Topological Groups

Topological Groups

An Introduction

Nelson G. Markley

A JOHN WILEY & SONS, INC., PUBLICATION

Published by John Wiley & Sons, Inc., Hoboken, New Jersey.

Published simultaneously in Canada.

Library of Congress Cataloging-in-Publication Data:

Markley, Nelson G., 1940–
  Topological groups : an introduction / Nelson G. Markley.
    p. cm.
  Includes bibliographical references and index.
  ISBN 978-0-470-62451-7 (cloth)
  1. Topological groups. I. Title.
  QA387.M364 2010
  512'.55—dc22   2010013147

10 9 8 7 6 5 4 3 2 1

To my wife

Pat

and our children

Susan

David

Kirk

Rebecca

Contents

Preface

The definition of a topological group is a marriage of the axioms of topological spaces and groups that provided fertile ground for an abstract theory. More importantly, it is a powerful unifying concept that links together diverse mathematical areas. It is a subject that many mathematicians encounter at one time or another. Some of us make more use of the theory of topological groups than others, but the potential to benefit from an understanding of topological groups is rather wide-spread in mathematics and related sciences. The aim of this book is to provide a user-friendly introduction to topological groups for the larger mathematical community.

What constitutes a user-friendly introductory book to a technical subject? Our perspective is first, that the prerequisites should be kept to a minimum. Second, the content should be broad, but not shallow. Third, the chapters should fit together in a way that provides a cohesive picture of a central portion of the subject. Fourth, it should provide a good foundation for exploring the more advanced literature on the subject. Of course, achieving these goals requires some compromises, but is also an opportunity to think about the subject from a different point of view.

The primary compromise made in this book is at key junctures to restrict attention to metric spaces. Without this restriction, there would have to be a substantial graduate-level topological prerequisite or a large amount of advanced general topology would have to be included. On the other side of the coin, many important topological groups, like matrix groups, are metric by definition, and metric group results can be put to good use developing the more general theory.

The perspective of the book is topological. Metric spaces are viewed as a particularly accessible class of topological spaces because sequences suffice in metric space proofs, but not in topological space proofs. The characterizations of different types of topologies in terms of their open sets is always stressed and used. We try to present the material in way that helps the reader see how the results fit into a more general picture.

The use of metric groups does make a largely self-contained book possible. In the way of prerequisites, what is required more than anything else is the mathematical maturity that is gained from upper-level undergraduate courses such as advanced calculus or real analysis, linear algebra, abstract algebra, and topology. Advanced calculus or real analysis and one or two of the others along with the motivation to do exercises should be sufficient prerequisites.

The exercises are an integral part of the contents and intended to facilitate the use of the book by a wide audience. Of course, there are exercises of the standard type that provide an opportunity to test the readers grasp of the material. In addition, the exercises contain many elementary propositions that are referenced and used in the text. There are, in fact, over 250 references to specific exercises in the text. The page numbers are always provided for these referenced exercises. In many cases, the more advanced reader will know what is in a referenced exercise without even looking at it and can continue reading without interruption, while the beginner will gain understanding by looking up the exercise and either recalling their earlier solution or writing out a solution to it. There are also exercises that are intended for the more dedicated reader to extend the theory beyond what is presented.

Although the chapter titles may seem diverse, there is more interconnection between them than one might expect. To help the reader more easily understand points in proofs that depend on prior material, there are many internal references to earlier theorems, propositions, and corollaries. We have tried to use good judgment on how often a specific result is referenced, and how it is done. Many of the references are a simple reference location of the result being used in parenthesis, such as (*Theorem* **??**), to be consulted as needed. There is also an index of special symbols located between the Bibliography and the Index.

Beyond this book is a vast body of knowledge about topological groups of all kinds. The bibliography and citations indicate some main avenues for further study of topological groups broadly interpreted. The abstract study of topological groups requires substantially more general topology than developed here, and we have provided references and brief comments on this necessary topological material. Although over half of the items in the bibliography bear dates before 1975, at this writing they are all still readily available. Several have been reprinted a number of times. In our judgement they have stood the test of time and are appropriate resources for initiating further study.

The primary motivation for writing this book was to fill a perceived gap in the literature on topological groups, namely the absence of a broad, accessible, and self-contained introduction to the subject. The content of this book is based on our intuitive understanding of the core ideas, having, as a user of topological groups for many years, acquired a knowledge of the subject from a wide range of sources. The sparkplug for the book, however, was the idea that focusing more on metric groups than general topological groups would not impoverish the subject. And it did not.

Without the intriguing idea that a worthwhile introduction to topological groups could be based primarily on metric groups, this book would not have been written. Determining whether or not it was a feasible idea quickly became a fascinating mental exercise. As the prospect of such a book seemed more and more possible, the questions became "What topics should be included?" and "How do they best fit together?" These questions lead to a constant and critical examination of ideas and proofs that was the pleasure of writing this book. There were interesting mathematical questions to be explored every day.

Among those who made valuable contributions to the preparation of the manuscript were the anonymous reviewers. Their thoughtful comments and suggestions were very useful to me in shaping the final version of the manuscript. This is my unique opportunity to extend my personal thanks to each of you.

Three other people deserve special thanks for their help making this book a reality. Robert Burckel, having heard about the project near its completion, offered to read portions of it. Besides catching careless errors on my part, he made some wonderful suggestions that improved the overall quality of the book. This is my second book with John Wiley and Sons, and both times I had the privilege and pleasure of working with Susanne Steitz-Filler. She did a masterful job guiding both books through the editorial process and into production. During this book's rather long embryonic state, my wife Pat's patience, support, and encouragement were invaluable and deeply appreciated.

Nelson G. Markley

Chapter 1

Groups and Topologies

During the twentieth century enormous progress was made abstracting the mathematical structures critical to nineteenth-century analysis and using them to attack classical problems in new ways. The simple algebraic operations of addition and multiplication of real numbers became part of the more general algebraic structures of groups, rings, fields, and algebras that included many common binary operations in mathematics. Distance functions of various kinds that were frequently used in analytic arguments lead to the concept of a metric, the study of metric spaces, and eventually general topology. These new areas of algebra and topology thrived as independent mathematical subjects. More importantly for us, their cross fertilization proved to be very productive ground for mathematical research, particularly, in what may be called modern analysis. Topological groups provide an ideal area for an introductory exploration of the interaction between algebraic and topological ideas.

In simple terms, a topological group is a group with a continuous multiplication function and a continuous inverse function. This book is a concrete self-contained introduction to topological groups. Since continuity is a topological concept, algebraic and topological ideas must be developed together. The primary topological context will be metric spaces, and the groups studied in Chapters 2, 3 and 4 are intrinsically metric groups. They will reappear many times in the last four chapters.

This chapter begins with introductions to group theory (Section 1.1) and metric space theory (Section 1.2), and then brings them together to begin the study of topological groups (Section 1.3). From that point on, algebraic and topological ideas will be constantly reinforcing each other. New concepts and results about groups and topological spaces will be introduced as needed to build a theory of topological groups.

Coset decompositions of a group and quotient groups are fundamental in all aspects of group theory. The metric-space side of these fundamental group constructions is developed in Section 1.4. The final section (1.5) in the chapter is devoted to the key concepts of compact and locally compact metric groups.

The exercises are an important part of the text. A variety of calculations, elementary results, and supplementary facts have been placed in the exercises and referenced in the text. The intent here is to help readers work through the material at very different speeds and levels depending on their mathematical background. A reference to a specific exercise always includes the page number for that exercise. There is also an index of special symbols on page 362.

1.1 Groups

Let G be a set, finite or infinite, and consider the set $G \times G$ of ordered pairs from G, in other words,

$$G \times G = \{(x, y) : x \in G \text{ and } y \in G\}.$$

(The notation $x \in G$ means that x is an element of the set G, and the colon is read as "such that".) A function $\mu : G \times G \to G$ is called a *binary operation* because it amalgamates two elements of G into one. Binary operations can also be thought of as a kind of abstract multiplication (or addition) of elements of G with $\mu(x, y)$ being the product of two elements x and y in G. Without requiring the function μ to satisfy some algebraic conditions from the familiar world of numbers, it is unlikely that μ would be of interest from an algebraic perspective.

A natural algebraic condition to impose on the function μ is the associative property. Specifically, the function $\mu : G \times G \to G$ is called an *associative binary operation* on G provided that

$$\mu(x, \mu(y, z)) = \mu(\mu(x, y), z)$$

for all x, y, and z in G. At this point it should be obvious that it would be a lot more convenient to write $\mu(x, y)$ like ordinary multiplication or addition as xy or $x + y$. Then, the associative property is simply

$$x(yz) = (xy)z$$

or

$$x + (y + z) = (x + y) + z$$

for all x, y, and z in G.

A set G with an associative binary operation $\mu : G \times G \to G$ is called a *semigroup*. Semigroups are fairly common, but can also be trivial because the constant function $\mu(x, y) = a \in G$ for all x and y in G defines a semigroup structure on G.

To construct a few more illustrative examples, let $\mathbb{R}$ denote the real numbers and let a be a positive real number. If $a \geq 1$, then $G_a = \{x \in \mathbb{R} : x \geq a\}$ is a semigroup with respect to ordinary multiplication of real numbers. There is a significant difference, however, between the properties of G_1 and G_2.

An element e of a semigroup G is called an *identity element* for G provided that $ex = xe = x$ for all $x \in G$. The semigroup G_1 has an identity element, namely, the number 1, and G_2 does not. The set of all positive real numbers,

$\mathbb{R}^+$ is also a semigroup under the multiplication of real numbers and has an even richer algebraic structure than G_1.

Let G be a semigroup with an identity element e, and let x be an element of G. An *inverse* of x, written x^{-1}, is an element of G such that $xx^{-1} = x^{-1}x = e$. Every real number x in the semigroup $\mathbb{R}^+$ has an inverse, namely, $1/x$, but only 1 has an inverse in G_1.

When all three properties (associativity, an identity element, and inverses) occur together, the structure is called a group. The semigroup $\mathbb{R}^+$ is an example of a group. Specifically, a *group* is a set G with a binary operation $\mu : G \times G \to G$, written $\mu(x, y) = xy$, that satisfies the following conditions:

(a) For all x, y, and z in G,
$$x(yz) = (xy)z.$$

(b) There exists an element e in G such that
$$ex = x = xe$$
for all x in G.

(c) For each x in G there exists x^{-1} such that
$$xx^{-1} = e = x^{-1}x.$$

Although there are algebraic structures like rings that are more complex than groups and ones that are simpler, like semigroups, the concept of a group captures a large middle ground that includes many familiar mathematical environments and serves as a fundamental algebraic building block.

The first example of a group was the positive real numbers, $\mathbb{R}^+$, under multiplication of real numbers. The nonzero real numbers are also a group under multiplication of real numbers. The real numbers, $\mathbb{R}$, are a group under addition of real numbers. Under multiplication $\mathbb{R}$ is a semigroup, but it is not a group because 0 does not have an inverse. Similarly, the complex numbers, $\mathbb{C}$, are a group under complex addition and the nonzero complex numbers are a group under complex multiplication. The integers, $\mathbb{Z}$, are a group under addition, but the nonzero integers are only a semigroup under multiplication.

On a set consisting of exactly one element, say a, there is a unique binary operation, namely, $\mu(a, a) = a$, and with this operation $\{a\}$ is a group called the *trivial group*. A *nontrivial group* is a group with more than one element. As we progress in the study of topological groups, more and more interesting groups will appear.

A basic dichotomy in the study of groups is the distinction between finite and infinite groups. A simple example of a finite group is $\mathbb{Z}_n = \{0, 1, \ldots, n-1\}$, where n is a positive integer. Addition is defined by $i + j = r$ if and only if r is the remainder of the usual sum $i + j$ divided by n. Clearly, 0 is the identity element and $i + (n - i) = 0$. Checking the associative property is an exercise. The group $\mathbb{Z}_n$ is a called the *integers mod n*.

Another basic distinction is between the Abelian groups and all the others. A group G is *Abelian* or *commutative* provided that

$$xy = yx$$

or

$$x + y = y + x$$

for all x and y in G. The additive groups $\mathbb{Z}$, $\mathbb{Z}_n$, and $\mathbb{R}$ are Abelian. In fact, when the binary operation of a group is written additively, the group is understood to be an Abelian group, but the converse is not true.. There are plenty of Abelian groups, like $\mathbb{R}^+$, for which the binary operation is written as xy.

For x in a group G written multiplicatively, x^k can be defined for all k in $\mathbb{Z}$. When $k > 0$, set

$$x^k = \underbrace{xx\ldots x}_{k-\text{times}}.$$

Set $x^0 = e$. When $k < 0$, set $x^k = (x^{-1})^{-k}$. In this context, the usual rules for exponents hold. Specifically, $x^m x^n = x^{m+n}$ and $(x^m)^n = x^{mn}$ for all m, n in $\mathbb{Z}$. Similarly, one defines kx when G is written additively and obtains $mx + nx = (m+n)x$ and $m(nx) = (mn)x$ for all m, n in $\mathbb{Z}$.

A group G is a *cyclic group* provided that there exists an element a, called a *generator*, in G such that $G = \{a^k : k \in \mathbb{Z}\}$ or $G = \{ka : k \in \mathbb{Z}\}$ in additive notation. The groups $\mathbb{Z}$ and $\mathbb{Z}_m$ are examples of infinite and finite cyclic groups, respectively. Cyclic groups are clearly Abelian.

Two elements x and y in a group G are said to *commute*, if $xy = yx$. Even when a group is not Abelian, it will contain elements that commute. For example, powers of x always commute because $x^m x^n = x^{m+n} = x^{n+m} = x^n x^m$. *Proposition 1.1.1* below provides some less obvious examples.

Functions are another important source of semigroups and groups. Given a set X, there is a natural associative binary operation on the set of functions from X to X, namely , the composition of functions. Specifically, if $f, g : X \to X$ are functions from X to X, then $f \circ g(x) = f(g(x))$ defines another function from X to X. Clearly,

$$f \circ (g \circ h)(x) = f\left(g\left(h(x)\right)\right) = (f \circ g) \circ h(x)$$

for all x in X and

$$f \circ (g \circ h) = (f \circ g) \circ h.$$

Thus the set F_X of all functions from X to X with the binary operation of composition of functions denoted by $\circ$ is a semigroup. The map $\iota : X \to X$ defined by $\iota(x) = x$ is an identity element for F_X. To obtain groups of functions additional restrictions must be placed on the set of functions.

We begin with some basic definitions. Let X and Y be sets. A function $f : X \to Y$ is *onto* provided that for every y in Y there exists x in X such that $f(x) = y$. And f is *one-to-one* provided that $f(x) = f(x')$ implies that $x = x'$. If f is both one-to-one and onto, then given y in Y there exists a unique x in X such that $f(x) = y$. Therefore, a function $f^{-1} : Y \to X$ can be defined unambiguously by setting $f^{-1}(y) = x$ if and only if $f(x) = y$. Clearly, $f^{-1}(f(x)) = x$ for all x in X and $f(f^{-1}(y)) = y$ for all y in Y. It follows that f^{-1} is also one-to-one and onto, when f is one-to-one and onto.

Given a set X, let

$$S_X = \{f : X \to X : f \text{ is one-to-one and onto}\} \subset F_X.$$

(The notation $A \subset B$ means that the set A is contained in the set B, or in other words every element of A is in B. It does not exclude the possibility that $A = B$.) If f is in S_X, then $f \circ f^{-1} = \iota = f^{-1} \circ f$. Conversely, if there exists g in S_X such that $f \circ g = \iota = g \circ f$, then $g = f^{-1}$. Therefore, S_X is the group of units of F_X by *Exercise* 3, p. 11.

Given α in S_X, the *α-orbit* of $x \in X$ is defined by

$$\mathcal{O}(x) = \{\alpha^k(x) : k \in \mathbb{Z}\}.$$

A subset E of X is an *α-invariant set* if $\alpha(E) = E$. The α-orbits are the smallest α-invariant sets, and α-invariant sets are simply unions of orbits. Notice that $\mathcal{O}(x) = \mathcal{O}(\alpha^m(x))$ for m in $\mathbb{Z}$ because $\alpha^k(x) = \alpha^{k-m}(\alpha^m(x))$. If z is in $\mathcal{O}(x) \cap \mathcal{O}(y)$, then $\alpha^m(x) = z = \alpha^n(y)$ for some m and n, and it follows that $\mathcal{O}(x) = \mathcal{O}(\alpha^m(x)) = \mathcal{O}(\alpha^n(y)) = \mathcal{O}(y)$. Consequently, given x and y in X, either $\mathcal{O}(x) = \mathcal{O}(y)$ or $\mathcal{O}(x) \cap \mathcal{O}(y) = \phi$. (The symbol ϕ denotes the set that contains no elements and is called the *empty set.*) Thus the α-orbits decompose X into disjoint α-invariant sets.

Now, consider the finite set $X = \{1, 2, 3, \ldots, n\}$, and set $S_n = S_X$. The elements of S_n are called *permutations of n symbols*, and the group S_n is called the *symmetric group on n symbols.* (For convenience, the symbols used here are the integers $1, 2, \ldots, n$, but they could just as well be letters or other distinct symbols.) The group S_n will play an important role in later chapters and will be the primary group studied in this section.

One way to express a permutation in S_n is to write the symbols on a line and the image of each symbol directly underneath it with parentheses as delimiters. For example, the following specifies a particular permutation in S_9:

$$\beta = \begin{pmatrix} 1 & 2 & 3 & 4 & 5 & 6 & 7 & 8 & 9 \\ 9 & 6 & 7 & 8 & 3 & 4 & 5 & 2 & 1 \end{pmatrix} \tag{1.1}$$

This permutation will be used to illustrate several basic permutation concepts.

Given a permutation α in S_n, the α-orbits are necessarily finite, and repeats must occur in the sequence $j, \alpha(j), \alpha^2(j), \ldots, \alpha^m(j), \ldots$. The first repeat is particularly useful. Specifically, given a symbol j, $1 \le j \le n$, there exists a smallest positive integer m such that $\alpha^m(j) = \alpha^k(j)$ for some k satisfying $0 \le k \le m - 1$. Applying α^{-k} to both sides of $\alpha^m(j) = \alpha^k(j)$ shows that $\alpha^{m-k}(j) = j$. If $k > 0$, then $0 < m - k < m$ making $\alpha^{m-k}(j) = j$ an earlier repeat. Therefore, $k = 0$ and the α-orbit of j consists precisely of the m distinct points $j, \alpha(j), \ldots, \alpha^{m-1}(j)$. Consequently, the α-orbits and the order of the points in each α-orbit completely describe the permutation α.

For example, the orbits of β defined by equation (1.1) written in the order in which β maps one symbol to another with the last going back to the first are:

$$\{1,9\}, \{2,6,4,8\}, \text{ and } \{3,7,5\}.$$

Because we know these orbits are written in the order in which β maps one symbol to another, it is easy to reconstruct equation (1.1) for β from the above information. Using the wrong order in which β maps one symbol to another will, however, produce a permutation that is not equal to β.

The discussion in the previous paragraphs leads naturally to the concept of a special class of permutations called cycles. The idea of a cycle incorporates both an orbit and the order in which the points occur in that orbit. Let $a_1, \ldots, a_k$ be distinct elements of $\{1, 2, \ldots, n\}$. The *cycle* $(a_1 a_2 \ldots a_k)$ is the permutation that maps a_1 to a_2, a_2 to a_3, ..., a_k to a_1, and maps j to j for every $j \neq a_i$ for $i = 1, \ldots, k$. In S_9, the cycle (2648) is the same as the permutation

$$\begin{pmatrix} 1 & 2 & 3 & 4 & 5 & 6 & 7 & 8 & 9 \\ 1 & 6 & 3 & 8 & 5 & 4 & 7 & 2 & 9 \end{pmatrix},$$

which agrees with β on the set $\{2, 4, 6, 8\}$. Although the order of a cycle is critical, the first element of a cycle is not. For example, $(375) = (753) = (537)$.

The cycle (a) is the identity map ι. A cycle of the form (ab) is called a *transposition*. If $\alpha = (a_1 a_2 \ldots a_k)$ is a cycle in S_n, then it is easy to see that $\alpha, \alpha^2, \ldots, \alpha^{k-1}$ are distinct permutations and $\alpha^k = \iota$, the identity permutation that fixes every j. In particular, for a transposition $\alpha^2 = \iota$.

Consider an α in S_n. If the α-orbit of the symbol j is

$$\mathcal{O}(j) = \{j, \alpha(j), \ldots, \alpha^{m-1}(j)\},$$

then the cycle $(j\,\alpha(j) \ldots \alpha^{m-1}(j))$ equals α on $\mathcal{O}(j)$. For example, the cycle (2468) agrees with the permutation β given by (1.1) on $\mathcal{O}(2) = \{2, 6, 4, 8\}$. So each α-orbit determines a unique cycle that equals α on that α-orbit.

The following three propositions about cycles will be used to prove an important theorem about S_n at the end of this section.

Proposition 1.1.1 *If $\alpha = (a_1 \ldots a_k)$ and $\beta = (b_1 \ldots b_m)$ are disjoint cycles, that is, $a_i \neq b_j$ for $1 \leq i \leq k$ and $1 \leq j \leq m$, then $\alpha \circ \beta = \beta \circ \alpha$.*

Proof. Consider some a_j, $1 \leq j \leq k$. Since $\beta(a_i) = a_i$ for $1 \leq i \leq k$ and $\alpha(a_j) = a_q$ for some q, it follows that

$$\alpha(\beta(a_j)) = \alpha(a_j) = a_q = \beta(a_q) = \beta(\alpha(a_j)).$$

Of course, the same argument applies to the points b_j. Obviously, $\alpha(\beta(k)) = k = \beta(\alpha(k))$, when k is neither an a_i nor a b_j. Thus $\alpha \circ \beta = \beta \circ \alpha$. □

Proposition 1.1.2 *If α is a permutation in S_n, then there exist disjoint cycles $\alpha_1, \ldots, \alpha_m$ such that $\alpha = \alpha_1 \circ \alpha_2 \circ \ldots \circ \alpha_m$.*

Proof. Let $\alpha_1, \ldots, \alpha_m$ be the cycles determined by the distinct α-orbits. They are disjoint cycles because distinct orbits are disjoint subsets of $\{1, \ldots, n\}$.

For each i, $\alpha_i = \alpha$ on exactly one of the m distinct α-orbits. Write $\alpha_i = (a_{(i,1)} \cdots a_{(i,k_i)})$ for $1 \leq i \leq m$. Hence, every symbol equals precisely one $a_{(i,j)}$. If $a_{(i,j)}$ is an arbitrary symbol, then

$$\begin{aligned} \alpha_1 \circ \alpha_2 \circ \ldots \circ \alpha_m(a_{(i,j)}) &= \\ \alpha_i \circ \alpha_1 \circ \ldots \circ \alpha_{i-1} \circ \alpha_{i+1} \circ \ldots \circ \alpha_m(a_{(i,j)}) &= \\ \alpha_i(a_{(i,j)}) = \alpha(a_{(i,j)}) \end{aligned}$$

for the following reasons: $\alpha_p \circ \alpha_q = \alpha_q \circ \alpha_p$ by *Proposition 1.1.1*, $\alpha_q(a_{(i,j)}) = a_{(i,j)}$ for $q \neq i$, and $\alpha_i(a_{(i,j)}) = \alpha(a_{(i,j)})$. Thus $\alpha_1 \circ \alpha_2 \circ \ldots \circ \alpha_m = \alpha$. □

Proposition 1.1.3 *Every permutation α in S_n is the composition of transpositions.*

Proof. By *Proposition 1.1.2*, it suffices to show that a cycle is a composition of transpositions. Applying the transpositions from right to left, it follows that

$$(a_1 \ldots a_k) = (a_1 a_k) \circ (a_1 a_{k-1}) \circ \ldots \circ (a_1 a_3) \circ (a_1 a_2)$$

to complete the proof. □

Returning to the permutation β in S_9 defined by equation (1.1), it follows from the preceding propositions that:

$$\begin{aligned} \beta &= (19) \circ (375) \circ (2648) \\ &= (19) \circ (35) \circ (37) \circ (28) \circ (24) \circ (26). \end{aligned}$$

Let G and G' be arbitrary groups. A *algebraic homomorphism* of G to G' is a function $\varphi : G \to G'$ such that

$$\varphi(xy) = \varphi(x)\varphi(y)$$

for all x and y in G. In other words, you can multiply before or after applying the function, and algebraic properties of G are preserved by φ. In addition, a homomorphism preserves identities and inverses (*Exercise* 11, p. 12). Of course, if the group operation in one or both of the groups is written additively, then the right or left side of the equation would be written as $\varphi(x+y)$ or $\varphi(x)+\varphi(y)$.

The map $\varphi(k) = 2k$ is an example of a algebraic homomorphism of $\mathbb{Z}$ into $\mathbb{Z}$. It is one-to-one, but not onto. To construct an example of a algebraic homomorphism that is not one-to-one define $\psi : \mathbb{Z} \to \mathbb{Z}_n$ by setting $\psi(k) = r$ if and only if q and r are the unique integers such that $k = qn + r$ and $0 \leq r < n$. In other words, r is the *remainder* of k divided by n.

When an algebraic homomorphism, $\varphi : G \to G'$, is both one-to-one and onto, there is a one-to-one onto inverse function $\varphi^{-1} : G' \to G$. If y and y' are elements of G', then, because φ is onto, $y = \varphi(x)$ and $y' = \varphi(x')$ for some x and x' in G. It follows that

$$\varphi^{-1}(yy') = \varphi^{-1}(\varphi(x)\varphi(x')) = \varphi^{-1}(\varphi(xx')) = xx' = \varphi^{-1}(y)\varphi^{-1}(y')$$

and φ^{-1} is also an algebraic homomorphism.

A one-to-one onto algebraic homomorphism $\varphi : G \to G'$ is called an *algebraic isomorphism*. As a consequence of the previous paragraph, the inverse of an algebraic isomorphism is an algebraic isomorphism. Two groups are said to be *algebraically isomorphic* if there exists an algebraic isomorphism mapping one group onto the other. Algebraically isomorphic groups have exactly the same algebraic properties. For example, they are either both Abelian or neither is Abelian. So S_3 is not algebraically isomorphic to $\mathbb{Z}_6$, even though they both contain 6 elements, because S_3 is not Abelian by *Exercise* 8, p. 12.

Let n be a positive integer and consider the complex number

$$a = \exp(2\pi i/n) = \cos(2\pi/n) + i\sin(2\pi/n).$$

Then

$$G = \{a^0, a^1, a^2, a^3, \ldots, a^{n-1}\},$$

the set of solutions of $z^n = 1$ in $\mathbb{C}$, is an example of a finite cyclic group. The function $\varphi : \mathbb{Z}_n \to G$ defined by $\varphi(r) = a^r$ is an algebraic isomorphism between $\mathbb{Z}_n$ and G. In fact, any two cyclic groups containing n elements are algebraically isomorphic, see *Exercise* 16, p. 12. Thus from a group-theoretic point of view there is one infinite cyclic group and for each positive integer n exactly one cyclic group containing n elements.

Suppose H is a nonempty subset of a group G. Then xy is defined for x and y in H. What conditions will guarantee that H is a group for this multiplication? Certainly, xy must be in H when x and y are in H so that $(x, y) \to xy$ defines a binary operation on H. The associative property holds for any three elements of G, so it automatically holds for any three elements of H. Suppose x^{-1} is also in H whenever x is in H. Since H is assumed to be nonempty, there exists $x \in H$, and hence $e = xx^{-1}$ is in H. It follows that H has an identity, and for every x in H there exists x^{-1} in H such that $xx^{-1} = x^{-1}x = e$. To summarize, if H is a nonempty subset of a group G such that x and y in H implies that both xy and x^{-1} are also in H, then H is also a group under the multiplication of G. This leads naturally to a definition. A nonempty subset H of a group G is a *subgroup of* G provided that for all x and y in H:

(a) $xy \in H$

(b) $x^{-1} \in H$.

For example, $\mathbb{Z}$ is a subgroup of the additive group $\mathbb{R}$, but the positive integers and the odd integers are not subgroups of $\mathbb{R}$. A more interesting example can be found in the complex numbers $\mathbb{C}$. The *conjugate of a complex number* $z = x + iy$ is defined by $\overline{z} = x - iy$. Note that $z\overline{z} = x^2 + y^2$, which is a non-negative real number. The *absolute value of a complex number* is defined by $|z| = \sqrt{z\overline{z}}$. Consider

$$\mathbb{K} = \{z \in \mathbb{C} : |z| = 1\} = \{z = x + iy : x^2 + y^2 = 1\}.$$

Obviously, $\mathbb{K} \subset \mathbb{C} \setminus \{0\}$. If z and w are in $\mathbb{K}$, then an easy calculation shows that $|zw| = 1$ and zw is in $\mathbb{K}$. Similarly, $1/z = \overline{z}$ when $|z| = 1$. So $1/z$ is in $\mathbb{K}$ when z is in $\mathbb{K}$. Thus, $\mathbb{K}$, the familiar unit circle in the complex plane, is a subgroup of the multiplicative group $\mathbb{C} \setminus \{0\}$, and the set of solutions of $z^n = 1$ in $\mathbb{C}$ is a subgroup of both $\mathbb{K}$ and of $\mathbb{C}$.

There are two simple remarks to be made about the subgroups of any group G with identity e. First, there are two extreme subgroups of G, namely, G itself and the trivial group $\{e\}$. All other subgroups will be referred to as *proper subgroups*. Second, if a is in G, then $\{a^k : k \in \mathbb{Z}\}$ is an Abelian subgroup of G called the *cyclic subgroup generated by a*.

If $\varphi : G \to G'$ is a algebraic homomorphism, then $K = \{x \in G : \varphi(x) = e'\}$ is a subgroup of G called the *kernel of* φ. Furthermore, $K = \{e\}$ if and only if φ is one-to-one (*Exercise* 12, p. 12). This is the initial observation in an important connection between subgroups and algebraic homomorphisms that will be explored more fully in Section 4.

The roots of $z^2 = 1$ in $\mathbb{C}$ are $\{1, -1\}$, which is a cyclic group containing 2 elements generated by -1. The final goal of this section is to prove the existence of an onto algebraic homomorphism $\operatorname{sgn} : S_n \to \{1, -1\}$. In other words, a sign can be assigned to permutations that is multiplicative when permutations are composed. This sign function will play a significant role in the construction and properties of another homomorphism, the determinant function (Section 2.5).

First, we define a function $N : S_n \to \mathbb{Z}$. Let $|\mathcal{O}(j)|$ denote the cardinality of the orbit $\mathcal{O}(j)$ for a permutation α in S_n, and set

$$N(\alpha) = \sum_{j=1}^{n} \frac{|\mathcal{O}(j)| - 1}{|\mathcal{O}(j)|}.$$

Notice that when $\alpha(j) = j$, the orbit of j is just $\{j\}$ and j contributes nothing to the above sum. It follows that $N(\iota) = 0$. If $\alpha = (a_1 a_2 \ldots a_k)$ is a cycle, then the only terms contributing to the sum are $a_1, \ldots, a_k$ and $\mathcal{O}(a_i)$ consists of the k points $\{a_1, \ldots, a_k\}$ for each i. Hence,

$$N(\alpha) = \sum_{i=1}^{k} \frac{k-1}{k} = k - 1.$$

More generally, if $\mathcal{O}_1 = \{a_1, \ldots, a_k\}$ is an orbit of α, then $\mathcal{O}(a_j) = \mathcal{O}_1$ for $j = 1, \ldots, k$ and

$$\sum_{j \in \mathcal{O}_1} \frac{|\mathcal{O}(j)| - 1}{|\mathcal{O}(j)|} = \sum_{i=1}^{k} \frac{k-1}{k} = k - 1 = |\mathcal{O}_1| - 1,$$

and the orbit $\mathcal{O}_1$ of α contributes $|\mathcal{O}_1| - 1$ to $N(\alpha)$. If the distinct orbits of α are $\mathcal{O}_1, \ldots, \mathcal{O}_p$, then

$$N(\alpha) = \sum_{j=1}^{p} (|\mathcal{O}_j| - 1).$$

The function sgn : $S_n \to \{1, -1\}$ is defined by setting

$$\mathrm{sgn}(\alpha) = (-1)^{N(\alpha)}. \tag{1.2}$$

The permutation α is said to be an *even permutation* when $\mathrm{sgn}(\alpha) = 1$, and an *odd permutation* when $\mathrm{sgn}(\alpha) = -1$.

Returning to our standard example β defined by equation (1.1), recall that the orbits of β are $\{1, 9\}, \{2, 6, 4, 8\}$, and $\{3, 7, 5\}$. Hence, $N(\beta) = (2 - 1) + (4 - 1) + (3 - 1) = 6$ and $\mathrm{sgn}(\beta) = 1$ making β an even permutation. The cycle (2648) is clearly odd. By using *Propositions 1.1.1 and 1.1.2*, it is easy to show that $\beta \circ (2648) = (19) \circ (24) \circ (68) \circ (375)$. Consequently, $N(\beta \circ (2648)) = 5$ and $\mathrm{sgn}(\beta \circ (2648)) = -1 = \mathrm{sgn}(\beta)\mathrm{sgn}((2648))$. The next task is to prove that this is a general phenomenon.

Understanding when $N(\alpha)$ is odd and even will be critical in showing that sgn is a algebraic homomorphism. The next lemma provides the key.

Lemma 1.1.4 *If $\alpha = (a_1 b_1) \circ \ldots \circ (a_m b_m)$ is the composition of m transpositions, then $N(\alpha)$ is odd or even according as m is odd or even.*

Proof. Observe that the statement is certainly true when $m = 1$ because $N((ab)) = 1$ and then proceed by induction. Assuming it holds for m, consider $\alpha = (a_1 b_1) \circ \ldots \circ (a_m b_m) \circ (a_{m+1} b_{m+1})$ and set $\beta = (a_2 b_2) \circ \ldots \circ (a_m b_m) \circ (a_{m+1} b_{m+1})$.

By *Proposition 1.1.2*, there exist disjoint cycles $\beta_1, \ldots, \beta_k$ such that $\beta = \beta_1 \circ \beta_2 \circ \ldots \circ \beta_k$. Because these cycles are disjoint, a_1 and b_1 each occur in exactly one of the cycles $\beta_1, \ldots, \beta_k$. There are two possibilities: either they occur in the same cycle or in different ones. By *Proposition 1.1.1*, we can assume that the cycle containing a_1 is β_1 and b_1 is in either β_1 or β_2. So either

$$\alpha = (a_1 b_1) \circ (a_1 c_1 \ldots c_p\, b_1 d_1 \ldots d_q) \circ \beta_2 \circ \ldots \circ \beta_k$$

or

$$\alpha = (a_1 b_1) \circ (a_1 c_1 \ldots c_p) \circ (b_1 d_1 \ldots d_q) \circ \beta_3 \circ \ldots \circ \beta_k$$

where p and q are allowed to be zero.

In the first case, it is easy to check that

$$(a_1 b_1) \circ (a_1 c_1 \ldots c_p\, b_1 d_1 \ldots d_q) = (a_1 c_1 \ldots c_p) \circ (b_1 d_1 \ldots d_q). \tag{1.3}$$

Thus one orbit of β splits into two orbits in α and the other orbits remain unchanged. The orbit $\{a_1, c_1, \ldots, c_p, b_1, d_1, \ldots, d_q\}$ contributes $p + q + 1$ to

$N(\beta)$ and the orbits $\{a_1, c_1, \ldots, c_p\}$ and $\{b_1, d_1, \ldots, d_q\}$ contribute $p+q$ to $N(\alpha)$. Since the other orbits are unchanged, $N(\alpha) = N(\beta) - 1$. It follows that $N(\alpha)$ is odd if and only if $N(\beta)$ is even. Since by induction $N(\beta)$ is even if and only if m is even, it follows that $N(\alpha)$ is odd if and only if $m+1$ is odd. Similarly, $N(\alpha)$ is even if and only if $m+1$ is even, and the first case is done.

Since $(a_1b_1) \circ (a_1b_1) = \iota$, it follows from equation (1.3) that

$$(a_1b_1) \circ (a_1c_1 \ldots c_p) \circ (b_1d_1 \ldots d_q) = (a_1c_1 \ldots c_p b_1 d_1 \ldots d_q),$$

and two orbits of β combine to form one orbit of α. Thus $N(\alpha) = N(\beta) + 1$ in the second case, and as in the first case, $N(\alpha)$ is odd or even according as $m+1$ is odd or even. □

There are, of course, many ways that a permutation can be written as a composition of cycles. What the lemma tells us is that no matter how one writes a given permutation as a composition of cycles, the number of cycles used will either always be even or always be odd.

Theorem 1.1.5 *The function* sgn : $S_n \to \{1, -1\}$ *defined by equation (1.2) is an algebraic homomorphism that is onto for* $n \geq 2$.

Proof. Let α and β be elements of S_n. By *Proposition 1.1.3*, α and β are compositions of p and q transpositions respectively. Hence, $\alpha \circ \beta$ is the composition of $p+q$ transpositions. It then follows from the lemma that $\text{sgn}(\alpha) = (-1)^p$, $\text{sgn}(\beta) = (-1)^q$, and $\text{sgn}(\alpha \circ \beta) = (-1)^{p+q}$. Therefore, $\text{sgn}(\alpha \circ \beta) = \text{sgn}(\alpha)\text{sgn}(\beta)$.

When $n \geq 2$, There is a transposition (ab) in S_n and sgn is onto because $N(\iota) = 0$ and $N((ab)) = 1$. □

The permutations that can be written as the composition of an even number of transpositions are the kernel of the sgn, and hence form a subgroup of S_n. This subgroup of S_n is called the *alternating group* and is denoted by A_n.

EXERCISES

1. Let G be a group. Show that G has only one identity element and each element of G has exactly one inverse.

2. Let x be an element of a group G. Show that $x^m x^n = x^{m+n}$ for all integers m and n.

3. Let G be a semigroup with identity e. An element x in G is a *unit*, if there exists y in G such that $xy = e = yx$. Show that the set of units of the semigroup G is a group. This group is called the *group of units*.

4. Let G be a group and let a be an element of G. Show that the functions $x \mapsto ax$ and $x \mapsto xa$ are one-to-one functions of G onto G. (The arrow $\mapsto$ is read "maps to" and is used to define a function.)

5. For X a finite set, show that a function $f : X \to X$ is one-to-one if and only if it is onto.

6. Show that S_n contains $n! = n(n-1)(n-2)\dots 1$ elements. How many functions are in F_n, the set of functions from $\{1, \dots, n\}$ to itself?

7. Show that S_n contains subgroups algebraically isomorphic to S_k and to $\mathbb{Z}_k$ for $k = 1, 2, 3, \dots, n$.

8. Show that S_n is Abelian if and only if $n \leq 2$.

9. Show that every element of the group S_n can be written as a composition of transpositions of the form $(k\,k+1)$.

10. Let H be a nonempty subset of a group G. Show that H is a subgroup of G if and only if $ab^{-1} \in H$ for all $a, b \in H$.

11. Let G and G' be groups with identity elements e and e', respectively, and let $\varphi : G \to G'$ be an algebraic homomorphism. Show that $\varphi(e) = e'$ and $\varphi(x^{-1}) = \varphi(x)^{-1}$.

12. Let G and G' be groups with identity elements e and e', respectively, and let $\varphi : G \to G'$ be an algebraic homomorphism. Show that $K = \{x \in G : \varphi(x) = e'\}$ is a subgroup of G. Show that φ is one-to-one if and only if $K = \{e\}$. Also show that $\varphi(G)$ is a subgroup of G'.

13. Let G be a group and let a be an element of G. Show that $\varphi_a(x) = axa^{-1}$ defines an algebraic isomorphism of G onto G. Also show that $\{a : \varphi_a = \iota\}$ is a subgroup of G.

14. Show that a group G is algebraically isomorphic to a subgroup of S_G.

15. Show that if $f : X \to Y$ is one-to-one and onto then S_X and S_Y are algebraically isomorphic groups.

16. Show that if G is a cyclic group, then G is algebraically isomorphic to $\mathbb{Z}$ or $\mathbb{Z}_n$ for some integer $n \geq 1$.

17. Show that the cyclic subgroup of $\mathbb{K}$ generated by $a = \exp(2\pi i\alpha)$ is finite if and only if α is a rational number.

18. Prove that a group G has no proper subgroups if and only if G is algebraically isomorphic to $\mathbb{Z}_n$ and n is a prime.

19. Calculate the inverse of a cycle $(a_1 \dots a_k)$ in S_n.

1.2 Metric and Topological Spaces

The starting point is again a set, but now distance replaces the concept of a binary operation as the central idea. Let X be a set. A function that measures the distance between two points of X is called a metric. Metrics should reflect enough of the basic properties of measuring distance in the world around us to make them worth studying, but not to the extent that metrics are uncommon.

The real numbers, denoted by $\mathbb{R}$, will play a prominent role in this chapter and subsequent ones. We are assuming that the reader is familiar with all the basic properties of the real numbers including: Every nonempty subset E of $\mathbb{R}$ that is bounded above has a *supremum*, written $\sup E$. To be more specific, a nonempty subset E is *bounded above* provided there exists a real number b such that $x \leq b$ for all x in E, and b is called an *upper bound* of E. If E is a nonempty subset of $\mathbb{R}$ that is bounded above, then there exists a real number $\sup E$ with the following properties:

(a) If x is in E, then $x \leq \sup E$.

(b) If $x \leq a$ for all x in E, then $\sup E \leq a$.

In other words, $\sup E$ is a smallest or *least upper bound* of E. The fact that every nonempty subset E of $\mathbb{R}$ that is bounded above has a supremum is one of the fundamental properties of the real numbers that distinguishes them from the rational numbers.

A *metric* on a set X is a real-valued function $d : X \times X \to \mathbb{R}$ that has the following properties for all x, y, and z in X:

(a) $d(x, y) = d(y, x)$

(b) $d(x, y) \geq 0$

(c) $d(x, y) = 0$ if and only if $x = y$

(d) $d(x, y) \leq d(x, z) + d(z, y)$.

The last property is called the *triangle inequality* and generalizes the fact that in Euclidean geometry the length of one side of a triangle is always less than the sum of the other two sides. Obviously, $d(x, y) = |x - y|$ is a metric on $\mathbb{R}$.

A *metric space* is simply a set X with a specified metric d. Because there is always more than one metric defined on a set X containing more than one point, it is critical at times to make sure there is no ambiguity about the metric being used. The question of when two metrics are or are not fundamentally different must also be addressed.

Among the most basic of all metric spaces are the *n-dimensional Euclidean spaces*, denoted by $\mathbb{R}^n$. The set $\mathbb{R}^n$ is the set of all n-tuples of real numbers and is defined by

$$\mathbb{R}^n = \underbrace{\mathbb{R} \times \ldots \times \mathbb{R}}_{n} = \{(x_1, \ldots, x_n) : x_j \in \mathbb{R} \text{ for } 1 \leq j \leq n\}.$$

It is convenient to use boldface type to denote points in $\mathbb{R}^n$, that is, $\mathbf{x} = (x_1, \ldots, x_n)$. Note that $\mathbb{R}^n$ is an Abelian group under the binary operation

$$\mathbf{x} + \mathbf{y} = (x_1 + y_2, \ldots, x_n + y_n)$$

with identity $\mathbf{0} = (0, \ldots, 0)$.

The definition of the familiar *dot product* on $\mathbb{R}^3$ can be extended to $\mathbb{R}^n$ by setting

$$\mathbf{x} \cdot \mathbf{y} = x_1 y_1 + \ldots + x_n y_n = \sum_{j=1}^{n} x_j y_j.$$

As usual for c, a real number, $c\mathbf{x} = (cx_1, \ldots, cx_n)$. Observe that

(a) $\mathbf{x} \cdot \mathbf{y} = \mathbf{y} \cdot \mathbf{x}$

(b) $(\mathbf{x} + \mathbf{x}') \cdot \mathbf{y} = \mathbf{x} \cdot \mathbf{y} + \mathbf{x}' \cdot \mathbf{y}$

(c) $c(\mathbf{x} \cdot \mathbf{y}) = (c\mathbf{x}) \cdot \mathbf{y} = \mathbf{x} \cdot (c\mathbf{y})$

hold for all c in $\mathbb{R}$ and for all $\mathbf{x}$, $\mathbf{x}'$, and $\mathbf{y}$ in $\mathbb{R}^n$.

Set

$$\|\mathbf{x}\| = \sqrt{\mathbf{x} \cdot \mathbf{x}} = \left(\sum_{j=1}^{n} x_j^2 \right)^{1/2} \tag{1.4}$$

and note that $\|\mathbf{x}\| = 0$ if and only if $\mathbf{x} = \mathbf{0}$. Then,

$$\|\mathbf{x} - \mathbf{y}\| = \left(\sum_{j=1}^{n} (x_j - y_j)^2 \right)^{1/2} \tag{1.5}$$

is the usual Euclidean distance function on $\mathbb{R}^n$, and we will show that it is a metric on $\mathbb{R}^n$. Obviously, $\|\mathbf{x} - \mathbf{y}\| = \|\mathbf{y} - \mathbf{x}\|$ and $\|\mathbf{x} - \mathbf{y}\| \geq 0$. Also $\|\mathbf{x} - \mathbf{y}\| = 0$ if and only if $\mathbf{x} = \mathbf{y}$ is self evident. Proving the triangle property requires a fundamental inequality that will prove to be very useful in general.

Proposition 1.2.1 (Cauchy-Schwarz Inequality) *If* $\mathbf{x}$ *and* $\mathbf{y}$ *are in* $\mathbb{R}^n$, *then*

$$|\mathbf{x} \cdot \mathbf{y}| \leq \|\mathbf{x}\| \, \|\mathbf{y}\|. \tag{1.6}$$

If $\mathbf{y} \neq \mathbf{0}$, *then equality holds if and only if*

$$\mathbf{x} = \frac{\mathbf{x} \cdot \mathbf{y}}{\mathbf{y} \cdot \mathbf{y}} \mathbf{y}.$$

Proof. When $\mathbf{y} = \mathbf{0}$, both sides of the inequality are 0 and there is nothing to prove. So for the rest of the proof $\mathbf{y} \neq \mathbf{0}$.

Suppose $\|\mathbf{v}\|^2 = \mathbf{v} \cdot \mathbf{v} = 1$ and set $\mathbf{u} = \mathbf{x} - (\mathbf{x} \cdot \mathbf{v})\mathbf{v}$. Then,

$$\begin{aligned} 0 &\leq \mathbf{u} \cdot \mathbf{u} \\ &= \mathbf{x} \cdot \mathbf{x} - 2(\mathbf{x} \cdot \mathbf{v})^2 + (\mathbf{x} \cdot \mathbf{v})^2(\mathbf{v} \cdot \mathbf{v}) \\ &= \mathbf{x} \cdot \mathbf{x} - (\mathbf{x} \cdot \mathbf{v})^2, \end{aligned}$$

and $0 = \mathbf{u} \cdot \mathbf{u}$ if and only if $\mathbf{u} = \mathbf{0}$ or equivalently $\mathbf{x} = (\mathbf{x} \cdot \mathbf{v})\mathbf{v}$.

For $\mathbf{y} \neq \mathbf{0}$, the above inequality applies to

$$\mathbf{v} = \frac{1}{\|\mathbf{y}\|}\mathbf{y}$$

and

$$0 \leq \mathbf{x} \cdot \mathbf{x} - \frac{(\mathbf{x} \cdot \mathbf{y})^2}{\|\mathbf{y}\|^2}.$$

The latter can be rewritten as

$$(\mathbf{x} \cdot \mathbf{y})^2 \leq \|\mathbf{x}\|^2 \|\mathbf{y}\|^2$$

or by taking the positive square root as

$$|\mathbf{x} \cdot \mathbf{y}| \leq \|\mathbf{x}\| \, \|\mathbf{y}\|.$$

It also follows from the previous paragraph that holds if and only if

$$\mathbf{x} = \left(\mathbf{x} \cdot \frac{1}{\|\mathbf{y}\|}\mathbf{y}\right) \frac{1}{\|\mathbf{y}\|}\mathbf{y} = \frac{\mathbf{x} \cdot \mathbf{y}}{\|\mathbf{y}\|^2}\mathbf{y} = \frac{\mathbf{x} \cdot \mathbf{y}}{\mathbf{y} \cdot \mathbf{y}}\mathbf{y}$$

to complete the proof. (Note that $\mathbf{x}$ occurs on both sides of this equation. And it holds for $c\mathbf{x}$ when it holds for $\mathbf{x}$.) □

Corollary 1.2.2 *For all* $\mathbf{x}$, $\mathbf{y}$ *and* $\mathbf{z}$ *in* $\mathbb{R}^n$,

$$\|\mathbf{x} - \mathbf{y}\| \leq \|\mathbf{x} - \mathbf{z}\| + \|\mathbf{z} - \mathbf{y}\|,$$

and $d(\mathbf{x}, \mathbf{y}) = \|\mathbf{x} - \mathbf{y}\|$ *is a metric on* $\mathbb{R}^n$.

Proof. To prove the triangle inequality observe that it suffices to prove

$$\|\mathbf{u} + \mathbf{v}\| \leq \|\mathbf{u}\| + \|\mathbf{v}\|, \tag{1.7}$$

and then set $\mathbf{u} = \mathbf{x} - \mathbf{z}$ and $\mathbf{v} = \mathbf{z} - \mathbf{y}$. Next, an easy calculation shows that

$$\|\mathbf{u} + \mathbf{v}\|^2 = (\mathbf{u} + \mathbf{v}) \cdot (\mathbf{u} + \mathbf{v}) = \|\mathbf{u}\|^2 + 2\mathbf{u} \cdot \mathbf{v} + \|\mathbf{v}\|^2.$$

The Cauchy-Schwarz Inequality can be applied here because $\mathbf{u} \cdot \mathbf{v} \leq |\mathbf{u} \cdot \mathbf{v}|$. Specifically,

$$\begin{aligned} \|\mathbf{u} + \mathbf{v}\|^2 &= \mathbf{u} \cdot \mathbf{u} + 2\mathbf{u} \cdot \mathbf{v} + \mathbf{v} \cdot \mathbf{v} \\ &\leq \|\mathbf{u}\|^2 + 2|\mathbf{u} \cdot \mathbf{v}| + \|\mathbf{v}\|^2 \\ &\leq \|\mathbf{u}\|^2 + 2\|\mathbf{u}\| \, \|\mathbf{v}\| + \|\mathbf{u}\|^2 \\ &= (\|\mathbf{u}\| + \|\mathbf{v}\|)^2 \end{aligned}$$

which implies that $\|\mathbf{u}+\mathbf{v}\| \leq \|\mathbf{u}\| + \|\mathbf{v}\|$. Hence, the triangle inequality holds and equation (1.5) defines a metric on $\mathbb{R}^n$. $\square$

The "metric space $\mathbb{R}^n$" will always mean $\mathbb{R}^n$ with this standard metric. When $n = 1$, the metric given by (1.5) equals the metric $d(x, y) = |x - y|$ on $\mathbb{R}$ and, unless stated otherwise, the metric on $\mathbb{R}$.

The *open ball* of radius $r > 0$ with center y in a metric space X is defined by

$$B_r(y) = \{x \in X : d(x, y) < r\}.$$

Of course, the open ball $B_r(y)$ also depends on the space X and the metric d, but making the specific X and d clear from the context is preferable to using more cumbersome notation like $B_r(y, X, d)$.

Suppose x and y are distinct points in a metric space X with metric d. If $0 < r < d(x, y)/2$, then it is easy to see that $B_r(x) \cap B_r(y) = \phi$ because $z \in B_r(x) \cap B_r(y)$ would imply that $d(x, y) \leq d(x, z) + d(z, y) < r + r < d(x, y)$, which is impossible. The existence of disjoint open balls centered at any two distinct points of a metric space is called the *Hausdorff property* of metric spaces.

When X is a metric space with metric d and Y is a nonempty subset of X, the restriction of d to $Y \times Y$ defines a metric on Y. For y in Y, the open ball of radius $r > 0$ with center y in Y is just $B_r(y) \cap Y$. So every nonempty subset of a metric space is a metric space in its own right.

It is easy to extend the definitions of convergent sequences and continuous functions from $\mathbb{R}$ to metric spaces. A sequence x_k, $k = 1, 2, 3, \ldots$, in a metric space X with metric d *converges* to y in X, if given $\varepsilon > 0$, there exists $K > 0$ such that $d(x_k, y) < \varepsilon$, whenever $k \geq K$. The Hausdorff property guarantees that the limit of a convergent sequence in a metric space is unique (*Exercise* 1, p. 27). We will usually say simply "Let x_k be a sequence in ...." with it understood that for each k in $\mathbb{Z}^+$, the positive integers, there is an x_k in a prescribed set.

Now, let X and Y be metric spaces with metrics d_X and d_Y, respectively, and let $f : X \to Y$ be a function from X to Y. The function f is *continuous* at z in X, if given $\varepsilon > 0$, there exists $\delta > 0$ such that $d_Y(f(x), f(z)) < \varepsilon$, whenever $d_X(x, z) < \delta$. Convergent sequences can be used to characterize continuity as follows:

Proposition 1.2.3 *Let X and Y be metric spaces with metrics d_X and d_Y and let $f : X \to Y$ be a function from X to Y. The following are equivalent:*

(a) The function f is continuous at z in X.

(b) Whenever x_k is a sequence in X converging to z, the sequence $f(x_k)$ converges to $f(z)$ in Y.

Proof. First, suppose f is continuous at z, and let x_k be a sequence in X converging to z. To show that $f(x_k)$ converges to $f(z)$, consider any $\varepsilon > 0$. Since f is continuous at z, there exists $\delta > 0$ such that $d_Y(f(x), f(z)) < \varepsilon$

whenever $d_X(x,z) < \delta$. Because x_k converges to z, there exists $K > 0$ such that $d_X(x_k, z) < \delta$ whenever $k \geq K$. It follows that $d_Y(f(x_k), f(z)) < \varepsilon$, whenever $k \geq K$, and $f(x_k)$ converges to $f(z)$.

For the second half of the proof, suppose that whenever x_k is a sequence in X converging to z, the sequence $f(x_k)$ converges to $f(z)$ in Y. The proof will proceed by contradiction, that is, it will be shown that the assumption that the conclusion is false leads to a contradiction. Therefore, the conclusion must be true.

Let $\varepsilon > 0$. Suppose there does not exist $\delta > 0$ such that $d_Y(f(x), f(z)) < \varepsilon$ whenever $d_X(z,x) < \delta$. Then for every positive integer k there exists an $x_k \in X$ such that $d_X(x_k, z) < 1/k$ and $d_Y(f(x_k), f(z)) \geq \varepsilon$. Since $1/k$ converges to 0 as k goes to infinity, it follows that x_k converges to z. It is obvious that $f(x_k)$ does not converge to $f(z)$ because $d_Y(f(x_k), f(z)) \geq \varepsilon$ for all k, contradicting the starting assumption that whenever x_k is a sequence in X converging to z, the sequence $f(x_k)$ converges to $f(z)$ in Y. Therefore, there must exist δ such that $d_Y(f(x), f(z)) < \varepsilon$ whenever $d_X(x,z) < \delta$, and f is continuous z. $\square$

A slightly different, but equivalent, way of stating the definition of $f : X \to Y$ being continuous at z is that for every $\varepsilon > 0$ there exists $\delta > 0$ such that

$$f\big(B_\delta(z)\big) \subset B_\varepsilon\big(f(z)\big).$$

For a subset E of Y, set $f^{-1}(E) = \big\{x \in X : f(x) \in E\big\}$. Using this notation, the above can be restated as the function $f : X \to Y$ is continuous at z if and only if for every $\varepsilon > 0$ there exists $\delta > 0$ such that

$$B_\delta(z) \subset f^{-1}\left[B_\varepsilon(f(z))\right]. \tag{1.8}$$

This point of view is helpful when working with functions that are continuous at every point of X and is linked to the definition of an open set.

Let X be a metric space with metric d. A subset U of X is an *open set* in X provided that for every x in U there exists $r > 0$ such that $B_r(x) \subset U$. The next proposition uses open sets to characterize functions that are continuous at every point in a metric space, such functions will be called simply *continuous functions*.

Proposition 1.2.4 *Let X and Y be metric spaces and let $f : X \to Y$ be a function from X to Y. The function f is continuous if and only if $f^{-1}(U)$ is an open set in X, whenever U is an open set in Y.*

Proof. Suppose f is continuous at every x in X, and let U be an open set in Y. It must be shown that for any x in $f^{-1}(U)$ there exists $r > 0$ such that $B_r(x) \subset f^{-1}(U)$. Since U is open and $f(x)$ is in U, there exists $\varepsilon > 0$ such that $B_\varepsilon(f(x)) \subset U$. Now using the continuity of f at x, there exists $\delta > 0$ such that (1.8) holds. Clearly, $f^{-1}\left[B_\varepsilon(f(x))\right] \subset f^{-1}(U)$, and hence

$$B_\delta(x) \subset f^{-1}\left[B_\varepsilon(f(x))\right] \subset f^{-1}(U).$$

Therefore, $f^{-1}(U)$ is an open set in X.

Next, suppose that whenever U is an open set in Y, then $f^{-1}(U)$ is an open set in X. Let x be an arbitrary point in X and let $\varepsilon > 0$. Because $B_\varepsilon(f(x))$ is an open subset of Y by *Exercise* 2, p. 27, the set $f^{-1}(B_\varepsilon(f(x))$ is an open subset of X by hypothesis, and it contains x. Hence, there exists $\delta > 0$ such that $B_\delta(x) \subset f^{-1}(B_\varepsilon(f(x))$. Thus (1.8) holds and f is continuous at x. □

Corollary 1.2.5 *Let X, Y and Z be metric spaces. If $f : X \to Y$ and $g : Y \to Z$ are continuous functions, then $g \circ f : X \to Z$ is a continuous function.*

Proof. $(g \circ f)^{-1}(U) = f^{-1}(g^{-1}(U))$ □

Since *Proposition 1.2.4* shows that the open sets completely determine the continuous functions from one metric space into another without specific reference to the metric, what role does the metric really play? To address this question, we first need to examine the union and intersection properties of open sets. To do this we work with collections of sets or sets whose elements are sets. Let $\mathcal{S}$ be a collection of subsets of a set X. The *union* and *intersection* of the sets in $\mathcal{S}$ are subsets of X defined by

$$\bigcup_{E \in \mathcal{S}} E = \{x : x \in E \text{ for some } E \in \mathcal{S}\}$$

and

$$\bigcap_{E \in \mathcal{S}} E = \{x : x \in E \text{ for all } E \in \mathcal{S}\},$$

respectively. Sometimes it is more convenient to define a finite collection of sets in X by indexing them as in $E_1, \ldots, E_m$.

Proposition 1.2.6 *Let X be a metric space with metric d. Then the open sets have the following properties:*

(a) Both X and the empty set ϕ are open sets.

(b) If $\mathcal{S}$ is any collection of open sets in X, then

$$\bigcup_{U \in \mathcal{S}} U = \{x : x \in U \text{ for some } U \in \mathcal{S}\}$$

is an open set in X.

(c) If $U_1, \ldots, U_m$ is any finite collection of open sets in X, then

$$\bigcap_{j=1}^{m} U_j = \{x : x \in U_j \text{ for } j = 1, \ldots, m\}$$

is an open set in X.

Proof. Obviously, X is an open set. The empty set, ϕ, is open because it is impossible to find x in ϕ such that $B_r(x)$ is not contained in ϕ for all $r > 0$. In other words, the empty set satisfies the definition of an open set vacuously.

For part *(b)*, let x be any point in

$$\bigcup_{U \in \mathcal{S}} U.$$

Then x is in at least one particular V in $\mathcal{S}$. Since V is open, there exists $r > 0$ such that $B_r(x) \subset V$. It follows that

$$B_r(x) \subset \bigcup_{U \in \mathcal{S}} U,$$

and the union of the sets in $\mathcal{S}$ is open because x was an arbitrary point in it.

For the third part, let x be an arbitrary point in

$$\bigcap_{j=1}^{m} U_j.$$

Consequently, x is in every U_j for $j = 1, \ldots, m$. Since U_j is open, for each j, there exists $r_j > 0$ such that $B_{r_j}(x) \subset U_j$. Set $r = \min\{r_j : j = 1, \ldots, m\}$, which is positive. It follows that $B_r(x) \subset U_j$ for $j = 1, \ldots, m$ and

$$B_r(x) \subset \bigcap_{j=1}^{m} U_j.$$

Therefore, the intersection of a finite collection of open sets is open. □

Let X be a nonempty set. A collection $\mathcal{T}$ of subsets of X such that (a) both X and ϕ are in $\mathcal{T}$; (b) if $\mathcal{S}$ is any collection of sets in $\mathcal{T}$, that is, $\mathcal{S} \subset \mathcal{T}$, then $\bigcup_{U \in \mathcal{S}}$ is in $\mathcal{T}$; (c) if $U_1, \ldots, U_m$ is any finite collection of sets in $\mathcal{T}$, then $\bigcap_{j=1}^{m} U_j$ is in $\mathcal{T}$. is called a *topology* on X, and the sets in $\mathcal{T}$ are called *open sets.* A *topological space* is simply a nonempty set with a specified topology. Thus the content of *Proposition 1.2.6* is that a metric on X determines a topology on X, making X is a topological space.

A topology on a set X has the *Hausdorff property* provided that given distinct points x and y in X there exist disjoint open sets U and V in the topology such that $x \in U$ and $y \in V$. When the topology on X has the Hausdorff property, X is called a *Hausdorff topological space.* It has already been shown that every metric topology has the Hausdorff property. Not every topological space has the Hausdorff property, and hence not every topological space is a metric space. For example, $\mathcal{T} = \{X, \phi\}$ is the smallest topology for any nonempty set X. If X contains more than one point, this topology does not have the Hausdorff property and does not come from a metric.

For any nonempty set X, the collection of all subsets of X is obviously a topology on X and the largest topology possible on X. This topology is called the *discrete topology*. The discrete topology on any nonempty set is always a metric topology. For example, on any nonempty set X

$$d(x,y) = \begin{cases} 0 & \text{if } x = y \\ 1 & \text{if } x \neq y \end{cases}, \tag{1.9}$$

is a metric that will produce the discrete topology on X because $B_{1/2}(x) = \{x\}$ for all $x \in X$. So on every nonempty set there is at least one metric.

Let X be a topological space with topology $\mathcal{T}_X$, and let Y be a nonempty subset of X. It is easy to verify that

$$\mathcal{T}_Y = \{U \cap Y : U \in \mathcal{T}_X\} \tag{1.10}$$

is a topology on Y. The topology $\mathcal{T}_Y$ is called the *relative topology* and is the default topology on any nonempty subset of a topological space. Note that if Z is a nonempty subset of Y, then the relative topology on Z from Y is the same as the relative topology on Z from X because $(U \cap Y) \cap Z = U \cap Z$.

A nonempty subset Y of a topological space X is *discrete* provided that for every $y \in Y$ there exists an open set U of X such that $U \cap Y = \{y\}$. Thus for every y in Y, the set $\{y\}$ is an open subset of Y with the relative topology. It follows that every subset of Y is an open subset of Y with the relative topology because the union of open sets is an open set. So the relative topology of a nonempty discrete subset of a topological space is the discrete topology.

If X is a metric space with metric d, then $\mathcal{T}(d)$ will denote the topology on X determined by d, that is, the collection of open sets determined by d. Two metrics d_1 and d_2 on X will be called *equivalent metrics* provided $\mathcal{T}(d_1) = \mathcal{T}(d_2)$. When working with open sets, equivalent metrics are interchangeable.

Theorem 1.2.7 *If d is a metric on X, then*

$$d'(x,y) = \frac{d(x,y)}{1 + d(x,y)}$$

is a metric on X that is equivalent to d and satisfies $d'(x,y) < 1$ for all x and y in X.

Proof. It is easy to see that $d'(x,y) < 1$ and satisfies the first three conditions for being a metric.

To prove the triangle inequality, first observe that the real-valued function

$$f(t) = \frac{t}{1+t} = \frac{1}{1+\frac{1}{t}}$$

is increasing for $t \geq 0$. Hence $d(x,y) \leq d(x,z) + d(z,y)$ implies that

$$\frac{d(x,y)}{1 + d(x,y)} \leq \frac{d(x,z) + d(z,y)}{1 + d(x,z) + d(z,y)},$$

and it follows that

$$\begin{aligned} d'(x,y) &= \frac{d(x,y)}{1+d(x,y)} \\ &\le \frac{d(x,z)+d(z,y)}{1+d(x,z)+d(z,y)} \\ &= \frac{d(x,z)}{1+d(x,z)+d(z,y)} + \frac{d(z,y)}{1+d(x,z)+d(z,y)} \\ &\le \frac{d(x,z)}{1+d(x,z)} + \frac{d(z,y)}{1+d(z,y)} \\ &= d'(x,z)+d'(z,y). \end{aligned}$$

Thus $d'(x,y) \le d'(x,z) + d'(z,y)$, and $d'(x,y)$ is a metric on X.

It remains to be shown that $\mathcal{T}(d) = \mathcal{T}(d')$. Actually, more is true; the two metrics have the same open balls, just the radii change. Let $B'_r(x)$ denote an open ball of radius r in the d' metric. It is easy to verify that for $r > 0$

$$B_r(x) = B'_{\frac{r}{1+r}}(x)$$

and that for $0 < r < 1$

$$B'_r(x) = B_{\frac{r}{1-r}}(x).$$

It now follows from the definition of an open set that $\mathcal{T}(d) = \mathcal{T}(d')$. □

The complementary behavior of open sets is also an important topological matter. If E is a subset of X, then the *complement* of E is by definition $X \setminus E = \{x \in X : x \notin E\}$. A subset E of a topological space X is a *closed set* provided that $X \setminus E$ is an open set. For metric spaces, we have the following useful characterization of closed sets:

Proposition 1.2.8 *Let X be a metric space. A subset F of X is closed if and only if the limit of every convergent sequence x_k in F is also in F.*

Proof. Suppose F is closed and x_k is a sequence in F converging to x. Since F is closed, $X \setminus F$ is open. If x is not in F, then x is in the open set $X \setminus F$ and there exists $r > 0$ such that $B_r(x) \subset X \setminus F$. Now there exists K such that $x_k \in B_r(x)$ for $k \ge K$, contradicting the fact that x_k is in F for all k.

For the second half of the proof, suppose that the limit of every convergent sequence x_k in F is also in F. It suffices to show that $X \setminus F$ is open. Let x be a point in $X \setminus F$. If $B_{1/k}(x) \subset X \setminus F$ does not hold for any k, then for every k there exists $x_k \in F \cap B_{1/k}(x)$. Clearly, x_k is a sequence in F converging to $x \in X \setminus F$, contradicting the hypothesis that the limit of every convergent sequence x_k in F is also in F. Therefore, $B_{1/k}(x) \subset X \setminus F$ for some k, and $X \setminus F$ is open. □

Proposition 1.2.9 *Let X be a topological space. Then the closed sets of X have the following properties:*

(a) Both X and the ϕ are closed sets.

(b) If $\mathcal{S}$ is any collection of closed sets in X, then

$$\bigcap_{F \in \mathcal{S}} F = \{x : x \in F \text{ for all } F \in \mathcal{S}\}$$

is a closed set in X.

(c) If $F_1, \ldots F_m$ is any finite collection of closed sets in X, then

$$\bigcup_{j=1}^{m} F_j = \{x : x \in F_j \text{ for some } j\}$$

is a closed set in X.

Proof. Since $X \setminus \phi = X$ and $X \setminus X = \phi$, both X and ϕ are closed because ϕ and X are open sets.

For *(b)* and *(c)*, it suffices to show that

$$X \setminus \bigcap_{F \in \mathcal{S}} F \text{ and } X \setminus \bigcup_{j=1}^{m} F_j$$

are open sets. Using de Morgan's formulas from *Exercise* 6, p. 27, these sets can be expressed as follows:

$$X \setminus \bigcap_{F \in \mathcal{S}} F = \bigcup_{F \in \mathcal{S}} (X \setminus F)$$

and

$$X \setminus \bigcup_{j=1}^{m} F_j = \bigcap_{j=1}^{m} (X \setminus F_j).$$

It follows from the definition of a topology that the right-hand sides are open sets because the sets $X \setminus F$ for $F \in \mathcal{S}$ and $X \setminus F_j$ for $j = 1, \ldots, m$ are open sets. Therefore,

$$\bigcap_{F \in \mathcal{S}} F \text{ and } \bigcup_{j=1}^{m} F_j$$

are closed sets. □

Let E be a subset of a topological space X. Define the *closure* of E by

$$E^{-} = \{x : E \cap U \neq \phi \text{ for all open } U \text{ containing } x\}. \tag{1.11}$$

Note that x is not in E^{-} if and only if there exists an open set V containing x such that $E \cap V = \phi$. In particular, if V is an open set such that $E \cap V = \phi$, then $V \subset X \setminus E^{-}$. The closure construction has a number of useful properties that are worth stating as a proposition.

Proposition 1.2.10 *Let E be a subset of a topological space X. Then the following hold:*

(a) The set E^- is a closed set containing E.

(b) The set E is closed if and only if $E = E^-$.

(c) If F is a closed set such that $E \subset F$, then $E^- \subset F$.

(d) $E^- = \bigcap\{F : E \subset F = F^-\}$.

Proof. Obviously, $E \subset E^-$. If x is not in E^-, then there exists an open set V containing x such that $E \cap V = \phi$. It follows that $V \subset X \setminus E^-$ and $X \setminus E^-$ is open by *Exercise* 8, p. 28, a basic fact that will be used frequently. Thus E^- is a closed set and *(a)* holds.

If E is closed, then $X \setminus E$ is open, and a point in $X \setminus E$ cannot be in E^- because $E \cap (X \setminus E) = \phi$. Hence, $E^- \subset E \subset E^-$ and $E = E^-$. The converse is obvious.

For part *(c)*, observe that $X \setminus F$ is an open set such that $E \cap (X \setminus F) = \phi$ and hence $E^- \cap (X \setminus F) = \phi$.

Finally, part *(d)* follows from parts *(a)*, *(b)*, and *(c)*. □

Given sets X_1, X_2, ..., X_m, their *Cartesian product* is defined by

$$X_1 \times \ldots \times X_m = \{(x_1, \ldots, x_m) : x_j \in X_j \text{ for } j = 1, \ldots, m\},$$

and for $1 \leq j \leq m$ the projections

$$p_j : X_1 \times \ldots \times X_m \to X_j$$

are defined by $p_j(x_1, \ldots, x_m) = x_j$.

Suppose X_1, ..., X_m are also metric spaces with metrics d_1, ..., d_m. Then

$$d((x_1, \ldots, x_m), (y_1, \ldots, y_m)) = \sum_{j=1}^{m} d_j(x_j, y_j) \tag{1.12}$$

defines a metric on $X_1 \times \ldots \times X_m$ called the *product metric.* An important property of the product metric is that its open sets can be characterized in terms of the open sets of X_1, ..., X_m.

Theorem 1.2.11 *Let X_1, ..., X_m be metric spaces with metrics d_1, ..., d_m, and let d be the metric on $X_1 \times \ldots \times X_m$ defined by (1.12). The following are equivalent for a subset U of $X_1 \times \ldots \times X_m$:*

(a) U is an open subset for the metric d.

(b) For every point $(x_1, \ldots, x_m) \in U$ there exist open sets U_j of X_j for $j = 1, \ldots, m$ such that

$$(x_1, \ldots, x_m) \in U_1 \times \ldots \times U_m \subset U.$$

In particular, $U_1 \times \ldots \times U_m$ is an open set for the metric d, when U_j is an open set in X_j for $j = 1, \ldots, m$.

Proof. To begin the proof, first observe that for $r > 0$

$$B_r(x_1) \times \ldots \times B_r(x_m) \subset B_{mr}\big((x_1, \ldots, x_m)\big) \tag{1.13}$$

and

$$B_r\big((x_1, \ldots, x_m)\big) \subset B_r(x_1) \times \ldots \times B_r(x_m). \tag{1.14}$$

If U is an open set in $X_1 \times \ldots \times X_m$ and $(x_1, \ldots, x_m)$ is in U, then there exists $r > 0$ such that

$$B_{mr}\big((x_1, \ldots, x_m)\big) \subset U,$$

and *(b)* follows from (1.13) because open balls are open sets.

If condition *(b)* holds, then, given $(x_1, \ldots, x_m) \in U$, for each j there exists $r_j > 0$ such that $B_{r_j}(x_j) \subset U_j$. Set $r = \min\{r_j : j = 1, \ldots, m\}$, and note that

$$B_r(x_1) \times \ldots \times B_r(x_m) \subset U_1 \times \ldots \times U_m \subset U.$$

Now(1.14) implies that

$$B_r\big((x_1, \ldots, x_m)\big) \subset U,$$

and U is an open set for the metric d. □

Corollary 1.2.12 *Let X_1, ..., X_m be metric spaces. If d_j and d'_j are equivalent metrics on X_j for $j = 1, \ldots, m$, then $d\big((x_1, \ldots, x_m), (y_1, \ldots, y_m)\big) = \sum_{j=1}^m d_j(x_j, y_j)$ and $d'\big((x_1, \ldots, x_m), (y_1, \ldots, y_m)\big) = \sum_{j=1}^m d'_j(x_j, y_j)$ are equivalent metrics on $X_1 \times \ldots \times X_m$.*

Condition *(b)* in *Theorem 1.2.11* now leads us to a topology on any product of a finite number of topological spaces $X_1, \ldots, X_m$. Let $\mathcal{T}$ be the collection of subsets U of $X_1 \times \ldots \times X_m$ such that for every point $(x_1, \ldots, x_m) \in U$ there exist open sets U_j of X_j satisfying

$$(x_1, \ldots, x_m) \in U_1 \times \ldots \times U_m \subset U.$$

We need to show that $\mathcal{T}$ is a topology. Obviously, $X_1 \times \ldots \times X_m$ is in $\mathcal{T}$ because each X_j is an open set in X_j. The empty set satisfies the required condition vacuously and is in $\mathcal{T}$. It is also clear that the union of any collection of sets in $\mathcal{T}$ is in $\mathcal{T}$.

To show that the finite intersection of sets in $\mathcal{T}$ is in $\mathcal{T}$, it suffices to show that $U \cap V$ is in $\mathcal{T}$ when U and V are in $\mathcal{T}$. If $(x_1, \ldots, x_m)$ is in $U \cap V$, then there exist open sets U_j and V_j of X_j for $j = 1, \ldots, m$ such that $(x_1, \ldots, x_m) \in U_1 \times \ldots \times U_m \subset U$ and $(x_1, \ldots, x_m) \in V_1 \times \ldots \times V_m \subset V$. It follows that

$$(x_1, \ldots, x_m) \in (U_1 \times \ldots \times U_m) \cap (V_1 \times \ldots \times V_m) \subset U \cap V.$$

Since

$$(U_1 \times \ldots \times U_m) \cap (V_1 \times \ldots \times V_m) = (U_1 \cap V_1) \times \ldots \times (U_m \cap V_m),$$

$U \cap V$ is in $\mathcal{T}$ when U and V are in $\mathcal{T}$. Thus $\mathcal{T}$ is a topology on $X_1 \times \ldots \times X_m$, and it is called the *product topology*. From *Theorem 1.2.11*, we know that the product topology is a metric topology when $X_1, \ldots, X_m$ are metric spaces

In the remainder of this section, we will establish a few useful properties of functions from one topological space to another. Let X and Y be topological spaces. A function $f : X \to Y$ is *continuous* at $z \in X$ provided that for every open set U of Y containing $f(z)$ there exists an open set V of X such that $z \in V \subset f^{-1}(U)$. It is easy to verify that for metric spaces this definition is equivalent to the definition of continuity at a point z found on page 16. As before, a *continuous function* means a function that is continuous at every point of its domain. We also have the following topological versions of *Proposition 1.2.4* and its corollary:

Proposition 1.2.13 *Let X and Y be topological spaces and let $f : X \to Y$ be a function from X to Y. The function f is continuous if and only if $f^{-1}(U)$ is an open subset of X, whenever U is an open subset of Y.*

Proof. Suppose f is continuous at every x in X, and let U be an open subset of Y. Consider an arbitrary x in $f^{-1}(U)$. Then, $f(x)$ is in U, and by the definition of continuity there exists an open subset V of X such that $x \in V \subset f^{-1}(U)$. It now follows from *Exercise* 8, p. 28 that $f^{-1}(U)$ is open. The other direction is trivial. $\square$

Corollary 1.2.14 *Let X, Y and Z be topological spaces. If $f : X \to Y$ and $g : Y \to Z$ are continuous functions, then $g \circ f : X \to Z$ is a continuous function.*

Corollary 1.2.15 *Let X and Y be topological spaces and let $f : X \to Y$ be a function. If the topology on X is the discrete topology, then f is continuous.*

Let X and Y be topological spaces. A function $f : X \to Y$ is said to be open if $f(U)$ is an open subset of Y, whenever U is an open subset of X.

Proposition 1.2.16 *If $X_1 \times \ldots \times X_m$ is a product of topological spaces with the product topology, then the projections are continuous open functions.*

Proof. If V is an open set in X_j, then

$$p_j^{-1}(V) = X_1 \times \ldots \times X_{j-1} \times V \times X_{j+1} \times \ldots \times X_m$$

which is an open set in the product topology. So the projections are continuous.

Given an open set U of $X_1 \times \ldots \times X_m$, it must be shown that $p_j(U)$ is open in X_j. It suffices by *Exercise* 8, p. 28, to show that for each y in $p_j(U)$ there exist an open set V in X_j such that $y \in V \subset p_j(U)$. If y is in $p_j(U)$, then there exists $(x_1, \ldots, x_m)$ in U such that $x_j = y$. So there exist open sets U_i of X_i for $i = 1, \ldots, m$ such that $(x_1, \ldots, x_m) \in U_1 \times \ldots \times U_m \subset U$. It follows that $y = x_j \in U_j = p_j(U_1 \times \ldots \times U_m) \subset p_j(U)$. Thus U_j is the required open set V of X_j and the projection p_j is an open function. $\square$

Proposition 1.2.17 *Let $X_1 \times \ldots \times X_m$ be a product of topological spaces with the product topology, and let Y be a topological space. A function $f : Y \to X_1 \times \ldots \times X_m$ is continuous if and only if the functions $f_j = p_j \circ f$ are continuous for $j = 1, \ldots, m$.*

Proof. If f is continuous, then the functions $f_j = p_j \circ f$ are all continuous because the projections are continuous and the composition of continuous functions is continuous by *Corollary 1.2.14*.

Suppose the functions $f_j = p_j \circ f$ are all continuous. These functions are called *coordinate functions* because $f(y) = (f_1(y), \ldots, f_m(y))$. Let U be an open set in $X_1 \times \ldots \times X_m$. To show that $f^{-1}(U)$ is open it suffices to show that for every point y in $f^{-1}(U)$, there exists an open set V of Y such that $y \in V \subset f^{-1}(U)$.

By *Theorem 1.2.11*, there exist open sets U_j of X_j such that $f(y) \in U_1 \times \ldots \times U_m \subset U$, and so

$$y \in f^{-1}(U_1 \times \ldots \times U_m) \subset f^{-1}(U).$$

Observe that

$$\bigcap_{j=1}^{m} (p_j \circ f)^{-1}(U_j) = f^{-1}(U_1 \times \ldots \times U_m).$$

Therefore, $V = f^{-1}(U_1 \times \ldots \times U_m)$ is the required open set because the functions $p_j \circ f$ are continuous. □

Let X and Y be topological spaces. A function $f : X \to Y$ is a *homeomorphism* provided that f is a one-to-one and onto function such that both f and f^{-1} are continuous functions. Suppose $f : X \to Y$ is a one-to-one and onto function. Since $(f^{-1})^{-1} = f$, the function f^{-1} is continuous if and only if f is an open function. Therefore, a one-to-one and onto function is a homeomorphism if and only if it is both continuous and open. It follows that for a homeomorphism f, a subset U of X is open if and only if $f(U)$ is an open set in Y. Since $f(X \setminus U) = Y \setminus f(U)$ because f is one-to-one and onto, a subset E of X is closed if and only if $f(E)$ is an closed set in Y. Consequently, two topological spaces are regarded as equal when they are homeomorphic.

A property of topological spaces is a *topological invariant* if given two homeomorphic topological spaces either they both have the property or neither one has it. A topological space X with topology $\mathcal{T}$ is said to be *metrizable* provided that there exists a metric d for X such that $\mathcal{T}(d) = \mathcal{T}$. Being metrizable is a topological invariant (*Exercise* 20, p. 29).

We close this section with a word of caution about the differences between metric spaces and topological spaces. *Propositions 1.2.3 and 1.2.8* are not true for all topological spaces. Convergence is a more delicate issue in general topological spaces and requires using *nets* instead of sequences. Many results like *Propositions 1.2.3 and 1.2.8* can be generalized to topological spaces by replacing sequences with nets, but this requires a substantial detour into the theory of nets. The interested reader can find a comprehensive treatment of nets in

Kelley [7], Chapter 2 or Willard [16], Chapter 4. These two books both with the same title, *General Topology*, will be our standard general topology references. Although not recent publications, they have stood the test of time and are still readily available.

EXERCISES

1. Show that a sequence in a metric space cannot converge to two distinct points.
2. Prove that an open ball of a metric space X is an open set in X.
3. Let X and Y be sets, let $f : X \to Y$ be a function, and let $\mathcal{S}$ be a collection of subsets of Y. Prove that
$$f^{-1}\left(\bigcup_{E\in\mathcal{S}} E\right) = \bigcup_{E\in\mathcal{S}} f^{-1}(E),$$
$$f^{-1}\left(\bigcap_{E\in\mathcal{S}} E\right) = \bigcap_{E\in\mathcal{S}} f^{-1}(E),$$
and
$$f^{-1}(Y \setminus E) = X \setminus f^{-1}(E).$$
(The usual way to prove two sets are equal is to show that each point in the set on the left is in the set on the right and visa versa.)
4. Let X be a metric space with metric d, and let y be a point in X. Show that the function $d_y(x) = d(x, y)$ is continuous on X. Given $\varepsilon > \delta > 0$, show that there exists a function $g : \mathbb{R} \to \mathbb{R}$ such that the function $f = g \circ d_y$ has the following properties:
   (a) f is continuous
   (b) $0 \leq f(x) \leq 1$ for all $x \in X$
   (c) $f(x) = 1$ if and only if $d(x, y) \leq \delta$,
   (d) $f(x) = 0$ if and only if $d(x, y) \geq \varepsilon$.
5. Show that $d_1(x, y) = |x^3 - y^3|$ is a metric on $\mathbb{R}$ equivalent to $d(x, y) = |x - y|$.
6. Let $\mathcal{S}$ be a collection of subsets of a set X. Prove *de Morgan's formulas*:
$$X \setminus \bigcap_{E\in\mathcal{S}} E = \bigcup_{E\in\mathcal{S}} (X \setminus E)$$
and
$$X \setminus \bigcup_{E\in\mathcal{S}} E = \bigcap_{E\in\mathcal{S}} (X \setminus E).$$

7. Prove that in a metric space X with metric d, the closed balls defined by $\{x \in X : d(x, y) \leq r\}$ are closed sets and that
$$B_r(y)^- \subset \{x \in X : d(x, y) \leq r\}.$$
Prove that in $\mathbb{R}^n$ the closure of an open Euclidean ball $B_r(\mathbf{y})$ is the closed ball $\{\mathbf{x} : d(\mathbf{x}, \mathbf{y}) \leq r\}$, and show by example that this is not true in all metric spaces.

8. Let X be a topological space. Prove that a subset U of X is an open set if and only if for every x in U there exists an open set V such that $x \in V \subset U$.

9. Let E be a nonempty subset of a metric space X. A point $x \in X$ is in E^- if and only if there exists a sequence x_k in E converging to x.

10. Let U be an open set in a metric space X, and let x be a point in U. Prove that there exists an open set V such that $x \in V \subset V^- \subset U$. A topological space with this property is said to be *regular* .

11. Let X be a Hausdorff topological space. Show that if x is in X, then $\{x\}$ is a closed set in X.

12. Let F be a closed subset of a metric space X with metric d. Define the *distance from the point x in X to the closed set F* by $d(x, F) = \inf\{d(x, y) : y \in F\}$ for any x in X. (If E is a set of real numbers that is bounded below, then the *infinum* of E, written $\inf E$, is the greatest lower bound and $\inf E = -\sup(-E)$.) For $r > 0$, show that set $\{x \in X : d(x, F) \geq r\}$ is a closed set or equivalently that the set $\{x \in X : d(x, F) < r\}$ is open.

13. Show that if E is a nonempty subset of $\mathbb{R}$, then $\sup E$ or $\inf E$ is in E^- according as E is bounded above or below.

14. Let x_k and y_k be sequences in the metric spaces X and Y. Show that the sequence (x_k, y_k) in $X \times Y$ with the product topology converges to (x, y) if and only if x_k converges to x in X and y_k converges to y in Y.

15. Show that the metrics $d(\mathbf{x}, \mathbf{y}) = \|\mathbf{x} - \mathbf{y}\|$ and $d'(\mathbf{x}, \mathbf{y}) = \sum_{j=1}^{n} |x_1 - y_2|$ on $\mathbb{R}^n$ are equivalent.

16. Let X be a metric space with metric d and let Y be a nonempty subset of X. Show that the relative topology on Y is the same the metric topology from restricting d to Y.

17. Show that the product topology on $\mathbb{R}^m \times \mathbb{R}^n$ coincides with the usual topology on $\mathbb{R}^{n+m}$. (One approach is to make use of *Exercise 15* above.)

18. Let X and Y be topological spaces, and let $f : X \to Y$ be a continuous function. Show that $f(A^-) \subset f(A)^-$. Then show that $f(A)^- = f(A^-)$, when f is a homeomorphism of X onto Y.

19. Let X and Y be metric spaces. An onto function $f : X \to Y$ such that $d_Y(f(u), f(v)) = d_X(u, v)$ for all u and v in X is called an *isometry*. Show that an isometry is a homeomorphism. Prove that the isometries of a metric space onto itself are a group under composition of functions. Give an example of a distance preserving function of a metric space to itself that is not onto.

20. Let X and Y be homeomorphic topological spaces. Prove that X is metrizable if and only if Y is metrizable.

1.3 Continuous Group Operations

Let G be both a group and a topological space. So we are assuming there is a specified binary operation on G that satisfies the group axioms on page 3 and a collection of open sets $\mathcal{T}$ satisfying the axioms of a topology on page 19. No assumption, however, is being made of any connection between these two structures on G. In general, there is no reason to believe that multiplying two elements in G or taking the inverse of an element in G are continuous functions. When these functions are continuous, however, algebraic and analytic ideas come together in a rich theory.

A *topological group* is a group that is also a topological space such that the function $(x, y) \to xy^{-1}$ is a continuous function from $G \times G$ with the product topology to G. When the topology on G comes from a metric, G will be called a *metric group*.

Proposition 1.3.1 *Let G be both a group and a topological space. Then G is a topological group if and only if both the function $x \to x^{-1}$ from G to G and the function $(x, y) \to xy$ from $G \times G$ with the product metric to G are continuous.*

Proof. Suppose G is a topological group and e is the identity element of G. The function $y \to (e, y)$ from G into $G \times G$ is continuous by *Proposition 1.2.17*. Since the composition of continuous maps is continuous (*Corollary 1.2.14*), the function $y \to (e, y) \to ey^{-1} = y^{-1}$ is continuous.

To prove that the function $(x, y) \to xy$ is continuous, consider the function from $G \times G$ with the product topology to itself defined by $f(x, y) = (x, y^{-1})$. Then $p_1 \circ f = p_1$ is obviously continuous, and $p_2 \circ f(x, y) = (p_2(x, y))^{-1}$ is continuous because p_2 is continuous and $y \to y^{-1}$ was just shown to be continuous. Thus f is also continuous by *Proposition 1.2.17*. Consequently, the composition of f followed by $(x, y) \to xy^{-1}$ is continuous, that is, $(x, y) \to (x, y^{-1}) \to x(y^{-1})^{-1} = xy$ is continuous.

For the second half of the proof, suppose the functions $x \to x^{-1}$ and $(x, y) \to xy$ are continuous. Then as above $f(x, y) = (x, y^{-1})$ is continuous because $y \to y^{-1}$ is continuous. It follows that the composition $(x, y) \to f(x, y) = (x, y^{-1}) \to xy^{-1}$ is continuous. □

Of course, if G is a topological group with the group operation written additively, then the function $(x, y) \to x - y$ is continuous by definition and equivalent to the functions $x \to -x$ and $(x, y) \to x + y$ being continuous.

Until Chapter 5, all of our examples of topological groups will be metric groups. The most obvious example of a metric group is the additive real numbers $\mathbb{R}$ with the usual metric $d(x, y) = |x - y|$. The product metric on $\mathbb{R} \times \mathbb{R}$ is $d'((x, y), (u, v)) = |x - u| + |y - v|$. Since

$$|(x + y) - (a + b)| \leq |x - a| + |y - b| = d'((x, y), (a, b)),$$

addition is a continuous function from $\mathbb{R} \times \mathbb{R}$ to $\mathbb{R}$ at every (a, b) in $\mathbb{R}^2$. The inverse functions, $x \to -x$, is an isometry of $\mathbb{R}$ onto itself and obviously continuous. So $\mathbb{R}$ is a metric group by *Proposition 1.3.1.* In fact, the same argument with $d(\mathbf{x}, \mathbf{y}) = \|\mathbf{x} - \mathbf{y}\|$ replacing $d(x, y) = |x - y|$ shows that $\mathbb{R}^n$ under addition is also a metric group for $n > 1$.

With more careful estimates, it is not difficult to show that the groups $\mathbb{R}^+$ and $\mathbb{R} \setminus \{0\}$ are also metric groups with the metric $d(x, y) = |x - y|$ (*Exercise* 1, p. 38). Similarly, one shows that $\mathbb{C} \setminus \{0\}$ and $\mathbb{K}$ are metric groups using the metric $d(z, w) = |z - w|$, where $|z - w|$ is the complex absolute value (*Exercise* 2, p. 38). *In the future,* $\mathbb{R}$, $\mathbb{R}^n$, $\mathbb{R}^+$, $\mathbb{R} \setminus \{0\}$, $\mathbb{C}$, $\mathbb{C} \setminus \{0\}$ *and* $\mathbb{K}$ *will always refer to these specific metric groups.*

Any group can be made into a metric group by using the discrete metric defined by (1.9) because the product topology of two spaces with discrete topologies is also a discrete topology and all functions from one space with the discrete topology to another metric space are continuous. Occasionally it will be useful to consider $\mathbb{R}_d$, the real numbers with the discrete topology. It will be understood that $\mathbb{Z}$ is always the metric group with the discrete topology on the additive group $\mathbb{Z}$, which is the same as its relative topology as a subset of $\mathbb{R}$. It is also convenient to assume that every finite group is a discrete metric space and hence a metric group (see *Exercise* 3, p. 38).

Proposition 1.3.2 *Let G be a topological group and let a and b be elements of G. Then the following functions are all homeomorphisms of G onto G:*

(a) $x \to xb$

(b) $x \to ax$

(c) $x \to axb$

(d) $x \to axa^{-1}$,

(e) $x \to x^{-1}$

Proof. The function $x \to xb$ is one-to-one and onto by *Exercise* 4, p. 11. It is continuous because it is the composition of the continuous functions $x \to (x, b)$ and $(x, y) \to xy$. The function $x \to xb^{-1}$ is thus also continuous, and clearly the inverse of $x \to xb$, making $x \to xb$ a homeomorphism. Similar arguments work in the remaining parts. □

Let A and B be subsets of a group G. The following notations will be convenient:

$$A^{-1} = \{a^{-1} : a \in A\},$$

and

$$AB = \{ab : a \in A \text{ and } b \in B\}.$$

Notice that if the identity element e is in A, then $B \subset AB$.

It will also be helpful to use the concept of a neighborhood. Let X be a topological space and let x be in X. A *neighborhood* of x is a subset U of X such that there exist and open set V satisfying $x \in V \subset U$. A neighborhood need not be an open set. A set U is open if and only if it is a neighborhood of each of its points. Also, a function $f : X \to Y$ is continuous at z if and only if $f^{-1}(U)$ is a neighborhood of z whenever U is a neighborhood of $f(z)$.

Lemma 1.3.3 *Let G be a topological group. If U is an open neighborhood of the identity e in G, then there exists an open neighborhood V of the identity such that $VV \subset U$.*

Proof. Because multiplication is continuous, the set $V' = \{(x, y) : xy \in U\}$ is an open set in $G \times G$ with the product topology. Obviously, (e, e) is in V'. It follows from the definition of the product topology on page 24 that there exist open neighborhoods V_1 and V_2 of e such that $V_1 \times V_2 \subset V'$. Hence, $V_1V_2 \subset U$. Set $V = V_1 \cap V_2$ to get $VV \subset U$. □

For a topological space X, a *countable neighborhood base at x* is a sequence of neighborhoods U_m of x such that for every neighborhood U of x there exists m such that $U_m \subset U$. (The concept of a set or collection of sets being countable will be examined in more detail in Section 3.4 and its exercises. For now, countable is just an adjective that occurs in the terminology.)

Every metric space X has a countable neighborhood base at any point x; just let $U_m = B_{1/m}(x)$. When there is a countable neighborhood base at every point of topological space, the space is said to be *first countable.* Metric spaces are first countable, but not all topological spaces are first countable.

If G is a topological group and U_m is a countable neighborhood base at the identity e, then *Proposition 1.3.2* implies that xU_m and U_mx are countable neighborhood bases at x for any x in G. Thus a topological group is first countable, if it has a countable neighborhood base at the identity.

For metric groups, there are countable neighborhood bases at e that reflect the algebraic structure, as well as the metric structure.

Proposition 1.3.4 *Let G be a metric group. There exists a countable neighborhood base U_m at the identity element e with the following properties:*

(a) U_m is open for all m.

(b) $U_m U_m U_m \subset U_{m-1}$ *for all* $m \geq 2$.

(c) $U_m^{-1} = U_m$ *for all* m.

(d) $\bigcap_{m=1}^{\infty} U_m = \{e\}$.

Proof. If V is an open set containing e, then so is V^{-1} because $x \to x^{-1}$ is a homeomorphism. So $U = V \cap V^{-1}$ is an open neighborhood of e such that $U^{-1} = U$.

Let d be a metric for G. The construction of the neighborhood base will proceed by induction. To start the process, let $V_1 = B_1(e)$ and set $U_1 = V_1 \cap V_1^{-1}$, making U_1 is an open neighborhood of e such that $U_1^{-1} = U_1 \subset B_{1/1}(e)$.

Suppose $U_1, \ldots, U_m$ have been constructed satisfying conditions *(a)*, *(b)*, *(c)*, and $U_m \subset B_{1/m}(e)$. Note that $U_m \cap B_{1/(m+1)}(e)$ is an open neighborhood of e. By applying *Lemma 1.3.3* twice, there exists an an open neighborhood V_{m+1} of e such that

$$V_{m+1}V_{m+1}V_{m+1}V_{m+1} \subset U_m \cap B_{1/(m+1)}(e).$$

Since e is in V_{m+1}, it clearly follows that $V_{m+1}V_{m+1}V_{m+1} \subset U_m$ and $V_{m+1} \subset B_{1/(m+1)}(e)$. Set $U_{m+1} = V_{m+1} \cap V_{m+1}^{-1}$, so U_{m+1} satisfies condition *(a)* and *(c)*. Since $U_{m+1} \subset V_{m+1}$, it follows that U_{m+1} satisfies conditions *(b)* and $U_{m+1} \subset B_{1/(m+1)}(e)$.

Therefore by induction there exists a sequence of open neighborhoods U_m of e satisfying conditions *(a)*, *(b)*, *(c)*, and $U_m \subset B_{1/m}(e)$. The last inclusion guarantees that the sequence of sets U_m is a neighborhood basis at e and

$$\{e\} \subset \bigcap_{m=1}^{\infty} U_m \subset \bigcap_{m=1}^{\infty} B_{1/m}(e) = \{e\}.$$

It follows that *(d)* holds. □

The usual metric $d(\mathbf{x}, \mathbf{y}) = \|\mathbf{x} - \mathbf{y}\|$ on $\mathbb{R}^n$ has the additional invariant property that $d(\mathbf{u} + \mathbf{x}, \mathbf{u} + \mathbf{y}) = d(\mathbf{x}, \mathbf{y})$. In fact, every metric group has such a metric. These invariant metrics will be very useful, but the proof of their existence is delicate.

Theorem 1.3.5 *Let G be a metric group with metric d. There exists a metric ρ on G that is equivalent to d and satisfies*

$$\rho(ax, ay) = \rho(x, y)$$

for all x, y, and a in G.

Proof. Let U_m be a countable neighborhood base at e given by *Proposition 1.3.4*. The proof will depend only on the properties of this sequence of sets. For technical reasons, set $U_0 = G$. Define a function $f : G \times G \to \mathbb{R}$ by

$$f(x, y) = \begin{cases} 1/2^m & \text{if } x^{-1}y \in U_m \setminus U_{m+1} \\ 0 & \text{if } x^{-1}y \in U_m \text{ for all } m \geq 0. \end{cases}$$

The function f has the following properties:

(a) $0 \leq f(x,y) \leq 1$

(b) $f(x,y) = f(y,x)$ because $y^{-1}x = (x^{-1}y)^{-1}$ and $U_m^{-1} = U_m$

(c) $f(ax,ay) = f(x,y)$ because $(ax)^{-1}ay = x^{-1}y$

(d) $f(x,y) > 0$ if $x \neq y$ because $\bigcap_{m=1}^{\infty} U_m = \{e\}$

(e) $f(x,y) \leq 1/2^m$ implies that $x^{-1}y \in U_m$ because $U_{k+1} \subset U_k$ for all k.

For each positive integer n define

$$F_n : \underbrace{G \times \ldots \times G}_{n+1 \text{ copies}} \to \mathbb{R}$$

by

$$F_n(x_0, x_1, \ldots, x_n) = \sum_{j=1}^{n} f(x_{j-1}, x_j).$$

In particular, $F_1(x,y) = f(x,y)$.

Set

$$\rho(x,y) = \inf\{F_n(x_0, \ldots, x_n) : x_0 = x, x_n = y, n \geq 1\}$$

The function ρ has the following elementary properties:

(a) $0 \leq \rho(x,y) \leq f(x,y) \leq 1$

(b) $\rho(x,y) = \rho(y,x)$ because $f(x,y) = f(y,x)$ and consequently

$$F_n(x_0, x_1, \ldots, x_{n-1}, x_n) = F_n(x_n, x_{n-1}, \ldots, x_1, x_0) \tag{1.15}$$

(c) $\rho(ax,ay) = \rho(x,y)$ because

$$F_n(ax, x_1, \ldots, x_{n-1}, ay) = F_n(x, a^{-1}x_1, \ldots, a^{-1}x_{n-1}, y).$$

The next step is to show that ρ satisfies the triangle inequality. Given x, y, and z in G, it suffices to show that $\rho(x,z) + \rho(z,y) + 2\varepsilon \geq \rho(x,y)$ for every $\varepsilon > 0$. From the definition of ρ, there exist $x_1, \ldots, x_{n-1}$ such that

$$\rho(x,z) + \varepsilon \geq F_n(x, x_1 \ldots, x_{n-1}, z),$$

and there exist $y_1, \ldots, y_{m-1}$ such that

$$\rho(z,y) + \varepsilon \geq F_m(z, y_1 \ldots, y_{m-1} y).$$

Consequently,

$$\begin{aligned}
\rho(x,z) + \rho(z,y) + 2\varepsilon &\geq F_n(x, x_1, \ldots, x_{n-1}, z) + F_m(z, y_1, \ldots, y_{m-1}, y) \\
&= F_{n+m}(x, x_1 \ldots, x_{n-1}, z, y_1, , \ldots, y_{m-1}, y) \\
&\geq \rho(x,y).
\end{aligned}$$

To show that ρ is a metric, it remains only to show that $\rho(x, y) > 0$ when $x \neq y$. It has already been shown that ρ has the required invariance property $\rho(ax, ay) = \rho(x, y)$ for all x, y, and a in G. Completing the proof that ρ is a metric will not, however, complete the proof of the theorem. It must also be shown that d and ρ produce the same topology on G, that is, $\mathcal{T}(d) = \mathcal{T}(\rho)$. The following lemma will be used to complete these last two steps of the proof:

Lemma 1.3.6 *For all $x, x_1, \ldots, x_{n-1}, y$ in G, and $n \geq 1$*

$$f(x, y) \leq 2F_n(x, x_1, \ldots, x_{n-1}, y).$$

Proof. Without loss of generality we can assume that consecutive terms in the finite sequence $x, x_1, \ldots, x_{n-1}, y$ are distinct. (For example, if $x_1 = x_2$, then $F_n(x, x_1, \ldots, x_{n-1}, y) = F_{n-1}(x, x_2, \ldots, x_{n-1}, y)$, and it suffices to prove that $f(x, y) \leq 2F_{n-1}(x, x_2, \ldots, x_{n-1}, y)$.)

The $n = 1$ case is trivial. For $n = 2$, let m be the smallest positive integer such that $1/2^m \leq F_2(x, x_1, y) = f(x, x_1) + f(x_1, y)$. If $xx_1^{-1} \notin U_m$, then the definition of f implies that $1/2^{m-1} \leq f(x, x_1) \leq F_2(x, x_1, y)$, contradicting the choice of m. Thus $xx_1^{-1} \in U_m$ and likewise $x_1y^{-1} \in U_m$. It follows that $xy^{-1} \in U_mU_m \subset U_{m-1}$, and then $f(x, y) \leq 1/2^{m-1} = 2/2^m \leq 2F_1(x, x_1, y)$. Similarly, $f(x, y) \leq 2F_3(x, x_1, x_2, y)$, using $U_mU_mU_m \subset U_{m-1}$.

Proceeding by induction for $n \geq 4$, assume that the conclusion of the lemma holds for all integers from 1 to n. To show that it is also true for $n + 1$, consider $F_{n+1}(x, x_1, \ldots, x_n, y)$. Using $f(x, y) = f(y, x)$ and equation (1.15), it suffices to prove either $f(x, y) \leq 2F_{n+1}(x, x_1, \ldots, x_n, y)$ or $f(y, x) \leq 2F_{n+1}(y, x_n, \ldots, x_1, x)$. We can assume without loss of generality that

$$f(x, x_1) = F_1(x, x_1) \leq \frac{1}{2}F_{n+1}(x, \ldots, y).$$

Now there exists k, $1 \leq k \leq n$, such that the following inequalities hold:

$$\begin{aligned} F_k(x, \ldots, x_k) &\leq \frac{1}{2}F_{n+1}(x, \ldots, y) \\ F_{k+1}(x, \ldots, x_{k+1}) &\geq \frac{1}{2}F_{n+1}(x, \ldots, y). \end{aligned}$$

We will assume that $1 \leq k < n$ and leave the case $k = n$ to the reader. It follows that

$$F_{n-k}(x_{k+1}, \ldots, y) \leq \frac{1}{2}F_{n+1}(x, \ldots, y).$$

The induction assumption now implies that

$$f(x, x_k) \leq 2F_k(x, \ldots, x_k) \leq F_{n+1}(x, \ldots, y)$$

and

$$f(x_{k+1}, y) \leq 2F_{n-k}(x_{k+1}, \ldots, y) \leq F_{n+1}(x, \ldots, y)$$

It is also clear that

$$f(x_k, x_{k+1}) \leq F_{n+1}(x, \ldots, y)$$

These last three inequalities are the crucial ingredients for the finale of the proof.

Choose the smallest positive integer m such that

$$\frac{1}{2^m} \leq F_{n+1}(x, \ldots, y).$$

If xx_k^{-1} is not in U_m, then the definition of f implies that $1/2^{m-1} \leq f(x, x_k)$ and the first of our three crucial inequalities implies that

$$\frac{1}{2^{m-1}} \leq f(x, x_k) \leq F_{n+1}(x, \ldots, x_n, y),$$

contradicting the choice of m. Therefore, xx_k^{-1} must lie in U_m.

The same reasoning shows that $x_{k+1}y^{-1} \in U_m$ and $x_k x_{k+1}^{-1} \in U_m$. It follows that

$$xy^{-1} = (xx_k^{-1})(x_k x_{k+1}^{-1})(x_{k+1}y^{-1}) \in U_m U_m U_m,$$

and then part *(b)* of *Proposition 1.3.4* implies that $xy^{-1} \in U_{m-1}$. (In the $k = n$ case, $x_n y^{-1}$ is in U_m and $U_m U_m \subset U_{m-1}$ suffices.) Therefore,

$$f(x, y) \leq \frac{1}{2^{m-1}} = \frac{2}{2^m} \leq 2F_{n+1}(x, \ldots, y),$$

and the proof of the lemma is complete. □

As a consequence of the lemma

$$\rho(x, y) = \inf \{F_n(x_0, \ldots, x_n) : x_0 = x, x_n = y, n \geq 1\} \geq \frac{1}{2} f(x, y)$$

or simply

$$\rho(x, y) \geq \frac{1}{2} f(x, y)$$

In particular, $\rho(x, y) > 0$ when $x \neq y$ because $f(x, y) > 0$ when $x \neq y$. Hence, ρ is a metric. In fact, it is a bounded metric because $\rho(x, y) \leq 1$. Furthermore, setting $B'_r(y) = \{x : \rho(x, y) < r\}$, it is easy to see that $B'_r(y) = yB'_r(e)$ because $\rho(yx, y) = \rho(x, e)$.

Finally it must be shown that d and ρ generate the same open sets in G, that is $\mathcal{T}(d) = \mathcal{T}(\rho)$. For each y in G the open sets yU_m are a countable neighborhood base at y for $\mathcal{T}(d)$. For $\mathcal{T}(\rho)$, the open sets $B'_{1/2^m}(y)$ are a countable neighborhood base at y. So it suffices (using the definition of a countable base and *Exercise* 8, p. 28 again.) to show that

$$B'_{1/2^m}(y) \subset yU_m \subset B'_{1/2^{m-1}}(y)$$

for all $m \geq 1$. Since $B'_r(y) = yB'_r(e)$ for all y and $r > 0$, the above is further reduced to showing that

$$B'_{1/2^m}(e) \subset U_m \subset B'_{1/2^{m-1}}(e).$$

If $x \in B'_{1/2^m}(e)$, then

$$\frac{1}{2^m} > \rho(x, e) \geq \frac{1}{2} f(x, e)$$

and $1/2^{m-1} > f(x, e)$. Since the positive values of $f(x, y)$ are always of the form $1/2^k$, it follows that $1/2^m \geq f(x, e) = f(e, x)$. Thus $x = e^{-1}x$ must be in U_m by property (e) of $f(x, y)$. Therefore, and $B'_{1/2^m}(e) \subset U_m$.

Next, consider $y \in U_m$. Then, $f(e, y) \leq 1/2^m < 1/2^{m-1}$ and $\rho(e, y) \leq f(e, y) < 1/2^{m-1}$, proving that $U_m \subset B'_{1/2^{m-1}}(e)$. □

A metric ρ for a topological group G such that $\rho(ax, ay) = \rho(x, y)$ for all x, y, and a in G is called a *left-invariant metric.* For a left invariant metric the functions $\psi_a : G \to G$ defined by $\psi_a(x) = ax$ for $a \in G$ are not just homeomorphisms of G onto G but they are also isometries because they preserve the distance between points and are onto (see *Exercise* 19, p. 29).

A left-invariant metric need not be right-invariant, but the proof of the existence of left-invariant metrics can obviously be modified to prove the existence of right-invariant metrics. *As needed, we can assume the metric on a metric group is left- or right-invariant without changing the topology on the group.*

In Section 1.1 (page 7), we defined an algebraic homomorphism from one group to another. For metric groups, we want to consider algebraic homomorphisms that are also continuous. Since topological groups are the central topic, the adjective "algebraic" will be dropped and the assumption of continuity added for topological groups.

Let G and G' be topological groups. A function $\varphi : G \to G'$ is a *homomorphism* of G to G' provided that φ is continuous on G and that

$$\varphi(xy) = \varphi(x)\varphi(x) \tag{1.16}$$

for all x and y in G. *It is important to stress that a homomorphism will always be a continuous function unless the adjective "algebraic" precedes it.*

A homeomorphism φ of a topological group G onto a topological group G' that satisfies (1.16) for all x and y in G is a called an *isomorphism* and G and G' are *isomorphic topological groups.* The inverse of an isomorphism between two topological groups is an isomorphism of topological groups because it has already been shown (page 8) that the inverse of an algebraic isomorphism is an algebraic isomorphism and the inverse of a homeomorphism is a homeomorphism. Finally, an *automorphism* is an isomorphism of a topological group onto itself.

Here are some familiar examples of homomorphisms and isomorphisms. The function $f(x) = \exp(x) = e^x$ is an isomorphism from $\mathbb{R}$ onto $\mathbb{R}^+$ with $f^{-1}(x) = \ln(x)$. Similarly, $\varphi(x) = \exp(2\pi i x) = \cos(2\pi x) + i\sin(2\pi x)$ is a homomorphism of $\mathbb{R}$ onto $\mathbb{K}$, but it is not one-to-one. The functions $f(x) = ax$, where a is a nonzero real number are all automorphisms of $\mathbb{R}$ (see *Exercises 6 and 7,* p. 39). The simple complex polynomial $f(z) = z^n$, n a positive integer, is a homomorphism of $\mathbb{C} \setminus \{0\}$ onto itself that is not an isomorphism for $n > 1$.

Let $G_1, \ldots, G_m$ be groups. There is a natural group structure on $G_1 \times \ldots \times G_m$ called the *direct product.* Specifically,

$$(x_1, \ldots, x_m)(y_1, \ldots, y_m) = (x_1y_1, \ldots, x_my_m),$$
$$e = (e_1, \ldots, e_m),$$

and

$$(x_1, \ldots, x_m)^{-1} = (x_1^{-1}, \ldots, x_m^{-1}).$$

Proposition 1.3.7 *If $G_1, \ldots, G_m$ are topological groups, then the direct product $G_1 \times \ldots \times G_m$ with the product topology is a topological group.*

Proof. It must be shown that the function

$$f\big((x_1, \ldots, x_m), (y_1, \ldots, y_m)\big) = (x_1y_1^{-1}, \ldots, x_my_m^{-1})$$

from $(G_1 \times \ldots \times G_m) \times (G_1 \times \ldots \times G_m)$ to $G_1 \times \ldots \times G_m$ is continuous. By *Proposition 1.2.17*, it suffices to show that the functions

$$p_j \circ f\big((x_1, \ldots, x_m), (y_1, \ldots, y_m)\big) = x_jy_j^{-1}$$

from $(G_1 \times \ldots \times G_m) \times (G_1 \times \ldots \times G_m)$ to G_j are continuous at an arbitrary point $\big((a_1, \ldots, a_m), (b_1, \ldots, b_m)\big)$. For convenience, we will assume that $j = 1$; the proof is the same for all j but the notation is easier with $j = 1$.

Let U be an open neighborhood of $a_1b_1^{-1}$ in G_1. Because G_1 is a topological group, $\{(x_1, y_1) : x_1y_1^{-1} \in U\}$ is an open neighborhood of (a_1, b_1) in $G_1 \times G_1$ and there exist open neighborhoods V_1 and V_2 of a_1 and b_1, respectively, such that $V_1 \times V_2 \subset \{(x_1, y_1) : x_1y_1^{-1} \in U\}$ and hence $V_1V_2^{-1} \subset U$. Clearly,

$$(V_1 \times G_2 \ldots \times G_m) \times (V_2 \times G_2 \times \ldots \times G_m) \subset (p_1 \circ f)^{-1}(U)$$

and

$$(V_1 \times G_2 \ldots \times G_m) \times (V_2 \times G_2 \times \ldots \times G_m)$$

is an open neighborhood of $\big((a_1, \ldots, a_m), (b_1, \ldots, b_m)\big)$ by *Theorem 1.2.11.* □

The list of topological groups has just grown significantly and now includes $\mathbb{K}^n$ for all positive integers and metric groups like $\mathbb{R}^m \times \mathbb{K}^n \times \mathbb{Z}_p$. There are also some new connections worth mentioning. The function $\varphi : \mathbb{R}^+ \times \mathbb{R} \to \mathbb{C} \setminus \{0\}$ defined by $\varphi(r, \theta) = re^{i\theta}$ is a homomorphism of the product metric group $\mathbb{R}^+ \times \mathbb{R}$ onto $\mathbb{C} \setminus \{0\}$, and there is a homomorphism of metric groups behind the polar form of a nonzero complex number. Similarly, The function $\psi(r, z) = e^r z$ defines an isomorphism from the product metric group $\mathbb{R} \times \mathbb{K}$ onto $\mathbb{C} \setminus \{0\}$, and $\mathbb{C} \setminus \{0\}$ can be thought of as a product group that looks like an infinite cylinder.

We close this section with a topological characterization of when a topological group is metrizable. It provides a different viewpoint of what a metric group is and is not.

Theorem 1.3.8 *A topological group G is metrizable if and only if G is first countable and the points of G are closed sets.*

Proof. If G is a metric group, then it is first countable because every metric space is first countable and points are closed sets by *Exercise* 11, p. 28, because metric spaces have the Hausdorff property.

Conversely, suppose G is first countable and the points of G are closed sets. Let W_m be a countable neighborhood base at e the identity of G. Without loss of generality we can assume that $W_{m+1} \subset W_m$ by replacing W_m with $\bigcap_{k=1}^{m} W_k$.

If $x \neq e$, then $G \setminus \{x\}$ is an open neighborhood of e because $\{x\}$ is a closed set by hypothesis. Hence, $W_m \subset G \setminus \{x\}$ for some m, and for every $x \neq e$ in G, there exists some W_m such that x is not in W_m. Therefore, $\bigcap_{m=1}^{\infty} W_m = \{e\}$.

The proof of *Proposition 1.3.4* can be applied here by replacing $B_{1/m}(e)$ with W_m to obtain a countable neighborhood base U_m at the identity element e with the following properties:

(a) U_m is open for all m,

(b) $U_m U_m U_m \subset U_{m-1}$ for all $m \geq 2$

(c) $U_m^{-1} = U_m$ for all m

(d) $\bigcap_{m=1}^{\infty} U_m = \{e\}$.

Now the proof of *Theorem 1.3.5* can be reused word for word to prove that there exists a left-invariant metric ρ such that $\mathcal{T}(\rho)$ equals the original topology on G. Thus G is metrizable by definition. □

Metric groups are a rich class of topological groups that cut a wide swath across the theory of topological groups. They are certainly ample for a solid introduction to topological groups and have the advantage of requiring a lot less mathematical machinery. Although we will frequently restrict our attention to metric groups, we do not want to loose sight of how they fit into the larger picture. To help provide the reader with a broader perspective than metric groups, a variety of basic ideas and introductory results will be presented in the general context of topological groups and used to discuss results beyond the scope of this book.

EXERCISES

1. Show that the groups $\mathbb{R}^+$ and $\mathbb{R} \setminus \{0\}$ are metric groups with the metric $d(x, y) = |x - y|$.

2. Prove that the groups $\mathbb{C} \setminus \{0\}$ and $\mathbb{K}$ are metric groups.

3. Show that if G is a finite metric group, then the topology on G is the discrete topology.

4. Let G be a group and a topological space such that the function $(x, y) \to xy$ is a continuous function from $G \times G$ with the product topology to G. Prove that if the inverse function $x \to x^{-1}$ from G to G is continuous at the identity e, then it is continuous on G and G is a topological group.

5. Let G and G' be topological groups. Show that if $\varphi : G \to G'$ is an algebraic homomorphism that is continuous at e, then φ is a homomorphism.

6. Let $\varphi : \mathbb{R} \to \mathbb{R}$ be a homomorphism of $\mathbb{R}$ to itself. Prove the following:

   (a) $(1/q)\varphi(x) = \varphi(x/q)$ for all $x \in \mathbb{R}$ and $q \in \mathbb{Z} \setminus \{0\}$.

   (b) $\varphi(rx) = r\varphi(x)$ for all $x \in \mathbb{R}$ $r \in \mathbb{Q}$, the rational numbers.

   (c) $\varphi(sx) = s\varphi(x)$ for all s and x in $\mathbb{R}$.

   (d) There exists $a \in \mathbb{R}$ such that $\varphi(x) = ax$.

7. Show that the group of automorphisms of the metric group $\mathbb{R}$ is algebraically isomorphic to the group $\mathbb{R} \setminus \{0\}$.

8. Let G and G' be metric groups with left-invariant metrics d and d', respectively. If $\varphi : G \to G'$ is a homomorphism, then given $\varepsilon > 0$, there exists $\delta > 0$ such that $d'(\varphi(x), \varphi(y)) < \varepsilon$ whenever $d(x, y) < \delta$.

9. Let G be a metric group. Given a neighborhood U of the identity of G and a positive integer n, show that there exists a neighborhood V of the identity such that $V^n \subset U$.

10. Let G be a topological group. Prove the following:

    (a) If V is a neighborhood of the identity and $x \in V^-$, then $xV \cap V \neq \phi$.

    (b) If U and V are neighborhoods of the identity such that $VV^{-1} \subset U$, then $V^- \subset U$.

    (c) G is a *regular* topological space, that is, given an open set U and $y \in U$, there exists an open set V such that $y \in V \subset V^- \subset U$.

11. Let G be a metric group. Prove that there exists a two-sided invariant metric if and only if there exist a neighborhood base of open sets at e such that $xU_m x^{-1} = U_m$ for all $x \in G$ and $m \geq 1$.

12. Let F be a closed subset of a topological group G. Prove that if x is in $G \setminus F$, then there exists an open neighborhood of e such that $xV \cap FV = \phi$.

1.4 Subgroups and Their Quotient Spaces

Let G be a topological group, and let H be a subgroup of G as defined in Section 1.1 on page 8. Then H is a group under the same binary operation that defines the group structure on G. If H has the relative topology from G, then the product topology on $H \times H$ is the same as the relative topology from the product topology on $G \times G$ because

$$(U \cap H) \times (V \cap H) = (U \times V) \cap (H \times H).$$

Hence, the function $(x, y) \to xy^{-1}$ on $H \times H$ with the product topology is a continuous function to H. *Therefore, with the relative topology a subgroup of a topological group is a topological group. It will be understood henceforth that a subgroup of a topological group is the topological group obtained by using the relative topology on the subgroup.*

Unless H is a closed subset of G, however, the topological group H can be a strange beast. Consequently, the focus will be on subgroups that are closed subsets of G or simply *closed subgroups.*

Let G be a topological group. When the topology on G has the Hausdorff property defined on page 19, G is called a Hausdorff topological group. They are the most common topological groups that one encounters in the literature. A set consisting of a single point is always a closed set in a Hausdorff topological group by *Exercise* 11, p. 28. When a subgroup H of G is also a discrete subset of G (page 20), H is said to be a *discrete subgroup* of G. For example, $\mathbb{Z}$ is a discrete subgroup of $\mathbb{R}$. Recall that the relative topology of a discrete subset of a topological space is the discrete topology.

Proposition 1.4.1 *Let H be a subgroup of a Hausdorff topological group G. The following hold:*

(a) If there exists an open set in G such that $U \cap H = \{e\}$, then H is a discrete subgroup of G.

(b) If H is a discrete subgroup of G, then H is a closed subgroup of G.

Proof. Given an open set U such that $U \cap H = \{e\}$, let a be an element of H. Then Ua is an open set containing a because $x \to xa$ is a homeomorphism of G onto itself by *Proposition 1.3.2.* Suppose x is in $Ua \cap H$. Then $x = ua$ for some $u \in U$. Since x is in H, it follows that $xa^{-1} = u \in U \cap H$, and consequently $xa^{-1} = e$. Therefore, $x = a$, $Ua \cap H = \{a\}$, and H is discrete.

Let H be a discrete subgroup of G, and consider $x \in G \setminus H$. There exists an open set containing e such that $U \cap H = \{e\}$, and another open set V containing e such that $VV^{-1} \subset U$ because $(x, y) \to xy^{-1}$ is continuous. Suppose there are two points a and b in $Vx \cap H$. Then $a = v_1 x$ and $b = v_2 x$ with v_1 and v_2 in V. So $ab^{-1} = (v_1 x)(v_2 x)^{-1} = v_1 v_2^{-1} \in VV^{-1} \subset U$. Hence $ab^{-1} \in U \cap H = \{e\}$ and $a = b$. Thus either $Vx \cap H$ is empty or contains exactly one point. If $Vx \cap H = \{a\}$, then $a \neq x$ because $x \notin H$, and there exists an open set W containing x such that $a \notin W$ because G is Hausdorff. It follows that

$(Vx \cap W) \cap H = \phi$. Thus every x in $G \setminus H$ has an open neighborhood that is contained in $G \setminus H$. Therefore, and $G \setminus H$ is open, and H is closed. □

Proposition 1.4.2 *If H is a subgroup of a topological group G, then H^{-} is also a subgroup of G.*

Proof. Let a and b be arbitrary points in H^{-}. It suffices by *Exercise* 10, p. 12, to show that ab^{-1} is in H^{-}. Let U be an open neighborhood of ab^{-1}. There exist open neighborhoods V and W of a and b, respectively, such that $VW^{-1} \subset U$. Because a and b are in H^{-} there exist $x \in V \cap H$ and $y \in W \cap H$. It follows that xy^{-1} is in $U \cap H$. As U is an arbitrary neighborhood of ab^{-1}, this means that ab^{-1} is in H^{-}. □

Proposition 1.4.3 *If H is an Abelian subgroup of a Hausdorff topological group G, then H^{-} is also an Abelian subgroup of G.*

Proof. Suppose a and b are elements of H^{-} and $ab \neq ba$. Then there exist disjoint open neighborhoods U_1 and U_2 of ab and ba, respectively, because G is Hausdorff. Since multiplication is continuous, there exist open neighborhoods V_1, V_2 of a and W_1, W_2 of b such that $V_1W_1 \subset U_1$ and $W_2V_2 \subset U_2$. Set $V = V_1 \cap V_2$ and $W = W_1 \cap W_2$. Because a and b are in H^{-}, there exists $x \in V \cap H$ and $y \in W \cap H$. It follows that xy is in U_1 and yx is in U_2. Therefore, $xy \neq yx$ and H is not Abelian, a contradiction. □

Quotient spaces are constructed by decomposing a space into disjoint sets and then letting the sets of the decomposition be the points of the quotient space. To understand better how a subgroup naturally decomposes a group into disjoint sets, we return to the idea of the orbits from Section 1.1 (page 5).

Let X be a nonempty set and let L be a subgroup of the group S_X. Then there is a natural *L-orbit* or simply *orbit*, when unambiguous, for each a in X defined by $\mathcal{O}(a) = \{\psi(a) : \psi \in L\}$. This definition extends the idea of an orbit used in Section 1.1 by replacing the cyclic group generated by an element α of S_n with any subgroup L of S_X. Like the permutation case, orbits decompose a set X into disjoint pieces.

Proposition 1.4.4 *Let L be a subgroup of S_X and let a and b be points in X. Then either $\mathcal{O}(a) = \mathcal{O}(b)$ or $\mathcal{O}(a) \cap \mathcal{O}(b) = \phi$.*

Proof. It suffices to show that $\mathcal{O}(a) \cap \mathcal{O}(b) \neq \phi$ implies $\mathcal{O}(a) = \mathcal{O}(b)$.. If $\mathcal{O}(a) \cap \mathcal{O}(b) \neq \phi$, then there exist ψ_1 and ψ_2 in L such that $\psi_1(a) = \psi_2(b)$ or equivalently $a = \psi_1^{-1} \circ \psi_2(b)$. For any ψ in L, it follows that $\psi \circ \psi_1^{-1} \circ \psi_2 \in L$ because L is a subgroup of S_X, and hence $\psi(a) = \psi \circ \psi_1^{-1} \circ \psi_2(b)$ is in $\mathcal{O}(b)$. Thus $\mathcal{O}(a) \subset \mathcal{O}(b)$. Similarly, $\mathcal{O}(b) \subset \mathcal{O}(a)$ and $\mathcal{O}(a) = \mathcal{O}(b)$. □

Now, let H be a subgroup of a group G, and as before, let $\psi_y : G \to G$ be defined by $\psi_y(x) = yx$. Each ψ_y is one-to-one, onto, and an element of S_G. The set $L = \{\psi_y : y \in H\}$ is a subgroup of S_G because $\psi_y \circ \psi_{y'} = \psi_{yy'}$ and

$\psi_y^{-1} = \psi_{y^{-1}}$. (In fact, $y \to \psi_y$ is algebraic isomorphism of H onto L. With $H = G$, this is the solution to *Exercise* 14, p. 12.) The L-orbit of a in G is

$$\begin{aligned} \mathcal{O}(a) &= \{\psi_y(a) : y \in H\} \\ &= \{ya : y \in H\} \\ &= Ha \end{aligned}$$

and is called a *left coset*. By the previous proposition, left cosets are either disjoint or equal. Each $a \in G$ is in the left coset Ha because e is in H. The subgroup itself is always a coset because $He = H$. Thus the left cosets decompose G into disjoint subsets, one of which is H itself. There is an important criterion for determining when two cosets are equal.

Proposition 1.4.5 *Let H be a subgroup of a group G, and let a and b be in G. Then $Ha = Hb$ if and only if ab^{-1} is in H.*

Proof. If $Ha = Hb$, then $a = yb$ for some $y \in H$ and $ab^{-1} = y \in H$. Conversely, $ab^{-1} = y \in H$ implies that $Ha = Hab^{-1}b = Hyb = Hb$ because $Hy = H$ for any $y \in H$. □

For topological groups, left cosets have additional topological properties.

Proposition 1.4.6 *Let G be a topological group.*

(a) If H is a closed subgroup, then every coset Ha is a closed set in G.

(b) If H is an open subgroup of G, then every coset Ha is an open set in G.

(c) If H is an open subgroup of G, then H is also a closed subgroup of G.

Proof. Since $x \to xa$ is a homeomorphism of G onto G, the left coset Ha is closed if and only if H is closed, and Ha is open if and only if H is open, proving parts *(a)* and *(b)*. (See the discussion following the definition of homeomorphism on page 26.)

If H is an open subgroup, then the cosets are all open sets and

$$\bigcup_{a \notin H} Ha$$

is an open subset of G. The fact that $Ha = H$ if and only if $a \in H$ implies that

$$H = G \setminus \bigcup_{a \notin H} Ha.$$

Thus H is the complement of an open set and closed. □

As an example consider the multiplicative group $\mathbb{R} \setminus \{0\}$. Clearly, $\mathbb{R}^+$ is a subgroup of $\mathbb{R} \setminus \{0\}$ and there are two cosets, $\mathbb{R}^+$ and $\mathbb{R}^- = \mathbb{R}^+(-1)$. Note that as in part *(c)*, $\mathbb{R}^+$ is both an open and a closed set in $\mathbb{R} \setminus \{0\}$. Open-and-closed subgroups will play an important role in Chapter 5.

Let G/H denote the collection of left cosets. So G/H is a set whose elements are the orbits of the points of G under the subgroup $L = \{\psi_y : y \in H\}$ of S_G. In other words,

$$G/H = \{E \subset G : E = Ha \text{ for some } a \in G\}.$$

There is a natural function π from G onto G/H defined by $\pi(a) = Ha$. We will restrict our study of quotient spaces to closed subgroups of metric groups. The choice of left cosets is arbitrary, and every result has an analogue for right cosets.

Theorem 1.4.7 *Let G be a metric group, and let d be a left-invariant metric for G. If H is a closed subgroup of G, then*

$$\rho(Ha, Hb) = \inf\{d(xa, yb) : x, y \in H\} \tag{1.17}$$

is a metric on G/H. Furthermore, a subset U of G/H is an open subset of G/H if and only if $\pi^{-1}(U)$ is an open subset of G.

Proof. Obviously, $\rho(Ha, Hb) = \rho(Hb, Ha)$, and $\rho(Ha, Hb) \geq 0$ for all Ha and Hb in G/H. Since d is a left-invariant metric, $d(xa, yb) = d(a, x^{-1}yb)$ and

$$\rho(Ha, Hb) = \inf\{d(a, wb) : w \in H\} = \inf\{d(wa, b) : w \in H\} \tag{1.18}$$

are alternative formulas for the function ρ.

Clearly, $Ha = Hb$ implies $\rho(Ha, Hb) = 0$. If $\rho(Ha, Hb) = 0$, then equation (1.18) implies that there exists a sequence x_k in H such that $d(a, x_k b)$ converges to zero, which is equivalent to saying that $x_k b$ converges to a. Since H is closed, its cosets are closed sets by *Proposition 1.4.6.* Hence a is in Hb and $Ha = Hb$.

To prove the triangle inequality, consider three cosets Ha, Hb, and Hc. Let $\varepsilon > 0$ be arbitrary. By using equation (1.18), there exist x and y in H such that

$$\begin{aligned} d(xa, c) &\leq \rho(Ha, Hc) + \varepsilon \\ d(c, yb) &\leq \rho(Hc, Hb) + \varepsilon. \end{aligned}$$

Equation (1.17) and these inequalities imply that

$$\begin{aligned} \rho(Ha, Hb) &\leq d(xa, yb) \\ &\leq d(xa, c) + d(c, yb) \\ &\leq \rho(Ha, Hc) + \rho(Hc, Hb) + 2\varepsilon. \end{aligned}$$

Since the inequality $\rho(Ha, Hb) \leq \rho(Ha, Hc) + \rho(Hc, Hb) + 2\varepsilon$ holds for all $\varepsilon > 0$, it must be true that $\rho(Ha, Hb) \leq \rho(Ha, Hc) + \rho(Hc, Hb)$. This completes the proof that ρ defines a metric on G/H.

Let $B'_r(Ha)$ denote the open ball around Ha in G/H determined by the metric ρ. In preparation for proving that U, a subset of G/H, is an open set in G/H if and only if $\pi^{-1}(U)$ is an open set in G, it will be shown that

$$\pi^{-1}(B'_r(Ha)) = HB_r(a). \tag{1.19}$$

Observe that x is in $\pi^{-1}(B'_r(Ha))$ if and only if $\rho(Hx, Ha) < r$ if and only if $d(x, ya) < r$ for some $y \in H$ if and only if $x \in B_r(ya) = yB_r(a)$ for some $y \in H$ if and only if $x \in HB_r(a)$. This proves the validity of equation (1.19).

Suppose that U is an open set in G/H and let a be in $\pi^{-1}(U)$. Then $B'_r(Ha) \subset U$ for some $r > 0$ and $B_r(a) \subset HB_r(a) = \pi^{-1}(B'_r(Ha)) \subset \pi^{-1}(U)$ by (1.19). So $\pi^{-1}(U)$ is open.

Suppose that $\pi^{-1}(U)$ is open and Ha is in U. Then a is in $\pi^{-1}(U)$ and $B_r(a) \subset \pi^{-1}(U)$ for some $r > 0$. Since $\pi^{-1}(U)$ is a union of cosets, $HB_r(a) \subset \pi^{-1}(U)$. Using (1.19) again, it follows that $B'_r(Ha) \subset U$ and U is open. □

Once again we are in the enviable position of having a formula for a metric and a characterization of the open sets that is independent of the metric. Given a left-invariant metric d for G and a closed subgroup H of G, the metric given by equation (1.17) or (1.18) on G/H is called a *quotient metric*. The topology determined by a quotient metric is a *quotient topology* on G/H and does not depend on the choice of the invariant metric d. With the quotient topology G/H is called a *quotient space*.

Corollary 1.4.8 *If H is a closed subgroup of the metric group G, then the function $\pi : G \to G/H$ from G to the quotient space G/H defined by $\pi(g) = Hg$ is continuous and open.*

Proof. The function π is obviously continuous because $\pi^{-1}(U)$ is an open set if and only if U is an open set in G by the theorem.

To prove that π is an open function, first observe that for any subset V of G

$$\begin{aligned} \pi^{-1}(\pi(V)) &= HV \\ &= \bigcup_{y \in H} yV \\ &= \bigcup_{y \in H} \psi_y(V). \end{aligned}$$

If V is open, then $\psi_h(V)$ is open because ψ_h is a homeomorphism of G onto itself, and then $\pi^{-1}(\pi(V))$ is open because it is a union of open sets. It follows from the theorem that $\pi(V)$ is open, when V is open □

Corollary 1.4.9 *Let H be a closed subgroup of a metric group G. If H' is a subgroup of G containing H, then H'/H is a subset of G/H and the relative topology on H'/H is the same as its quotient topology.*

Proof. Clearly H is a closed subgroup of H' and H'/H is a subset of G/H. Let d be a left-invariant metric for G, and let ρ be the metric on G/H given by equation (1.17). Let d' be the restriction of d to H'. Because d' is a left-invariant metric on H', it can be used to construct a quotient metric ρ' on H'/H by setting $\rho'(Ha, Hb) = \inf\{d'(xa, yb) : x, y \in H\}$ for a and b in H'. It is now obvious that ρ' is just the restriction of ρ to H'/H. Hence the relative topology on H'/H is the same as its quotient topology by *Exercise* 16, p. 28. □

Corollary 1.4.10 *Let H be a closed subgroup of a metric group. The quotient topology on G/H is the discrete topology if and only if H is an open subgroup of G.*

Proof. If the quotient topology is discrete, then the points of G/H are open sets, and, in particular, $\{H\}$ is an open set in G/H. It follows from the theorem that $\pi^{-1}(\{H\}) = H$ is an open set in G.

Conversely, if H is an open set in G, then every coset Ha is an open subset of G, and each point of G/H is an open set by *Corollary 1.4.8.* So the quotient topology is discrete. □

The next proposition provides a criterion for the convergence of a sequence in G/H in terms of the convergence of sequences in the metric group G. It will be used in the proofs of several results in the remainder of this section.

Proposition 1.4.11 *Let H be a closed subgroup of a metric group G. A sequence Hx_k in G/H converges in the quotient topology to Hx if and only if there exists a sequence y_k in H such that the sequence y_kx_k converges to x in the group G.*

Proof. Let d be a left-invariant metric on G, and let $\rho(Ha, Hb)$ be the metric on G/H defined by equation (1.18). Suppose Hx_k converges to Hx in G/H, so the sequence of numbers $\rho(Hx_k, Hx)$ converges to 0. It follows from equation (1.18) that for each k there exists y_k in H such that $d(y_kx_k, x) < \rho(Hx_k, Hx) + 1/k$. Hence $d(y_kx_k, x)$ converges to 0, and y_kx_k converges to x in G.

Conversely, if there exists a sequence y_k in H such that y_kx_k converges to x in G, then $Hx_k = Hy_kx_k = \pi(y_kx_k)$ converges to $\pi(x) = Hx$ in G/H because π is continuous. □

For a group G, defining a group structure on G/H by $(Ha, Hb) \mapsto HaHb$ does not work in general because $HaHb$ need not equal any Hc. An additional hypothesis is required to produce a natural group structure on G/H. A subgroup H of G is a *normal subgroup* of G provided $aHa^{-1} = H$ for all $a \in G$ or equivalently $aH = Ha$ for all $a \in G$.

Recall that the function $\varphi_a(x) = axa^{-1}$ is an algebraic automorphism of a group G. These automorphisms are called the *inner automorphisms.* Alternatively, a subgroup H of G is normal if and only if every inner automorphism maps H onto itself. In particular, the subgroups G and $\{e\}$ of a group G are normal subgroups, and every subgroup of an Abelian group is normal.

Theorem 1.4.12 *If H is a closed normal subgroup of a metric group G, then G/H with the quotient topology and the binary operation $(Ha, Hb) \mapsto HaHb = Hab$ is a metric group. Furthermore, $\pi(a) = Ha$ is an open homomorphism of G onto G/H.*

Proof. Routine calculations show that $(Ha)(Hb) = H(aHa^{-1})ab = HHab = Hab$ is an associative binary operation on G/H, the coset H is an identity element, and $(Ha)^{-1} = Ha^{-1}$, giving G/H a group structure. Clearly, $\pi(a)\pi(b) = (Ha)(Hb) = Hab = \pi(ab)$ and π is an onto algebraic homomorphism.

We already know that when H is closed, G/H is a metric space and π is an open continuous function. To complete the proof, it remains to show that $(Ha, Hb) \to Hab^{-1}$ is a continuous function from $G/H \times G/H$ to G/H .

Let d be a left-invariant metric on G, and let ρ be the metric on G/H given by (1.17). By *Proposition 1.2.3*, it suffices to prove that the sequence $Ha_kb_k^{-1}$ converges to Hab^{-1}, whenever the sequence (Ha_k, Hb_k) converges to (Ha, Hb) in $G/H \times G/H$. It follows from *Exercise* 14, p. 28, that (Ha_k, Hb_k) converges to (Ha, Hb) if and only if Ha_k and Hb_k converge to Ha and Hb, respectively, in G/H.

By *Proposition 1.4.11*, there exist sequences x_k and y_k in H such that x_ka_k and y_kb_k converge to a and b respectively in G. Because G is a metric group, $x_ka_k(y_kb_k)^{-1}$ converges to ab^{-1}. Note that $x_ka_k(y_kb_k)^{-1} = x_ka_kb_k^{-1}y_k^{-1} = x_kw_ka_kb_k^{-1}$ for some w_k in H because $a_kb_k^{-1}H = Ha_kb_k^{-1}$ for all k by normality. Using *Proposition 1.4.11* again, it follows that $Ha_kb_k^{-1}$ converges to Hab^{-1} because x_kw_k is in H for all k and $x_kw_ka_kb_k^{-1}$ converges to ab^{-1}. □

When G is a metric group and H is a closed normal subgroup of G, the open homomorphism $\pi : G \to G/H$ defined by $\pi(x) = Hx$ will be called the *canonical homomorphism*.

Proposition 1.4.13 *Let G and G' be metric groups. If $\varphi : G \to G'$ is homomorphism of G to G', then the kernel $K = \{g \in G : \varphi(g) = e'\} = \varphi^{-1}(e')$ is a closed normal subgroup of G.*

Proof. By *Exercise* 12, p. 12, the kernel $K = \varphi^{-1}(e')$ is a subgroup, and it is closed by *Exercise* 3, p. 50, because φ is continuous and $\{e'\}$ is a closed subset of G'.

If x is in K, then

$$\begin{aligned} \varphi(axa^{-1}) &= \varphi(a)\varphi(x)\varphi(a^{-1}) \\ &= \varphi(a)e'\varphi(a)^{-1} \\ &= e'. \end{aligned}$$

Hence $aKa^{-1} \subset K$ for all a in G, and K is normal by *Exercise* 5, p. 50. □

Theorem 1.4.14 (First Isomorphism Theorem) *Let G and G' be metric groups. If $\varphi : G \to G'$ is homomorphism of G onto G' with kernel K, then $\tilde{\varphi}(Ka) = \varphi(a)$ defines a one-to-one homomorphism of G/K onto G' satisfying $\tilde{\varphi} \circ \pi = \varphi$. If φ an open function, then $\tilde{\varphi}$ is an isomorphism of G/K onto G'.*

Proof. If $Ka = Kb$, then ab^{-1} is in K and $\varphi(ab^{-1}) = e'$ or $\varphi(a) = \varphi(b)$. Thus setting $\tilde{\varphi}(Ka) = \varphi(a)$ unambiguously defines a function from G/K to G' which is onto because φ is onto. Clearly, $\tilde{\varphi} \circ \pi = \varphi$. Moreover,

$$\begin{aligned} \tilde{\varphi}(KaKb) &= \tilde{\varphi}(Kab) \\ &= \varphi(ab) \\ &= \varphi(a)\varphi(b) \\ &= \tilde{\varphi}(Ka)\tilde{\varphi}(Kb) \end{aligned}$$

and φ is an algebraic homomorphism.

To show that $\tilde{\varphi}$ is continuous, it suffices to show that $\pi(\varphi^{-1}(U)) = \tilde{\varphi}^{-1}(U)$ because φ is continuous and π is open. An element of $\pi(\varphi^{-1}(U))$ has the form Ka with $a \in \varphi^{-1}(U)$. Hence, $\tilde{\varphi}(Ka) = \varphi(a) \in U$ and $\pi(\varphi^{-1}(U)) \subset \tilde{\varphi}^{-1}(U)$. If Ka is in $\tilde{\varphi}^{-1}(U)$, then $Ka = \pi(a)$ and $\varphi(a) = \tilde{\varphi}(Ka) \in U$, proving that $\tilde{\varphi}^{-1}(U) \subset \pi(\varphi^{-1}(U))$. Thus $\pi(\varphi^{-1}(U)) = \tilde{\varphi}^{-1}(U)$, and φ is continuous.

If $\tilde{\varphi}(Ka) = \varphi(a) = e'$, then a is in K and $Ka = K$, the identity element of G/K. So the kernel of $\tilde{\varphi}$ is the trivial subgroup of G/K, and $\tilde{\varphi}$ is one-to-one.

Although it has now been shown that $\tilde{\varphi}$ is an algebraic isomorphism of G/K onto G', it does not follow that it is an isomorphism of metric groups just because $\tilde{\varphi}$ is continuous. The difficulty is that $\tilde{\varphi}^{-1}$ need not be continuous (see *Exercise* 12, p. 51).

The function $\tilde{\varphi}^{-1}$ is continuous if and only if $\tilde{\varphi}$ is an open function because

$$(\tilde{\varphi}^{-1})^{-1}(U) = \tilde{\varphi}(U).$$

If the function φ is an open function and U is an open set in G/K, then $\pi^{-1}(U)$ is an open subset of G, and $\varphi(\pi^{-1}(U))$ is an open subset of G'. Since

$$\varphi(\pi^{-1}(U)) = \tilde{\varphi} \circ \pi(\pi^{-1}(U)) = \tilde{\varphi}(U),$$

the homomorphism $\tilde{\varphi}$ is an open function and an isomorphism of the metric group G/K onto G'. □

Theorem 1.4.15 (Second Isomorphism Theorem) *Let G and G' be metric groups and let φ be an open homomorphism of G onto G' with kernel K. If H' is a closed normal subgroup of G' and $H = \varphi^{-1}(H')$, then H is a closed normal subgroup of G containing K, and H/K is a closed normal subgroup of G/K. Furthermore, the three metric groups G/H, G'/H' and $(G/K)/(H/K)$ are all isomorphic.*

Proof. It follows from *Exercises 7 and 8*, p. 50, that H is a closed normal subgroup of G. Since K is a normal subgroup of G, it is obviously a normal subgroup of any subgroup of G that contains K. Consequently, G'/H', G/H, G/K, and H/K are all well-defined metric groups.

Clearly, H/K is a subgroup of G/K. If $a \in G$, then $KaKxKa^{-1} = Kaxa^{-1}$ lies in H/K for all $x \in H$ because H normal. It follows that H/K is a normal subgroup of G/K. From *Corollary 1.4.9*, we know that the relative topology on H/K is the same as its quotient topology. (See *Exercises 9 and 10*, p. 51, for a full description of subgroups of quotient groups.)

To show that H/K is a closed set in G/K, it suffices to show that $(H/K)^{-} \subset H/K$. Consider Ka in $(H/K)^{-}$. Let U be any open neighborhood of a in G. Then $\pi(U)$ is an open neighborhood of Ka, and there exists x in H such that Kx is in $\pi(U) \cap H/K$ because Ka is in $(H/K)^{-}$. There exists y in U such that $Ky = \pi(y) = Kx$. It follows that y is in H because $K \subset H$ and $x \in H$. Thus every open neighborhood of a contains point y in H, proving that a is in $H^{-} = H$. Therefore, Ha is in H/K.

Let $\pi' : G' \to G'/H'$ be the canonical open homomorphism. By *Theorem 1.4.14*, there exists an isomorphism $\tilde{\varphi}$ of G/K onto G'. Consider the composition $\pi' \circ \tilde{\varphi} : G/K \to G'/H'$. It is an open homomorphism because both $\tilde{\varphi}$ and π' are open homomorphisms. Note that $\pi' \circ \tilde{\varphi}(Ka) = H'\varphi(a) = H'$ if and only if $\varphi(a)$ is in H' if and only if a is in H. Therefore, the kernel of $\pi' \circ \tilde{\varphi}$ is H/K and $(G/K)/(H/K)$ is isomorphic to G'/H' by *Theorem 1.4.14*.

Lastly, consider the composition $\pi' \circ \varphi : G \to G'/H'$, which is an open homomorphism because both φ and π' are open homomorphisms. Its kernel is clearly H. Therefore, G/H and G'/H' are isomorphic by *Theorem 1.4.14*. □

Corollary 1.4.16 *Let H and K be closed normal subgroups of the metric group G. If $K \subset H$, then the metric groups $(G/K)/(H/K)$ and G/H are isomorphic.*

Proof. Set $G' = G/K$ and $H' = H/K$. Then as in the proof of the theorem, H' is a closed subgroup of G'. Now the theorem can be applied with $\varphi = \pi$, the canonical homomorphism of G onto G', and $G'/H' = (G/K)/(H/K)$. □

If H is a subgroup of a Hausdorff topological group G, then *Theorem 1.4.7* provides the definition of a *quotient topology* on G/H by letting $\mathcal{T} = \{U \subset G/H : \pi^{-1}(U) \text{ is open}\}$ and applying *Exercise* 3, p. 27, to show that $\mathcal{T}$ is a topology. This is a Hausdorff topology when H is closed, and G/H is a topological group when H is normal by *Exercises 18, 19, and 20* beginning on page 193, which can be worked without further prerequisites. Then the proofs of the First and Second Isomorphism Theorems for Hausdorff topological groups are essentially the same as the proofs just presented for metric groups.

To illustrate the use of the First and Second Isomorphism Theorems (*Theorems 1.4.14 and 1.4.15*), we will determine all the closed subgroups of $\mathbb{R}$, their quotients, and the quotients of the quotients. The first proof requires the *greatest integer function* denoted by $[x]$ for $x \in \mathbb{R}$ and defined by $[x]$ equals the largest integer less than or equal to x. Note that $0 \leq x - [x] < 1$, and that x is an integer if and only if $x - [x] = 0$.

Theorem 1.4.17 *If H is a closed subgroup of $\mathbb{R}$, then exactly one of the following holds:*

(a) $H = \mathbb{R}$

(b) $H = \{0\}$

(c) there exists a positive number a such that $H = \{ma : m \in \mathbb{Z}\}$, making H a discrete subgroup of $\mathbb{R}$ isomorphic to $\mathbb{Z}$.

Proof. The first step is to show that either $H = \mathbb{R}$ or H is discrete. If H is not discrete, then by *Proposition 1.4.1* for every positive integer k there exists $x_k \in H$ with $0 < |x_k| < 1/k$. Observe that every interval of length $1/k$ contains an element of H of the form mx_k with $m \in \mathbb{Z}$. Given any real number x, for each positive integer k choose m_k such that $|m_k x_k - x| < 1/k$. Clearly, $m_k x_k$

converges to x, and x is in the closed subgroup H. Since x was an arbitrary real number, $H = \mathbb{R}$.

If H is discrete, then either $H = \{0\}$ or H contains a nonzero real number. Suppose H contains a nonzero real number, and hence its inverse. Thus H contains a positive number. Set $a = \inf\{x \in H : x > 0\}$. Then *Exercise* 13, p. 28, implies that a must lie in H because H is closed. Moreover, a is positive because H is discrete. Let x be an arbitrary element of H, and set $m = [x/a]$. Then $x - ma$ is in H. It follows from $0 \leq x/a - [x/a] < 1$ that $0 \leq x - ma < a$, contradicting the choice of a unless $0 = x - ma$. Therefore, $x = ma$ and $H = \{ma : m \in \mathbb{Z}\}$. Obviously, $m \mapsto ma$ defines an isomorphism from $\mathbb{Z}$ onto H to complete the proof. □

Corollary 1.4.18 *If H is a proper closed subgroup of $\mathbb{R}$, then $\mathbb{R}/H$ is isomorphic to $\mathbb{K}$.*

Proof. The function $\varphi(x) = \exp(2\pi i x/a)$ is a homomorphism of $\mathbb{R}$ onto $\mathbb{K}$ and its kernel is precisely $H = \{ma : m \in \mathbb{Z}\}$. Since it maps open intervals of $\mathbb{R}$ to open circular arcs, it is an open function. (In Section 1.5, we will prove a more general result, *Theorem 1.5.18*, that also implies that φ is open.) An application of the First Isomorphism Theorem completes the proof. □

The next corollary is usually proved by purely algebraic means, but we will take advantage of the theorem.

Corollary 1.4.19 *If H is a subgroup of $\mathbb{Z}$, then exactly one of the following holds:*

(a) $H = \mathbb{Z}$

(b) $H = \{0\}$

(c) there exists a positive integer $a > 1$ such that $H = \{ma : m \in \mathbb{Z}\}$.

Proof. Every subgroup of $\mathbb{Z}$ is a discrete subgroup of $\mathbb{R}$, and hence a closed subgroup of $\mathbb{R}$ by *Proposition 1.4.1.* □

Theorem 1.4.20 *If H' is a closed subgroup of $\mathbb{K}$ other than $\mathbb{K}$, then*

(a) there exists a positive integer n such that $H' = \{z \in \mathbb{K} : z^n = 1\}$

(b) H' is a finite cyclic group

(c) H' is a discrete subgroup of $\mathbb{K}$

(d) $\mathbb{K}/H'$ is isomorphic to $\mathbb{K}$.

Proof. Consider the homomorphism $\varphi(x) = \exp(2\pi i x)$ of $\mathbb{R}$ onto $\mathbb{K}$. Then, as noted in the proof of *Corollary 1.4.18*, φ is an open homomorphism with kernel $\mathbb{Z}$. Set $H = \varphi^{-1}(H')$. It follows that H cannot be $\{0\}$ because H contains $\mathbb{Z}$, and H cannot be $\mathbb{R}$ because $H' \neq \mathbb{K}$. Thus $H = \{ma : m \in \mathbb{Z}\}$ for some $a > 0$.

Since $\mathbb{Z} \subset H$, there must exist a positive integer n such that $1 = na$ or $a = 1/n$. Therefore,

$$\begin{aligned} H' &= \{\exp(2\pi ima) : m \in \mathbb{Z}\} \\ &= \{\exp(2\pi im/n) : m \in \mathbb{Z}\} \\ &= \{\exp(2\pi i/n)^m : m \in \mathbb{Z}\} \\ &= \{\exp(2\pi i/n)^m : 1 \leq m \leq n\} \end{aligned}$$

and completes the proof of parts *(a)* and *(b)*. Since a finite subgroup of a metric group is always a discrete subgroup, *(c)* holds.

For the last part, The Second Isomorphism Theorem implies that $\mathbb{R}/H$ and $\mathbb{K}/H'$ are isomorphic. It follows form *Corollary 1.4.18* that $\mathbb{R}/H$ is isomorphic to $\mathbb{K}$. □

EXERCISES

1. Prove that a finite subgroup of a Hausdorff topological groups is a discrete subgroup.
2. Let H be a subgroup of a metric group G with identity e. Show that if H is a neighborhood of the identity, then H is an open-and-closed subgroup of G.
3. Let X and Y be topological spaces. Prove that $f : X \to Y$ is a continuous function if and only if $f^{-1}(E)$ is a closed set in X when E is a closed set in Y.
4. Show that the quotient space for right cosets is homeomorphic to the quotient space for left cosets. (One approach is to use the function $gH \to (gH)^{-1} = Hg^{-1}$.)
5. Let H be a subgroup of a group G. Prove that H is a normal subgroup of G if and only if $aHa^{-1} \subset H$ for all a in G.
6. Prove that if H is a normal subgroup of a topological group G, then H^- is a normal subgroup of G.
7. Let G and G' be groups, and let $\varphi : G \to G'$ be an algebraic homomorphism. Prove the following:
   (a) If H' is a subgroup of G', then $\varphi^{-1}(H')$ is a subgroup of G containing the kernel of φ.
   (b) If H' is a normal subgroup of G', then $\varphi^{-1}(H')$ is a normal subgroup of G.
   (c) If H is a subgroup of G, then $\varphi(H)$ is a subgroup of G'.
   (d) If H is a normal subgroup of G and φ is onto, then $\varphi(H)$ is a normal subgroup of G'.

8. Let G and G' be topological groups, and let $\varphi : G \to G'$ be a homomorphism. Prove the following:

   (a) If H' is a closed subgroup of G', then $\varphi^{-1}(H')$ is a closed subgroup of G.

   (b) If φ is an open onto homomorphism and $\varphi^{-1}(H')$ is a closed subgroup of G, then H' is a closed subgroup of G'.

9. Let K be a normal subgroup of the group G, and let H' be a subset of G/K. Show that H' is a subgroup of G/K if and only if there exists a subgroup H of G such that $K \subset H$ and $H' = H/K$. In addition, show that H' is a normal subgroup of G/K if and only if there exists a normal subgroup H of G such that $K \subset H$ and $H' = H/K$.

10. Let G be a metric group, and let K be a closed normal subgroup of G. Show that H' is a closed subgroup of G/K if and only if there exists a closed subgroup H of G containing K such that $H' = H/K$ and that H' is normal if and only if H is normal.

11. Let H be a closed normal subgroup of a metric group G with left-invariant metric d. Show that the metric ρ defined by equation (1.17) is a left-invariant metric on G/H, and that if d is a left-invariant and right-invariant metric for G, then ρ is a left- and right-invariant metric.

12. Show that $\mathbb{K}$ is the continuous image of $\mathbb{R}_d$, the real numbers with the discrete topology, with kernel $\mathbb{Z}$, but the metric groups $\mathbb{K}$ and $\mathbb{R}_d/\mathbb{Z}$ are not isomorphic.

13. Let α be a real number and let H be the cyclic subgroup of the circle generated by $a = \exp(2\pi i\alpha)$. Show that $H^{-} = \mathbb{K}$ if and only if α is irrational.

14. Let $\varphi : G \to G'$ be a homomorphism of the metric group G onto the metric group G'. Prove that φ is open if and only if $\varphi(V)$ is a neighborhood of e' whenever V is a neighborhood of e.

15. Let G and G' be metric groups, let $\varphi : G \to G'$ be a homomorphism, and let H be a closed normal subgroup of G. Prove that there exists a homomorphism $\tilde{\varphi} : G/H \to G'$ such that $\varphi = \tilde{\varphi} \circ \pi$ if and only if H is contained in the kernel K of φ.

1.5 Compactness and Metric Groups

A particularly important class of subsets of topological spaces are the compact subsets. Although the definition may not seem to be an intuitively natural one, it turns out to be a useful property in the analytical study of topological spaces.

A subset C of a topological space X is a *compact set* provided that whenever $\mathcal{S}$ is a collection of open sets such that

$$C \subset \bigcup_{U \in \mathcal{S}} U, \tag{1.20}$$

there exists a finite collection $U_1, \ldots, U_k$ of sets in $\mathcal{S}$ such that

$$C \subset \bigcup_{i=1}^{k} U_i. \tag{1.21}$$

Obviously, any finite subset of a topological space is compact. When X itself is compact, X is called a *compact topological space.*

A collection of open sets $\mathcal{S}$ of X that satisfies condition (1.20) is called an *open cover of* C, and $U_1, \ldots, U_k$ such that (1.21) holds is called a *finite subcover.* An alternative way of stating the definition of a compact set is that C is compact provided that every open cover of C contains a finite subcover of C.

Proposition 1.5.1 *Let X and Y be topological spaces, and let $f : X \to Y$ be a continuous function. If C is a compact subset of X, then $f(C)$ is a compact subset of Y.*

Proof. Let $\mathcal{S}$ be an open covering of $f(C)$. Then $\mathcal{S}' = \{f^{-1}(U) : U \in \mathcal{S}\}$ is an open covering of C because f is continuous. Hence, there exists a finite subcover $f^{-1}(U_1), \ldots, f^{-1}(U_k)$ of C. It follows that $U_1, \ldots, U_k$ is a finite subcover of $f(C)$. $\square$

Corollary 1.5.2 *Let $f : X \to Y$ be a homeomorphism of the topological space X onto the topological space Y and let C be a subset of X. Then C is a compact subset of X if and only if $f(C)$ is a compact subset of Y.*

Proposition 1.5.3 *A closed set contained in a compact set of a topological space is compact.*

Proof. Let F be a closed set contained in the compact set C, and let $\mathcal{S}$ be an open cover of F. Note that $X \setminus F$ is an open set and

$$C \subset \left(\bigcup_{U \in \mathcal{S}} U\right) \cup (X \setminus F).$$

So $\mathcal{S}' = \{U : U \in \mathcal{S} \text{ or } U = X \setminus F\}$ is an open cover for C. Because C is compact, there exist a finite subcover $U_1, \ldots, U_k$ of C from $\mathcal{S}'$. Without loss of generality we can assume that $U_k = X \setminus F$ and U_j is in $\mathcal{S}$ for $j = 1, \ldots, k-1$. It follows that

$$F \subset \bigcup_{i=1}^{k-1} U_i,$$

and F is compact. $\square$

Proposition 1.5.4 *A compact set in a Hausdorff topological space is a closed set.*

Proof. Let C be a compact subset of a Hausdorff topological space. It suffices to show that $X \setminus C$ is open. Let y be in $X \setminus C$. For each $x \in C$, there exist disjoint open neighborhoods U_x and V_y of x and y, respectively, because X is Hausdorff. Obviously, $\{U_x : x \in X\}$ is an open cover of C. Since C is compact, there exist a finite set of points $x_1, \ldots, x_m$ in C such that $C \subset \bigcup_{k=1}^{m} U_{x_k} = U$. Then $V = \bigcap_{k=1}^{m} V_{x_k}$ is an open neighborhood of y such that $V \cap U = \phi$. Since $C \subset U$, it follows that $V \cap C = \phi$ and $y \in V \subset X \setminus C$. Thus $X \setminus C$ is open.

Corollary 1.5.5 *A compact subset of a metric space is closed.*

Proof. Metric spaces are Hausdorff.

A subset E of $\mathbb{R}$ is *bounded* provided that it is both bounded above and below or equivalently there exists $R > 0$ such that E is contained in the closed interval $[-R, R] = \{x : -R \leq x \leq R\}$.

Theorem 1.5.6 (Heine-Borel) *A subset of* $\mathbb{R}$ *is compact if and only if it is both closed and bounded.*

Proof. First, suppose that C is a compact subset of $\mathbb{R}$. Then C is closed by *Corollary 1.5.5.* If C is not bounded, then the open intervals $(-n, n)$, $n = 1, 2, \ldots$ are an open cover of C that does not have a finite subcover.

To prove that a closed and bounded subset of $\mathbb{R}$ is compact it suffices by *Proposition 1.5.3* to show that the closed interval $[-R, R]$ is compact for positive R.

Let $\mathcal{S}$ be an open cover of $[-R, R]$. Clearly $-R$ is in U for some $U \in \mathcal{S}$, and because U is open, there exists $r > 0$ such that $(-R - r, -R + r) \subset U$. So there exists a (very simple) finite subcover of $[-R, -R + r/2]$. The idea of the proof is to push the property that there is finite subcover for $[-R, x]$ for as large an x as possible using the same ideas.

Set

$$A = \left\{ x > -R : [-R, x] \subset \bigcup_{i=1}^{k} U_i \textit{ for some } U_1, \ldots, U_k \in \mathcal{S} \right\}.$$

From the previous paragraph, we know A is not the empty set. Moreover, if $-R < y < x$ and x is in A, then y is in A.

It suffices to prove that R is in A. Suppose R is not in A. Then, A must be contained in the open interval $(-R, R)$. Hence, $-R < c = \sup A \leq R$, and c is in U' for some U' in $\mathcal{S}$. Because U' is open, there exist $\varepsilon > 0$ such that the open interval $(c - \varepsilon, c + \varepsilon)$ is contained in U'. By the construction of c, there exists $a \in A$ such that $c - \varepsilon < a \leq c$, and there exist $U_1, \ldots, U_k$ such that $[-R, a] \subset \bigcup_{j=1}^{k} U_j$. It follows that

$$[-R, c + \varepsilon/2] \subset [-R, a] \cup (c - \varepsilon, c + \varepsilon) \subset \bigcup_{j=1}^{k} U_j \cup U'$$

and $c + \varepsilon/2$ is in A, which is impossible because c is an upper bound for A. Therefore R must be in A and $[-R, R]$ is compact. □

Corollary 1.5.7 *The metric group $\mathbb{K}$ is compact.*

Proof. The function $f(t) = e^{2\pi i t} = \cos(2\pi t) + i\sin(2\pi t)$ maps $\mathbb{R}$ continuously onto $\mathbb{K}$ such that $f([0,1]) = \mathbb{K}$. Since $[0,1]$ is compact by the theorem, $\mathbb{K}$ is compact by *Proposition 1.5.1.* □

Proposition 1.5.8 *Let X be a topological space, and let $f : X \to \mathbb{R}$ be a continuous function. If C is a compact subset of X, then $f(C)$ is a closed and bounded subset of $\mathbb{R}$ and there exist x_m and x_M in C such that*

$$f(x_m) \le f(x) \le f(x_M)$$

for all x in C.

Proof. By *Proposition 1.5.1*, $f(C)$ is a compact subset of $\mathbb{R}$, and hence closed and bounded by *Theorem 1.5.6*. It follows that $a = \inf\{t \in \mathbb{R} : t \in f(C)\}$ and $b = \sup\{t \in \mathbb{R} : t \in f(C)\}$ are finite. Because $f(C)$ is closed, a and b are in $f(C)$ by *Exercise* 13, p. 28. So there exist x_m and x_M such that $f(x_m) = a$ and $f(x_m) = b$. □

The remainder of this section examines the basic properties of compact subsets of metric spaces with an emphasis on compact and locally compact metric groups.

Proposition 1.5.9 *If x_k is a sequence in a compact subset C of a metric space X, then x_k has a convergent subsequence.*

Proof. If for some integer k, the set $\{j : x_j = x_k\}$ is infinite, then x_k has a constant subsequence that obviously converges. Thus it can be assumed without loss of generality that for all k the set $\{j : x_j = x_k\}$ is finite. It follows that the set $E = \{y \in C : y = x_j \text{ for some } j\}$ is infinite.

Suppose x_k does not have a convergent subsequence. Then for each y in C there exists an integer $n(y)$ depending on y such that $\{k : x_k \in B_{1/n(y)}(y)\}$ is the empty set when $y \ne x_k$ for all k and a nonempty finite set when $y = x_k$ for some k. Clearly,

$$\mathcal{S} = \{B_{1/n(y)}(y) : y \in C\}$$

is an open cover of C. Since C is compact, there exist $y_1, \ldots, y_m$ such that

$$C \subset \bigcup_{i=1}^{m} B_{1/n(y_i)}(y_i).$$

Because E is infinite, it follows that for some y_i the set $\{k : x_k \in B_{1/n(y_i)}(y_i)\}$ is infinite, a contradiction. Therefore, x_k has a convergent subsequence. □

Corollary 1.5.10 *Every bounded sequence of real numbers contains a convergent subsequence.*

Proof. Apply the proposition to a closed and bounded interval containing the sequence. □

The product of a collection of compact sets is compact in a very general setting. The next result, *Theorem 1.5.11*, is a finite version of this major theorem for metric spaces that will suffice for the next three chapters. We will return to this subject in Chapter 5.

Theorem 1.5.11 *Let $X_1, \ldots, X_m$ be metric spaces. If $C_1, \ldots, C_m$ are compact subsets of $X_1, \ldots, X_m$, respectively, then $C_1 \times \ldots \times C_m$ is a compact subset of $X_1 \times \ldots \times X_m$ with the product topology.*

Proof. The proof proceeds by induction with the theorem being trivial when $m = 1$. Assume the theorem is true for $m - 1$ metric spaces with $m \geq 2$ and consider compact subsets $C_1, \ldots, C_m$ of the metric spaces $X_1, \ldots, X_m$, respectively. Given x in X_1, the function $(x_2, x_3, \ldots, x_m) \to (x, x_2, x_3, \ldots, x_m)$ is a continuous function from $X_2 \times \ldots \times X_m$ to $X_1 \times \ldots \times X_m$. By induction, $C_2 \times \ldots \times C_m$ is compact and the set $C_x = \{x\} \times C_2 \times \ldots \times C_m$ is compact set because it is the continuous image of a compact set.

Let $\mathcal{S}$ be an open covering of $C_1 \times \ldots \times C_m$. For each x in C_1, the collection of open sets $\{U \in \mathcal{S} : x \in p_1(U)\}$ from $\mathcal{S}$ is an open cover of the compact set C_x. Therefore, for each x in C_1, there exists a finite collection of open sets, $\mathcal{S}_x$ from $\mathcal{S}$ such that $x \in p_1(U)$ for each $U \in \mathcal{S}_x$ and

$$C_x \subset \bigcup_{U \in \mathcal{S}_x} U.$$

Although this shows that $\mathcal{S}$ contains a finite subcover $\mathcal{S}_x$ of C_x for each x in C_1, there are possibly infinitely many points in C_1. To complete the proof, the compactness of C_1 must be used to prove that there exists a finite number of points $x_1, \ldots, x_k$ in C_1 such that together the collections of open sets $\mathcal{S}_{x_1}, \ldots, \mathcal{S}_{x_k}$ cover $C_1 \times \ldots \times C_m$.

The next step is to show that for $x \in C_1$ there exists an open ball $B_{1/k}(x)$ in X_1 such that

$$B_{1/k}(x) \times C_2 \times \ldots \times C_m \subset \bigcup_{U \in \mathcal{S}_x} U.$$

Suppose no such $B_{1/k}(x)$ exists. Then for each positive integer k there exists $x_k \in B_{1/k}(x) \subset X_1$ and $\mathbf{y}_k \in C_2 \times \ldots \times C_m$ such that

$$(x_k, \mathbf{y}_k) \notin \bigcup_{U \in \mathcal{S}_x} U.$$

(The boldface notation $\mathbf{y}$ is being used for points in $X_2 \times \ldots \times X_m$ to remind us that they are points in a product space with coordinates.) By applying

Proposition 1.5.9, there exists a subsequence $\mathbf{y}_{k_i}$ converging to $\mathbf{y}$ in $C_2 \times \ldots \times C_m$. It follows that $(x_{k_i}, \mathbf{y}_{k_i})$ converges to $(x, \mathbf{y})$ in C_x. Thus $(x, \mathbf{y})$ must be in some U in $\mathcal{S}_x$. Consequently, $(x_{k_i}, \mathbf{y}_{k_i})$ is in the same U for large i, contradicting the construction of $(x_k, \mathbf{y}_k)$. Therefore, the required $B_{1/k}(x)$ exists. Let $k(x)$ be the smallest positive integer k such that $B_{1/k}(x) \times C_2 \times \ldots \times C_m \subset \bigcup_{U \in \mathcal{S}_x} U$, and set $V_x = B_{1/k(x)}(x)$.

Clearly, $\{V_x : x \in C_1\}$ is an open cover of C_1, and there exist $y_1, \ldots, y_k$ in X_1 such that

$$C_1 \subset \bigcup_{j=1}^{k} V_{y_j}.$$

To complete the proof it suffices to show that

$$\mathcal{S}' = \bigcup_{j=1}^{k} \mathcal{S}_{y_j}$$

is a finite subcover of $C_1 \times \ldots \times C_m$. By construction, $\mathcal{S}'$ is a finite collection of open sets in the original cover $\mathcal{S}$.

Let $(w_1, \ldots, w_m)$ be an arbitrary point in $C_1 \times \ldots \times C_m$. Then w_1 is in some V_{y_j}. It follows that

$$(w_1, w_2, \ldots, w_m) \in V_{y_j} \times C_2 \times \ldots \times C_m \subset \bigcup_{U \in \mathcal{S}_{y_j}} U \subset \bigcup_{U \in \mathcal{S}'} U$$

and the proof is complete. □

Corollary 1.5.12 *The metric groups $\mathbb{K}^n$ are compact.*

A subset of $\mathbb{R}^n$ is bounded provided all the coordinates of all the points in it is a bounded set of real numbers. In other words, a subset E of $\mathbb{R}^n$ is *bounded* provided there exists $R > 0$ such that

$$E \subset [-R, R]^n.$$

Theorem 1.5.13 *A subset of $\mathbb{R}^n$ is compact if and only if it is both closed and bounded.*

Proof. Suppose E is a closed and bounded subset of $\mathbb{R}^n$. Then $E \subset [-R, R]^n$ for some $R > 0$. By *Theorems 1.5.6 and 1.5.11* the set $[-R, R]^n$ is a compact subset of $\mathbb{R}^n$. Hence, E is a closed set contained in a compact set and is compact by *Proposition 1.5.3*.

Suppose E is a compact subset of $\mathbb{R}^n$. Then it is closed by *Proposition 1.5.4*. Because the projections $p_j : \mathbb{R}^m \to \mathbb{R}$ are continuous, the sets $p_j(E)$ are compact by *Proposition 1.5.1*. The Heine-Borel Theorem now implies that $p_j(E)$ is bounded for $j = 1, \ldots, m$, and hence E is bounded. □

Small compact neighborhoods can be as important as large compact sets. When every point of a topological space has a compact neighborhood, the space is said to be *locally compact.* Every compact topological space is, of course, locally compact, and $\mathbb{R}^n$ is an example of a locally compact space that is not compact. Discrete metric spaces are locally compact because points are both compact and open sets. Not all topological spaces, however, are locally compact (see *Exercise* 14, p. 61). Local compactness is a critical property in the study of Hausdorff topological groups.

If C is a compact neighborhood of x in a metric space X, then there exists $r > 0$ such that $B_\varepsilon(x) \subset C$ for $0 < \varepsilon \leq r$. Because C is closed by *Corollary 1.5.5*, it follows that $B_\varepsilon(x)^- \subset C$ and $B_\varepsilon(x)^-$ is compact by *Proposition 1.5.3.* Since $B_\varepsilon(x)^- \subset \{y : d(x, y) \leq \varepsilon\}$ by *Exercise* 7, p. 28, the sequence of sets $B_{r/k}(x)^-$ is a countable neighborhood base of compact sets at x. The point of this discussion can be summarized as follows:

Proposition 1.5.14 *A metric space X is locally compact if and only if for every $x \in X$ there exists a countable neighborhood base of compact sets.*

Proposition 1.5.15 *Let G be a metric group. Then G is locally compact if and only if there is a compact neighborhood of the identity.*

Proof. Let C be a compact neighborhood of the identity e of G. Since the map $\psi_a(x) = ax$ is a homeomorphism of G onto itself mapping e to a, the set $\psi_a(C)$ is a compact neighborhood of a, proving that G is locally compact. The other direction is trivial. □

A subset E of a metric space X is *dense* in X, provided $E^- = X$. Equivalently, E is dense if $E \cap B_r(x) \neq \phi$ for all $x \in X$ and $r > 0$. Both the rational numbers and the irrational numbers are dense subsets of $\mathbb{R}$. When α is an irrational number, $\{e^{2\pi i k\alpha} : k \in \mathbb{Z}\}$ is a dense subgroup of $\mathbb{K}$ (*Exercise 13*, p. 51).

Theorem 1.5.16 (Baire) *Let X be a locally compact metric space. If D_k is a sequence of open dense subsets in X, then*

$$\bigcap_{k=1}^{\infty} D_k$$

is a dense subset of X.

Proof. Set $D = \bigcap_{k=1}^{\infty} D_k$ and let x be an arbitrary point in X. It suffices to show that $B_{r'}(x) \cap D \neq \phi$ for all $r' > 0$. Because X is a locally compact metric space, for every $r' > 0$ there exists $r > 0$ such that $B_r(x)^- \subset B_{r'}(x)$ and $B_r(x)^-$ is compact. This further reduces the proof to showing that $B_r(x)^- \cap D \neq \phi$ when $B_r(x)^-$ is compact and $r > 0$.

Since D_1 is both open and dense, there exists $r_1 > 0$ and x_1 such that $B_{r_1}(x_1)^- \subset B_r(x) \cap D_1$. In fact, we can define, r_k and x_k inductively so that

$$B_{r_k}(x_k)^- \subset B_{r_{k-1}}(x_{k-1}) \cap \bigcap_{j=1}^{k} D_j \subset B_r(x)$$

because $\bigcap_{j=1}^{k} D_j$ is also both open and dense (*Exercise* 8, p. 61). It follows that

$$B_{r_k}(x_k)^- \subset B_{r_{k-1}}(x_{k-1})^- \subset B_r(x)^-$$

for $k > 1$ and

$$\bigcap_{k=1}^{\infty} B_{r_k}(x_k)^- \subset \bigcap_{k=1}^{\infty} D_k = D.$$

Thus,

$$\bigcap_{k=1}^{\infty} B_{r_k}(x_k)^- \subset B_r(x)^- \cap D,$$

and it suffices to show that

$$\bigcap_{k=1}^{\infty} B_{r_k}(x_k)^- \neq \phi.$$

Suppose it is empty. Then by de Morgan's formulas (*Exercise* 6, p. 27),

$$\bigcup_{k=1}^{\infty} (X \setminus B_{r_k}(x_k)^-) = X$$

and the open sets

$$U_k = X \setminus B_{r_k}(x_k)^-$$

form an open cover of X and of $B_r(x)^-$ in particular. So there exists a finite subcover of $B_r(x)^-$ because it is compact. Observe that $B_{r_k}(x_k)^- \subset B_{r_{k-1}}(x_{k-1})^-$ implies that $U_{k-1} \subset U_k$ for all $k > 1$. Thus any finite subcover reduces to $B_r(x)^- \subset U_k$ for some k. It follows that

$$X \setminus B_r(x)^- \supset X \setminus U_k = B_{r_k}(x_k)^-,$$

which contradicts the property that

$$B_{r_k}(x_k)^- \subset B_r(x)^-$$

holds for all $k \geq 1$. Therefore, D is a dense subset of X. □

Theorem 1.5.16 is usually referred to as the Baire Category Theorem. (For a purely topological version of this theorem see *Exercise* 20, p. 62.) Because of the complementary relationship between open and closed sets, The Baire Category Theorem has an equivalent version stated in terms of closed sets The complement of an open dense subset is a closed set that is not the neighborhood of any point in the metric space. Such a set is called a *closed nowhere dense set.* The Baire Category Theorem can also be stated using closed nowhere dense sets instead of open dense sets.

Corollary 1.5.17 *Let X be a locally compact metric space. If F_k is a sequence of closed sets such that*

$$\bigcup_{k=1}^{\infty} F_k = X,$$

then some F_k contains a nonempty open set. In particular, a locally compact metric space cannot be the union of a sequence of closed nowhere dense subsets.

Proof. Suppose none of the sets F_k contains an open set. Set $D_k = X \setminus F_k$, which is open for all k. Since no open set U is contained in F_k, the set $U \cap D_k \neq \phi$ and D_k is dense.

The theorem implies that $\bigcap_{k=1}^{\infty} D_k$ is dense, and hence not empty. By de Morgan's formulas

$$\phi = X \setminus \bigcup_{k=1}^{\infty} F_k = \bigcap_{k=1}^{\infty} (X \setminus F_k) = \bigcap_{k=1}^{\infty} D_k \neq \phi.$$

This contradiction completes the proof. □

There is also a version of the Baire Category Theorem based on a purely metric hypothesis instead of a topological one. A sequence x_k in a metric space X with metric d is a *Cauchy sequence*, provided that given $\varepsilon > 0$ there exists a positive integer K such that $d(x_n, x_m) < \varepsilon$ when both $m \geq K$ and $n \geq K$. A metric space X with metric d is said to be *complete* if every Cauchy sequence in X converges to a point in X. The hypothesis that X is a "locally compact metric space" can be replaced by "complete metric space" and the theorem is still valid. This result (*Exercise* 19, p. 62) and several other basic facts about complete metric spaces are contained in the exercises.

A metric space X is *σ-compact or sigma-compact* provided that there exists a sequence of compact sets C_k in X such that

$$X = \bigcup_{k=1}^{\infty} C_k.$$

An example of a σ-compact metric space is $\mathbb{R}^n$. If H is a closed subgroup of a σ-compact metric group G, then clearly H and G/H are σ-compact.

Theorem 1.5.18 (Open Homomorphism Criterion) *Let G and G' be locally compact metric groups, and let $\varphi : G \to G'$ be an onto homomorphism. If G is σ-compact, then φ is an open homomorphism.*

Proof. It suffices to show that φ maps neighborhoods of e to neighborhoods of e' by *Exercise* 14, p. 51. Let U be a neighborhood of e. Because G is locally compact, there exists a compact neighborhood V of e such that $VV^{-1} \subset U$ and an open set W such that $e \in W \subset V$.

The open sets Wx, $x \in G$, are an open cover of G. Since G is σ-compact, there exists a sequence x_k such that $G = \bigcup_{k=1}^{\infty} Wx_k$ by *Exercise* 9, p. 61. It follows that $G = \bigcup_{k=1}^{\infty} Vx_k$ and

$$G' = \bigcup_{k=1}^{\infty} \varphi(V)\varphi(x_k).$$

Because V is compact, $\varphi(V)$ is compact and hence closed by *Corollary 1.5.5*. Now *Proposition 1.3.2* implies that the sets $\varphi(V)\varphi(x_k)$ are closed for all k. By *Corollary 1.5.17*, there exists k such that $\varphi(V)\varphi(x_k)$ contains an nonempty open set, and then $\varphi(V)$ must contain an open set V'.

Let y be a point in V' and pick x in V such that $\varphi(x) = y$. Then $V'y^{-1}$ is an open set in G' such that

$$e' \in V'y^{-1} \subset \varphi(V)\varphi(x^{-1}) = \varphi(Vx^{-1}) \subset \varphi(VV^{-1}) \subset \varphi(U).$$

Thus φ maps neighborhoods of e to neighborhoods of e'. □

Corollary 1.5.19 *Let G and G' be locally compact metric groups and let $\varphi : G \to G'$ be an onto homomorphism. If G is σ-compact, then G' is isomorphic to G/K where K is the kernel of φ.*

Proof. Since φ is open by the theorem, *Theorem 1.4.14* applies. □

Theorem 1.5.18 and its corollary provide another proof that $x \to e^{2\pi i x}$ is an open homomorphism of $\mathbb{R}$ onto $\mathbb{K}$, and the metric groups $\mathbb{R}/\mathbb{Z}$ and $\mathbb{K}$ are isomorphic. Actually, we can say more. Suppose G' is a locally compact group and $\varphi : \mathbb{R} \to G'$ is an onto homomorphism with kernel K. *Corollary 1.5.19* implies that $\mathbb{R}/K$ is isomorphic to G'. If K equals $\mathbb{R}$ or $\{0\}$, then G' is isomorphic to the trivial group or to $\mathbb{R}$. What happens when $K = \{ka : k \in \mathbb{Z}\}$ for some $a > 0$? Note that the function $x \to e^{2\pi i x/a}$ is a homomorphism of $\mathbb{R}$ onto $\mathbb{K}$ with kernel K. It follows that $\mathbb{K}$ is isomorphic to $\mathbb{R}/K$ and hence to G'. Thus we have determined all the locally compact metric groups that are homomorphic images of $\mathbb{R}$.

Since the Baire Category Theorem holds for locally compact regular topological spaces (*Exercise* 20, p. 62) and every topological group is regular (*Exercise* 10, p. 39), the Baire Category Theorem can be applied to sequences of open dense sets or sequences of closed nowhere dense sets in a locally compact topological group. For example, the proof of the Open Homomorphism Criterion can be used to prove a more general version of this theorem, with one caveat. The one step in the proof of *Theorem 1.5.18* that uses the metric hypothesis is the inference that $\varphi(V)$ is closed because it is compact. If the "locally compact metric groups" hypothesis is replaced with the hypothesis that the groups are locally compact Hausdorff topological groups, then *Proposition 1.5.4* can be applied to show that $\varphi(V)$ is closed because it is compact. Consequently, the Open Homomorphism Criterion holds with the words "metric groups" replaced by the words "Hausdorff topological groups".

EXERCISES

1. Let X be a compact metric space with metric d. Show that there exists a constant M such that $d(x,y) \leq M$ for all x and y in X.

2. Let X and Y be Hausdorff topological spaces and let $f : X \to Y$ is a continuous one-to-one onto function. Prove that if X is compact, then f^{-1} is continuous and f is a homeomorphism. (One approach is to use *Exercise* 3, p. 50.)

3. Let G be a metric group. Prove that if C is a compact subset of G and F is a closed subset of G such that $C \cap F = \phi$, then there exists an open neighborhood V of the identity such that $(CV) \cap (FV) = \phi$. (One approach is to use *Exercise* 12, p. 39.)

4. Let C_1 and C_2 be disjoint compact subsets of a metric space X with metric d. Prove that $0 < \inf\{d(x,y) : x \in C_1 \text{ and } y \in C_2\}$.

5. Prove that with the relative topology, open and closed subsets of locally compact metric spaces are locally compact metric spaces.

6. Let X and Y be topological spaces and let $f : X \to Y$ be a continuous onto open function. Prove that Y is locally compact, if X is locally compact.

7. Let $X_1, \ldots, X_n$ be metric spaces. Prove that $X_1, \ldots, X_n$ are locally compact if and only if $X_1 \times \ldots \times X_n$ is locally compact.

8. Prove that in a metric space the intersection of a finite number of open dense sets is open and dense.

9. Let $\mathcal{S}$ be an open cover of a topological space X. Show that if X is σ-compact, then there exists a sequence of open sets U_k in $\mathcal{S}$ such that $X = \bigcup_{k=1}^{\infty} U_k$.

10. Show that a closed subset of a σ-compact metric space is σ-compact.

11. Show that for a σ-compact space there exists a sequence of compact subsets C_k such that $C_k \subset C_{k+1}$ for all $k \geq 1$ and $X = \bigcup_{k=1}^{\infty} C_k$.

12. Prove that a finite product of σ-compact spaces is σ-compact.

13. Let U be an open subset of a σ-compact metric space X. Show that U with the relative topology is also a σ-compact metric space. (One approach is to use *Exercise* 12, p. 28, with $F = X \setminus U$.)

14. Show that $\mathbb{Q}$, the rational numbers, with the relative topology from $\mathbb{R}$, is not locally compact.

15. Use *Theorem 1.5.16* to prove that there does not exist a one-to one function from $\mathbb{Z}^+ = \{k \in \mathbb{Z} : k \geq 1\}$ onto $\mathbb{R}$.

16. Let $\mathbb{R}_d$ denote the reals with the discrete topology. Show that $\mathbb{R}_d$ is not σ-compact. (One approach is to use *Exercise 15* above.)

17. Let X be a metric space with metric d. Prove the following:

    (a) Every convergent sequence in X is a Cauchy sequence.

    (b) A Cauchy sequence converges if and only if it has a convergent subsequence.

    (c) If X is compact, then it is a complete metric space.

18. Prove that $\mathbb{R}^n$ is a complete metric space for the metric $d(\mathbf{x}, \mathbf{y}) = \|\mathbf{x} - \mathbf{y}\|$.

19. Prove the Baire category theorem for complete metric spaces. (One approach is to modify the proof in the text by choosing $r_k \leq 1/2^k$ and showing that x_k is Cauchy.)

20. Prove the Baire Category Theorem for regular (see *Exercise* 10, p. 39 for the definition of regular) locally compact topological spaces by showing that the sets $B_{r_k}(x_k)$ in the proof of *Theorem 1.5.16* can be replaced with open sets V_k such that V_k^- is compact.

Chapter 2

Linear Spaces and Algebras

A group can have additional algebraic structures that are linked in some way to the underlying group multiplication. For example, the additive group $\mathbb{R}$ has a multiplicative structure, and the multiplication is linked to the additive group structure by the distribution formula $a(b + c) = ab + ac$. This chapter is about metric groups with additional algebraic structures.

Linear spaces or vector spaces are Abelian groups with a linear structure. A linear structure on a group G is a function, called scalar multiplication, from the Cartesian product of $\mathbb{R}$ or $\mathbb{C}$ with G to G that satisfies certain algebraic conditions. In the Sections 2.1 and 2.2, the basic algebraic properties of linear structures and linear functions are established. Although this material is part of what is usually called linear algebra, the presentation here emphasizes the underlying group structure of a linear space. Only the ideas that are pertinent to studying metric groups with continuous scalar multiplication are included.

Norms provide natural metrics on linear spaces. They are introduced in Section 2.3. These metrics are particularly interesting because they are invariant metrics for the underlying group and behave linearly with respect to the stretching and shrinking produced by scalar multiplication. In this context, some homomorphisms preserve the scalar multiplication. They are discussed in Section 2.4. These linear homomorphisms form new linear spaces with norms.

In addition to a linear structure, some Abelian groups have another associative binary operation, and all three algebraic entities are linked by specified formulas. They are called algebras. For example, square matrices of a given size form an algebra. Algebras with norms lead to important families of metric groups, and basic results about them are included in Sections 2.2 and 2.4.

Throughout the chapter, special attention will be paid to finite-dimensional linear spaces culminating in the construction of the determinant function in Section 2.5. The determinant plays an essential role in the study of metric groups of square matrices under matrix multiplication in Chapter 4, and it will be used to prove a key result in the determination of the subgroups of $\mathbb{R}^n$ in Chapter 3. In fact, ideas from this chapter on linear spaces and algebras will be used in all the remaining chapters.

2.1 Linear Structures on Groups

The group $\mathbb{R}^n$ has an additional algebraic structure. There is a natural scalar multiplication of $c \in \mathbb{R}$ times $\mathbf{x} \in \mathbb{R}^n$ defined by $c\mathbf{x} = (cx_1, \ldots, cx_m)$ called a *real linear structure.* Similarly, $\mathbb{C}^n$ has a natural complex linear structure defined by $c\mathbf{z} = (cz_1, \ldots, cz_n)$ for $c \in \mathbb{C}$. In this section, the algebraic properties of such linear structures will be explored.

For this discussion of linear structures, there is no essential difference between the real numbers and the complex numbers. For example, the solution of $aw + b = 0$ with $a \neq 0$ is $w = b/a$ if a and b are real or complex numbers. Henceforth, the term scalar will refer to a real or complex number. For convenience, we will let $\mathbb{S}$ denote either $\mathbb{R}$ or $\mathbb{C}$, and will refer to $\mathbb{S}$ as the scalar field. The letters c and d will typically be used to denote scalars.

To develop the connection between a group and a linear structure, a basic fact from systems of linear equations will be used. Consider the system of linear equations

$$\begin{array}{ccccccc} c_{11}x_1 & + & \ldots & + & c_{1q}x_q & = & 0 \\ \vdots & & & & \vdots & & \vdots \\ c_{p1}x_1 & + & \ldots & + & c_{pq}x_q & = & 0 \end{array} \tag{2.1}$$

where c_{ij} are scalars, x_1 through x_q are unknowns, and, of course, p and q are positive integers. It is understood that each equation has at least one nonzero coefficient and each unknown has a least one nonzero coefficient. Otherwise, (2.1) is not really a system of p equations in q unknowns.

Note that $x_1 = \ldots = x_q = 0$ is a solution of (2.1). The basic result about systems of linear equations that will be used in this section is the following: *If there are more unknowns than equations in (2.1), that is if $q > p$, then there exists a solution $\gamma_1, \ldots, \gamma_q$ of (2.1) such that $\gamma_k \neq 0$ for some k.* Although this fact is usually embedded in a more general discussion of systems of linear equations and their solutions, it is not difficult to prove it as an independent result (see *Exercise* 1, p. 68).

A *linear structure* on an Abelian group V is a function $(c, \mathbf{v}) \mapsto c\mathbf{v}$ from $\mathbb{S} \times V$ to V satisfying the following conditions for all $c, d \in \mathbb{S}$ and $\mathbf{u}, \mathbf{v} \in V$:

(a) $1\mathbf{u} = \mathbf{u}$

(b) $(cd)\mathbf{u} = c(d\mathbf{v})$

(c) $(c + d)\mathbf{u} = c\mathbf{u} + d\mathbf{u}$

(d) $c(\mathbf{u} + \mathbf{v}) = c\mathbf{u} + c\mathbf{v}$.

Property (d) is equivalent to saying that given a scalar c, the function defined by $\mathbf{v} \to c\mathbf{v}$ is an algebraic homomorphism of the Abelian group V to itself. These properties of scalar multiplication imply that $0\mathbf{u} = \mathbf{0}$ and $(-1)\mathbf{u} = -\mathbf{u}$.

A *real or complex linear space* or *vector space* is an Abelian group V and a specific linear structure. Alternatively, a real or complex linear space will some times be referred to as a linear space over $\mathbb{R}$ or $\mathbb{C}$. The elements of a linear space or vector space are being denoted with boldface type like the elements of $\mathbb{R}^n$ and $\mathbb{C}^n$ to help distinguish them from scalars. Likewise, the identity element of the underlying Abelian group will be denoted by $\mathbf{0}$ to avoid confusion with the scalar 0. Most of the time we will use the term linear space instead of vector space, but the elements of a linear space will at times be referred to as vectors.

It is easy to check that $\mathbb{R}^n$ and $\mathbb{C}^n$ are linear spaces. They are both also metric spaces, but the discussion of linear spaces in this section will be strictly algebraic and will not use any metric-space concepts.

It is also worth noting that the underlying Abelian group for $\mathbb{C}^n$ is the same as $\mathbb{R}^{2n}$ because the complex plane $\mathbb{C}$ is just $\mathbb{R}^2$ with $(x, y) = x + iy$ and $(x+iy)+(x'+iy') = (x+x')+i(y+y')$, which is the same as $(x, y)+(x', y') = (x+x', y+y')$ in $\mathbb{R}^2$. The real and complex linear structures on $\mathbb{R}^{2n}$ are, however, quite different.

A *subspace* of a linear space V is a subgroup W of V such that $c\mathbf{u}$ is in W for all $c \in \mathbb{S}$ and $\mathbf{u} \in W$. Or equivalently, a nonempty subset W of V is a subspace of V if and only if $c\mathbf{u} + d\mathbf{v}$ is in W for all $\mathbf{u}$ and $\mathbf{v}$ in W and scalars c and d. The linear space V itself is a subspace of V, as is the trivial subgroup $\{\mathbf{0}\}$. Subspaces other than these two extremes will be referred to as proper subspaces. Of course, a subspace of a linear space is a linear space in its own right.

It is worth noting that not all subgroups of a linear space are subspaces. For example, $\mathbb{Z}^2 = \{(p, q) : p, q \in \mathbb{Z}\}$ is a subgroup of $\mathbb{R}^2$, but it is not a subspace because $1/2(1, 1) = (1/2, 1/2)$ is not in $\mathbb{Z}^2$.

Given a finite set of vectors $\mathbf{v}_1, \ldots, \mathbf{v}_n$ in the linear space V, the *span* of $\mathbf{v}_1, \ldots, \mathbf{v}_n$ is defined by

$$\text{Span}\,[\mathbf{v}_1, \ldots, \mathbf{v}_n] = \{c_1\mathbf{v}_1 + \ldots + c_n\mathbf{v}_n : c_j \in \mathbb{S}\}$$

and is a subspace of V. In fact, $\text{Span}\,[\mathbf{v}_1, \ldots, \mathbf{v}_n]$ is the smallest subspace containing $\mathbf{v}_1, \ldots, \mathbf{v}_n$. The smallest subgroup of V containing $\mathbf{v}_1, \ldots, \mathbf{v}_n$ is given by

$$\{c_1\mathbf{v}_1 + \ldots + c_n\mathbf{v}_n : c_j \in \mathbb{Z}\},$$

and is a proper subset of $\text{Span}\,[\mathbf{v}_1, \ldots, \mathbf{v}_n]$ unless $\mathbf{v}_1 = \ldots = \mathbf{v}_n = \mathbf{0}$.

A finite set of vectors $\mathbf{v}_1, \ldots, \mathbf{v}_n$ in a linear space V is *linearly independent* provided

$$c_1\mathbf{v}_1 + \ldots + c_n\mathbf{v}_n = \mathbf{0}$$

implies that $c_1 = \ldots = c_n = 0$. When $\mathbf{v}_1, \ldots, \mathbf{v}_n$ are not linearly independent they are said to be *linearly dependent.* If $\mathbf{v}_1, \ldots, \mathbf{v}_n$ are linearly independent in V and $\text{Span}\,[\mathbf{v}_1, \ldots, \mathbf{v}_n] = V$, then V is said to be a *finite-dimensional linear space* and $\mathbf{v}_1, \ldots, \mathbf{v}_n$ is a *basis* for V.

Proposition 2.1.1 *If* $\mathbf{v}_1, \ldots, \mathbf{v}_n$ *is a basis for a linear space* V*, then for each* $\mathbf{v}$ *in* V *there exists unique scalars* $c_1, \ldots, c_n$ *such that* $\mathbf{v} = c_1\mathbf{v}_1 + \ldots + c_n\mathbf{v}_n$.

Proof. Because $\mathbf{v}_1, \ldots, \mathbf{v}_n$ is a basis, Span $[\mathbf{v}_1, \ldots, \mathbf{v}_n] = V$ and there exist scalars $c_1, \ldots, c_n$ such that $\mathbf{v} = c_1\mathbf{v}_1 + \ldots + c_n\mathbf{v}_n$. If $d_1, \ldots, d_n$ are also scalars such that $\mathbf{v} = d_1\mathbf{v}_1 + \ldots + d_n\mathbf{v}_n$, then $\mathbf{0} = (c_1 - d_1)\mathbf{v}_1 + \ldots + (c_n - d_n)\mathbf{v}_n$. Now the linear independence of $\mathbf{v}_1, \ldots, \mathbf{v}_n$ implies that $c_i - d_i = 0$ for $i = 1, \ldots, n$. Hence, $c_i = d_i$ for $i = 1, \ldots, n$, and the coefficients $c_1, \ldots, c_n$ are uniquely determined by $\mathbf{v}$. □

Proposition 2.1.1 shows that every element of a linear space has uniquely determined coordinates with respect to a given basis. These coordinates will be used repeatedly in a variety ways throughout this chapter.

Theorem 2.1.2 *Let $\mathbf{v}_1, \ldots, \mathbf{v}_n$ be a basis for a finite-dimensional linear space V. If $\mathbf{w}_1, \ldots, \mathbf{w}_m$ is a set of linearly independent vectors in V, then $m \leq n$.*

Proof. Since Span $[\mathbf{v}_1, \ldots, \mathbf{v}_n] = V$, there exist scalars c_{ij} such that

$$\mathbf{w}_j = \sum_{i=1}^{n} c_{ij}\mathbf{v}_i$$

for $j = 1, \ldots, m$.

Consider the system of equations:

$$\begin{array}{ccccccc} c_{11}x_1 & + & \cdots & + & c_{1m}x_m & = & 0 \\ \vdots & & & & \vdots & & \vdots \\ c_{n1}x_1 & + & \cdots & + & c_{nm}x_m & = & 0 \end{array}$$

If $m > n$, there exists a solution $\gamma_1, \ldots, \gamma_m$ such that $\gamma_k \neq 0$ for at least one k. Since

$$\sum_{j=1}^{m} c_{ij}\gamma_j = 0$$

for $i = 1, \ldots, n$, it follows that

$$\begin{aligned} \sum_{j=1}^{m} \gamma_j \mathbf{w}_j &= \sum_{j=1}^{m} \gamma_j \sum_{i=1}^{n} c_{ij}\mathbf{v}_i \\ &= \sum_{j=1}^{m}\sum_{i=1}^{n} c_{ij}\gamma_j\mathbf{v}_i \\ &= \sum_{i=1}^{n} \left(\sum_{j=1}^{m} c_{ij}\gamma_j \right) \mathbf{v}_i \\ &= \sum_{i=1}^{n} 0\mathbf{v}_i \\ &= \mathbf{0} \end{aligned}$$

This contradicts the linear independence of the vectors $\mathbf{w}_1, \ldots, \mathbf{w}_m$ because some $\gamma_k \neq 0$. Thus $m \leq n$. □

Corollary 2.1.3 *If V is a finite-dimensional linear space, then any two bases for V have the same cardinality.*

Proof. Suppose both $\mathbf{v}_1, \ldots, \mathbf{v}_n$ and $\mathbf{w}_1, \ldots, \mathbf{w}_m$ are a basis for V. Then by the proposition $n \leq m$ and $m \leq n$. □

By definition, the *dimension* of a finite-dimensional linear space V is the number of elements in a basis for V and is denoted by $\text{Dim}\,[V]$. This definition is unambiguous by *Corollary 2.1.3*. Since there are no linearly independent elements in the linear space $\{\mathbf{0}\}$, its dimension is zero.

Let $\mathbf{e}_j$ denote the element of $\mathbb{R}^n$ that has a 1 in the jth coordinate and zeros in all the other coordinates. Then it is easy to see that $\mathbf{e}_1, \ldots, \mathbf{e}_n$ is a basis for $\mathbb{R}^n$ and the dimension of $\mathbb{R}^n$ is n for $n > 0$. Similarly, the dimension of $\mathbb{C}^n$ (as a complex linear space) is n. The basis $\mathbf{e}_1, \ldots, \mathbf{e}_n$ of $\mathbb{S}^n$ is called the *standard basis*.

There are a number of essential results relating linear independence and dimension that flow out of *Theorem 2.1.2*. They are presented next in a series of propositions.

Proposition 2.1.4 *Let $\mathbf{v}_1, \ldots, \mathbf{v}_n$ be a linearly independent set of vectors in the linear space V. If* $\text{Span}\,[\mathbf{v}_1, \ldots, \mathbf{v}_n] \neq V$, *then there exists a $\mathbf{v}_{n+1} \in V$ such that $\mathbf{v}_1, \ldots, \mathbf{v}_n, \mathbf{v}_{n+1}$ is a linearly independent set in V.*

Proof. Since $\text{Span}\,[\mathbf{v}_1, \ldots, \mathbf{v}_n]$ does not equal V by hypothesis, there exists a nonzero $\mathbf{v}_{n+1} \in V$ that is not in $\text{Span}\,[\mathbf{v}_1, \ldots, \mathbf{v}_n]$. Suppose

$$c_1\mathbf{v}_1 + \ldots + c_n\mathbf{v}_n + c_{n+1}\mathbf{v}_{n+1} = \mathbf{0}.$$

If $c_{n+1} \neq 0$, then

$$\mathbf{v}_{n+1} = \frac{-1}{c_{n+1}} \sum_{j=1}^{n} c_j \mathbf{v}_j \in \text{Span}\,[\mathbf{v}_1, \ldots, \mathbf{v}_n]$$

contrary to the choice of $\mathbf{v}_{n+1}$. Hence, $c_{n+1} = 0$, and the linear independence of $\mathbf{v}_1, \ldots, \mathbf{v}_n$ forces $c_1 = \ldots = c_n = 0$. Therefore, $\mathbf{v}_1, \ldots, \mathbf{v}_n, \mathbf{v}_{n+1}$ is a linearly independent set. □

Corollary 2.1.5 *If $\mathbf{v}_1, \ldots, \mathbf{v}_n$ is a linearly independent set of vectors in a linear space V of dimension n, then $\mathbf{v}_1, \ldots, \mathbf{v}_n$ is a basis for V.*

Proof. It must be shown that $\text{Span}\,[\mathbf{v}_1, \ldots, \mathbf{v}_n] = V$. If this is not the case, then by *Proposition 2.1.4* there exists $\mathbf{v}_{n+1} \in V$ such that $\mathbf{v}_1, \ldots, \mathbf{v}_n, \mathbf{v}_{n+1}$ is a linearly independent set, which is impossible by *Theorem 2.1.2* because $\text{Dim}\,[V] = n$. □

Proposition 2.1.6 *A subspace of a finite-dimensional linear space is finite-dimensional.*

Proof. Let W be a subspace of a finite-dimensional linear space V. If $W = \{\mathbf{0}\}$, then $\text{Dim}\,[W] = 0$. Assume $W \neq \{0\}$. A nonzero element $\mathbf{v}_1$ of W is linearly independent. If $\text{Span}\,[\mathbf{v}_1] \neq W$, then by *Proposition 2.1.4* there exist $\mathbf{v}_2$ in W such that $\mathbf{v}_1, \mathbf{v}_2$ is a linearly independent set in W. If $\text{Span}\,[\mathbf{v}_1, \mathbf{v}_2] \neq W$, this argument can be repeated, but the process must terminate in a finite basis for W by *Theorem 2.1.2* because V is a finite-dimensional linear space. □

Proposition 2.1.7 *Let $\mathbf{v}_1, \ldots, \mathbf{v}_m$ be a linearly independent set of vectors in a linear space V of dimension n. If $m < n$, then there exist $\mathbf{v}_{m+1}, \ldots, \mathbf{v}_n$ in V such that $\mathbf{v}_1, \ldots, \mathbf{v}_m, \mathbf{v}_{m+1}, \ldots, \mathbf{v}_n$ is a basis for V.*

Proof. Since $m < n$, the vectors $\mathbf{v}_1, \ldots, \mathbf{v}_m$ cannot be a basis for V by *Corollary 2.1.3*. It follows that $\text{Span}\,[\mathbf{v}_1, \ldots, \mathbf{v}_m] \neq V$ because $\mathbf{v}_1, \ldots, \mathbf{v}_m$ are linearly independent. Then, by *Proposition 2.1.4* there exists a vector $\mathbf{v}_{m+1} \notin \text{Span}\,[\mathbf{v}_1, \ldots, \mathbf{v}_m]$ such that $\mathbf{v}_1, \ldots, \mathbf{v}_m, \mathbf{v}_{m+1}$ are linearly independent. Repeating this argument $n - m$ times produces the linearly independent set

$$\mathbf{v}_1, \ldots, \mathbf{v}_m, \mathbf{v}_{m+1}, \ldots, \mathbf{v}_n$$

which is a basis for V by *Corollary 2.1.5*. □

Not all linear spaces, however, are finite-dimensional. For example, $\mathcal{C}([0,1])$, the set of all real-valued continuous functions defined on the closed interval $[0,1]$, is perfectly good linear space. Just define the sum of two functions f and g in $\mathcal{C}([0,1])$ by $(f+g)(x) = f(x) + g(x)$. Then $f + g$ is in $\mathcal{C}([0,1])$ because the sum of continuous functions is continuous. The additive identity or zero is the function that is identically zero. Similarly, set $(cf)(x) = cf(x)$. In this context $c_1 f_1 + \ldots + c_m f_m = \mathbf{0}$ means the function $c_1 f_1(x) + \ldots + c_m f_m(x) = 0$ for all x in the closed interval $[0,1]$. In other words, the function is identically zero. Since polynomials only have a finite set of zeros, the functions $f_1(x) = x, f_2(x) = x^2, \ldots, f_k(x) = x^k$ are a linearly independent set in $\mathcal{C}([0,1])$ for all $k > 0$. Therefore, $\mathcal{C}([0,1])$ is not a finite-dimensional linear space. We will see in Section 2.3 that $\mathcal{C}([0,1])$ is an interesting metric group and a useful source of examples.

EXERCISES

1. Consider the system of linear equations (2.1) with the standard assumption that each equation has at least one nonzero coefficient and each unknown has a least one nonzero coefficient. Prove the following:

   (a) Adding β times the pth equation to the jth equation does not change the set of solutions.

   (b) If $q > p$, then there exists a solution $\gamma_1, \ldots, \gamma_q$ of (2.1) such that $\gamma_k \neq 0$ for some k. (One approach is to use induction on q and part(a).)

2. Let V be a linear space. Show that $0\mathbf{v} = \mathbf{0}$ for all $\mathbf{v} \in V$ and $c\mathbf{0} = \mathbf{0}$ for all $c \in \mathbb{S}$. Prove that $c\mathbf{v} = \mathbf{0}$ if and only if $c = 0$ or $\mathbf{v} = \mathbf{0}$.

3. Show that $\mathcal{C}(X) = \{f : X \to \mathbb{R} : f \text{ is continuous}\}$ with $(f+g)(x) = f(x) + g(x)$ and $(cf)(x) = cf(x)$ is a linear space when X is a topological space. Prove that for a metric space, $\mathcal{C}(X)$ is a finite-dimensional linear space if and only if X is a finite set with the discrete topology. (One approach is to use *Exercise* 4, p. 27 to show that X must be finite.)

4. Let G be an Abelian group written additively and let $g_1, \ldots, g_k$ be in G. Show that the smallest subgroup H of G containing $g_1, \ldots, g_k$, which is called the *subgroup generated by* $g_1, \ldots, g_k$, is given by

$$H = \{n_1 g_1 + \ldots + n_k g_k : n_1, \ldots, n_k \in \mathbb{Z}\}.$$

5. Let V be a finite-dimensional linear space, and let W be a subspace of V. Show that if $\text{Dim}\,[W] = \text{Dim}\,[V]$, then $W = V$.

6. Let $\mathbf{v}_1, \ldots, \mathbf{v}_n$ be elements of a linear space V. Prove that $\mathbf{v}_1, \ldots, \mathbf{v}_n$ are linearly dependent if and only if for some i there exist scalars such that

$$\mathbf{v}_i = \sum_{j \neq i} c_i \mathbf{v}_j.$$

7. Let $\mathbf{u}_1, \ldots, \mathbf{u}_n$ be elements of a linear space V. Show that

$$\text{Dim}\,[\text{Span}\,[\mathbf{u}_1, \ldots, \mathbf{u}_n]] \leq n.$$

8. Let $\mathbf{u}_1, \ldots, \mathbf{u}_n$ be elements of a linear space V. Show that if

$$\text{Dim}\,[\text{Span}\,[\mathbf{u}_1, \ldots, \mathbf{u}_n]] = n,$$

then $\mathbf{u}_1, \ldots, \mathbf{u}_n$ are linear independent and a basis for $\text{Span}\,[\mathbf{u}_1, \ldots, \mathbf{u}_n]$.

9. Let V_1 and V_2 be linear spaces over the same scalar field and show that $V_1 \times V_2$ is also a linear space. Prove that $V_1 \times V_2$ is a finite-dimensional linear space if and only if both V_1 and V_2 are finite-dimensional. Determine the dimension of $V_1 \times V_2$, when V_1 and V_2 are finite-dimensional?

10. Let V be a finite-dimensional complex linear space. Since $\mathbb{R}$ is contained in $\mathbb{C}$, the complex linear space V is also a real linear space. Determine the dimension of V as a real linear space in terms of its dimension as a complex linear space.

2.2 Linear Functions

In this section, we study linear functions from one linear space to another and introduce algebras. Again the discussion is strictly algebraic, and the emphasis is on finite-dimensional linear spaces.

Let V and V' be linear spaces over the same scalar field. Recall that a function $T : V \to V'$ is an algebraic homomorphism of the Abelian group V to the Abelian group V' provided that $T(\mathbf{u} + \mathbf{v}) = T(\mathbf{u}) + T(\mathbf{v})$ for all $\mathbf{u}$ and $\mathbf{v}$ in V. An algebraic homomorphism that preserves the scalar multiplication is called a linear function. To be precise, a function $T : V \to V'$ is a *linear function* or *linear map* provided that it is an algebraic homomorphism of V to V' and $T(c\mathbf{u}) = cT(\mathbf{u})$ for all $\mathbf{u}$ in V and $c \in \mathbb{S}$. Equivalently, a function $T : V \to V'$ is a linear function if and only if $T(c\mathbf{u} + d\mathbf{v}) = cT(\mathbf{u}) + dT(\mathbf{v})$ for all $\mathbf{u}, \mathbf{v} \in V$ and $c, d \in \mathbb{S}$. In other words, a linear function preserves the group structure and the scalar multiplication.

If V and V' are linear spaces and $T : V \to V'$ is a one-to-one onto linear function, then T is an algebraic group isomorphism of the Abelian group V onto V' and its inverse is an algebraic group isomorphism of the Abelian group V' onto V as shown on page 8. Not surprisingly, T^{-1} is also linear as the following calculation shows:

$$T^{-1}(c\mathbf{v}) = T^{-1}\big(cT(T^{-1}(\mathbf{v}))\big) = T^{-1}\big(T(cT^{-1}(\mathbf{v}))\big) = cT^{-1}(\mathbf{v}).$$

Two linear spaces are said to be *algebraically isomorphic* when there exists a one-to-one linear function from one onto the other.

Constructing linear functions is easy when the dimension of the domain is finite, as the first proposition shows.

Proposition 2.2.1 *Let $\mathbf{v}_1, \ldots, \mathbf{v}_n$ be a basis for a finite-dimensional linear space V, and let V' be another linear space over the same scalar field. If $\mathbf{w}_1, \ldots, \mathbf{w}_n$ are elements of V', then there exists a unique linear function $T : V \to V'$ such that $T(\mathbf{v}_j) = \mathbf{w}_j$ for $j = 1, \ldots, n$.*

Proof. By *Proposition 2.1.1*, for each $\mathbf{u}$ in V there exist unique $c_1, \ldots, c_n$ in the scalar field such that $\mathbf{u} = c_1\mathbf{v}_1 + \ldots + c_n\mathbf{v}_n$. Therefore, $T : V \to W$ can be unambiguously defined by setting

$$T(\mathbf{u}) = T(c_1\mathbf{v}_1 + \ldots + c_n\mathbf{v}_n) = c_1\mathbf{w}_1 + \ldots + c_n\mathbf{w}_n.$$

Let $\mathbf{u} = c_1\mathbf{v}_1 + \ldots + c_n\mathbf{v}_n$ and $\mathbf{v} = d_1\mathbf{v}_1 + \ldots + d_n\mathbf{v}_n$ be two vectors in V. Then $\mathbf{u} + \mathbf{v} = (c_1 + d_1)\mathbf{v}_1 + \ldots + (c_n + d_n)\mathbf{v}_n$ and

$$\begin{aligned} T(\mathbf{u} + \mathbf{v}) &= T\big((c_1 + d_1)\mathbf{v}_1 + \ldots + (c_n + d_n)\mathbf{v}_n\big) \\ &= (c_1 + d_1)\mathbf{w}_1 + \ldots + (c_n + d_n)\mathbf{w}_n) \\ &= (c_1\mathbf{w}_1 + \ldots + c_n\mathbf{w}_n) + (d_1\mathbf{w}_1 + \ldots + d_n\mathbf{w}_n) \\ &= T(\mathbf{u}) + T(\mathbf{v}). \end{aligned}$$

Therefore, T is an algebraic homomorphism of V to V'. Similarly, it is easy to verify that $T(c\mathbf{v}) = cT(\mathbf{v})$ to complete the proof that T is linear.

Let $T' : V \to V'$ be a linear function such that $T'(\mathbf{v}_j) = \mathbf{w}_j$ for $j = 1, \ldots, n$. If $\mathbf{u} = c_1\mathbf{v}_1 + \ldots + c_n\mathbf{v}_n$ is an arbitrary element of V, then

$$\begin{aligned} T'(\mathbf{u}) &= T'(c_1\mathbf{v}_1 + \ldots + c_n\mathbf{v}_n) \\ &= c_1\mathbf{w}_1 + \ldots + c_n\mathbf{w}_n \\ &= c_1T(\mathbf{v}_1) + \ldots + c_nT(\mathbf{v}_n) \\ &= T(c_1\mathbf{v}_1 + \ldots + c_n\mathbf{v}_n) \\ &= T(\mathbf{u}) \end{aligned}$$

shows that T is uniquely defined. □

Let $T : V \to V'$ be a linear function. Since T is an algebraic homomorphism of groups, the kernel K of T is a subgroup of V. The kernel is also a subspace because $T(c\mathbf{v}) = cT(\mathbf{v}) = c\mathbf{0} = \mathbf{0}$, when $T(\mathbf{v}) = \mathbf{0}$. The kernel of a linear function T is also often called the *null space* of T. Of course, a linear function is one-to-one if and only if the kernel or null space $K = \{\mathbf{0}\}$.

Similarly, the image $T(V) = \{T(\mathbf{v}) : \mathbf{v} \in V\}$ of a linear function $T : V \to V'$ is a subgroup and a subspace because $cT(\mathbf{v}) = T(c\mathbf{v})$. When V is finite-dimensional there is an important relationship between the dimension of V and the dimensions of the kernel and image of a linear function $T : V \to V'$.

Theorem 2.2.2 *Let V and V' be linear spaces. If V is finite-dimensional and $T : V \to V'$ is a linear function with kernel K, then*

$$\text{Dim}\,[V] = \text{Dim}\,[K] + \text{Dim}\,[T(V)].$$

Proof. Let $\mathbf{v}_1, \ldots, \mathbf{v}_m$ be a basis for the kernel K. By *Proposition 2.1.7*, there exist $\mathbf{v}_{m+1}, \ldots, \mathbf{v}_n$ such that $\mathbf{v}_1, \ldots, \mathbf{v}_m, \mathbf{v}_{m+1}, \ldots, \mathbf{v}_n$ is a basis for V. So $\text{Dim}\,[V] = \text{Dim}\,[K] + n - m$ and it suffices to show that $T(\mathbf{v}_{m+1}), \ldots, T(\mathbf{v}_n)$ is a basis for $T(V)$.

Given $\mathbf{v} \in V$, there exist scalars $c_1, \ldots, c_n$ such that $\mathbf{v} = \sum_{j=1}^{n} c_j\mathbf{v}_j$. It follows that

$$\begin{aligned} T(\mathbf{v}) &= \sum_{j=1}^{n} T(c_j\mathbf{v}_j) \\ &= \sum_{j=1}^{m} c_jT(\mathbf{v}_j) + \sum_{j=m+1}^{n} c_jT(\mathbf{v}_j) \\ &= \sum_{j=m+1}^{n} c_jT(v_j) \end{aligned}$$

because $T(\mathbf{v}_j) = \mathbf{0}$ for $j = 1, \ldots, m$. Hence,

$$\text{Span}\,[T(\mathbf{v}_{m+1}), \ldots, T(\mathbf{v}_n)] = T(V).$$

Suppose

$$\sum_{j=m+1}^{n} c_j T(\mathbf{v}_j) = \mathbf{0}$$

or equivalently

$$T\left(\sum_{j=m+1}^{n} c_j \mathbf{v}_j\right) = \mathbf{0}.$$

It follows that

$$\sum_{j=m+1}^{n} c_j \mathbf{v}_j$$

is in the kernel K and there exist scalars $c_1, \ldots, c_m$ such that

$$\sum_{j=1}^{m} c_j \mathbf{v}_j = \sum_{j=m+1}^{n} c_j \mathbf{v}_j$$

or

$$\sum_{j=1}^{m} c_j \mathbf{v}_j - \sum_{j=m+1}^{n} c_j \mathbf{v}_j = \mathbf{0}.$$

The linear independence of $\mathbf{v}_1, \ldots, \mathbf{v}_n$ implies that $c_j = 0$ for $j = 1, \ldots, n$. Thus $T(\mathbf{v}_{m+1}), \ldots, T(\mathbf{v}_n)$ are linearly independent and a basis for $T(V)$. □

Corollary 2.2.3 *Let V be a finite-dimensional linear space and let $T : V \to V$ be a linear function. Then T is one-to-one if and only if it is onto.*

Proof. Just observe that T is one-to-one if and only if its kernel K is $\{\mathbf{0}\}$ if and only if $\text{Dim}\,[K] = 0$ if and only if the $\text{Dim}\,[T(V)] = \text{Dim}\,[V]$ if and only if $T(V) = V$, using *Exercise* 5, p. 69 for the last step. □

Theorem 2.2.4 *Two finite-dimensional linear spaces over the same scalar field are algebraically isomorphic if and only if their dimensions are equal.*

Proof. Let V and V' be two finite-dimensional linear spaces. If V and V' are algebraically isomorphic, there exists a one-to-one onto linear function $T : V \to V'$. Consequently, the kernel of T is $\{\mathbf{0}\}$ and $T(V) = V'$. *Theorem 2.2.2* now implies that $\text{Dim}\,[V] = 0 + \text{Dim}\,[T(V)] = \text{Dim}\,[V']$.

For the converse, suppose that $\text{Dim}\,[V] = \text{Dim}\,[V'] = n$. Let $\mathbf{v}_1, \ldots, \mathbf{v}_n$ and $\mathbf{w}_1, \ldots, \mathbf{w}_n$ be a bases for V and V' respectively. By *Proposition 2.2.1*, there exists a unique linear map $T : V \to V'$ such that $T(\mathbf{v}_i) = \mathbf{w}_i$ for $i = 1, \ldots, n$. Clearly, $V' = \text{Span}\,[\mathbf{w}_1, \ldots, \mathbf{w}_n] \subset T(V) \subset V'$ and T is onto. Hence $\text{Dim}\,[V] = \text{Dim}\,[K] + \text{Dim}\,[T(V)] = \text{Dim}\,[K] + \text{Dim}\,[V']$, by *Theorem 2.2.2*, and $\text{Dim}\,[K] = 0$. Thus T is also one-to-one. □

As a consequence of this theorem, finite-dimensional vector spaces are algebraically indistinguishable if and only if their dimensions are equal. This will play a key role in understanding finite-dimensional linear spaces as metric spaces in the next section.

Given two linear spaces V and V' over the same scalar field, let $\mathcal{L}(V, V')$ denote the set of linear functions from V to V'. It has a natural Abelian group structure. Just define a binary operation $S + T$ on $\mathcal{L}(V, V')$ by setting $(S + T)(\mathbf{v}) = S(\mathbf{v}) + T(\mathbf{v})$ and checking that $S + T$ is linear. This binary operation is easily seen to be associative and Abelian. The zero element is the linear function that sends every $\mathbf{v}$ in V to $\mathbf{0}$ in V', and $(-T)(\mathbf{v}) = -T(\mathbf{v})$.

Furthermore, there is a natural linear structure on $\mathcal{L}(V, V')$. For c in the common scalar field for V and V', define $(cT)(\mathbf{v}) = cT(\mathbf{v})$ and observe that it satisfies the scalar multiplication conditions in the definition of a linear structure on page 64, making $\mathcal{L}(V, V')$ a linear space.

When V is a finite-dimensional linear space with $\mathrm{Dim}\,[V] = n$, a basis $\mathbf{v}_1, \mathbf{v}_2 \ldots, \mathbf{v}_n$ is indexed by the set $\{1, 2, \ldots, n\}$, that is, there is a first element $\mathbf{v}_1$, a second element $\mathbf{v}_2$, and so forth, until the last element $\mathbf{v}_n$. Note that there can be no repeats in a basis. Formally, we have a function $\theta : \{1, 2, \ldots, n\} \to V$ such that $\theta(j) = \mathbf{v}_j$. If α is an element of S_n, that is, a permutation on n symbols, then $\theta \circ \alpha$ or $\mathbf{u}_1 = \mathbf{v}_{\theta(1)}, \mathbf{u}_2 = \mathbf{v}_{\theta(2)}, \ldots, \mathbf{u}_n = \mathbf{v}_{\theta(n)}$ is also a basis for V. *The order of the vectors does not effect linear independence or the span.*

Every element $\mathbf{v}$ of V can be expressed uniquely in the form $c_1\mathbf{v}_1 + c_2\mathbf{v}_2 + \ldots + c_n\mathbf{v}_n$ using scalars $c_1, \ldots, c_n$, when $\mathbf{v}_1, \mathbf{v}_2 \ldots, \mathbf{v}_n$ is basis. Now the order is important for both the basis elements and the scalars. Having a set of n vectors and a set of n scalars does not work. First it may or may not be necessary to repeat one or more scalars and that requires an indexed list of n scalars or a function from $\{1, 2, \ldots, n\}$ to $\mathbb{S}$. Second, if $\mathbf{v}_j$ is not multiplied by the scalar c_j the result need not be the desired vector $\mathbf{v}$.

There are, however, situations where using a set of consecutive integers like $\{1, 2, \ldots, n\}$ as an indexing set is not the most convenient choice. One such situation is $\mathcal{L}(V, V')$ when V and V' are finite-dimensional linear spaces. Here it is more convenient to use indexing sets like

$$\begin{aligned} \Lambda &= \{1 \le i \le n\} \times \{1 \le j \le m\} \\ &= \{(i, j) : 1 \le i \le n \text{ and } 1 \le j \le m\} \end{aligned} \tag{2.2}$$

as will become apparent in the proof of the following proposition:

Proposition 2.2.5 *If V and V' are finite-dimensional linear spaces over the same scalar field, then $\mathcal{L}(V, V')$ is also finite-dimensional and its dimension is* $\mathrm{Dim}\,[V]\mathrm{Dim}\,[V']$.

Proof. Let m and n denote the dimensions of V and V', respectively. We need to construct a set mn linear functions in $\mathcal{L}(V, V')$ that are linearly independent and whose span equals $\mathcal{L}(V, V')$. Rather than using the integers $1, \ldots, mn$ to index them, it is preferable to use the set Λ defined above by equation (2.2) as the indexing set.

Let $\mathbf{u}_1, \ldots, \mathbf{u}_m$ and $\mathbf{v}_1, \ldots, \mathbf{v}_n$ be bases for V and V', respectively. Using *Proposition 2.2.1*, for each (i,j) in Λ there exists $T_{(i,j)}$ in $\mathcal{L}(V, V')$ such that

$$T_{(i,j)}(\mathbf{u}_k) = \begin{cases} \mathbf{v}_i & \text{if } k = j \\ \mathbf{0} & \text{if } k \neq j \end{cases} \tag{2.3}$$

for $k = 1, \ldots, m$. It will be shown that $\{T_{(i,j)} : (i,j) \in \Lambda\}$ is a basis for $\mathcal{L}(V, V')$.

Let T be an arbitrary element of $\mathcal{L}(V, V')$. Then there exist unique scalars $B(i,j)$ indexed by Λ such that

$$T(\mathbf{u}_k) = \sum_{i=1}^{n} B(i,k)\mathbf{v}_i$$

for $k = 1, \ldots, m$. Set

$$S = \sum_{(i,j)\in\Lambda} B(i,j)T_{(i,j)}.$$

To prove that

$$\text{Span}\left[\{T_{(i,j)} : (i,j) \in \Lambda\}\right] = \mathcal{L}(V, V'),$$

it suffices to show that $T = S$.

To show that $T = S$ it is sufficient by linearity to show that $T(\mathbf{u}_k) = S(\mathbf{u}_k)$ for $k = 1, \ldots, m$. The first step is to observe that

$$\sum_{j=1}^{m} B(i,j)T_{(i,j)}(\mathbf{u}_k) = B(i,k)\mathbf{v}_i$$

using the definition of $T_{(i,j)}(\mathbf{u}_k)$. Then,

$$\begin{aligned} S(\mathbf{u}_k) &= \sum_{(i,j)\in\Lambda} B(i,j)T_{(i,j)}(\mathbf{u}_k) \\ &= \sum_{i=1}^{n}\sum_{j=1}^{m} B(i,j)T_{(i,j)}(\mathbf{u}_k) \\ &= \sum_{i=1}^{n} B(i,k)\mathbf{v}_i \\ &= T(\mathbf{u}_k). \end{aligned}$$

Hence, $T = S$ and $\text{Span}\,[T_{(i,j)} : (i,j) \in \Lambda] = \mathcal{L}(V, V')$.

To prove the linear independence suppose that there are scalars $A(i,j)$ for (i,j) in Λ such that

$$\sum_{(ij)\in\Lambda} A(i,j)T_{(i,j)} = \mathbf{0}.$$

Since this is an equation in $\mathcal{L}(V, V')$, it means that

$$\sum_{(ij)\in\Lambda} A(i,j)T_{(i,j)}(\mathbf{u}) = \mathbf{0}$$

for all $\mathbf{u} \in V$. In particular,

$$\sum_{(ij)\in\Lambda} A(i,j)T_{(i,j)}(\mathbf{u}_k) = \mathbf{0}$$

for $k = 1, \ldots, m$. As above this reduces to

$$\sum_{i=1}^{n} A(i,k)\mathbf{v}_i = \mathbf{0}$$

which implies that $A(i,k) = 0$ for $i = 1, \ldots, n$ because $\mathbf{v}_1, \ldots, \mathbf{v}_n$ are linearly independent. Since k was arbitrary, it follows that $A(i,k) = 0$ for all $(i,k) \in \Lambda$ and $\{T_{(i,j)} : (i,j) \in \Lambda\}$ is a linear independent set in $\mathcal{L}(V, V')$. $\square$

This proof has led us naturally to the idea of a matrix. An $n \times m$ *matrix* is a function from the set Λ defined by (2.2) to a scalar field $\mathbb{S}$. Just as it is natural to write $\mathbf{x}$ in $\mathbb{R}^m$ as the row of real numbers $(x_1, \ldots, x_m)$, it is natural to write a matrix as a rectangular array of scalars. The conventional way of doing this is to let $A(i,j)$ denote the jth element in the ith row. So an $n \times m$ matrix A is written as

$$A = \begin{pmatrix} A(1,1) & A(1,2) & \ldots & A(1,m) \\ \vdots & \vdots & & \vdots \\ A(n,1) & A(n,2) & \ldots & A(n,m) \end{pmatrix}$$

Let $\mathcal{M}_{n\times m}(\mathbb{S})$ denote the set of $n \times m$ matrices for the scalar field $\mathbb{S}$. The notation $\mathcal{M}_{n\times m}(\mathbb{S})$ primarily indicates the use of the index set Λ defined by (2.2); otherwise $\mathcal{M}_{n\times m}(\mathbb{S})$ is a thinly disguised version of $\mathbb{S}^{nm}$. In particular, $\mathcal{M}_{n\times m}(\mathbb{S})$ is a linear space with $(A+B)(i,j) = A(i,j)+B(i,j)$ and $(cA)(i,j) = cA(i,j)$, and the dimension of $\mathcal{M}_{n\times m}(\mathbb{S})$ is nm.

Consider again two finite-dimensional linear spaces V and V' over $\mathbb{S}$ of dimensions m and n, respectively. Let $\mathbf{u}_1, \ldots, \mathbf{u}_m$ and $\mathbf{v}_1, \ldots, \mathbf{v}_n$ be bases for V and V' respectively. In the proof of *Proposition 2.2.5* it was shown that the linear functions $\{T_{(i,j)} : (i,j) \in \Lambda\}$ defined by (2.3) are a basis for $\mathcal{L}(V, V')$. Hence for each T in $\mathcal{L}(V, V')$ there exists a unique matrix B called the *matrix of the linear function S* with respect to the bases $\mathbf{u}_1, \ldots, \mathbf{u}_m$ and $\mathbf{v}_1, \ldots, \mathbf{v}_n$ such that

$$T = \sum_{(ij)\in\Lambda} B(i,j)T_{(i,j)}. \tag{2.4}$$

It follows that

$$\begin{aligned} T(\mathbf{u}_k) &= \sum_{(ij)\in\Lambda} B(i,j)T_{(i,j)}(\mathbf{u}_k) \\ &= \sum_{i=1}^{n}\sum_{j=1}^{m} B(i,j)T_{(i,j)}(\mathbf{u}_k) \\ &= \sum_{i=1}^{n} B(i,k)\mathbf{v}_i \end{aligned}$$

or simply

$$T(\mathbf{u}_k) = \sum_{i=1}^{n} B(i,k)\mathbf{v}_i. \tag{2.5}$$

Consequently, the kth *column of the matrix* B *consists of the coefficients of* $T(\mathbf{u}_k)$ *with respect to the basis* $\mathbf{v}_1, \ldots, \mathbf{v}_n$ *and the* i^{th} *row consists of the coefficients of* $\mathbf{v}_i$ *in the vectors* $T(\mathbf{u}_1), \ldots, T(\mathbf{u}_m)$. *The number of rows in the matrix is the dimension of* V' *and the number of columns is the dimension of* V.

Since equation (2.4) uniquely determines the matrix B in $\mathcal{M}_{n\times m}(\mathbb{S})$ for a given linear function T in $\mathcal{L}(V, V')$, it defines a function $\Phi : \mathcal{L}(V, V') \to \mathcal{M}_{n\times m}(\mathbb{S})$. It is easy to see that Φ is linear and onto. Hence, it is one-to-one by *Theorem 2.2.2* and has a linear inverse Φ^{-1}. The linear isomorphism Φ will reappear with more details in the last theorem of this section.

When $V = \mathbb{S}^m$ and $V' = \mathbb{S}^n$, the matrix B of a linear function $T : \mathbb{S}^m \to \mathbb{S}^n$ with respect to the standard bases $\mathbf{e}_1, \ldots, \mathbf{e}_m$ and $\mathbf{e}_1, \ldots, \mathbf{e}_n$ can be used to explicitly calculate $T(\mathbf{x})$ for $\mathbf{x} = (x_1, \ldots, x_m)$. (Admittedly this notation appears somewhat ambiguous, but it is to be understood that $\mathbf{e}_1, \ldots, \mathbf{e}_m$ are m-tuples because they are in $\mathbb{S}^m$ and $\mathbf{e}_1, \ldots, \mathbf{e}_n$ are n-tuples because they are in $\mathbb{S}^n$. When $m = n$, the problem disappears.) Specifically,

$$\begin{aligned} T(\mathbf{x}) &= T\left(\sum_{k=1}^{m} x_k \mathbf{e}_k\right) \\ &= \sum_{k=1}^{m} x_k T(\mathbf{e}_k) \\ &= \sum_{k=1}^{m} x_k \sum_{i=1}^{n} B(i,k)\mathbf{e}_i \\ &= \sum_{i=1}^{n} \left(\sum_{k=1}^{m} B(i,k)x_k\right)\mathbf{e}_i. \end{aligned}$$

Thus the ith coordinate of $T(\mathbf{x})$ is just

$$T(\mathbf{x})_i = \sum_{k=1}^{m} B(i,k)x_k \tag{2.6}$$

Now let V, V', and V'' be three finite-dimensional linear spaces over the same scalar field with dimensions m, n, and p, respectively. Let $\mathbf{u}_1, \ldots, \mathbf{u}_m$; $\mathbf{v}_1, \ldots, \mathbf{v}_n$; and $\mathbf{w}_1, \ldots, \mathbf{w}_p$ be bases for V, V', and V'', respectively. Consider linear functions $T : V \to V'$ and $S : V' \to V''$. Let B be the matrix of T with respect to the bases $\mathbf{u}_1, \ldots, \mathbf{u}_m$ and $\mathbf{v}_1, \ldots, \mathbf{v}_n$, and let A be the matrix of S with respect to the bases $\mathbf{v}_1, \ldots, \mathbf{v}_n$ and $\mathbf{w}_1, \ldots, \mathbf{w}_p$. In this context, can the matrix C of $S \circ T : V \to V''$ with respect to the bases $\mathbf{u}_1, \ldots, \mathbf{u}_m$ and $\mathbf{w}_1, \ldots, \mathbf{w}_p$ be calculated from the matrices A and B?

This is not as complicated as it sounds. The key is to use equation (2.5) and its counterparts for S and $S \circ T$ to calculate $S \circ T(\mathbf{u}_k)$ as follows:

$$\begin{aligned}
S \circ T(\mathbf{u}_k) &= S(T(\mathbf{u}_k)) \\
&= S\left(\sum_{i=1}^{n} B(i,k)\mathbf{v}_i\right) \\
&= \sum_{i=1}^{n} B(i,k)S(\mathbf{v}_i) \\
&= \sum_{i=1}^{n} B(i,k)\sum_{j=1}^{p} A(j,i)\mathbf{w}_j \\
&= \sum_{i=1}^{n}\sum_{j=1}^{p} B(i,k)A(j,i)\mathbf{w}_j \\
&= \sum_{j=1}^{p}\left(\sum_{i=1}^{n} A(j,i)B(i,k)\right)\mathbf{w}_j.
\end{aligned}$$

Therefore,

$$S \circ T(\mathbf{u}_k) = \sum_{j=1}^{p}\left(\sum_{i=1}^{n} A(j,i)B(i,k)\right)\mathbf{w}_j \tag{2.7}$$

and

$$C(j,k) = \sum_{i=1}^{n} A(j,i)B(i,k), \tag{2.8}$$

which is the standard definition of matrix multiplication. To be more specific, if A is a $p \times n$ matrix and B is a $n \times m$ matrix, then AB is the $p \times m$ matrix C whose (j,k) entry is given by equation (2.8). In other words, the (j,k) entry of AB is the usual dot product of the jth row of A and the kth column of B. *The matrix C of $S \circ T$ with respect to the bases $\mathbf{u}_1, \ldots, \mathbf{u}_m$ and $\mathbf{w}_1, \ldots, \mathbf{w}_p$ is the matrix product AB.*

If we restrict our attention to a single linear space V, not necessarily finite-dimensional, then $\mathcal{L}(V,V)$ has even richer algebraic properties than V. If S and T are elements of $\mathcal{L}(V,V)$, it is easy to see that $S \circ T$ is also in $\mathcal{L}(V,V)$. So $\mathcal{L}(V,V)$ is a subsemigroup of the semigroup F_V of all functions from V to V with the binary operation of composition of functions introduced on page 4. The identity function $\iota(\mathbf{v}) = \mathbf{v}$ is obviously a linear function. So $\mathcal{L}(V,V)$ is a semigroup with identity under composition of functions.

The addition, composition, and scalar multiplication of linear functions are connected by the following routinely verified formulas:

(a) $S \circ (T_1 + T_2) = S \circ T_1 + S \circ T_2$

(b) $(S_1 + S_2) \circ T = S_1 \circ T + S_2 \circ T$

(c) $c(S \circ T) = (cS) \circ T = S \circ (cT)$.

This amounts to saying that $\mathcal{L}(V, V)$ is an algebra with identity, but that concept must be defined first.

An *algebra* is an Abelian group $\mathcal{A}$ with a linear structure and an associative binary operation written as $\mathbf{uv}$ on $\mathcal{A}$ such that for all $\mathbf{u}, \mathbf{v}, \mathbf{w}$ in $\mathcal{A}$ and c in $\mathbb{S}$ the following hold:

(a) $\mathbf{u}(\mathbf{v} + \mathbf{w}) = \mathbf{uv} + \mathbf{uw}$

(b) $(\mathbf{u} + \mathbf{v})\mathbf{w} = \mathbf{uw} + \mathbf{vw}$

(c) $c(\mathbf{uv}) = (c\,\mathbf{u})\mathbf{v} = \mathbf{u}(c\,\mathbf{v})$

Thus there are two interrelated binary operations called addition and multiplication in an algebra such that $\mathcal{A}$ is an Abelian group with respect to addition and a semigroup with respect to multiplication. In addition, there is a linear structure for the Abelian group that is interrelated with the multiplication operation. The three formulas specifying these interrelations are equivalent to requiring that multiplication on either the right or left by a fixed element of the algebra is a linear map of the algebra to itself.

In general, two algebraic objects are algebraically isomorphic if there exists a one-to-one function from one onto the other such that the algebraic calculations can be carried out before or after applying the function. Applying this general principle one more time leads to the definition of algebraically isomorphic algebras. Specifically, two algebras $\mathcal{A}$ and $\mathcal{A}'$ are *algebraically isomorphic* if there exists a one-to-one onto function $\Phi : \mathcal{A} \to \mathcal{A}'$ satisfying the following conditions for all $\mathbf{u}, \mathbf{v}$ in $\mathcal{A}$ and c in $\mathbb{S}$:

$$\begin{aligned} \Phi(\mathbf{u} + \mathbf{v}) &= \Phi(\mathbf{u}) + \Phi(\mathbf{v}) \\ \Phi(c\,\mathbf{u}) &= c\Phi(\mathbf{u}) \\ \Phi(\mathbf{uv}) &= \Phi(\mathbf{u})\Phi(\mathbf{v}). \end{aligned}$$

Not surprisingly Φ^{-1} is an algebraic isomorphism of the algebra $\mathcal{A}'$ onto the algebra $\mathcal{A}$. (Again, the same technique used on page 8 to prove that the inverse of an algebraic isomorphism of groups was also an algebraic isomorphism of groups works here.)

An algebra $\mathcal{A}$ is an *algebra with an identity* provided that there exists $\mathbf{e} \neq \mathbf{0}$ in $\mathcal{A}$ such that $\mathbf{eu} = \mathbf{ue}$. An algebra is a *commutative algebra* provided that $\mathbf{uv} = \mathbf{vu}$ for all $\mathbf{u}, \mathbf{v}$ in $\mathcal{A}$. Obviously, $\mathcal{L}(V, V)$ is an algebra with identity, but it is rarely commutative (see *Exercise* 11, p. 82).

If $\mathcal{A}$ is an algebra with identity $\mathbf{e}$, then the set of units,

$$G = \{\mathbf{u} \in \mathcal{A} : \text{ there exists } \mathbf{v} \in \mathcal{A} \text{ such that } \mathbf{uv} = \mathbf{e} = \mathbf{vu}\}, \tag{2.9}$$

is a group by *Exercise* 3, p. 11.

If V is a linear space, and T is an element of $\mathcal{L}(V, V)$, then T is a unit in $\mathcal{L}(V, V)$ if and only if there exists a linear function $T^{-1} \in \mathcal{L}(V, V)$ such that

$T \circ T^{-1} = \iota = T^{-1} \circ T$. It follows that a unit of $\mathcal{L}(V, V)$ is one-to-one and onto, and hence a linear algebraic isomorphism of V onto itself. The units in $\mathcal{L}(V, V)$ are often called *invertible linear functions.* When V is finite-dimensional, it suffices to check that a linear function is either one-to-one or onto to show that it is invertible by *Corollary 2.2.3.*

The $n \times n$ matrices are of special interest and will be denoted by $\mathcal{M}_n(\mathbb{S})$ instead of $\mathcal{M}_{n \times n}(\mathbb{S})$. Now equation (2.8) for matrix multiplication defines a binary operation on $\mathcal{M}_n(\mathbb{S})$ and routine calculations show that $\mathcal{M}_n(\mathbb{S})$ is an algebra with identity. The *identity matrix* I is defined by $I(i, j)$ equals 1 or 0 according as $i = j$ or $i \neq j$.

An $n \times n$ matrix A is said to be an *invertible matrix* provided there exists an $n \times n$ matrix A^{-1} such that $AA^{-1} = I = A^{-1}A$. In other words, a matrix is invertible if and only if it is a unit in $\mathcal{M}_n(\mathbb{S})$. The set of invertible matrices in $\mathcal{M}_n(\mathbb{S})$ is the group of units of $\mathcal{M}_n(\mathbb{S})$ and a particularly important metric group. A matrix that is not invertible is called a *singular matrix*

The linear isomorphism $\Phi : \mathcal{L}(V, V') \to \mathcal{M}_{n \times m}(\mathbb{S})$ briefly discussed earlier in this section is an algebraic isomorphism of algebras when $V = V'$.

Theorem 2.2.6 *If V is a finite-dimensional linear space of dimension n over the scalar field $\mathbb{S}$, then there exists an algebraic isomorphism Φ of the algebra $\mathcal{L}(V, V)$ onto the algebra $\mathcal{M}_n(\mathbb{S})$ such that $\Phi(\iota) = I$ and such that T is invertible if and only if $\Phi(T)$ is invertible.*

Proof. Let $\mathbf{v}_1, \ldots, \mathbf{v}_n$ be a basis of V, and let

$$\begin{aligned} \Lambda &= \{1 \leq i \leq n\} \times \{1 \leq j \leq n\} \\ &= \{(i, j) : 1 \leq i \leq n \text{ and } 1 \leq j \leq n\}. \end{aligned}$$

Define $T_{(i,j)}$ for $(i, j) \in \Lambda$ by

$$T_{(i,j)}(\mathbf{v}_k) = \begin{cases} \mathbf{v}_i & \text{if } k = j \\ \mathbf{0} & \text{if } k \neq j \end{cases} \tag{2.10}$$

for $k = 1, \ldots, m$. [Equation (2.10) is just a special case of equation (2.3).] The proof of *Proposition 2.2.5* shows that $\{T_{(i,j)} : (i, j) \in \Lambda\}$ is a basis for $\mathcal{L}(V, V)$.

Let S be an element of $\mathcal{L}(V, V)$. There exists a matrix A uniquely determined by S such that

$$S = \sum_{(i,j) \in \Lambda} A(i, j) T_{(i,j)}.$$

Set $\Phi(S) = A$, the matrix of S with respect to the basis $\mathbf{v}_1, \ldots, \mathbf{v}_n$ and itself or simply with respect to the basis $\mathbf{v}_1, \ldots, \mathbf{v}_n$. Then Φ maps $\mathcal{L}(V, V)$ onto $\mathcal{M}_n(\mathbb{S})$ because every matrix A in $\mathcal{M}_n(\mathbb{S})$ determines an S in $\mathcal{L}(V, V)$ by setting

$$S = \sum_{(i,j) \in \Lambda} A(i, j) T_{(i,j)}.$$

Clearly,

$$cS = \sum_{(i,j)\in\Lambda} cA(i,j)T_{(i,j)}$$

and $\Phi(cS) = c\Phi(S)$. If T is another element of $\mathcal{L}(V,V)$, then

$$T = \sum_{(i,j)\in\Lambda} B(i,j)T_{(i,j)},$$

and

$$S + T = \sum_{(i,j)\in\Lambda} (A(i,j) + B(i,j))T_{(i,j)}.$$

It follows that $\Phi(S+T) = \Phi(S) + \Phi(T)$ and Φ is a linear function. Since Φ is onto, it is also one-to-one by *Theorem 2.2.2* because

$$\text{Dim}\,[\mathcal{L}(V,V)] = n^2 = \text{Dim}\,[\mathcal{M}_n(\mathbb{S})].$$

It was shown on page 77 that

$$S \circ T = \sum_{(i,j)\in\Lambda} C(i,j)T_{(i,j)}$$

where $C = AB$ is defined by (2.8). Hence, $\Phi(S \circ T) = AB$ and Φ is an algebraic isomorphism of the algebra $\mathcal{L}(V,V)$ onto the algebra $\mathcal{M}_n(\mathbb{S})$.

Observe $\iota = \sum_{i=1}^{n} T_{(i,i)}$ and hence $\Phi(\iota) = I$. From this it follows easily that T is invertible if and only if $\Phi(T)$ is invertible. □

When $V = \mathbb{S}^n$ and the basis is the standard one given by $\mathbf{e}_1, \ldots, \mathbf{e}_n$, there exists $\Phi : \mathcal{L}(\mathbb{S}^n, \mathbb{S}^n) \to \mathcal{M}_n(\mathbb{S})$ satisfying the conclusion of *Theorem 2.2.6*. In this case one can give a useful and more explicit formula for $\Phi^{-1} : \mathcal{M}_n(\mathbb{S}) \to \mathcal{L}(\mathbb{S}^n, \mathbb{S}^n)$. The details appear in *Exercise* 6, p. 82. We can also derive a useful formula for changing basis.

Proposition 2.2.7 *Let V be a finite-dimensional linear space. Let $\mathbf{u}_1, \ldots, \mathbf{u}_n$ and $\mathbf{v}_1, \ldots, \mathbf{v}_n$ be two bases for V.*

(a) There exist unique matrices B and C in $\mathcal{M}_n(\mathbb{S})$ such that

$$\mathbf{u}_k = \sum_{i=1}^{n} B(i,k)\mathbf{v}_i \quad \text{and} \quad \mathbf{v}_k = \sum_{j=1}^{n} C(j,k)\mathbf{u}_j. \tag{2.11}$$

(b) Moreover, $C = B^{-1}$.

(c) If A is the matrix of T in $\mathcal{L}(V,V)$ with respect to the basis $\mathbf{u}_1, \ldots, \mathbf{u}_n$, then the matrix of T with respect to the basis $\mathbf{v}_1, \ldots, \mathbf{v}_n$ is BAB^{-1}.

Proof. The matrix B is just the matrix of ι with respect to the bases $\mathbf{u}_1, \ldots, \mathbf{u}_n$ and $\mathbf{v}_1, \ldots, \mathbf{v}_n$ given by (2.5), and C is the matrix of ι with the roles of the bases reversed. These matrices are unique by (2.4).

Using equations (2.7) and (2.8), it follows that CB is the matrix of $\iota \circ \iota = \iota$ with respect to $\mathbf{u}_1, \ldots, \mathbf{u}_n$ and itself. Therefore, $CB = I$ and similarly $BC = I$, proving that $C = B^{-1}$.

Suppose A is the matrix of $T \in \mathcal{L}(V, V)$ with respect to $\mathbf{u}_1, \ldots, \mathbf{u}_n$ and itself. Then using (2.7) and (2.8) as above, AC is the matrix of $T \circ \iota = T$ with respect to $\mathbf{v}_1, \ldots, \mathbf{v}_n$ and $\mathbf{u}_1, \ldots, \mathbf{u}_n$. Similarly, $B(AC) = BAB^{-1}$ is the matrix of $\iota \circ T = T$ with respect to $\mathbf{v}_1, \ldots, \mathbf{v}_n$ and itself. (Alternatively, one can use A, B, and C to calculate directly that $T(\mathbf{v}_k) = \sum_{j=1}^{n} BAC(j, k)\mathbf{v}_j$.) □

The results in this section provide an algebraic classification of all finite-dimensional linear spaces and their algebras of linear functions for the scalar fields $\mathbb{R}$ and $\mathbb{C}$. Every finite-dimensional linear space over $\mathbb{R}$ or $\mathbb{C}$ is linearly isomorphic to $\mathbb{R}^n$ or $\mathbb{C}^n$, respectively, by *Theorem 2.2.4*. Furthermore, every algebra $\mathcal{L}(V, V)$ for finite-dimensional linear spaces over $\mathbb{R}$ or $\mathbb{C}$ is algebraically isomorphic to the matrix algebra $\mathcal{M}_n(\mathbb{R})$ or $\mathcal{M}_n(\mathbb{C})$. These algebraic classifications will be very useful in the study of topological properties of finite-dimensional normed linear spaces, which are another class of metric groups.

EXERCISES

1. Let V and V' be finite-dimensional linear spaces and let K be a subspace of V. Establish necessary and sufficient conditions for the existence of a linear function $T : V \to V'$ such that K is the kernel of T.

2. Let V be a finite-dimensional linear space and let $T : V \to V$ be a linear function with kernel K. Prove that if $K \cap T(V) = \{\mathbf{0}\}$, then given $\mathbf{v} \in V$ there exist unique vectors $\mathbf{u}_1 \in K$ and $\mathbf{u}_2 \in T(V)$ such that $\mathbf{v} = \mathbf{u}_1 + \mathbf{u}_2$.

3. Let V be linear space of dimension $n > 1$ and let K be a subspace of dimension 1. Show that there exists a linear function $T : V \to V$ with kernel K such that
$$\underbrace{T \circ \ldots \circ T}_{n-1}(\mathbf{v}) = \mathbf{0}$$
for all $\mathbf{v}$ in V.

4. The *transpose of a matrix* A is defined by $A^{\mathfrak{t}}(i, j) = A(j, i)$. Show that the function $A \to A^{\mathfrak{t}}$ is a one-to-one linear function from $\mathcal{M}_{m \times n}(\mathbb{S})$ onto $\mathcal{M}_{n \times m}(\mathbb{S})$ such that $(AB)^{\mathfrak{t}} = B^{\mathfrak{t}}A^{\mathfrak{t}}$. When $m = n$, also show that A is invertible if and only if $A^{\mathfrak{t}}$ is invertible.

5. Let $T : \mathbb{S}^m \to \mathbb{S}^n$ be a linear function and let B be in $\mathcal{M}_{n \times m}(\mathbb{S})$. Thinking of $\mathbf{x} = (x_1, \ldots, x_m)$ in $\mathbb{S}^m$ as a $1 \times m$ matrix, show that $T(\mathbf{x}) = \mathbf{x}B^{\mathfrak{t}}$ for all $\mathbf{x}$ in $\mathbb{S}^m$ if and only if B is the matrix of T with respect to the standard bases $\mathbf{e}_1, \ldots, \mathbf{e}_m$ and $\mathbf{e}_1, \ldots, \mathbf{e}_n$.

6. Define $\Psi : \mathcal{M}_n(\mathbb{S}) \to \mathcal{L}(\mathbb{S}^n, \mathbb{S}^n)$ by $\Psi(A) = T_A$, where $T_A(\mathbf{x}) = \mathbf{x}A^{\mathsf{t}}$. (See *Exercises 4 and 5* above for background information.) Prove the following:
   (a) The function $\Psi(A) = T_A$ is an algebraic isomorphism of the algebra $\mathcal{M}_n(\mathbb{S})$ onto the algebra $\mathcal{L}(\mathbb{S}^n, \mathbb{S}^n)$ such that $\Psi(I) = \iota$.
   (b) If $\Phi : \mathcal{L}(\mathbb{S}^n, \mathbb{S}^n) \to \mathcal{M}_n(\mathbb{S})$ is the algebraic isomorphism given by *Theorem 2.2.6* using the basis $\mathbf{e}_1, \ldots, \mathbf{e}_n$, then $\Psi = \Phi^{-1}$.

7. Show that if A is an $n \times n$ real matrix, then $(\mathbf{x}A) \cdot \mathbf{y} = \mathbf{x} \cdot (\mathbf{y}A^{\mathsf{t}})$ for all $\mathbf{x}$ and $\mathbf{y}$ in $\mathbb{R}^n$. (Here, $\mathbf{x}A$ is viewed as an $1 \times n$ matrix times and $n \times n$ matrix.)

8. For $\mathbf{z}$ and $\mathbf{w}$ in $\mathbb{C}^n$, define
$$\mathbf{z} \cdot \mathbf{w} = \sum_{j=1}^{n} z_j \overline{w}_j \tag{2.12}$$
   where $\overline{w}$ is the conjugate of the complex number w. This extension of the usual dot product to $\mathbb{C}^n$ is called the *Hermitian product.* Show that
   (a) $(\mathbf{z}_1 + \mathbf{z}_2) \cdot \mathbf{w} = \mathbf{z}_1 \cdot \mathbf{w} + \mathbf{z}_2 \cdot \mathbf{w}$ for all $\mathbf{z}_1$ $\mathbf{z}_2$, and $\mathbf{w}$ in $\mathbb{C}^n$
   (b) $\mathbf{z} \cdot \mathbf{w} = \overline{\mathbf{w} \cdot \mathbf{z}}$ for all $\mathbf{z}$ and $\mathbf{w}$ in $\mathbb{C}^n$
   (c) $c(\mathbf{z} \cdot \mathbf{w}) = (c\mathbf{z}) \cdot \mathbf{w} = \mathbf{z} \cdot (\overline{c}\mathbf{w})$ for all c in $\mathbb{C}$ and for all $\mathbf{z}$ and $\mathbf{w}$ in $\mathbb{C}^n$.

9. Using the Hermitian product from *Exercise 8* above, prove the following generalization of the Cauchy-Schwarz Inequality for $\mathbb{C}^n$: If $\mathbf{z}$ and $\mathbf{w}$ are in $\mathbb{C}^n$ and $\|\mathbf{z}\|^2 = \mathbf{z} \cdot \mathbf{z}$, then
$$|\mathbf{z} \cdot \mathbf{w}| \leq \|\mathbf{z}\| \, \|\mathbf{w}\|. \tag{2.13}$$
   If $\mathbf{w} \neq \mathbf{0}$, then equality holds if and only if
$$\mathbf{z} = \frac{\mathbf{z} \cdot \mathbf{w}}{\mathbf{w} \cdot \mathbf{w}} \mathbf{w}.$$
   (One approach is to follow the proof of *Proposition 1.2.1* remembering that $|z|^2 = z\overline{z}$ for a complex number z.)

10. Let A be an $n \times n$ complex matrix, and define $\overline{A}$ by taking the complex conjugate of the entries of A, that is, $\overline{A}(j,k) = \overline{A(j,k)}$. For convenience, set $A^* = \left(\overline{A}\right)^{\mathsf{t}}$. Using $\mathbf{z} \cdot \mathbf{w}$ defined by (2.12), show that $(\mathbf{z}A) \cdot \mathbf{w} = \mathbf{z} \cdot (\mathbf{w}A^*)$ for all $\mathbf{z}$ and $\mathbf{w}$ in $\mathbb{C}^n$.

11. Let V be a finite-dimensional linear space. Show that $\mathcal{L}(V, V)$ is a commutative algebra if and only if $\mathrm{Dim}\,[V] = 1$.

12. Let X be a topological space and consider the linear space $\mathcal{C}(X)$. (See *Exercise* 3, p. 69 for background information.) Show that setting $(fg)(x) = f(x)g(x)$ for f and g in $\mathcal{C}(X)$ makes $\mathcal{C}(X)$ a commutative algebra with identity. Determine the group of units of $\mathcal{C}(X)$.

2.3 Norms on Linear Spaces

Thus far, the discussion in this chapter has been purely algebraic. The linear spaces $\mathbb{R}^n$ and $\mathbb{C}^n$, however, are product metric spaces of n and $2n$ copies of $\mathbb{R}$, respectively. From Chapter 1, we know their topologies are both locally compact and sigma compact, and they are metric groups. Indeed they are very nice metric groups. In this section, we use the idea of a norm to introduce a similar metric group structure into linear spaces more generally.

A real-valued function $\|\mathbf{v}\|_a$ on a linear space V is called a *norm* if it satisfies the following conditions:

(a) $\|\mathbf{v}\|_a \neq 0$ when $\mathbf{v} \neq \mathbf{0}$

(b) $\|c\mathbf{v}\|_a = |c|\,\|\mathbf{v}\|_a$ for all $c \in \mathbb{S}$ and $\mathbf{v} \in V$

(c) $\|\mathbf{u}+\mathbf{v}\|_a \leq \|\mathbf{u}\|_a + \|\mathbf{v}\|_a$ for all $\mathbf{u}$, $\mathbf{v} \in V$.

The third condition

$$\|\mathbf{u}+\mathbf{v}\|_a \leq \|\mathbf{u}\|_a + \|\mathbf{v}\|_a \tag{2.14}$$

is called the *triangle inequality* and is usually the more difficult condition to realize. A *normed linear space* is a linear space with a norm defined on it.

Proposition 2.3.1 *If $\|\mathbf{v}\|_a$ is a norm on a linear space V, then the following hold:*

(a) $\|\mathbf{v}\|_a = 0$ if and only if $\mathbf{v} = \mathbf{0}$.

(b) For all $\mathbf{u}$ and $\mathbf{v}$ in V

$$\big|\,\|\mathbf{u}\|_a - \|\mathbf{v}\|_a\,\big| \leq \|\mathbf{u}-\mathbf{v}\|_a. \tag{2.15}$$

(c) $0 \leq \|\mathbf{u}\|_a$ for all $\mathbf{u}$ in V.

Proof. Since $0\mathbf{v} = \mathbf{0}$ for all $\mathbf{v}$, it follows from part (b) of the definition of a norm that $\|\mathbf{0}\|_a = \|0\,\mathbf{0}\|_a = |0|\,\|\mathbf{0}\|_a = 0$. Hence, $\|\mathbf{v}\|_a = 0$ if and only if $\mathbf{v} = \mathbf{0}$ by (a) of the definition. This proves the first part of the proposition.

By the triangle inequality

$$\|\mathbf{u}\|_a = \|\mathbf{u}-\mathbf{v}+\mathbf{v}\|_a \leq \|\mathbf{u}-\mathbf{v}\|_a + \|\mathbf{v}\|_a$$

or

$$\|\mathbf{u}\|_a - \|\mathbf{v}\|_a \leq \|\mathbf{u}-\mathbf{v}\|_a.$$

Since the same calculation works with $\mathbf{u}$ and $\mathbf{v}$ interchanged, it follows that

$$\|\mathbf{v}\|_a - \|\mathbf{u}\|_a \leq \|\mathbf{v}-\mathbf{u}\|_a = \|\mathbf{u}-\mathbf{v}\|.$$

Consequently,

$$\big|\,\|\mathbf{u}\|_a - \|\mathbf{v}\|_a\,\big| \leq \|\mathbf{u}-\mathbf{v}\|_a,$$

and (2.15) holds.

Using this inequality, we get

$$0 \leq \big|\, \|\mathbf{u}\|_a - \|\mathbf{0}\|_a \,\big| \leq \|\mathbf{u} - \mathbf{0}\|_a = \|\mathbf{u}\|_a$$

and $0 \leq \|\mathbf{u}\|_a$ for all $\mathbf{u}$ in V. □

The real-valued function defined on $\mathbb{R}^n$ by

$$\|\mathbf{x}\| = \sqrt{\mathbf{x} \cdot \mathbf{x}} = \left(\sum_{j=1}^{n} x_j^2 \right)^{1/2}$$

first appeared as equation (1.4) on page 14. It defines a norm called the *Euclidean norm* on $\mathbb{R}^n$. This norm is sufficiently important that we have reserved the notation $\|\mathbf{x}\|$ without a subscript for it. It obviously possesses the first two properties of a norm. The triangle inequality (1.7) was proved using the Cauchy-Schwarz Inequality as part of the proof of *Corollary 1.2.2* beginning on page 15. It followed from the triangle inequality that $d(\mathbf{x}, \mathbf{y}) = \|\mathbf{x} - \mathbf{y}\|$ was a metric on $\mathbb{R}^n$. The same idea works here.

Proposition 2.3.2 *If $\|\mathbf{v}\|_a$ is a norm on a linear space V, then*

(a) $d(\mathbf{u}, \mathbf{v}) = \|\mathbf{u} - \mathbf{v}\|_a$ defines a left- and right-invariant metric on V such that V is a metric group.

(b) The function $\mathbf{u} \mapsto \|\mathbf{u}\|_a$ is a continuous function from V to $\mathbb{R}$.

(c) Scalar multiplication $(c, \mathbf{u}) \mapsto c\mathbf{u}$ is a continuous function from $\mathbb{S} \times V$ to V.

Proof. It follows readily from the definition of a norm and *Proposition 2.3.1* that $d(\mathbf{u}, \mathbf{v}) = \|\mathbf{u} - \mathbf{v}\|_a$ is a metric on V. Since V is an Abelian group, right-invariant metrics are left-invariant. Then $d(\mathbf{u}+\mathbf{w}, \mathbf{v}+\mathbf{w}) = \|\mathbf{u}+\mathbf{w}-(\mathbf{v}+\mathbf{w})\|_a = \|\mathbf{u} - \mathbf{v}\|_a = d(\mathbf{u}, \mathbf{v})$, and d is an invariant metric. The function $(\mathbf{u}, \mathbf{v}) \mapsto \mathbf{u} - \mathbf{v}$ is continuous because

$$\|(\mathbf{u} - \mathbf{v}) - (\mathbf{u}' - \mathbf{v}')\|_a \leq \|\mathbf{u} - \mathbf{u}'\|_a + \|\mathbf{v}' - \mathbf{v}\|_a.$$

Therefore, V is a metric group with the metric $d(\mathbf{u}, \mathbf{v}) = \|\mathbf{u} - \mathbf{v}\|_a$, and the first part of the conclusion is established.

It is an immediate consequence of inequality (2.15) that the norm is a continuous function.

For part *(c)*, it suffices to show that $\|c_k\mathbf{u}_k - c\mathbf{u}\|_a$ converges to zero when c_k converges to c in $\mathbb{S}$ and $\mathbf{u}_k$ converges to $\mathbf{u}$ in V. Note that $\|\mathbf{u}_k\|_a$ converges to $\|\mathbf{u}\|_a$ by part *(b)*. The continuity of $(c, \mathbf{u}) \mapsto c\mathbf{u}$ now follows from the estimate:

$$\begin{aligned}
\|c_k\mathbf{u}_k - c\mathbf{u}\|_a &= \|c_k\mathbf{u}_k - c\mathbf{u}_k + c\mathbf{u}_k - c\mathbf{u}\|_a \\
&\leq \|c_k\mathbf{u}_k - c\mathbf{u}_k\|_a + \|c\mathbf{u}_k - c\mathbf{u}\|_a \\
&= |c_k - c|\, \|\mathbf{u}_k\|_a + |c|\, \|\mathbf{u}_k - \mathbf{u}\|_a
\end{aligned}$$

because the right-hand side converges to $0\|\mathbf{u}\|_a + |c|0 = 0$. □

Consequently, normed linear spaces are a particular type of metric group. To be more specific, a normed linear space is an Abelian metric group with a continuous linear structure and metric of the form $d(\mathbf{u}, \mathbf{v}) = \|\mathbf{u} - \mathbf{v}\|_a$, where $\|\mathbf{u} - \mathbf{v}\|_a$ is a specified norm on V. The topology determined by such a metric will be referred to as a *norm linear topology.* These metric groups will be very useful in the study of more general metric groups.

Scalar multiplication uniformly expands distances or contracts them for the metric $d(\mathbf{u}, \mathbf{v}) = \|\mathbf{u} - \mathbf{v}\|_a$. Specifically,

$$\begin{aligned} d(c\mathbf{u}, c\mathbf{v}) &= \|c\mathbf{u} - c\mathbf{v}\|_a \\ &= |c|\,\|\mathbf{u} - \mathbf{v}\|_a \\ &= |c|\,d(\mathbf{u}, \mathbf{v}), \end{aligned}$$

and distances expand or contract according as $|c| > 1$ or $|c| < 1$. In this same vein for $c \neq 0$,

$$\begin{aligned} cB_r(\mathbf{0}) &= \{c\mathbf{u} : \mathbf{u} \in B_r(\mathbf{0})\} \\ &= \{c\mathbf{u} : \|\mathbf{u}\|_a < r\} \\ &= \{\mathbf{v} : \|\mathbf{v}\|_a < |c|r\} \\ &= B_{|c|r}(\mathbf{0}) \end{aligned}$$

and

$$cB_r(\mathbf{0}) = B_{|c|r}(\mathbf{0}). \tag{2.16}$$

Proposition 2.3.3 *Let V be a normed linear space, and let W be a subspace of V. If W is a closed set in V, then V/W is a normed linear space and the canonical homomorphism $\pi : V \to V/W$ is a continuous linear function.*

Proof. Since V is an Abelian group, W is a normal subgroup and V/W is an Abelian group with $(W + \mathbf{u}) + (W + \mathbf{v}) = W + \mathbf{u} + \mathbf{v}$.

Since W is closed, *Theorem 1.4.12* implies that V/W is a metric group and π is a homomorphism of metric groups. Furthermore, because the metric $d(\mathbf{u}, \mathbf{v}) = \|\mathbf{u} - \mathbf{v}\|_a$ is invariant, equation (1.18) provides the following formula for a metric on V/W

$$\begin{aligned} \rho(W + \mathbf{u}, W + \mathbf{v}) &= \inf\{d(\mathbf{u}, \mathbf{w} + \mathbf{v}) : \mathbf{w} \in W\} \\ &= \inf\{\|\mathbf{u} - (\mathbf{w} + \mathbf{v})\|_a : \mathbf{w} \in W\} \\ &= \inf\{\,\|\mathbf{u} - \mathbf{v} - \mathbf{w}\|_a : \mathbf{w} \in W\}. \end{aligned}$$

Next, a scalar multiplication function must be defined for V/W. If $W + \mathbf{u} = W + \mathbf{v}$ or equivalently $\mathbf{u} - \mathbf{v}$ is in W, then, because W is a subspace, $c(\mathbf{u} - \mathbf{v})$ is in W and $W + c\mathbf{u} = W + c\mathbf{v}$. Thus $(c, W + \mathbf{u}) \to W + c\mathbf{u}$ is a properly defined function. It is now easy to verify that it satisfies the conditions of a scalar multiplication function on page 64. Thus V/W is a linear space. Clearly $\pi(c\mathbf{u}) = W + c\mathbf{u} = c(W + \mathbf{u}) = c\pi(\mathbf{u})$, and π is a linear function.

The formula $\rho(W + \mathbf{u}, W + \mathbf{v}) = \inf \{ \|\mathbf{u} - \mathbf{v} - \mathbf{w}\|_a : \mathbf{w} \in W\}$ suggests setting

$$\|W + \mathbf{u}\|_q = \inf \{ \|\mathbf{u} + \mathbf{w}\|_a : \mathbf{w} \in W\}. \tag{2.17}$$

The next three steps show that (2.17) is a norm on V/W.

First, if $\|W + \mathbf{u}\|_q = 0$, then there exists a sequence $\mathbf{w}_k$ in W such that $\|\mathbf{u} + \mathbf{w}_k\| \to 0$. Hence, $\mathbf{u} + \mathbf{w}_k \to \mathbf{0}$ or equivalently $-\mathbf{w}_k \to \mathbf{u}$, implying that $\mathbf{u}$ is in the closed set W and $W + \mathbf{u} = W$. Therefore, $\|W + \mathbf{u}\|_q \neq 0$, if $W + \mathbf{u} \neq W$.

Second, using the assumption that W is a subspace it follows for $c \neq 0$ that

$$\begin{aligned}
\|c(W + \mathbf{u})\|_q &= \|W + c\mathbf{u}\|_q \\
&= \inf \{ \|c\mathbf{u} + \mathbf{w}\|_a : \mathbf{w} \in W\} \\
&= \inf \{ |c| \, \|\mathbf{u} + (1/c)\mathbf{w}\|_a : \mathbf{w} \in W\} \\
&= |c| \inf \{ \|\mathbf{u} + \mathbf{w}'\|_a : \mathbf{w}' \in W\} \\
&= |c| \, \|W + \mathbf{u}\|_q.
\end{aligned}$$

The $c = 0$ case is trivial.

Third, using *Exercise* 3, p. 91, we have

$$\begin{aligned}
\|(W + \mathbf{u}) + (W + \mathbf{v})\|_q &= \|W + \mathbf{u} + \mathbf{v}\|_q \\
&= \inf \{ \|\mathbf{u} + \mathbf{v} + \mathbf{w}\|_a : \mathbf{w} \in W\} \\
&= \inf \{ \|\mathbf{u} + \mathbf{w} + \mathbf{v} + \mathbf{w}'\|_a : \mathbf{w}, \mathbf{w}' \in W\} \\
&\leq \inf \{ \|\mathbf{u} + \mathbf{w}\|_a + \|\mathbf{v} + \mathbf{w}'\|_a : \mathbf{w}, \mathbf{w}' \in W\} \\
&= \inf \{ \|\mathbf{u} + \mathbf{w}\|_a : \mathbf{w} \in W\} + \inf \{\|\mathbf{v} + \mathbf{w}'\|_a : \mathbf{w}' \in W\} \\
&= \|W + \mathbf{u}\|_q + \|W + \mathbf{v}\|_q.
\end{aligned}$$

Therefore, $\|W + \mathbf{u}\|_q$ is a norm on V/W and V/W is a normed linear space.

Finally, it must be verified that the metric ρ and norm $\|W + \mathbf{u}\|_q$ produce the same topology on V/W. The following calculation shows that the metric ρ actually equals the metric coming from the norm

$$\begin{aligned}
\rho(W + \mathbf{u}, W + \mathbf{v}) &= \inf \{\|\mathbf{u} - \mathbf{v} - \mathbf{w}\|_a : \mathbf{w} \in W\} \\
&= \|W + \mathbf{u} - \mathbf{v}\|_q \\
&= \|(W + \mathbf{u}) - (W + \mathbf{v})\|_q,
\end{aligned}$$

which by definition (see *Proposition 2.3.2*) is the metric determined by the norm $\|\cdot\|_q$. Thus the quotient topology on V/W equals the norm linear topology on V/W and the proof is complete. □

There are other norms on $\mathbb{R}^n$ besides

$$\|\mathbf{x}\| = \sqrt{\mathbf{x} \cdot \mathbf{x}} = \left(\sum_{j=1}^{n} x_j^2 \right)^{1/2}$$

For example,

$$\|\mathbf{x}\|_s = \max\{|x_i| : 1 \leq i \leq n\} \tag{2.18}$$

is also a norm for $\mathbb{R}^n$, as is

$$\|\mathbf{x}\|_1 = \sum_{j=1}^{n} |x_j|.$$

Notice that the metric $d_1(\mathbf{x}, \mathbf{y}) = \|\mathbf{x} - \mathbf{y}\|_1$ on $\mathbb{R}^n$ is just the product metric on $\mathbb{R}^n$. So both the Euclidean metric and the product metric on $\mathbb{R}^n$ come from norms on $\mathbb{R}^n$.

Are the resulting topologies $\mathbb{R}^n$ given by different norms the same or are they fundamentally different? To address this question, we must first define when two norms are equivalent and understand what that means topologically.

Two norms, $\|\cdot\|_a$ and $\|\cdot\|_b$ on a linear space V are called *equivalent* if there exist positive real constants c and d satisfying

$$c\|\mathbf{v}\|_a \leq \|\mathbf{v}\|_b \leq d\|\mathbf{v}\|_a$$

for all $\mathbf{v}$ in V. (The dot $\cdot$ in the notation $\|\cdot\|_a$ or $g(\cdot)$ indicates an unnamed variable of a norm or a function. It will be convenient to use this notation from time to time.)

If two norms are equivalent, then for an arbitrary $\mathbf{v}$ in V and $r > 0$

$$\{\mathbf{u} : \|\mathbf{u} - \mathbf{v}\|_a < r/d\} \subset \{\mathbf{u} : \|\mathbf{u} - \mathbf{v}\|_b < r\} \subset \{\mathbf{u} : \|\mathbf{u} - \mathbf{v}\|_a < r/c\}.$$

Consequently, either norm will define the same family of open sets in V. *Therefore, $d_a(\mathbf{u}, \mathbf{v}) = \|\mathbf{u} - \mathbf{v}\|_a$ and $d_b(\mathbf{u}, \mathbf{v}) = \|\mathbf{u} - \mathbf{v}\|_b$ are equivalent metrics, when $\|\cdot\|_a$ and $\|\cdot\|_b$ are equivalent norms.*

The next theorem is at the heart of the difference between finite- and infinite-dimensional normed linear spaces.

Theorem 2.3.4 *Any two norms on $\mathbb{R}^n$ are equivalent.*

Proof. It suffices from *Exercise* 4, p. 91, to show that the Euclidean norm, $\|\cdot\|$, is equivalent to an arbitrary norm $\|\cdot\|_a$. It will be shown that $c\|\mathbf{x}\| \leq \|\mathbf{x}\|_a \leq d\|\mathbf{x}\|$.

Let $\mathbf{e}_1, \ldots, \mathbf{e}_n$ be the standard basis of $\mathbb{R}^n$. Set

$$\gamma = \max\{\|\mathbf{e}_j\|_a : 1 \leq j \leq n\}.$$

If $\mathbf{x} = (x_1, \ldots, x_n)$, then

$$\mathbf{x} = \sum_{j=1}^{n} x_j \mathbf{e}_j$$

and

$$\|\mathbf{x}\|_a \leq \sum_{j=1}^{n} |x_j| \, \|\mathbf{e}_j\|_a \leq \gamma \sum_{j=1}^{n} |x_j| \leq \gamma n \|\mathbf{x}\|$$

because $|x_j| \leq \|\mathbf{x}\|$. Thus $\|\mathbf{x}\|_a \leq d\|\mathbf{x}\|$ with $d = \gamma n$.

It follows from inequality (2.15) that

$$\big|\, \|\mathbf{x}\|_a - \|\mathbf{y}\|_a \,\big| \le \|\mathbf{x} - \mathbf{y}\|_a \le \gamma n \|\mathbf{x} - \mathbf{y}\|,$$

and hence $\|\mathbf{x}\|_a$ is a continuous function of $\mathbf{x}$ on $\mathbb{R}^n$.

Since $\{\mathbf{x} : \|\mathbf{x}\| = 1\}$ is compact by *Theorem 1.5.13* and $\|\mathbf{x}\|_a$ is continuous, *Proposition 1.5.8* implies that there exists $\mathbf{u}$ with $\|\mathbf{u}\| = 1$ such that

$$\|\mathbf{u}\|_a \le \|\mathbf{x}\|_a$$

whenever $\|\mathbf{x}\| = 1$. Set $c = \|\mathbf{u}\|_a$. Then, $c > 0$ because $\|\mathbf{u}\| = 1$ implies $\mathbf{u} \ne \mathbf{0}$.

For any $\mathbf{x} \ne \mathbf{0}$,

$$\left\| \tfrac{1}{\|\mathbf{x}\|}\, \mathbf{x} \right\| = 1,$$

and hence

$$c \le \left\| \tfrac{1}{\|\mathbf{x}\|}\, \mathbf{x} \right\|_a = \frac{1}{\|\mathbf{x}\|}\, \|\mathbf{x}\|_a$$

or

$$c\|\mathbf{x}\| \le \|\mathbf{x}\|_a,$$

which completes the proof. □

Using the Hermitian product defined in *Exercise* 8, p. 82, set

$$\|\mathbf{z}\| = \sqrt{\mathbf{z} \cdot \mathbf{z}} = \left(\sum_{j=1}^{n} z_j \bar{z}_j \right)^{1/2} = \left(\sum_{j=1}^{n} (x_j^2 + y_j^2) \right)^{1/2}$$

where $\mathbf{z} = (z_1, \ldots, z_n) \in \mathbb{C}^n$ and $z_j = x_j + iy_j$ with both x_j and y_j real. Thus $\|\mathbf{z}\|$ is just the Euclidean norm on $\mathbb{R}^{2n}$ expressed using complex multiplication and conjugates, and it satisfies conditions (a) and (c) in the definition of a norm. It is easy to verify that it also satisfies condition *(b)*.

Corollary 2.3.5 *Any two norms on $\mathbb{C}^n$ are equivalent.*

Proof. If $\|\cdot\|_a$ is a norm on $\mathbb{C}^n$, then it is also a norm on $\mathbb{R}^{2n}$ because the underlying Abelian groups of $\mathbb{C}^n$ and $\mathbb{R}^{2n}$ are the same and $\mathbb{R} \subset \mathbb{C}$. Since the scalars do not play a role in the definition of equivalent norms, *Theorem 2.3.4* can be applied. □

Proposition 2.3.6 *Any two norms on a finite-dimensional linear space are equivalent.*

Proof. Let $\|\cdot\|_a$ and $\|\cdot\|_b$ be norms on a linear space V of dimension n. By *Theorem 2.2.4* there exists a one-to-one onto linear function $T : \mathbb{S}^n \to V$. Define norms on $\mathbb{S}^n$ by setting $\|\mathbf{w}\|_{a'} = \|T(\mathbf{w})\|_a$ and $\|\mathbf{w}\|_{b'} = \|T(\mathbf{w})\|_b$. By either *Theorem 2.3.4* or *Corollary 2.3.5* the norms $\|\mathbf{w}\|_{a'}$ and $\|\mathbf{w}\|_{b'}$ are equivalent, and there exist positive real numbers c and d such that $c\|\mathbf{w}\|_{a'} \le \|\mathbf{w}\|_{b'} \le d\|\mathbf{w}\|_{a'}$.

For each $\mathbf{v}$ in V there exists a unique $\mathbf{w} \in \mathbb{S}^n$ such that $\mathbf{v} = T(\mathbf{w})$. It follows that

$$c\|\mathbf{v}\|_a = c\|T(\mathbf{w})\|_a = c\|\mathbf{w}\|_{a'} \leq \|\mathbf{w}\|_{b'} = \|T(\mathbf{w})\|_b = \|\mathbf{v}\|_b$$

and

$$\|\mathbf{v}\|_b = \|T(\mathbf{w})\|_b = \|\mathbf{w}\|_{b'} \leq d\|\mathbf{w}\|_{a'} = d\|T(\mathbf{w})\|_a = d\|\mathbf{v}\|_a.$$

Therefore, the two norms $\|\cdot\|_a$ and $\|\cdot\|_b$ on V are equivalent. □

Theorem 2.3.4 is only true for finite-dimensional linear spaces. In fact, its failure in the infinite-dimensional case is one of the key differences between finite- and infinite-dimensional normed vector spaces. Recall that at the end of Section 2.1 (page 68) it was shown that the linear space $C([0,1])$ is infinite-dimensional. The following are easily seen to be norms on $C([0,1])$:

$$\|f\|_s = \sup\{|f(x)| : 0 \leq x \leq 1\} \tag{2.19}$$

and

$$\|f\|_1 = \int_0^1 |f(x)|dx, \tag{2.20}$$

but they are not equivalent. To see why they are not equivalent, consider $f_k(x) = x^k$. Observe that $\|f_k\|_1$ converges to 0 as k goes to infinity while $\|f_k\|_s = 1$ for all k. Hence, there is no constant $c > 0$ such that $c\|f\|_s \leq \|f\|_1$ for all f in $C([0,1])$.

Let V be a finite-dimensional linear space. *Exercise* 6, p. 91 shows that there exist norms on V. Because V is finite-dimensional, all norms on V are equivalent. Therefore, every finite-dimensional linear space V is a normed linear space and there is exactly one norm linear topology on V. Henceforth, we will always assume that finite-dimensional linear spaces are normed linear spaces with this unique norm linear topology.

A subset E of $\mathbb{R}^n$ is bounded (page 56) if there exists $R > 0$ such that

$$E \subset [-R, R]^n = \{\mathbf{x} \in \mathbb{R}^n : \|\mathbf{x}\|_s \leq R\},$$

with $\|\cdot\|_s$ defined by equation (2.18). Since all norms on $\mathbb{R}^n$ are equivalent, it follows that a subset E is bounded if and only if given any norm $\|\cdot\|_a$ on $\mathbb{R}^n$ there exists $R_a > 0$ such that $E \subset \{\mathbf{x} \in \mathbb{R}^n : \|\mathbf{x}\|_a \leq R_a\}$.

Let V be a normed linear space with norm $\|\cdot\|_a$. A subset E of V is *bounded* if there exists $R_a > 0$ such that $E \subset \{\mathbf{v} \in V : \|\mathbf{v}\|_a \leq R_a\}$. As in the case of $\mathbb{R}^n$, a subset that is bounded with respect to one norm is bounded with respect to all equivalent norms on V. Moreover, *Theorem 1.5.13* generalizes to finite-dimensional normed linear spaces.

Theorem 2.3.7 *Let V is a finite-dimensional normed linear space. A subset E of V is compact if and only if it is a closed and bounded set in the normed linear topology.*

Proof. Let $\|\cdot\|_a$ be norm on V and let E be a subset of V. By using *Theorem 2.2.4* again, there exists a one-to-one onto linear function $T : V \to \mathbb{S}^n$ and $\|\mathbf{w}\|_{a'} = \|T^{-1}(\mathbf{w})\|_a$ defines a norm on $\mathbb{S}^n$.

Observe that $\|T(\mathbf{u}) - T(\mathbf{v})\|_{a'} = \|T(\mathbf{u} - \mathbf{v})\|_{a'} = \|\mathbf{u} - \mathbf{v}\|_a$ and T is a metric space isometry and hence a homeomorphism between V and $\mathbb{S}^n$ (see *Exercise 19*, p. 29). It follows that:

(a) E is bounded in V if and only if $T(E)$ is bounded in $\mathbb{S}^n$.

(b) E is closed in V if and only if $T(E)$ is closed in $\mathbb{S}^n$.

(c) E is compact in V if and only if $T(E)$ is compact in $\mathbb{S}^n$.

Therefore, by *Theorem 1.5.13*, E is compact in V if and only if E is closed and bounded in V. □

Corollary 2.3.8 *If V is a finite-dimensional normed linear space with norm $\|\cdot\|_a$, then*

$$\{\mathbf{v} : \|\mathbf{v}\|_a \leq r\}$$

is compact. Moreover, V is both locally compact and σ-compact.

Theorem 2.3.7 and its corollary are finite-dimensional results. For example, the closed and bounded subset $\{f \in \mathcal{C}([0,1]) : \|f\|_s \leq 1\}$ is not a compact subset of the normed linear space $\mathcal{C}([0,1])$ with the sup norm $\|f\|_s = \sup\{|f(x)| : 0 \leq x \leq 1\}$ by *Exercise* 9, p. 91.

Normed linear spaces V and W over the same scalar field are said to be *isomorphic normed linear spaces*, if there exists a linear homeomorphism from V onto W. In particular, isomorphic normed linear spaces are algebraically isomorphic linear spaces and isomorphic metric groups.

Theorem 2.3.9 *Two finite-dimensional normed linear spaces over the same scalar field are isomorphic normed linear spaces if and only if their dimensions are equal.*

Proof. Let V and V' be two finite-dimensional normed linear spaces over the same scalar field with norms $\|\cdot\|_a$ and $\|\cdot\|_b$. If they are isomorphic normed linear spaces, then their dimensions are equal by *Theorem 2.2.4*.

If $\text{Dim}\,[V] = \text{Dim}\,[V'] = n$, there exists a one-to-one onto linear map $T : V \to V'$ by *Theorem 2.2.4*, and $\|\mathbf{w}\|_{a'} = \|T^{-1}(\mathbf{w})\|_a$ defines a norm on V'. As in the proof of *Proposition 2.3.7*, T is an isometry between the metrics determined by the norms $\|\cdot\|_a$ and $\|\cdot\|_{a'}$, and therefore a homeomorphism between these metric spaces. Since all norms are equivalent on V', the two norms $\|\cdot\|_{a'}$ and $\|\cdot\|_b$ determine the same topology on V'. It follows that T is a linear homeomorphism between the original two normed linear spaces. □

Once it had been shown that all norms were equivalent on finite-dimensional vector spaces over $\mathbb{R}$ or $\mathbb{C}$, the algebra drove the topological results, as was evident by the repeated use of *Theorem 2.2.4*. Consequently, for each n there is essentially only one normed linear space of that dimension over each of the scalar fields $\mathbb{R}$ and $\mathbb{C}$, namely, $\mathbb{R}^n$ and $\mathbb{C}^n$. In other words, $\mathbb{S}^n$ is isomorphic as a normed linear space to any other normed linear space of dimension n over $\mathbb{S}$.

For example, if $\text{Dim}\,[V] = n$, then both $\mathcal{L}(V, V)$ and $\mathcal{M}_n(\mathbb{S})$ are isomorphic as normed linear spaces to $\mathbb{S}^{n^2}$. There are, however, instances where a particular norm is more desirable than others. This will be apparent in the further discussion of $\mathcal{L}(V, V)$ and $\mathcal{M}_n(\mathbb{S})$ in the Section 2.4.

EXERCISES

1. Let V be a normed linear space. Show that if W is a subspace of V, then W^- is also a subspace of V.

2. Let V be a normed linear space with norm $\|\cdot\|_a$. Show that $B_r(\mathbf{u})^- = \{\mathbf{v} : \|\mathbf{v} - \mathbf{u}\|_a \leq r\}$ for $r > 0$.

3. Let E and F be nonempty subsets of $\mathbb{R}$ that are bounded below. Show that
$$\inf\{x + y : x \in E \text{ and } y \in F\} = \inf E + \inf F.$$

4. Let $\|\cdot\|_a$, $\|\cdot\|_b$, and $\|\cdot\|_c$ be three norms on a linear space V. Show that if $\|\cdot\|_a$ and $\|\cdot\|_b$ are equivalent norms, and if $\|\cdot\|_b$ and $\|\cdot\|_c$ are equivalent norms, then $\|\cdot\|_a$ and $\|\cdot\|_c$ are equivalent norms.

5. Let $\mathbf{v}_k$ be a convergent sequence in a normed linear space V with norm $\|\cdot\|_a$. Prove there exists a constant $c > 0$ such that $\|\mathbf{v}_k\|_a \leq c$ for all k.

6. Let $\mathbf{u}_1, \ldots, \mathbf{u}_n$ be a basis for a finite-dimensional linear space V. Using the uniqueness of the scalars $c_1, \ldots, c_n$ such that $\mathbf{v} = c_1\mathbf{u}_1 + \ldots + c_n\mathbf{u}_n$, show that
$$\|\mathbf{v}\|_1 = \left\|\sum_{j=1}^{n} c_j \mathbf{u}_j\right\|_1 = \sum_{j=1}^{n} |c_j| \tag{2.21}$$
defines a norm on V.

7. Let $\mathbf{u}_1, \ldots, \mathbf{u}_n$ be a basis for a finite-dimensional linear space V and consider a sequence
$$\mathbf{v}_k = c_{1k}\mathbf{u}_1 + \ldots + c_{nk}\mathbf{u}_n.$$
Prove that $\mathbf{v}_k$ converges to $\mathbf{0}$ in the norm linear topology on V if and only if $lim_{k\to\infty} c_{jk} = 0$ for $j = 1, \ldots, n$.

8. Show that there exist functions f_k in $\mathcal{C}([0, 1])$ such that $\|f_j - f_k\|_s = 1$, when $j \neq k$.

9. Show that the closed and bounded set $B_1(\mathbf{0})^-$ is not a compact subset $\mathcal{C}([0, 1])$ with the sup norm $\|\cdot\|_s$. (One approach is to use *Exercise 8* above.) Then show that $\mathcal{C}([0, 1])$ with the sup norm is not locally compact.

2.4 Continuous Linear Functions

Let V and V' be two normed linear spaces with norms $\|\cdot\|_a$ and $\|\cdot\|_b$. By *Proposition 2.3.2*, they are metric groups with additional linear structures and with metrics obtained from their norms. In this section, we explore the homomorphisms of V to V'.

Proposition 2.4.1 *Let V and V' be normed linear spaces over $\mathbb{R}$. If $\varphi : V \to V'$ is a homomorphism of the metric group V to the metric group V', then φ is a linear map.*

Proof. Since φ is a homomorphism of metric groups, it is continuous and $\varphi(m\mathbf{v}) = m\varphi(\mathbf{v})$ for all $\mathbf{v} \in V$ and $m \in \mathbb{Z}$. Both these facts will play a crucial role in the proof.

Let p and q be integers with $q \neq 0$ and let $\mathbf{v}$ be in V. Then

$$\begin{aligned} \varphi\left(\frac{p}{q}\mathbf{v}\right) &= p\varphi\left(\frac{1}{q}\mathbf{v}\right) \\ &= \frac{pq}{q}\varphi\left(\frac{1}{q}\mathbf{v}\right) \\ &= \frac{p}{q}\varphi\left(\frac{q}{q}\mathbf{v}\right) \\ &= \frac{p}{q}\varphi(\mathbf{v}). \end{aligned}$$

Thus

$$\varphi(r\mathbf{v}) = r\varphi(\mathbf{v})$$

for all $r \in \mathbb{Q}$, the rational numbers, and $\mathbf{v} \in V$.

If s is an irrational real number, there exists a sequence r_k of rational numbers converging to s. *Proposition 1.2.3* will now be used several times. Because scalar multiplication is continuous, it follows that $r_k\mathbf{v}$ and $r_k\varphi(\mathbf{v})$ converge to $s\mathbf{v}$ and $s\varphi(\mathbf{v})$ respectively. The continuity of φ implies that

$$\begin{aligned} \varphi(s\mathbf{v}) &= \varphi\left(\lim_{k\to\infty} r_k\mathbf{v}\right) \\ &= \lim_{k\to\infty} \varphi(r_k\mathbf{v}) \\ &= \lim_{k\to\infty} r_k\varphi(\mathbf{v}) \\ &= s\varphi(\mathbf{v}) \end{aligned}$$

to complete the proof. □

To prove a similar result for $\mathbb{C}$, an additional hypothesis is required because $i\mathbf{v}$ does not lie on the line $\{t\mathbf{v} : t \in \mathbb{R}\}$ through the origin. Recall that multiplication by the imaginary number i in $\mathbb{C}$ is a counterclockwise rotation of $90°$ in $\mathbb{R}^2$. If, however, it is known that $\varphi(i\mathbf{v}) = i\varphi(\mathbf{v})$ for all $\mathbf{v} \in V$, then the above proof can be used twice - once for the real numbers and once for the pure imaginary numbers - to prove the following result:

Corollary 2.4.2 *Let V and V' be normed linear spaces over $\mathbb{C}$. If $\varphi : V \to V'$ is a homomorphism such that $\varphi(i\mathbf{v}) = i\varphi(\mathbf{v})$ for all $\mathbf{v} \in V$, then φ is a linear map.*

Continuing to let V and V' be two normed linear spaces with norms $\|\cdot\|_a$ and $\|\cdot\|_b$, a linear function $T : V \to V'$ is *bounded* provided that the set of real numbers

$$\{\|T(\mathbf{v})\|_b : \|\mathbf{v}\|_a \leq 1\}$$

is bounded above. (It is always bounded below by 0.) If M is an upper bound for $\{\|T(\mathbf{v})\|_b : \|\mathbf{v}\|_a \leq 1\}$ and $\mathbf{v} \neq \mathbf{0}$, then

$$\|T(\mathbf{v})\|_b = \|\mathbf{v}\|_a T\left(\frac{1}{\|\mathbf{v}\|_a}\mathbf{v}\right) \leq M\|\mathbf{v}\|_a$$

because

$$\left\|\frac{1}{\|\mathbf{v}\|_a}\mathbf{v}\right\|_a = 1.$$

Since $T(\mathbf{0}) = \mathbf{0}$, it follows that $\|T\mathbf{v}\|_b \leq M\|\mathbf{v}\|_a$ for all $\mathbf{v} \in V$. Consequently, a linear function $T : V \to V'$ is bounded if and only if there exists a positive real number M such that

$$\|T(\mathbf{v})\|_b \leq M\|\mathbf{v}\|_a \tag{2.22}$$

for all $\mathbf{v} \in V$.

Proposition 2.4.3 *Let V and V' be two normed linear spaces over the same scalar field. A linear function $T : V \to V'$ is bounded if and only if it is continuous.*

Proof. As usual, let $\|\cdot\|_a$ and $\|\cdot\|_b$ denote the norms on V and V'. Suppose T is bounded. So there exists a positive real number M such that $\|T(\mathbf{v})\|_b \leq M\|\mathbf{v}\|_a$ for all $\mathbf{v} \in V$. To prove that T is continuous, it suffices to show that T is continuous at $\mathbf{0}$ by *Exercise* 5, p. 39. Given $\varepsilon > 0$, just set $\delta = \varepsilon/M$. Then, $\|\mathbf{v}\|_a < \delta$ implies that $\|T(\mathbf{v})\|_b \leq M\|\mathbf{v}\|_a < M\delta = \varepsilon$, which shows that T is continuous at $\mathbf{0}$.

Suppose T is continuous. By the continuity of T at $\mathbf{0}$, there exists $\delta > 0$ such that $\|T(\mathbf{v})\|_b < 1$ when $\|\mathbf{v}\|_a < \delta$. If $\|\mathbf{v}\|_a \leq 1$, then the following calculation shows that T is bounded:

$$\begin{aligned} \|T(\mathbf{v})\|_b &= \frac{2}{\delta}\left\|T\left(\frac{\delta}{2}\mathbf{v}\right)\right\|_b \\ &< \frac{2}{\delta} \end{aligned}$$

because $\|(\delta/2)\mathbf{v}\|_a < \delta$. □

The next result is the First Isomorphism Theorem for Normed Linear Spaces.

Theorem 2.4.4 *Let V and V' be normed linear spaces. If $T : V \to V'$ is a bounded linear function of V onto V' with kernel K, then $\tilde{T}(K + \mathbf{v}) = T(\mathbf{v})$ defines a one-to-one bounded linear function of V/K onto V' satisfying $\tilde{T} \circ \pi = T$, where $\pi : V \to V/K$ is the canonical homomorphism. If T is an open function, then $\tilde{T}$ is a linear isomorphism of the normed linear space V/K onto the normed linear space V'.*

Proof. By *Proposition 2.3.3*, V/K is a normed linear space, and *Theorem 1.4.14* implies that $\tilde{T}$ is a one-to-one homomorphism of metric groups and an isomorphism of metric groups if T is an open function. It follows from the linearity of T and the definition of $\tilde{T}$ that $\tilde{T}$ is linear. □

To construct an example of an unbounded linear function, let V be the set of real-valued functions defined on $\mathbb{R}$ such that f satisfies the following conditions:

(a) $f^{(n)}$, the nth derivative of f, exists for all $n \in \mathbb{Z}^+$

(b) $f(x + 2\pi) = f(x)$ for all $x \in \mathbb{R}$

(c) $\int_0^{2\pi} f(x)dx = 0$.

Obviously, $\sin(x)$ and $\cos(x)$ are in V. Defining $(f + g)(x) = f(x) + g(x)$ and $(cf)(x) = cf(x)$, gives V a linear space structure, and $\|f\|_s = \sup\{|f(x)| : 0 \leq x \leq 2\pi\}$ defines a norm on V. Define a linear function $T : V \to V$ by setting $T(f) = f'$, the derivative of f. Clearly, f' can be differentiated infinitely often and $f'(x + 2\pi) = f'(x)$. Also $\int_0^{2\pi} f'(x)dx = f(2\pi) - f(0) = 0$. So $T(f) = f'$ is in V. To see that T is not bounded, note that for all positive integers k the function $f_k(x) = \sin(kx)$ is in V and $\|f_k\|_s = 1$, but $\|f_k'\|_s = k$ because $f_k'(x) = k\cos(kx)$.

Given two normed linear spaces V and V' over the same scalar field, let $\mathcal{B}(V, V')$ denote the set of continuous linear or equivalently bounded linear functions from V to V'. It is a subspace of $\mathcal{L}(V, V')$ because the sum of two bounded linear functions is bounded and a scalar multiple of a bounded linear function is bounded.

Proposition 2.4.5 *If V and V' are normed linear spaces with norms $\|\cdot\|_a$ and $\|\cdot\|_b$, then $\mathcal{B}(V, V')$ is a normed linear space with norm*

$$\|T\|_{\mathcal{B}} = \sup\{\|T(\mathbf{v})\|_b : \|\mathbf{v}\|_a \leq 1\}. \tag{2.23}$$

Moreover, the following inequality holds for all $\mathbf{v} \in V$ and $T \in \mathcal{B}(V, V')$:

$$\|T(\mathbf{v})\|_b \leq \|T\|_{\mathcal{B}} \|\mathbf{v}\|_a. \tag{2.24}$$

Proof. It was just pointed out that $\mathcal{B}(V, V')$ is a linear space. The main part of the proof is verifying that $\|T\|_{\mathcal{B}} = \sup\{\|T(\mathbf{v})\|_b : \|\mathbf{v}\|_a \leq 1\}$ satisfies the conditions for a norm on page 83.

If $T \neq \mathbf{0}$, then $T(\mathbf{u}) \neq \mathbf{0}$ for some $\mathbf{u} \in V$. Clearly, $\mathbf{u} \neq \mathbf{0}$. Set $\mathbf{v} = (1/\|\mathbf{u}\|_a)\mathbf{u}$. Then $\|\mathbf{v}\|_a = 1$ and $T(\mathbf{v}) \neq \mathbf{0}$. It follows that $\|T\|_{\mathcal{B}} \neq 0$, and the first condition for a norm is verified.

Next,

$$\begin{aligned} \|cT\|_{\mathcal{B}} &= \sup\{\|cT(\mathbf{v})\|_b : \|\mathbf{v}\|_a \leq 1\} \\ &= \sup\{|c|\,\|T(\mathbf{v})\|_b : \|\mathbf{v}\|_a \leq 1\} \\ &= |c| \sup\{\|T(\mathbf{v})\|_b : \|\mathbf{v}\|_a \leq 1\} \\ &= |c|\,\|T\|_{\mathcal{B}} \end{aligned}$$

shows that the second condition for a norm is satisfied by $\|T\|_{\mathcal{B}}$.

The triangle inequality or third condition is obtained from the following estimates:

$$\begin{aligned} \|S+T\|_{\mathcal{B}} &= \sup\{\|S(\mathbf{v}) + T(\mathbf{v})\|_b : \|\mathbf{v}\|_a \leq 1\} \\ &\leq \sup\{\|S(\mathbf{v})\|_b + \|T(\mathbf{v})\|_b : \|\mathbf{v}\|_a \leq 1\} \\ &\leq \sup\{\|S(\mathbf{v})\|_b : \|\mathbf{v}\|_a \leq 1\} \\ &+ \sup\{\|T(\mathbf{v})\|_b : \|\mathbf{v}\|_a \leq 1\} \\ &= \|S\|_{\mathcal{B}} + \|T\|_{\mathcal{B}}. \end{aligned}$$

Therefore, $\mathcal{B}(V, V')$ is a normed linear space with norm $\|\cdot\|_{\mathcal{B}}$. Finally, inequality (2.24) is just another version of inequality (2.22). □

In *Proposition 2.4.5*, we see for the first time homomorphisms of metric groups forming new metric groups in a natural way. In general, finding the right metric for a group of homomorphism is not as easy as in this case. We will pursue this subject for locally compact Abelian metric groups in the final chapters.

If we restrict our attention to a single normed linear space V with norm $\|\cdot\|_a$, then $\mathcal{B}(V, V)$ has richer algebraic and topological properties than V. If S and T are elements of $\mathcal{B}(V, V)$, then $S \circ T$ is also in $\mathcal{B}(V, V)$ because it follows from (2.24) that

$$\begin{aligned} \sup\{\|(S \circ T)(\mathbf{v})\|_a : \|\mathbf{v}\|_a \leq 1\} &= \sup\{\|S(T(\mathbf{v}))\|_a : \|\mathbf{v}\|_a \leq 1\} \\ &\leq \sup\{\|S\|_{\mathcal{B}}\,\|T(\mathbf{v})\|_a : \|\mathbf{v}\|_a \leq 1\} \\ &= \|S\|_{\mathcal{B}} \sup\{\|T(\mathbf{v})\|_a : \|\mathbf{v}\|_a \leq 1\} \\ &= \|S\|_{\mathcal{B}}\|T\|_{\mathcal{B}}. \end{aligned}$$

Therefore, $S \circ T$ is in $\mathcal{B}(V, V)$ and

$$\|S \circ T\|_{\mathcal{B}} \leq \|S\|_{\mathcal{B}}\|T\|_{\mathcal{B}}, \tag{2.25}$$

when S and T are in $\mathcal{B}(V, V)$. In addition, the identity function ι is in $\mathcal{B}(V, V)$ and $\|\iota\|_{\mathcal{B}} = 1$. Therefore, $\mathcal{B}(V, V)$ is also an algebra with identity. This kind of a structure is important enough to merit a name.

A *normed algebra with identity* is an algebra with an identity $\mathbf{e}$ and with a norm $\|\cdot\|_b$ which is *submultiplicative*, meaning that

$$\|\mathbf{uv}\|_b \leq \|\mathbf{u}\|_b \|\mathbf{v}\|_b. \tag{2.26}$$

for all $\mathbf{u}$ and $\mathbf{v}$ in the algebra and satisfies

$$\|\mathbf{e}\|_b = 1. \tag{2.27}$$

In particular, $\mathcal{B}(V, V)$ is a normed algebra with an identity, when V is a normed linear space.

The normed linear space $\mathcal{C}([0,1])$ with norm $\|f\|_s = \sup\{|f(x)| : 0 \leq x \leq 1\}$ provides another interesting example of a normed algebra with identity. For f and g in $\mathcal{C}([0,1])$, define $fg(x) = f(x)g(x)$. Because the product of continuous functions is continuous, $(f,g) \to fg$ is a binary operation on $\mathcal{C}([0,1])$. It is associative because the multiplication of real numbers is associative. Here the constant function 1 is the identity and the algebra is commutative.

Not surprisingly, $\mathcal{M}_n(\mathbb{S})$ is also normed algebra with identity. Since $\mathcal{M}_n(\mathbb{S})$ is an algebra with identity, all that is needed is a norm that satisfies (2.26) and (2.27).

Let A be an $n \times n$ matrix and think of $\mathbf{x}$ in $\mathbb{S}^n$ as a $1 \times n$ matrix. Then, $\mathbf{x}A$ is another $1 \times n$ matrix. Using the usual Euclidean norm $\|\cdot\|$, set

$$\|A\|_{\mathcal{B}} = \sup\{\|\mathbf{x}A\| : \|\mathbf{x}\| \leq 1\}. \tag{2.28}$$

The same kinds of estimates used to prove that $\|T\|_{\mathcal{B}}$ is a norm on $\mathcal{B}(V, V)$ that satisfies (2.26) and (2.27) can be used to show that $\|A\|_{\mathcal{B}}$ is a norm on $\mathcal{M}_n(\mathbb{S})$ satisfying (2.26) and (2.27). The norm defined by equation (2.28), called the *operator norm*, will be the preferred norm for $\mathcal{M}_n(\mathbb{S})$.

Proposition 2.4.6 *If $\mathcal{A}$ is a normed algebra with identity and norm $\|\cdot\|_a$, then the multiplication function $(\mathbf{u}, \mathbf{v}) \to \mathbf{uv}$ from $\mathcal{A} \times \mathcal{A} \to \mathcal{A}$ with the product topology on $\mathcal{A} \times \mathcal{A}$ is a continuous function.*

Proof. It suffices to show that if $\mathbf{u}_k$ and $\mathbf{v}_k$ are sequences in $\mathcal{A}$ converging to $\mathbf{u}$ and $\mathbf{v}$, respectively, then $\mathbf{u}_k\mathbf{v}_k$ converges to $\mathbf{uv}$ by *Proposition 1.2.3* and *Exercise* 14, p. 28. The triangle inequality (2.14) and the submultiplicative property of the norm (2.26) imply that

$$\begin{aligned}
\|\mathbf{u}_k\mathbf{v}_k - \mathbf{uv}\|_a &\leq \|\mathbf{u}_k\mathbf{v}_k - \mathbf{u}_k\mathbf{v}\|_a + \|\mathbf{u}_k\mathbf{v} - \mathbf{uv}\|_a \\
&\leq \|\mathbf{u}_k\|_a \|\mathbf{v}_k - \mathbf{v}\|_a + \|\mathbf{u}_k - \mathbf{u}\|_a \|\mathbf{v}\|_a \\
&\leq c\|\mathbf{v}_k - \mathbf{v}\|_a + \|\mathbf{u}_k - \mathbf{u}\|_a \|\mathbf{v}\|_a,
\end{aligned}$$

where c is from *Exercise* 5, p. 91. It follows that $\mathbf{u}_k\mathbf{v}_k$ converges to $\mathbf{uv}$. □

The study of normed linear spaces and bounded linear functions is the foundation of functional analysis. Our focus has been the concepts and results that will be useful in the further study of topological groups. For a broader more systematic introducion to functional analysis see [15]. It also contains a good introduction to topological spaces that lies between what is included here and books like [7] and [16].

Let G be the group of units of a normed algebra $\mathcal{A}$ with identity $\mathbf{e}$ as defined by (2.9) on page 78. When the topology on G is the relative norm topology from $\mathcal{A}$, it follows from *Proposition 2.4.6* that G is a group with continuous multiplication, but it does not automatically follow that G is a metric group. Recall (*Proposition 1.3.1*) that it must also be shown that the inverse function is continuous. By the end of this chapter, we will be able to prove that the group of the units of the finite-dimensional algebra $\mathcal{M}_n(\mathbb{S})$ with its unique norm topology is a metric group. Toward this end we restrict our attention to finite-dimensional linear spaces with their unique norm linear topologies.

Proposition 2.4.7 *Let V and V' be two normed linear spaces. If V is finite-dimensional, then every linear map $T : V \to V'$ is continuous.*

Proof. Let $\mathbf{v}_1, \ldots, \mathbf{v}_n$ be a basis for V. Then for every $\mathbf{v}$ in V there are uniquely determined scalars $c_1, \ldots, c_n$ such that $\mathbf{v} = c_1\mathbf{v} + \ldots + c_n\mathbf{v}_n$ and $\|\mathbf{v}\|_1 = \sum_{j=1}^n |c_j|$ is a norm on V by *Exercise* 6, p. 91. Because V is finite-dimensional, $\|\cdot\|_1$ is equivalent to every other norm on V. Letting $\|\cdot\|_b$ be a norm for V', it suffices by *Proposition 2.4.3* to prove that the set of real numbers $\{\|T(\mathbf{v})\|_b : \|\mathbf{v}\|_1 \leq 1\}$ is bounded above.

Let

$$M = \max\{\|T(\mathbf{v}_j)\|_b : j = 1, \ldots, n\}.$$

Given $\mathbf{v} = c_1\mathbf{v}_1 + \ldots + c_n\mathbf{v}_n$ such that $\|\mathbf{v}\|_1 = \sum_{j=1}^n |c_j| \leq 1$, it follows that

$$\begin{aligned}
\|T(\mathbf{v})\|_b &= \|c_1T(\mathbf{v}_1) + \ldots + c_nT(\mathbf{v}_n)\|_b \\
&\leq |c_1|\,\|T(\mathbf{v}_1)\|_b + \ldots + |c_n|\,\|T(\mathbf{v}_n)\|_b \\
&= (|c_1| + \ldots + |c_n|)M \\
&\leq M
\end{aligned}$$

Thus $\{\|T(\mathbf{v})\|_b : \|\mathbf{v}\|_1 \leq 1\}$ is bounded above by M. □

Corollary 2.4.8 *If V and V' are finite-dimensional linear spaces, then for any choice of norms on V and V'*

$$\mathcal{B}(V, V') = \mathcal{L}(V, V'). \tag{2.29}$$

Corollary 2.4.9 *Let V and V' be normed linear spaces. If the dimension of V is finite, then the kernel of every linear map $T : V \to V'$ is a closed set in V.*

Proof. The theorem implies that T is continuous and hence a homomorphism of metric groups. Therefore the kernel is closed. □

Proposition 2.4.10 *Every subspace W of a finite-dimensional normed linear space V is a closed set in V.*

Proof. If W equals either of the trivial subspaces, $\{\mathbf{0}\}$ or V, then it is clearly a closed set in V. So it can be assumed that $0 < \operatorname{Dim}[W] = m < \operatorname{Dim}[V] = n$. By the preceding corollary, it suffices to show that W is the kernel of a linear function $T : V \to V$.

By *Proposition 2.1.7* there exists a basis $\mathbf{v}_1, \ldots, \mathbf{v}_n$ of V such that $\mathbf{v}_1, \ldots, \mathbf{v}_m$ is a basis for W. *Proposition 2.2.1* implies that there exists a unique linear function $T : V \to V$ such that

$$T(\mathbf{v}_j) = \begin{cases} \mathbf{0} & \text{for } j = 1, \ldots, m \\ \mathbf{v}_j & \text{for } j = m+1, \ldots, n. \end{cases}$$

Obviously, W is contained in the kernel of T. If $\mathbf{v} = c_1\mathbf{v}_1 + \ldots + c_n\mathbf{v}_n$ is in the kernel of T, then $T(\mathbf{v}) = c_{m+1}\mathbf{v}_{m+1} + \cdots + c_n\mathbf{v}_n = \mathbf{0}$. It follows that $c_{m+1} = \ldots = c_n = 0$ because a nonempty subset of a set of linearly independent vectors is linearly independent. Thus the kernel of T is contained in W. Therefore, W is the kernel of T and a closed set by *Corollary 2.4.8*. □

Once again we are seeing that in finite dimensions the algebraic structure has significant topological implications. It is worth summarizing some of the topological information implied by finite dimensionality. Suppose V and V' are finite-dimensional linear spaces over the same scalar field and $T : V \to V'$ is an onto linear function with kernel K. Then the following have been established:

(a) K is a finite-dimensional subspace (*Proposition 2.1.6*).

(b) $\operatorname{Dim}[V] = \operatorname{Dim}[K] + \operatorname{Dim}[V']$ (*Theorem 2.2.2*).

(c) T is a continuous function (*Proposition 2.4.7*).

(d) K is a closed set in V (*Corollary 2.4.8*).

(e) V/K is normed linear space (*Proposition 2.3.3*).

(f) V, V', and K have unique norm linear topologies (*Theorem 2.3.6*).

(g) V, V', and K are locally compact and σ-compact (*Corollary 2.3.8*).

(h) T is an open function (*Theorem 1.5.18*)

(i) V/K and V' are linearly isomorphic normed linear spaces (*Theorem 2.4.4*).

Another consequence of *Proposition 2.4.7* is the following proposition:

Theorem 2.4.11 *If V is a finite-dimensional linear space of dimension n over the scalar field $\mathbb{S}$, then $\mathcal{L}(V,V)$ and $\mathcal{M}_n(\mathbb{S})$ are isomorphic normed algebras with identities.*

Proof. Because V is finite-dimensional, V has a unique norm linear topology and $\mathcal{L}(V,V) = \mathcal{B}(V,V)$. By *Theorem 2.2.6*, $\mathcal{L}(V,V)$ and $\mathcal{M}_n(\mathbb{S})$ are algebraically isomorphic algebras. Since an algebraic isomorphism of algebras and its inverse are linear, they are continuous by *Proposition 2.4.7*. □

EXERCISES

1. Let V and V' be normed linear spaces with norms $\|\cdot\|_a$ and $\|\cdot\|_b$, respectively. Prove that a linear function $T: V \to V'$ is bounded if and only if the image of every bounded set is a bounded set as defined on page 89.
2. Let V be a normed linear space, and let $T: V \to V$ be a bounded linear function. Prove that if W is a closed subspace of V such that $T(W) \subset W$, then $\tilde{T}(W+\mathbf{v}) = W+T(\mathbf{v})$ defines a bounded linear function $\tilde{T}: V/W \to V/W$ such that $\tilde{T} \circ \pi = \pi \circ T$ where, as usual, $\pi(\mathbf{v}) = W + \mathbf{v}$.
3. Show that $f \to \int_0^1 f(x)dx$ is a bounded linear function to $\mathbb{R}$ from $\mathcal{C}([0,1])$ with the norm $\|\cdot\|_s$ and that its norm in $\mathcal{B}(\mathcal{C}([0,1]),\mathbb{R})$ is 1.
4. Let X be a compact metric space and consider the linear space $\mathcal{C}(X)$ as defined in *Exercise* 3, p. 69. Prove the following:

   (a) $\mathcal{C}(X)$ with the norm $\|f\|_s = \sup\{|f(x)| : x \in X\}$ and $fg(x) = f(x)g(x)$ is a commutative normed algebra with identity.

   (b) Given x in X the linear function $T: \mathcal{C}(X) \to \mathbb{R}$ defined by $T(f) = f(x)$ is a bounded linear function with $\|T\|_{\mathcal{B}} = 1$.

   (c) The group of units of $\mathcal{C}(X)$ is a metric group.
5. Let V and V' be normed linear spaces with norms $\|\cdot\|_a$ and $\|\cdot\|_b$. Consider sequences $\mathbf{x}_k$ in V and T_k in $\mathcal{B}(V,V')$. Prove that if $\mathbf{x}_k$ converges to $\mathbf{x}$ in V and T_k converges to T in $\mathcal{B}(V,V')$ then the sequence $T_k(\mathbf{x}_k)$ converges to $T(\mathbf{x})$ in V'.
6. Let $\mathcal{A}$ be a normed algebra with identity, and set $T_{\mathbf{u}}(\mathbf{v}) = \mathbf{uv}$ for $\mathbf{u}$ and $\mathbf{v}$ in $\mathcal{A}$. Show that $T_{\mathbf{u}}$ is in $\mathcal{B}(\mathcal{A},\mathcal{A})$ and $\|T_{\mathbf{u}}\|_{\mathcal{B}} = \|\mathbf{u}\|_a$.

2.5 The Determinant Function

Many high-school and college students have encountered the formula for the determinant of a 2×2 matrix in some form or another. Even the formula for the determinant of a 3×3 matrix is well known to most engineering and physical science students. But these two formulas barely scratch the surface of the properties of the determinant. This section is devoted to developing the basic theory of the determinant for $n \times n$ matrices.

Let A be an $n \times n$ matrix in $\mathcal{M}_n(\mathbb{S})$. Recall from Section 1.1 that S_n denotes the symmetric group consisting of the permutations of n symbols and $\text{sgn} : S_n \to \{-1, 1\}$ is an algebraic homomorphism by *Theorem 1.1.5.* The *determinant* of A is defined by

$$\text{Det}\,[A] = \sum_{\alpha \in S_n} \text{sgn}(\alpha) \prod_{i=1}^{n} A(i, \alpha(i)). \tag{2.30}$$

The determinant is a function from either $\mathcal{M}_n(\mathbb{R})$ or $\mathcal{M}_n(\mathbb{C})$ to $\mathbb{R}$ or $\mathbb{C}$ with the property that $\text{Det}\,[cA] = c^n \text{Det}\,[A]$ for any $c \in \mathbb{S}$.

Proposition 2.5.1 *The determinant is a continuous function from the normed algebra $\mathcal{M}_n(\mathbb{S})$ to $\mathbb{S}$.*

Proof. The norm linear topology on $\mathcal{M}_n(\mathbb{S})$ is the product topology on $\mathbb{S}^{n^2}$ because all norms are equivalent on $\mathcal{M}_n(\mathbb{S})$ and the product metric comes from the norm $|A|_1 = \sum_{(j,k) \in \Lambda} |A(j,k)|$. The function $\mathcal{M}_n(\mathbb{S}) \to \mathbb{S}$ defined by $A \to A(i,j)$ is a projection function and continuous. Since the sums and products of continuous functions are continuous, the determinant is continuous. □

Given α in S_n and $1 \leq i \leq n$, let $j = \alpha(i)$. Then $A(i, \alpha(i)) = A(\alpha^{-1}(j), j)$ and

$$\prod_{i=1}^{n} A(i, \alpha(i)) = \prod_{j=1}^{n} A(\alpha^{-1}(j), j).$$

As α ranges through S_n, note that α^{-1} also ranges through S_n, because $\alpha \to \alpha^{-1}$ is a one-to-one function of S_n onto itself. Since every element in the group $\{1, -1\}$ is its own inverse and since sgn is a homomorphism, $\text{sgn}(\alpha^{-1}) = \text{sgn}(\alpha)^{-1} = \text{sgn}(\alpha)$. Putting the pieces together,

$$\begin{aligned} \text{Det}\,[A] &= \sum_{\alpha \in S_n} \text{sgn}(\alpha) \prod_{j=1}^{n} A(\alpha^{-1}(j), j) \\ &= \sum_{\alpha \in S_n} \text{sgn}(\alpha^{-1}) \prod_{j=1}^{n} A(\alpha(j), j) \\ &= \sum_{\alpha \in S_n} \text{sgn}(\alpha) \prod_{j=1}^{n} A(\alpha(j), j) \end{aligned}$$

and

$$\text{Det}\,[A] = \sum_{\alpha \in S_n} \text{sgn}(\alpha) \prod_{j=1}^{n} A(\alpha(j), j) \tag{2.31}$$

is an alternative formula for the determinant. Both are useful.

The *transpose of a matrix* A is defined by

$$A^{\mathfrak{t}}(i,j) = A(j,i).$$

In other words, the transpose of a matrix just interchanges the rows and columns of the matrix, and the transpose of an $m \times n$ matrix is an $n \times m$ matrix. For $n \times n$ matrices, the transpose is a one-to-one function of $\mathcal{M}_n(\mathbb{R})$ or $\mathcal{M}_n(\mathbb{C})$ onto itself (see *Exercise* 4, p. 81). Moreover,

$$\begin{aligned}\text{Det}\,[A^{\mathsf{t}}] &= \sum_{\alpha \in S_n} \text{sgn}(\alpha) \prod_{i=1}^{n} A^{\mathsf{t}}(i, \alpha(i)) \\ &= \sum_{\alpha \in S_n} \text{sgn}(\alpha) \prod_{i=1}^{n} A(\alpha(i), i) \\ &= \text{Det}\,[A]\end{aligned}$$

by equations (2.30) and (2.31). Therefore,

$$\text{Det}\,[A^{\mathsf{t}}] = \text{Det}\,[A]. \tag{2.32}$$

Since for any $\alpha \in S_n$ the expression

$$\prod_{i=1}^{n} A(i, \alpha(i)) = \prod_{j=1}^{n} A(\alpha^{-1}(j), j)$$

contains exactly one term from every row and exactly one term from every column, $\text{Det}\,[A] = 0$, when A has a row of all zeros or a column of all zeros.

It is also easy to calculate the determinant of the identity matrix I. If $\alpha \neq \iota$, then there exists an i such that $\alpha(i) \neq i$ and $I(i, \alpha(i)) = 0$. Thus $\prod_{i=1}^{n} A(i, \alpha(i)) = 0$ for $\alpha \neq \iota$ and

$$\text{Det}\,[I] = \text{sgn}(\iota) \prod_{i=1}^{n} I(i, \iota(i)) = \prod_{i=1}^{n} I(i, i) = 1.$$

If β is an element of S_n, then β can be used to permute the rows of an $n \times n$ matrix by setting ${}_{\beta}A(i, j) = A(\beta(i), j)$. Of course, the columns could also be permuted by β, but rows will suffice for our purposes.

Proposition 2.5.2 *If A is an $n \times n$ matrix and β is an element of S_n, then*

$$\text{Det}\,[{}_{\beta}A] = \text{sgn}(\beta)\text{Det}\,[A] \tag{2.33}$$

Proof. This proof makes critical use of the fact that sgn is an algebraic homomorphism of S_n to the multiplicative group $\{1, -1\}$ (*Theorem 1.1.5*). Because every element in the group $\{1, -1\}$ is its own inverse, $\text{sgn}(\beta)\text{sgn}(\beta) = 1$. Now use equation (2.31) to compute $\text{Det}\,[{}_{\beta}A]$ as follows:

$$\text{Det}\,[{}_{\beta}A] = \sum_{\alpha \in S_n} \text{sgn}(\alpha) \prod_{j=1}^{n} {}_{\beta}A(\alpha(j), j)$$

$$\begin{aligned}
&= \sum_{\alpha \in S_n} \operatorname{sgn}(\alpha) \prod_{j=1}^{n} A(\beta \circ \alpha(j), j) \\
&= \operatorname{sgn}(\beta) \sum_{\alpha \in S_n} \operatorname{sgn}(\beta)\operatorname{sgn}(\alpha) \prod_{j=1}^{n} A(\beta \circ \alpha(j), j) \\
&= \operatorname{sgn}(\beta) \sum_{\alpha \in S_n} \operatorname{sgn}(\beta \circ \alpha) \prod_{j=1}^{n} A(\beta \circ \alpha(j), j) \\
&= \operatorname{sgn}(\beta)\operatorname{Det}[A].
\end{aligned}$$

The final equality holds because $\alpha \to \beta \circ \alpha$ is a one-to-one function of S_n onto itself. □

If β is a transposition, then $\operatorname{sgn}(\beta) = -1$ and ${}_\beta A$ just interchanges two rows of A. Consequently, switching the rows of a matrix changes the sign of the determinant, a useful property of determinants.

Corollary 2.5.3 *Let A be an $n \times n$ matrix. If A has two equal rows, then* $\operatorname{Det}[A] = 0$. *If A has two equal columns, then* $\operatorname{Det}[A] = 0$.

Proof. Let β be the transposition that switches the identical rows. Then, ${}_\beta A = A$ and

$$\operatorname{Det}[A] = \operatorname{Det}[{}_\beta A] = -\operatorname{Det}[A].$$

Hence, $2\operatorname{Det}[A] = 0$, which implies that $\operatorname{Det}[A] = 0$.

If A has two equal columns, then A^t has two equal rows and $\operatorname{Det}[A] = \operatorname{Det}[A^t] = 0$ by the first part. □

Although the determinant is not a linear function, that is, $\operatorname{Det}[A + B] \neq \operatorname{Det}[A] + \operatorname{Det}[B]$ for all A and B, it is linear in a row when the other rows are held fixed. This property along with equation (2.33) and $\operatorname{Det}[I] = 1$ uniquely defines the determinant function. To obtain this result and make good use of it, we need to develop the idea of a multilinear function.

Let A_i denote the ith row of the $n \times n$ matrix A. If A is a matrix with entries from the scalar field $\mathbb{S}$, then A_i is an element of $\mathbb{S}^n$ and we can think of A as $A = (A_1, \ldots, A_n)$, an element of $\mathbb{S}^n \times \ldots \times \mathbb{S}^n$ and write $\operatorname{Det}[A_1, \ldots, A_n]$. This point of view has no effect on the additive group structure, the linear structure, or the norm topology for $n \times n$ real or complex matrices. But it does provide a different algebraic perspective for understanding the multilinear behavior of the determinant.

A function

$$\mu : \underbrace{\mathbb{S}^n \times \ldots \times \mathbb{S}^n}_{m \text{ times}} \to \mathbb{S}$$

is an *m-multilinear function* or simply a *multilinear function* provided that given any $(\mathbf{x}_1, \ldots, \mathbf{x}_m)$ in $\mathbb{S}^n \times \ldots \times \mathbb{S}^n$ and $1 \leq i \leq m$, the function from $\mathbb{S}^n$ to $\mathbb{S}$ defined by

$$\mathbf{x} \mapsto \mu(\mathbf{x}_1, \ldots, \mathbf{x}_{i-1}, \mathbf{x}, \mathbf{x}_{i+1}, \ldots, \mathbf{x}_m)$$

is linear. In other words, for all $\mathbf{x}, \mathbf{y} \in \mathbb{S}^n$ and $c, d \in \mathbb{S}$, we have

$$\mu(\mathbf{x}_1, \ldots, \mathbf{x}_{i-1}, c\mathbf{x} + d\mathbf{y}, \mathbf{x}_{i+1}, \ldots, \mathbf{x}_m) =$$

$$c\mu(\mathbf{x}_1, \ldots, \mathbf{x}_{i-1}, \mathbf{x}, \mathbf{x}_{i+1}, \ldots, \mathbf{x}_m) + d\mu(\mathbf{x}_1, \ldots, \mathbf{x}_{i-1}, \mathbf{y}, \mathbf{x}_{i+1}, \ldots, \mathbf{x}_m)$$

for $i = 1, \ldots, m$.

For any two positive integers m and n, let $F_{(m,n)}$ denote the set of functions from the set $\{1, \ldots, m\}$ to the set $\{1, \ldots, n\}$. There are exactly n^m functions in $F_{(m,n)}$.

Proposition 2.5.4 *A function*

$$\mu : \underbrace{\mathbb{S}^n \times \ldots \times \mathbb{S}^n}_{m \text{ times}} \to \mathbb{S}$$

is multilinear if and only if it has the form

$$\mu(\mathbf{x}_1, \ldots, \mathbf{x}_m) = \sum_{\gamma \in F_{(m,n)}} \mu(\mathbf{e}_{\gamma(1)}, \ldots, \mathbf{e}_{\gamma(m)}) \left(\prod_{j=1}^{m} x_{j\,\gamma(j)} \right) \tag{2.34}$$

where $\mathbf{x}_i = (x_{i1}, \ldots, x_{in})$.

Proof. Given $\gamma \in F_{(m,n)}$, it is obvious that the $\mu(\mathbf{x}_1, \ldots, \mathbf{x}_m) = \prod_{j=1}^{m} x_{j\,\gamma(j)}$ is multilinear. It is even more obvious that linear combinations of multilinear functions are multilinear; in fact, the multilinear functions on $\mathbb{S}^n \times \ldots \times \mathbb{S}^n$ are a linear space. Hence, any function of the form of equation (2.34) is multilinear.

Suppose $\mu : \mathbb{S}^n \times \ldots \times \mathbb{S}^n \to \mathbb{S}$ is multilinear and write $\mathbf{x}_i = \sum_{j=1}^{n} x_{ij}\mathbf{e}_j$. The argument precedes by induction on m. If $m = 1$, then μ is linear and equation (2.34) is the simple consequence of linearity that

$$\mu(\mathbf{x}_1) = \sum_{j=1}^{n} x_{1j}\mu(\mathbf{e}_j) = \sum_{\gamma \in F_{(1,n)}} \mu(\mathbf{e}_{\gamma(1)})x_{1\gamma(1)}.$$

Assume the result holds for m and consider

$$\mu : \underbrace{\mathbb{S}^n \times \ldots \times \mathbb{S}^n}_{m+1 \text{ times}} \to \mathbb{S}.$$

Two key observations are needed to finish the proof. First, the functions given by $\mu(\mathbf{x}_1, \ldots, \mathbf{x}_m, \mathbf{e}_k)$ for $k = 1 \ldots, n$ are clearly multilinear functions, and the induction assumption applies to them. Second, every function γ in $F_{(m,n)}$ can be extended to a function in $F_{(m+1,n)}$ in exactly n ways by setting $\gamma(m+1) = k$ for $k = 1, \ldots, n$, and every function β in $F_{(m+1,n)}$ can be obtained in this way from exactly one γ in $F_{(m,n)}$.

It follows from the these observations that

$$\begin{aligned}
\mu(\mathbf{x}_1, \ldots, \mathbf{x}_{m+1}) &= \sum_{k=1}^{n} \mu(\mathbf{x}_1, \ldots, \mathbf{x}_m, \mathbf{e}_k) x_{m+1k} \\
&= \sum_{k=1}^{n} \left(\sum_{\gamma \in F_{(m,n)}} \mu(\mathbf{e}_{\gamma(1)}, \ldots, \mathbf{e}_{\gamma(m)}, \mathbf{e}_k) \left(\prod_{j=1}^{m} x_{j\,\gamma(j)} \right) \right) x_{m+1k} \\
&= \sum_{\gamma \in F_{(m,n)}} \sum_{k=1}^{n} \mu(\mathbf{e}_{\gamma(1)}, \ldots, \mathbf{e}_{\gamma(m)}, \mathbf{e}_k) \left(\prod_{j=1}^{m} x_{j\,\gamma(j)} \right) x_{m+1k} \\
&= \sum_{\beta \in F_{(m+1,n)}} \mu(\mathbf{e}_{\beta(1)}, \ldots, \mathbf{e}_{\beta(m+1)}) \left(\prod_{j=1}^{m+1} x_{j\,\beta(j)} \right)
\end{aligned}$$

by setting $\beta(j) = \gamma(j)$ for $j = 1, \ldots, m$ and $\beta(m+1) = k$ for $k = 1, \ldots, n$ to obtain the right form for μ. $\square$

Proposition 2.5.5 *The determinant is multilinear in its rows.*

Proof. A notational shift is required here before *Proposition 2.5.4* can be applied. For the determinant of an $n \times n$ matrix A, the rows $A_1, \ldots, A_n$ of A are playing the role of the $\mathbf{x}_i$ in the formula for a general multilinear function. So $A(j, \alpha(j))$ is an $x_{i\alpha(j)}$ term. And the rows $I_1, \ldots, I_n$ of the identity matrix are just $\mathbf{e}_1, \ldots, \mathbf{e}_n$. In this context, it must be shown that $\text{Det}\,[A_1, \ldots, A_n]$ is an n-multilinear function.

Clearly, S_n is a subset of $F_{(n,n)}$, and equation (2.30) can be viewed as the sum over $F_{(n,n)}$ with $\text{sgn}(\gamma) = 0$ for γ in $F_{(n,n)} \setminus S_n$. To apply *Proposition 2.5.4* and complete the proof, it suffices to show that $\text{sgn}(\alpha) = \text{Det}\,[I_{\alpha(1)}, \ldots, I_{\alpha(n)}]$. Since $(I_{\alpha(1)}, \ldots, I_{\alpha(n)}) = {}_{\alpha}I$, it follows immediately from equation (2.33) that $\text{Det}\,[I_{\alpha(1)}, \ldots, I_{\alpha(n)}] = \text{Det}\,[{}_{\alpha}I] = \text{sgn}(\alpha)\text{Det}\,[I] = \text{sgn}(\alpha)$. $\square$

Proposition 2.5.6 *If the rows of the $n \times n$ matrix A are linearly dependent, then* $\text{Det}\,[A] = 0$.

Proof. If the rows are linearly dependent, then for some i there exist scalars c_j such that $A_i = \sum_{j \neq i} c_j A_j$ (*Exercise* 6, p. 69). Since the determinant is multilinear in its rows by the previous proposition, it follows that

$$\begin{aligned}
\text{Det}\,[A] &= \text{Det}\,[A_1, \ldots, A_n] \\
&= \text{Det}\,[A_1, \ldots, A_{i-1}, \underbrace{\sum_{j \neq i} c_j A_j}_{i^{\text{th}}\ row}, A_{i+1}, \ldots, A_n] \\
&= c_j \sum_{j \neq i} \text{Det}\,[A_1, \ldots, A_{i-1}, \underbrace{A_j}_{i^{\text{th}}\ row}, A_{i+1}, \ldots, A_n] \\
&= 0
\end{aligned}$$

because the ith row and the jth with $i \neq j$ are the same in each determinant appearing in the penultimate equation. □

An m-multilinear function

$$\mu : \underbrace{\mathbb{S}^n \times \ldots \times \mathbb{S}^n}_{m \text{ times}} \to \mathbb{S}$$

is an *alternating multilinear function* provided that

$$\mu(\mathbf{x}_{\alpha(1)}, \ldots, \mathbf{x}_{\alpha(m)}) = \operatorname{sgn}(\alpha)\mu(\mathbf{x}_1, \ldots, \mathbf{x}_m) \tag{2.35}$$

for all α in S_m. The determinant is an alternating multilinear function by *Proposition 2.5.2.*

The multilinear functions on the rows of $n \times n$ matrices obviously form a linear space V. Those that are alternating satisfy the equation

$$\mu(A_{\beta(1)}, \ldots, A_{\beta(n)}) = \operatorname{sgn}(\beta)\mu(A_1, \ldots, A_n) \tag{2.36}$$

for all β in S_n and form a linear subspace W of V. Clearly, the determinant function is in W. It will be shown (*Theorem 2.5.7*) that every element of W is a multiple of the determinant function and hence $\operatorname{Dim}[W] = 1$. Furthermore, the determinant function is the unique alternating multilinear function μ on the rows of $n \times n$ matrix such that $\mu(I) = 1$. This is valuable tool. For example, the formula $\operatorname{Det}[AB] = \operatorname{Det}[A]\operatorname{Det}[B]$ will follow readily from this property of the determinant function.

Theorem 2.5.7 *Let $\mu : \mathcal{M}_n(\mathbb{S}) \to \mathbb{S}$ be a multilinear function on the rows of a matrix. If for every matrix $A = (A_1, \ldots, A_n)$ in $\mathcal{M}_n(\mathbb{S})$ and $\beta \in S_n$,*

$$\mu(A_{\beta(1)}, \ldots, A_{\beta(n)}) = \operatorname{sgn}(\beta)\mu(A_1, \ldots, A_n),$$

then

$$\mu(A_1, \ldots, A_n) = \mu(I_1, \ldots, I_n)\operatorname{Det}[A_1, \ldots, A_n].$$

or more simply $\mu(A) = \mu(I)\operatorname{Det}[A]$ for all $A \in \mathcal{M}_n(\mathbb{S})$

Proof. By *Proposition 2.5.4*, the function μ can be expressed as

$$\mu(A_1, \ldots, A_n) = \sum_{\gamma \in F_{(n,n)}} \mu(I_{\gamma(j)}, \ldots, I_{\gamma(n)}) \left(\prod_{j=1}^{n} A(j, \gamma(j)) \right)$$

The value of μ is zero when A has two equal rows by *Exercise* 7, p. 114. If $\gamma \in F_{(n,n)}$ is not one-to-one, $\mu(I_{\gamma(j)}, \ldots, I_{\gamma(n)}) = 0$ because $I_{\gamma(j)} = I_{\gamma(k)}$ for some $j \neq k$. Since a function in $F_{(n,n)}$ is one-to-one if and only if it is onto (*Exercise* 5, p. 12), the above expression for μ simplifies to

$$\mu(A_1, \ldots, A_n) = \sum_{\gamma \in S_n} \mu(I_{\gamma(j)}, \ldots, I_{\gamma(n)}) \left(\prod_{j=1}^{n} A(j, \gamma(j)) \right).$$

Using the hypothesis $\mu(A_{\beta(1)}, \ldots, A_{\beta(n)}) = \operatorname{sgn}(\beta)\mu(A_1, \ldots, A_n)$ again,

$$\begin{aligned}\mu(A_1, \ldots, A_n) &= \sum_{\gamma \in S_n} \operatorname{sgn}(\gamma)\mu(I_1, \ldots, I_n)\left(\prod_{j=1}^{n} A(j, \gamma(j))\right) \\ &= \mu(I_1, \ldots, I_n)\sum_{\gamma \in S_n} \operatorname{sgn}(\gamma)\left(\prod_{j=1}^{n} A(j, \gamma(j))\right) \\ &= \mu(I_1, \ldots, I_n)\operatorname{Det}[A_1, \ldots, A_n],\end{aligned}$$

which completes the proof. □

Proposition 2.5.8 *If A and B are arbitrary $n \times n$ matrices, then* $\operatorname{Det}[AB] = \operatorname{Det}[A]\operatorname{Det}[B]$.

Proof. Fix a matrix B in $\mathcal{M}_n(\mathbb{S})$ and consider the function $\mu(A) = \operatorname{Det}[AB]$. Writing A in its row form as $A = (A_1, \ldots, A_n)$ and thinking of the rows as $1 \times n$ matrices, it is clear that the jth row of AB is A_jB and the row form of AB is $AB = (A_1B, \ldots, A_nB)$. Since the function $\mathbf{x} \mapsto \mathbf{x}B$ is linear, it follows that $\mu(A) = \operatorname{Det}[AB]$ is multilinear in the rows of A. Furthermore, $AB = (A_1B, \ldots, A_nB)$ also implies that $({}_\alpha A)B = {}_\alpha(AB)$ for $\alpha \in S_n$ and that

$$\mu({}_\alpha A) = \operatorname{Det}[({}_\alpha A)B] = \operatorname{Det}[{}_\alpha(AB)] = \operatorname{sgn}(\alpha)\operatorname{Det}[AB] = \operatorname{sgn}(\alpha)\mu(A).$$

Therefore, μ satisfies the hypothesis of *Theorem 2.5.7* and consequently

$$\begin{aligned}\mu(A) &= \mu(I)\operatorname{Det}[A] \\ &= \operatorname{Det}[IB]\operatorname{Det}[A] \\ &= \operatorname{Det}[B]\operatorname{Det}[A].\end{aligned}$$

Since B was an arbitrary $n \times n$ matrix, $\operatorname{Det}[AB] = \operatorname{Det}[A]\operatorname{Det}[B]$ for all A and B in $\mathcal{M}_n(\mathbb{S})$. □

Theorem 2.5.9 *Given A in $\mathcal{M}_n(\mathbb{S})$, the following are equivalent:*

(a) $\operatorname{Det}[A] \neq 0$.

(b) $\{\mathbf{x} \in \mathbb{S}^n : \mathbf{x}A = \mathbf{0}\} = \{\mathbf{0}\}$.

(c) A is invertible.

Proof. Suppose $\operatorname{Det}[A] \neq 0$. If there exists $\mathbf{x} \neq \mathbf{0}$ in $\mathbb{S}^n$ such that $\mathbf{x}A = \mathbf{0}$, then $x_1A_1 + \ldots + x_nA_n = \mathbf{0}$ and the rows of A are linearly dependent, implying that $\operatorname{Det}[A] = 0$ by *Proposition 2.5.6*. So *(a)* implies *(b)*.

Consider the linear function $T(\mathbf{x}) = \mathbf{x}A$ in $\mathcal{L}(\mathbb{S}^n, \mathbb{S}^n)$. If $\{\mathbf{x} \in \mathbb{S}^n : \mathbf{x}A = \mathbf{0}\} = \{\mathbf{0}\}$, then the kernel of T is $\{\mathbf{0}\}$ and T is one-to-one. Then *Corollary 2.2.3* implies that T is onto, and hence invertible.

Next observe that $T(\mathbf{e}_j) = \mathbf{e}_j A$ is the jth row of A. So the matrix of T with respect to the standard basis $\mathbf{e}_1, \ldots, \mathbf{e}_n$ is A^t (*Exercise* 5, p. 81). Let Φ be the isomorphism of $\mathcal{L}(\mathbb{S}^n, \mathbb{S}^n)$ onto $\mathcal{M}_n(\mathbb{S})$ given by *Theorem 2.2.6* using the standard basis. Then $\Phi(T)$ is the matrix of T with respect to the basis $\mathbf{e}_1, \ldots, \mathbf{e}_n$ and $\Phi(T) = A^t$ is invertible by *Theorem 2.2.6.* Now A is invertible by *Exercise* 4, p. 81, and *(b)* implies *(c)*.

Finally, if A is invertible, then $1 = \text{Det}\,[I] = \text{Det}\,[AA^{-1}] = \text{Det}\,[A]\,\text{Det}\,[A^{-1}]$. Hence neither $\text{Det}\,[A]$ nor $\text{Det}\,[A^{-1}]$ can be zero. In fact,

$$\text{Det}\,[A^{-1}] = \frac{1}{\text{Det}\,[A]}. \tag{2.37}$$

Thus *(c)* implies *(a)*. □

Given an $n \times n$ matrix A with real or complex entries, set $\rho(z) = \text{Det}\,[A - zI]$, where z is a complex variable. It follows from the definition of the determinant that $\rho(z)$ is a polynomial of degree n, called the *characteristic polynomial of A*. The Fundamental Theorem of Algebra states that every complex polynomial of positive degree has a zero, that is, a complex number λ exists such that $\rho(\lambda) = 0$.

The zeros of the characteristic polynomial for A are called the *characteristic values* of A, and every matrix has at least one characteristic value. Since a polynomial can have repeated zeros, a matrix may have only one characteristic value. The only characteristic value of the matrix

$$\begin{bmatrix} 2 & i \\ 0 & 2 \end{bmatrix}$$

is 2. The following corollary of *Theorem 2.5.9* is obvious.

Corollary 2.5.10 *Let A be in $\mathcal{M}_n(\mathbb{C})$. The matrix $A - \lambda I$ is not invertible if and only if λ is a characteristic value of A if and only if there exists a nonzero vector $\mathbf{v}$ in $\mathbb{C}^n$ such that $\mathbf{v}A = \lambda\mathbf{v}$. In particular, A is not invertible if and only if 0 is a characteristic value of A.*

If A is a real matrix, them the characteristic polynomial of A has real coefficients, but it need not have any real zeros. For example, the characteristic polynomial of

$$A = \begin{bmatrix} 0 & 1 \\ -1 & 0 \end{bmatrix}$$

is $z^2 + 1$ and its characteristic values are $\pm i$. So there are no real characteristic values. Consequently, for all real c the real matrix

$$\begin{bmatrix} 0 & 1 \\ -1 & 0 \end{bmatrix} - c \begin{bmatrix} 1 & 0 \\ 0 & 1 \end{bmatrix} = \begin{bmatrix} -c & 1 \\ -1 & -c \end{bmatrix}$$

is invertible, but the complex matrices

$$\begin{bmatrix} 0 & 1 \\ -1 & 0 \end{bmatrix} - \pm i \begin{bmatrix} 1 & 0 \\ 0 & 1 \end{bmatrix} = \begin{bmatrix} \mp i & 1 \\ -1 & \mp i \end{bmatrix}$$

are not invertible because $\pm i$ are characteristic values of A. For a real matrix A, there exists a nonzero vector $\mathbf{u}$ in $\mathbb{R}^n$ such that $\mathbf{v}A = \lambda\mathbf{v}$ if and only if λ is a real characteristic value of A.

Theorem 2.5.11 *Let V be a finite-dimensional linear space over $\mathbb{C}$ and let $T : V \to V$ be a linear function. There exists a basis $\mathbf{v}_1, \ldots, \mathbf{v}_n$ for V such that $T(\mathbf{v}_k) \in \text{Span}\,[\mathbf{v}_1, \ldots, \mathbf{v}_k]$ for $k = 1, \ldots, n$.*

Proof. The proof will proceed by induction on the dimension of V. The theorem is trivial if the dimension of V is 1. Assume the result holds when the dimension of the linear space is $n - 1$, and consider an arbitrary linear function $T : V \to V$ on a linear space V of dimension n.

Let $\mathbf{u}_1, \ldots, \mathbf{u}_n$ be a basis for V and let A be the matrix of T with respect to this basis. Then A has a characteristic value λ and $A - \lambda I$ is not invertible.

Since $A - \lambda I$ is clearly the matrix of $T - \lambda\iota$, the linear function $T - \lambda\iota$ is not invertible (*Theorem 2.2.6*) and thus not one-to-one (*Corollary 2.2.3*). So there exists $\mathbf{v}_1 \neq \mathbf{0}$ in V such that $T(\mathbf{v}_1) = \lambda\mathbf{v}_1$. It follows that $W = \text{Span}\,[\mathbf{v}_1]$ is a closed subspace of V such that $T(W) \subset W$. Obviously, $\text{Dim}\,[W] = 1$.

Using *Proposition 2.3.3*, form the quotient norm linear space V/W. It follows from *Theorem 2.2.2* that the dimension of V/W is $n - 1$ because the canonical homomorphism is linear and onto. By *Exercise* 2, p. 99, there exists a linear function $\tilde{T} : V/W \to V/W$ such that $\pi \circ T = \tilde{T} \circ \pi$. Now the induction assumption can be applied to $\tilde{T}$ and V/W. So there exists a basis $\tilde{\mathbf{v}}_2, \ldots, \tilde{\mathbf{v}}_n$ for V/W such that $\tilde{T}(\tilde{\mathbf{v}}_k) \in \text{Span}\,[\tilde{\mathbf{v}}_2, \ldots, \tilde{\mathbf{v}}_k]$ for $k = 2, \ldots, n$.

Choose $\mathbf{v}_k$ in V for $k = 2, \ldots, n$ in V such that $\pi(\mathbf{v}_k) = \tilde{\mathbf{v}}_k$. For $k > 1$, it follows that for appropriate scalars

$$\begin{aligned}
\pi(T(\mathbf{v}_k)) &= \tilde{T}(\pi(\mathbf{v}_k)) \\
&= \tilde{T}(\tilde{\mathbf{v}}_k) \\
&= \sum_{j=2}^{k} c_j \tilde{\mathbf{v}}_j \\
&= \pi\left(\sum_{j=2}^{k} c_j \mathbf{v}_j\right).
\end{aligned}$$

Therefore,

$$T(\mathbf{v}_k) - \sum_{j=2}^{k} c_j \mathbf{v}_j$$

is in W, the kernel of π, and so for some scalar c_1

$$T(\mathbf{v}_k) - \sum_{j=2}^{k} c_j \mathbf{v}_j = c_1 \mathbf{v}_1.$$

Consequently, $T(\mathbf{v}_k)$ is in $\text{Span}\,[\mathbf{v}_1, \ldots, \mathbf{v}_k]$.

To prove that $\mathbf{v}_1, \ldots, \mathbf{v}_n$ is a basis, it suffices to show that $\mathbf{v}_1, \ldots, \mathbf{v}_n$ are linearly independent because $\mathrm{Dim}\,[V] = n$. If $c_1\mathbf{v}_1 + \ldots + c_n\mathbf{v}_n = \mathbf{0}$, then

$$\pi(c_1\mathbf{v}_1 + \ldots + c_n\mathbf{v}_n) = \mathbf{0},$$

that is,

$$c_2\tilde{\mathbf{v}}_2 + \ldots + c_n\tilde{\mathbf{v}}_n = \mathbf{0},$$

implying that $c_2 = \ldots = c_n = 0$. So $c_1\mathbf{v}_1 = \mathbf{0}$ and $c_1 = 0$ because $\mathbf{v}_1 \neq \mathbf{0}$. □

Let $A(i|j)$ denote the $(n-1) \times (n-1)$ matrix obtained by deleting the ith row and j^{th} column from an $n \times n$ matrix A. The expression $(-1)^{i+j}\mathrm{Det}\,[A(i|j)]$ is called the i, j *cofactor* of A and used to calculate determinants.

Theorem 2.5.12 *If A is in $\mathcal{M}_n(\mathbb{S})$, then for each $j = 1, \ldots, n$*

$$\mathrm{Det}\,[A] = \sum_{i=1}^{n} A(i,j)(-1)^{i+j}\mathrm{Det}\,[A(i|j)] \tag{2.38}$$

Proof. Consider the function $\mu : \mathcal{M}_n(\mathbb{S}) \to \mathbb{S}$ defined by

$$\mu(A) = \sum_{i=1}^{n} A(i,j)(-1)^{i+j}\mathrm{Det}\,[A(i|j)].$$

It will be shown that *Theorem 2.5.7* applies so that $\mu(A) = \mu(I)\mathrm{Det}\,[A]$. Then showing that $\mu(I) = 1$ will complete the proof. The proof is broken into three distinct steps.

First, it must be shown that $\mu(A) = \mu(A_1, \ldots, A_n)$ is multilinear in its rows, $A_1, \ldots, A_n$. It suffices to show that each term $A(i,j)(-1)^{i+j}\mathrm{Det}\,[A(i|j)]$ is multilinear in the rows of A because a sum of multilinear functions is multilinear.

The rows of $A(i|j)$ are rows of A with the jth element deleted and the numbering changed. For example, the second row of $A(i|j)$ would be the third or second row of A with the jth element deleted according as $i \leq 2$ or $i \geq 3$.

Since the determinant function is multilinear, the function $\mathrm{Det}\,[A(i|j)]$ is clearly linear in the kth row of A, when $k \neq i$. Thus $A(i,j)(-1)^{i+j}\mathrm{Det}\,[A(i|j)]$ is linear in the kth row of A, when $k \neq i$ because $A(i,j)$ does not depend on the kth row of A.

It must now be shown that $A(i,j)(-1)^{i+j}\mathrm{Det}\,[A(i|j)]$ is linear in the ith row A. Here the roles are reversed. The expression $(-1)^{i+j}\mathrm{Det}\,[A(i|j)]$ does not depend on the ith row, and $A(i,j)$ is obviously linear on the ith. Hence, $A(i,j)(-1)^{i+j}\mathrm{Det}\,[A(i|j)]$ is also linear in the ith row of A. Therefore, each term $A(i,j)(-1)^{i+j}\mathrm{Det}\,[A(i|j)]$ is linear in the rows of A and $\mu(A)$ is a multilinear function on the rows of A. This completes the first step.

The second step is to show that $\mu({}_\alpha A) = \mathrm{sgn}(\alpha)\mu(A)$ for all A in $\mathcal{M}_n(\mathbb{S})$ and α in S_n. By *Exercise* 6, p. 113, it suffices to show that this formula holds for a set of generators of S_n. In *Exercise* 9, p. 12, it was shown that transpositions of the form $\alpha = (k\,k+1)$ generate S_n. Therefore, it suffices to show that $\mu({}_\alpha A) = \mathrm{sgn}(\alpha)\mu(A)$ holds for $\alpha = (k\,k+1)$ and each $1 \leq k \leq n-1$.

Assume that $\alpha = (k\,k+1)$. Then $\operatorname{sgn}(\alpha) = -1$ and it must be shown that $\mu({}_\alpha A) = -\mu(A)$. The analysis depends on breaking the terms in the formula for $\mu({}_\alpha A)$ into three sets, $i < k$, $i = k$ or $k+1$, and $i > k+1$. The first and third are the easiest and will be handled first. Before starting, it is necessary to be absolutely clear about the notation. Here ${}_\alpha A(i,j)$ denotes the (i,j) entry of ${}_\alpha A$ and ${}_\alpha A(i|j)$ denotes the $n-1 \times n-1$ matrix obtained by deleting the ith row and j^{th} column from ${}_\alpha A$.

When $i < k$ or when $i > k+1$, the matrix ${}_\alpha A(i|j)$ is obtained from $A(i|j)$ by switching two adjacent rows in $A(i|j)$ and $\operatorname{Det}[{}_\alpha A(i|j)] = -\operatorname{Det}[A(i|j)]$. Moreover, ${}_\alpha A(i,j) = A(i,j)$

When $i = k$, it follows that ${}_\alpha A(k,j) = A(k+1,j)$ and ${}_\alpha A(k|j) = A(k+1|j)$. Similarly, for $i = k+1$ we have ${}_\alpha A(k+1,j) = A(k,j)$ and ${}_\alpha A(k+1|j) = A(k|j)$. The sum of the k and $k+1$ terms in the formula for $\mu({}_\alpha A)$ can now be simplified as follows:

$$\begin{aligned}
{}_\alpha A(k,j)(-1)^{k+j}\operatorname{Det}[{}_\alpha A(k|j)] + {}_\alpha A(k+1,j)(-1)^{k+1+j}\operatorname{Det}[{}_\alpha A(k+1|j)] &= \\
A(k+1,j)(-1)^{k+j}\operatorname{Det}[A(k+1|j)] + A(k,j)(-1)^{k+1+j}\operatorname{Det}[A(k|j)] &= \\
-A(k+1,j)(-1)^{k+1+j}\operatorname{Det}[A(k+1|j)] - A(k,j)(-1)^{k+j}\operatorname{Det}[A(k|j)]&.
\end{aligned}$$

The critical observation is that the last line is just the negative of the sum of the $k+1$ and k terms in the formula for $\mu(A)$.

Substituting these calculations into the formula for $\mu({}_\alpha A)$ yields (with the first summation absent if $k = 1$ and the second summation absent if $k = n-1$)

$$\begin{aligned}
\mu({}_\alpha A) &= \sum_{i=1}^{k-1} -A(i,j)(-1)^{i+j}\operatorname{Det}[A(i|j)] \\
&- A(k+1,j)(-1)^{k+1+j}\operatorname{Det}[A(k+1|j)] \\
&- A(k,j)(-1)^{k+j}\operatorname{Det}[A(k|j)] \\
&+ \sum_{i=k+2}^{n} -A(i,j)(-1)^{i+j}\operatorname{Det}[A(i|j)] \\
&= -\sum_{i=1}^{n} A(i,j)(-1)^{i+j}\operatorname{Det}[A(i|j)] \\
&= -\mu(A).
\end{aligned}$$

As a result of the first two steps of the proof, *Theorem 2.5.7* applies to the function μ and $\mu(A) = \mu(I)\operatorname{Det}[A]$.

The third and easiest step is show that $\mu(I) = 1$, and hence that $\mu(A) = \operatorname{Det}[A]$. Clearly, $I(i|i)$ is just the $n-1 \times n-1$ identity matrix, and $I(i,j) = 0$ when $i \neq j$. Consequently,

$$\mu(I) = \sum_{i=1}^{n} I(i,j)(-1)^{i+j}\operatorname{Det}[I(i|j)] = I(j,j)(-1)^{2j}\operatorname{Det}[I(j|j)] = 1$$

and the proof is finished. □

Equation (2.38) provides a useful calculation tool for some determinants, but that is not its importance here. Rather it leads to an explicit formula for A^{-1} when A is invertible, and to the continuity of the function $A \to A^{-1}$.

Theorem 2.5.13 *If A is an invertible $n \times n$ matrix, then*

$$A^{-1}(j,i) = \frac{(-1)^{i+j}\mathrm{Det}\,[A(i|j)]}{\mathrm{Det}\,[A]} \tag{2.39}$$

Proof. The cofactors $(-1)^{i+j}\mathrm{Det}\,[A(i|j)]$ of an $n \times n$ matrix A form another $n \times n$ matrix. Since A is invertible, $\mathrm{Det}\,[A] \neq 0$ by *Theorem 2.5.9*, and we can multiply the matrix of cofactors by $1/\mathrm{Det}\,[A]$. Denote this matrix by B. According to the statement of the theorem, B is not the inverse of A but the transpose of the inverse. Thus to prove the theorem it must be shown that for every k

$$\sum_{i=1}^{n} B^{t}(j,i)A(i,k) = \sum_{i=1}^{n} B(i,j)A(i,k) = \begin{cases} 1 & \text{for } j = k \\ 0 & \text{for } j \neq k \end{cases}$$

or

$$\sum_{i=1}^{n} \frac{(-1)^{i+j}\mathrm{Det}\,[A(i|j)]}{\mathrm{Det}\,[A]} A(i,k) = \begin{cases} 1 & \text{for } j = k \\ 0 & \text{for } j \neq k \end{cases}.$$

The two cases will be handled separately.

When $j = k$,

$$\begin{aligned} \sum_{i=1}^{n} \frac{(-1)^{i+j}\mathrm{Det}\,[A(i|j)]}{\mathrm{Det}\,[A]} A(i,j) &= \\ \frac{1}{\mathrm{Det}\,[A]} \sum_{i=1}^{n} A(i,j)(-1)^{i+j}\mathrm{Det}\,[A(i|j)] &= \\ \frac{1}{\mathrm{Det}\,[A]}\mathrm{Det}\,[A] &= 1 \end{aligned}$$

by equation (2.38).

When $j \neq k$, let C denote the matrix obtained from A by replacing its jth column with a copy of the kth column. So the jth and k^{th} columns of C are equal and $\mathrm{Det}\,[C] = 0$ by *Corollary 2.5.3*. Moreover, $C(i|j) = A(i|j)$ for $i = 1, \ldots, n$. Now it follows from *Theorem 2.5.12* that

$$\begin{aligned} 0 &= \mathrm{Det}\,[C] \\ &= \sum_{i=1}^{n} C(i,j)(-1)^{i+j}\mathrm{Det}\,[C(i|j)] \\ &= \sum_{i=1}^{n} A(i,k)(-1)^{i+j}\mathrm{Det}\,[A(i|j)] \end{aligned}$$

It follows that

$$\sum_{i=1}^{n} \frac{(-1)^{i+j}\mathrm{Det}\,[A(i|j)]}{\mathrm{Det}\,[A]} A(i,k) = 0$$

when $j \neq k$, which completes the proof. $\square$

The groups of units for the matrix algebras $\mathcal{M}_n(\mathbb{R})$ and $\mathcal{M}_n(\mathbb{C})$ are called the *real and complex general linear groups*, respectively. They are denoted by $\mathrm{GL}(n, \mathbb{R})$ and $\mathrm{GL}(n, \mathbb{C})$, respectively. The units of the matrix algebra $\mathcal{M}_n(\mathbb{S})$ are the invertible matrices. It was shown in *Theorem 2.5.9* that an $n \times n$ matrix A is invertible if and only if $\mathrm{Det}\,[A] \neq 0$. Hence,

$$\mathrm{GL}(n, \mathbb{S}) = \mathrm{Det}^{-1}[\{c \in \mathbb{S} : c \neq 0\}].$$

Since $\{x \in \mathbb{R} : x \neq 0\}$ and $\{z \in \mathbb{C} : z \neq 0\}$ are open subsets of $\mathbb{R}$ and $\mathbb{C}$, respectively, the general linear groups $\mathrm{GL}(n, \mathbb{R})$ and $\mathrm{GL}(n, \mathbb{C})$ are open subsets of $\mathcal{M}_n(\mathbb{R})$ and $\mathcal{M}_n(\mathbb{C})$, respectively, because the determinant function is continuous..

Theorem 2.5.14 *The general linear groups* $\mathrm{GL}(n, \mathbb{R})$ *and* $\mathrm{GL}(n, \mathbb{C})$ *are locally compact* σ*-compact metric groups and the determinant is a homomorphism of* $\mathrm{GL}(n, \mathbb{R})$ *and* $\mathrm{GL}(n, \mathbb{C})$ *onto the metric (multiplicative) groups* $\mathbb{R} \setminus \{\mathbf{0}\}$ *and* $\mathbb{C} \setminus \{\mathbf{0}\}$*, respectively.*

Proof. There is no need to distinguish between $\mathrm{GL}(n, \mathbb{R})$ and $\mathrm{GL}(n, \mathbb{C})$ in the proof. The metric on $\mathrm{GL}(n, \mathbb{S})$ is, of course, the restriction of $d(A, B) = \|A - B\|_{\mathcal{B}}$ to $\mathrm{GL}(n, \mathbb{S})$, where $\|\cdot\|_{\mathcal{B}}$ is defined by equation (2.28).

Because $\mathcal{M}_n(\mathbb{S})$ is a normed algebra, the multiplication of matrices is a continuous function from $\mathcal{M}_n(\mathbb{S}) \times \mathcal{M}_n(\mathbb{S})$ to $\mathcal{M}_n(\mathbb{S})$ by *Proposition 2.4.6*. It follows that the restriction of matrix multiplication to $\mathrm{GL}(n, \mathbb{S}) \times \mathrm{GL}(n, \mathbb{S})$ is a continuous function, and its range is $\mathrm{GL}(n, \mathbb{S})$ because $\mathrm{GL}(n, \mathbb{S})$ is a group.

Now, it must be shown that the function $A \mapsto A^{-1}$ is continuous on $\mathrm{GL}(n, \mathbb{S})$. *Proposition 2.5.1* shows that the determinant function is continuous on $\mathcal{M}_n(\mathbb{S})$, and the same proof shows that the cofactors are also continuous. Since the determinant function is zero-free on $\mathrm{GL}(n, \mathbb{S})$, the function $1/\mathrm{Det}\,[\cdot]$ is continuous on $\mathrm{GL}(n, \mathbb{S})$ and the scalar-valued functions

$$A^{-1}(i, j) = \frac{(-1)^{i+j}\mathrm{Det}\,[A(j|i)]}{\mathrm{Det}\,[A]}.$$

are continuous for all i and j.

Recall that the norm topology on $\mathcal{M}_n(\mathbb{S})$ is just the product topology on $\mathbb{S}^{n^2}$. It has just been shown that the coordinate functions of the function $A \mapsto A^{-1}$ from $\mathrm{GL}(n, \mathbb{S})$ to $\mathcal{M}_n(\mathbb{S})$ are continuous. Therefore, *Proposition 1.2.17* applies and the function $A \mapsto A^{-1}$ defined by equation (2.39) is a continuous function from $\mathrm{GL}(n, \mathbb{S})$ into $\mathcal{M}_n(\mathbb{S})$. Obviously, A^{-1} is in $\mathrm{GL}(n, \mathbb{S})$ when A is in $\mathrm{GL}(n, \mathbb{S})$. In summary, $\mathrm{GL}(n, \mathbb{S})$ is a metric group.

The groups $\mathrm{GL}(n, \mathbb{S})$ are locally compact and σ-compact because they are open subsets of $\mathbb{R}^{n^2}$ or $\mathbb{R}^{(2n)^2}$, so *Exercises 5 and 13*, p. 61, apply.

Propositions 2.5.1 and 2.5.8 imply that the determinant is a homomorphism of metric groups. It is easy to check that it is onto by varying the (1,1) entry of the identity matrix I. $\square$

The *special linear groups* are defined by

$$\mathrm{SL}(n,\mathbb{S}) = \{A \in \mathrm{GL}(n,\mathbb{S}) : \mathrm{Det}\,[A] = 1\}$$

and are the kernels of the determinant homomorphisms from $\mathrm{GL}(n,\mathbb{S})$ to $\mathbb{S}\backslash\{0\}$.

Corollary 2.5.15 *The group* $\mathrm{SL}(n,\mathbb{S})$ *is a closed normal subgroup of* $\mathrm{GL}(n,\mathbb{S})$ *and* $\mathrm{GL}(n,\mathbb{S})/\mathrm{SL}(n,\mathbb{S})$ *is isomorphic to* $\mathbb{S}\setminus\{0\}$.

Proof. It follows from the theorem that *Corollary 1.5.19* applies. □

The closed subgroups of the general linear groups $\mathrm{GL}(n,\mathbb{R})$ and $\mathrm{GL}(n,\mathbb{C})$ are generally referred to as matrix groups. They are a surprisingly large class of metric groups. For example, there are matrix groups isomorphic to the permutation groups S_n, and others isomorphic to the group of rigid motions of Euclidean geometry. The study of matrix groups will resume in Chapter 4, but first we want to determine all the closed subgroups of $\mathbb{R}^n$. Since linear subspaces of $\mathbb{R}^n$ are closed subgroups of $\mathbb{R}^n$, it is inevitable that the theory of linear spaces including determinants plays a role in understanding the subgroups of $\mathbb{R}^n$.

EXERCISES

1. An element of S_n, say β, can also be used to permute the columns of an $n \times n$ matrix by setting ${}^{\beta}A(i,j) = A(i,\beta(j))$. Show that $\mathrm{Det}\,[{}^{\beta}A] = \mathrm{sgn}(\beta)\mathrm{Det}\,[A]$.

2. Let α be a permutation in S_n. Prove that if $\alpha \neq \iota$, then there exists j such that $\alpha(j) < j$.

3. Let A be an $n\times n$ matrix such that $A(j,k) = 0$ when $k < j$. These matrices are called *triangular matrices.* Use *Exercise 2* above to show that

$$\mathrm{Det}\,[A] = \prod_{j=1}^{n} A(j,j).$$

4. Let A be an $n \times n$ matrix and let i be an integer such that $1 \leq i \leq n$. Show that if c_j for $1 \leq j \leq n$ and $j \neq i$ are scalars, then

$$\mathrm{Det}\,[A_1,\ldots,A_{i-1},\underbrace{A_i + \sum_{j\neq i} c_j A_j}_{i\text{th row}},A_{i+1},\ldots,A_n] = \mathrm{Det}\,[A]. \tag{2.40}$$

5. Show that $\mathrm{Det}\,[A] = \sum_{j=1}^{n} A(i,j)(-1)^{i+j}\mathrm{Det}\,[A(i|j)]$.

6. Let μ be an m-multilinear function. Show that μ is alternating if equation (2.35) holds for $\alpha_1,\ldots,\alpha_p$, a set of *generators* for the group S_m that is, the smallest subgroup of S_n containing $\alpha_1,\ldots,\alpha_p$ is S_n.

7. Let μ be an m-multilinear function. Prove that the following are equivalent:

   (a) μ is alternating.

   (b) $\mu(\mathbf{x}_1, \ldots, \mathbf{x}_m) = 0$, when $\mathbf{x}_i = \mathbf{x}_j$ and $i \neq j$.

   (c) Equation (2.35) holds for all transpositions $\alpha = (ij)$.

8. Show that $A \to (A^*)^{-1}$ is an isomorphism of $\mathrm{GL}(n, \mathbb{S})$ onto itself.

9. Let A and B be $n \times n$ matrices with B invertible. Show that $\mathrm{Det}\,[A] = \mathrm{Det}\,[BAB^{-1}]$ and that A and BAB^{-1} have the same characteristic values.

10. Prove that the following statements are equivalent for an $n \times n$ matrix A:

    (a) $\mathrm{Det}\,[A] \neq 0$.

    (b) The rows of A are linearly independent.

    (c) The rows of A are a basis for $\mathbb{S}^n$.

    (d) The columns of A are linearly independent.

    (e) The columns of A are a basis for $\mathbb{S}^n$.

11. Let V be a normed linear space, and let $\mathcal{B}(V, V)$ be the normed algebra with identity of bounded linear functions of V into itself. Prove that if $\mathrm{Dim}\,[V] = n < \infty$, then the group of units of $\mathcal{B}(V, V)$ is a metric group isomorphic to $\mathrm{GL}(n, \mathbb{S})$.

12. Let V_m denote the linear space of all m-multilinear functions

$$\mu : \underbrace{\mathbb{S}^n \times \ldots \times \mathbb{S}^n}_{m \text{ times}} \to \mathbb{S}.$$

Prove that $\mathrm{Dim}\,[V_m] = n^m$. Let W_m be the subspace of alternating m-multilinear functions in V_m. Prove that $W_m = \{\mathbf{0}\}$ when $m > n$ and that

$$\mathrm{Dim}\,[W_m] = \binom{n}{m} = \frac{n!}{m!(n-m)!}$$

for $1 \leq m \leq n$. (One approach to the second part is to first construct an alternating m-multilinear function for a given subset of $\{\mathbf{e}_1, \ldots, \mathbf{e}_n\}$ containing m elements.)

Chapter 3

The Subgroups of $\mathbb{R}^n$

The only closed proper subgroups of $\mathbb{R}$ are the infinite cyclic subgroups generated by positive real numbers (*Theorem 1.4.17*). The quotient group of $\mathbb{R}$ by an infinite cyclic subgroup is always the circle group $\mathbb{K}$ (*Corollary 1.4.18*). Furthermore, the closed subgroups of $\mathbb{K}$ other than $\mathbb{K}$ itself are finite cyclic groups, and their quotients are again $\mathbb{K}$ (*Theorem 1.4.20*). The goal of this chapter is to determine the comparable results for $\mathbb{R}^n$.

We cannot, however, expect the results to be as clean and simple for $\mathbb{R}^n$ as they are for $\mathbb{R}$. There are two primary reasons for the increased complexity of the subgroups of $\mathbb{R}^n$ when $n > 1$. First, subspaces of $\mathbb{R}^n$ are closed subgroups of $\mathbb{R}^n$ (*Proposition 2.4.10*). The only subspaces of $\mathbb{R}$ are $\{0\}$ and $\mathbb{R}$, while $\mathbb{R}^2$ has infinitely many subspaces of dimension 1. Second, there are more degrees of freedom when $n > 1$. For example, there is room in $\mathbb{R}^2$ for infinite discrete subgroups with two linearly independent generators like $\mathbb{Z}^2$ instead of just one as in $\mathbb{R}$.

With discrete subgroups and subspaces occurring in $\mathbb{R}^n$, closed subgroups that are amalgamations of discrete subgroups and subspaces occur. For example, a subgroup of $\mathbb{R}^2$ can be constructed from a one-dimensional subspace or line through the origin in one direction and an infinite cycle subgroup in another direction. The result is a closed subgroup of $\mathbb{R}^2$ consisting of a sequence of equidistant parallel lines. Such a closed subgroup cannot occur in $\mathbb{R}$. The higher dimensional versions of such closed groups are the limit of what can occur in $\mathbb{R}^n$, but not in $\mathbb{R}$.

The key step in analyzing the closed subgroups of $\mathbb{R}^n$ in Section 3.1 is to determine the structure of the discrete subgroups of $\mathbb{R}^n$. Once this is accomplished it is possible to completely describe all the closed subgroups of $\mathbb{R}^n$. It turns out that every closed subgroup of $\mathbb{R}^n$ for fixed n is isomorphic to one of only a finite number of metric groups. The analysis necessary to prove these results will make essential use of the material in Chapter 2.

In Section 3.2, the structure of the closed subgroups of $\mathbb{R}^n$ is applied to related questions. It is used to determine the structure of the metric groups $\mathbb{R}^n/H$ where H is a closed subgroup of $\mathbb{R}^n$ and then to describe the closed subgroups of $\mathbb{K}^n$ and their quotients. The last topic in this section is a complete description of the homomorphisms of both $\mathbb{R}^m$ and $\mathbb{K}^m$ to $\mathbb{K}^n$, including the group of automorphisms of $\mathbb{K}^n$.

Section 3.3 is devoted to dense subgroups of $\mathbb{R}^n$ and $\mathbb{K}^n$. The main result is Kronecker's classical criterion for a finitely-generated subgroup of $\mathbb{K}^n$ to be dense in $\mathbb{K}^n$. One consequence of this theorem is that $\mathbb{K}^n$, like $\mathbb{K}$, contains dense infinite cyclic subgroups.

3.1 Closed Subgroups

If H is a closed subgroup of $\mathbb{R}$, then its structure is known. There are three possibilities. They are $H = \mathbb{R}$, $H = \{0\}$, or $H = \{ma : m \in \mathbb{Z}\}$ for some positive real number a. In the latter case, H is isomorphic to $\mathbb{Z}$. This section addresses the similar but more difficult question: If H is a closed subgroup of $\mathbb{R}^n$, what is its structure?

Subspaces of $\mathbb{R}^n$ are closed sets of $\mathbb{R}^n$ (*Proposition 2.4.10*). Since subspaces are subgroups of a linear space that are closed under scalar multiplication, the subspaces of $\mathbb{R}^n$ are a large family of closed subgroups of $\mathbb{R}^n$, when $n > 1$. There are other subgroups such as

$$H = \{(p_1, \ldots, p_m, 0, \ldots, 0) : p_j \in \mathbb{Z} \text{ for } j = 1, \ldots, m\}$$

where m is an integer such that $1 \leq m \leq n$. Or amalgamations of these two types like the following closed subgroup of $\mathbb{R}^2$:

$$H = \{(x, p) : x \in \mathbb{R} \text{ and } p \in \mathbb{Z}\}.$$

In this section, it will be shown that all the closed subgroups of $\mathbb{R}^n$ fit these three models. The key to establishing this result is an analysis of the basic dichotomy between discrete and non-discrete closed subgroups. Recall that discrete subgroups of a metric groups are closed by *Proposition 1.4.1*. Also, a subgroup H of $\mathbb{R}^n$ is discrete if and only if $B_r(\mathbf{0}) \cap H = \{\mathbf{0}\}$ for some $r > 0$. The analysis of the closed subgroups of $\mathbb{R}^n$ begins with the non-discrete ones.

Proposition 3.1.1 *If H is a closed subgroup of $\mathbb{R}^n$, then there exists a largest subspace L_H contained in H.*

Proof. Set

$$L_H = \{\mathbf{x} \in H : c\mathbf{x} \in H \text{ for all } c \in \mathbb{R}\}.$$

Obviously, $\mathbf{0}$ is in L_H and L_H is a subset H. Every subspace of $\mathbb{R}^n$ that is contained in H is clearly contained in L_H. If $\mathbf{x}$ and $\mathbf{y}$ are in L_H, then $c(\mathbf{x} - \mathbf{y}) = c\mathbf{x} - c\mathbf{y}$ is in H for all $c \in \mathbb{R}$ and $\mathbf{x} - \mathbf{y}$ is in L_H. Thus L_H is a subgroup of $\mathbb{R}^n$ and a subspace by definition. □

Of course, L_H can consist of just the zero vector $\mathbf{0}$ as is the case when $H = \mathbb{Z}^n$, which is a discrete subgroup of $\mathbb{R}^n$. As soon as H is not discrete, however, L_H is at least a one-dimensional subspace.

Theorem 3.1.2 *Every non-discrete closed subgroup of* $\mathbb{R}^n$ *contains a line passing through the origin.*

Proof. Because H is not discrete, there exists a sequence $\mathbf{x}_k$ in H converging to $\mathbf{0}$ such that $\mathbf{x}_k \neq \mathbf{0}$ for all k. Let

$$R = \{\mathbf{x} = (x_1, \ldots, x_n) : |x_j| < 1 \text{ for; } j = 1, \ldots, n\}.$$

Note that R is a bounded open set containing $\mathbf{0}$ and its closure is compact. Without loss of generality, $\mathbf{x}_k \in R$ for all k. Let q_k denote the largest positive integer such that $q_k\mathbf{x}_k \in R^-$. Since R^- is compact, by *Proposition 1.5.9* we can assume by passing to a subsequence that $q_k\mathbf{x}_k$ converges to $\mathbf{a} \in R^-$. Note that $\mathbf{a}$ is in H because $q_k\mathbf{x}_k \in H$ and H is closed.

Let $\varepsilon > 0$. There exists $N > 0$ such that $\|\mathbf{x}_k\| < \varepsilon$ and $\|q_k\mathbf{x}_k - \mathbf{a}\| < \varepsilon$ when $k > N$. Then, by the triangle inequality

$$\|(q_k + 1)\mathbf{x}_k - \mathbf{a}\| \leq \|q_k\mathbf{x}_k - \mathbf{a}\| + \|\mathbf{x}_k\| \leq 2\varepsilon$$

when $k > N$. Thus $(q_k + 1)\mathbf{x}_k$ also converges to $\mathbf{a}$. By the choice of q_k, we know that $(q_k + 1)\mathbf{x}_k$ is in the closed set $\mathbb{R}^n \setminus R$ and hence $\mathbf{a}$ is in $\mathbb{R}^n \setminus R$. In particular, $\mathbf{a} \neq \mathbf{0}$.

To complete the proof it suffices to show that $c\,\mathbf{a} \in H$ for all $c \in \mathbb{R}$. Recall that the greatest integer function $[x]$ equals the largest integer less than or equal to x and that $0 \leq x - [x] < 1$. Using the triangle inequality and $\big|[cq_k] - cq_k\big| = cq_k - [cq_k] < 1$,

$$\begin{aligned} \|[cq_k]\mathbf{x}_k - c\,\mathbf{a}\| &\leq \|[cq_k]\mathbf{x}_k - cq_k\mathbf{x}_k\| + \|cq_k\mathbf{x}_k - c\,\mathbf{a}\| \\ &\leq |[cq_k] - cq_k|\,\|\mathbf{x}_k\| + |c|\,\|q_k\mathbf{x}_k - \mathbf{a}\| \\ &\leq \|\mathbf{x}_k\| + |c|\,\|q_k\mathbf{x}_k - \mathbf{a}\|. \end{aligned}$$

Clearly, $\|\mathbf{x}_k\| + |c|\,\|q_k\mathbf{x}_k - \mathbf{a}\|$ goes to 0 as k goes to infinity, and hence the sequence $[cq_k]\mathbf{x}_k$ converges to $c\,\mathbf{a}$. Since $[cq_k]\mathbf{x}_k$ is in H for all k and H is closed, $c\,\mathbf{a}$ is in H. □

As a consequence of this theorem, $\text{Dim}\,[L_H] > 0$, if H is not discrete. Obviously, a closed subgroup H of $\mathbb{R}^n$ is not discrete, if $\text{Dim}\,[L_H] > 0$. Hence, we have the following corollary to the theorem:

Corollary 3.1.3 *A closed subgroup* H *of* $\mathbb{R}^n$ *is a discrete subgroup of* $\mathbb{R}^n$ *if and only if* $\text{Dim}\,[L_H] = 0$.

The span of a finite set of vectors was defined on page 65. That definition can be extended to an arbitrary subset E of a linear space V by allowing sums of any finite length. Specifically, set

$$\text{Span}\,[E] = \{c_1\mathbf{v}_1 + \ldots + c_k\mathbf{v}_k : k \in \mathbb{Z}^+, c_j \in \mathbb{R}, \text{ and } \mathbf{v}_j \in E \text{ for } 1 \leq j \leq k\}.$$

and check that it is a subspace of V containing E. In fact, it is the smallest subspace of V containing E.

The *rank* of a nonempty subset E of $\mathbb{R}^n$ is by definition the dimension of Span $[E]$. So the rank of E is an integer r satisfying $0 \leq r \leq n$. When H is a closed subgroup of $\mathbb{R}^n$, the rank of H is a useful tool for studying H.

Let H be a closed subgroup of $\mathbb{R}^n$ of rank r. If $r = 0$, then $H = \{\mathbf{0}\}$. Let $p = \text{Dim}\,[L_H]$ and set $q = r - p$. Obviously, $L_H \subset H \subset \text{Span}\,[H]$, and hence $0 \leq p \leq r$ and $q \geq 0$. These three numbers turn out to be complete descriptors of the structure of H as will be shown in the remainder of this section.

If $q = 0$, then $r = p$ and there are two possibilities. If $r = p = 0$, then Span $[H] = \{\mathbf{0}\}$ and $H = \{\mathbf{0}\} = L_H$. This is the trivial case. If $r = p > 0$, then *Corollary 2.1.5* implies that $L_H = \text{Span}\,[H]$ and $H = L_H$, a subspace of $\mathbb{R}^n$. Therefore, when $r = p > 0$, H is isomorphic to $\mathbb{R}^r$ by *Theorem 2.3.9*.

Proposition 3.1.4 *If H is a closed subgroup of $\mathbb{R}^n$ of rank $r > 0$, then there a basis $\mathbf{v}_1, \ldots, \mathbf{v}_n$ of $\mathbb{R}^n$ with the following properties:*

(a) $\mathbf{v}_1, \ldots, \mathbf{v}_p$ is a basis for L_H, when $p > 0$.

(b) $\mathbf{v}_k$ is in H for $1 \leq k \leq r$ and $\mathbf{v}_1, \ldots, \mathbf{v}_r$ is a basis for Span $[H]$.

(c) $H \cap \text{Span}\,[\mathbf{v}_{p+1}, \ldots, \mathbf{v}_r]$ is an infinite discrete subgroup of $\mathbb{R}^n$, when $q > 0$.

Proof. If $L_H = H$, then $p = r > 0$ and $q = 0$. Let $\mathbf{v}_1, \ldots, \mathbf{v}_p$ be a basis for L_H and use *Proposition 2.1.7* to extend $\mathbf{v}_1, \ldots, \mathbf{v}_p$ to a basis for $\mathbb{R}^n$.

Suppose $H \neq L_H$ and let $\mathbf{v}_1, \ldots, \mathbf{v}_p$ be a basis for L_H, when $p > 0$. There exists $\mathbf{v}_{p+1}$, an element of H, that is not in L_H. (This includes the case when $p = 0$ because $r > 0$ by hypothesis.) Obviously, $\mathbf{v}_{p+1} \neq \mathbf{0}$. Then as in the proof of *Proposition 2.1.4*, $\mathbf{v}_1, \ldots, \mathbf{v}_p, \mathbf{v}_{p+1}$ is a linearly independent set of vectors. Either Span $[\mathbf{v}_1, \ldots, \mathbf{v}_p, \mathbf{v}_{p+1}] = \text{Span}\,[H]$ and $r = p+1$ or there exists $\mathbf{v}_{p+2} \in H$ such that $\mathbf{v}_{p+2} \notin \text{Span}\,[\mathbf{v}_1, \ldots, \mathbf{v}_p, \mathbf{v}_{p+1}]$ and the argument can be repeated. But it can be used at most $n - p$ times. Hence, there exist $\mathbf{v}_1, \ldots, \mathbf{v}_r$ in H such that $\mathbf{v}_1, \ldots, \mathbf{v}_r$ is a basis for Span $[H]$. Now another application of *Proposition 2.1.7* completes the proof of parts *(a)* and *(b)*.

Since Span $[\mathbf{v}_{p+1}, \ldots, \mathbf{v}_r]$ is a closed subgroup of $\mathbb{R}^n$, it follows that $H' = H \cap \text{Span}\,[\mathbf{v}_{p+1}, \ldots, \mathbf{v}_r]$ is a closed subgroup of $\mathbb{R}^n$ contained in H. Clearly, $L_H \cap H' = \{\mathbf{0}\}$ by parts *(a)* and *(b)*. It is equally clear that $L_{H'} \subset L_H$ because $H' \subset H$. It follows that $L_{H'} = \{\mathbf{0}\}$ and H' is discrete by *Corollary 3.1.3*.

If $q > 0$, then $\mathbf{v}_r$ is a nonzero vector in H'. Hence, $\{k\mathbf{v}_r : k \in \mathbb{Z}\}$ is an infinite cyclic subgroup of H' and H' is infinite. □

Although we have encountered subgroups generated by subsets of a group, we have not introduced any notation for the smallest subgroup containing a specified subset of a group. To do so let G be a group and let S be a subset of G. The smallest subgroup of G containing S will be denoted by $\langle S \rangle$. A group G is *finitely generated* if there exists a finite subset set S of G such that $G = \langle S \rangle$. For example, $\mathbb{Z}^n = \langle \mathbf{e}_1, \ldots, \mathbf{e}_n \rangle$ is a finitely generated group that contains infinitely many elements. When G is an Abelian group, *Exercise* 4, p. 69, shows that every finitely generated subgroup of G has the form

$$\langle g_1, \ldots, g_m \rangle = \left\{ \sum_{j=1}^{m} k_j g_j : k_j \in \mathbb{Z} \text{ for } 1 \le j \le m \right\}.$$

The next task is to prove the existence of a finite set generators for an arbitrary discrete subgroup H of $\mathbb{R}^n$.

Proposition 3.1.5 *Let H be a discrete subgroup of $\mathbb{R}^m$ of rank r, let $\mathbf{a}_1, \ldots, \mathbf{a}_r$ be linearly independent vectors in H, and let*

$$P = \left\{ \sum_{j=1}^{r} c_j \mathbf{a}_j : 0 \le c_j \le 1 \text{ for } 1 \le j \le r \right\}. \tag{3.1}$$

Then $H \cap P$ is finite and $\langle H \cap P \rangle = H$. Moreover, every point of H can be expressed as a unique linear combination of $\mathbf{a}_1, \ldots, \mathbf{a}_r$ with rational coefficients.

Proof. Observe that $H \cap P$ is compact. Because H is discrete, there exists $\varepsilon > 0$ such that $B_\varepsilon(\mathbf{0}) \cap H = \{\mathbf{0}\}$. Using the invariance of the metric, it follows that $B_\varepsilon(\mathbf{b}) \cap H = \{\mathbf{b}\}$ for all $\mathbf{b} \in H$. Since the sets $B_\varepsilon(\mathbf{b})$ such that $\mathbf{b} \in H$ are an open cover of H, a finite number of them $B_\varepsilon(\mathbf{b}_1), \ldots, B_\varepsilon(\mathbf{b}_k)$ cover $H \cap P$. It follows that

$$H \cap P \subset H \cap \left(\bigcup_{j=1}^{k} B_\varepsilon(\mathbf{b}_j) \right) = \{\mathbf{b}_1, \ldots, \mathbf{b}_k\}$$

and hence $H \cap P$ is finite.

Since the rank of H is r, the vectors $\mathbf{a}_1, \ldots, \mathbf{a}_r$ are a basis for Span $[H]$. Let $\mathbf{x}$ be an arbitrary element of H. Then, $\mathbf{x} = \sum_{j=1}^{r} c_j \mathbf{a}_j$. Construct the sequence

$$\begin{aligned} \mathbf{x}_k &= k\mathbf{x} - \sum_{j=1}^{r} [kc_j]\, \mathbf{a}_j \\ &= \sum_{j=1}^{r} (kc_j - [kc_j])\mathbf{a}_j. \end{aligned}$$

Clearly, $\mathbf{x}_k \in H$ and $0 \le kc_j - [kc_j] < 1$. Hence, $\mathbf{x}_k$ is in $H \cap P$ for all k. Because $\mathbf{x}_1$ and $\mathbf{a}_j$, $1 \le j \le r$ are in $H \cap P$,

$$\mathbf{x} = \mathbf{x}_1 + \sum_{j=1}^{r} [c_j]\mathbf{a}_j \in \langle H \cap P \rangle.$$

and $\langle H \cap P \rangle = H$.

Because $H \cap P$ is finite, there exist integers k and k', such that $k \neq k'$ and $\mathbf{x}_k = \mathbf{x}_{k'}$. Consequently,

$$kc_j - [kc_j] = k'c_j - [k'c_j]$$

or

$$(k - k')c_j = [k'c_j] - [kc_j].$$

Since $k - k' \neq 0$, it follows that c_j is a rational number for $j = 1, \ldots, r$. These rational coefficients are unique because $\mathbf{a}_1, \ldots, \mathbf{a}_r$ are linearly independent. □

What we really want is unique integer coefficients not rational ones. The next step is to prove the existence of a better basis for Span $[H]$.

Theorem 3.1.6 *Let H be a discrete subgroup of $\mathbb{R}^n$ of rank r. Then there exist linearly independent vectors $\mathbf{b}_1, \ldots, \mathbf{b}_r$ in H such that*

$$H = \langle \mathbf{b}_1, \ldots, \mathbf{b}_r \rangle = \left\{ \sum_{j=1}^{r} m_j \mathbf{b}_j : m_j \in \mathbb{Z} \right\}.$$

Proof. By *Proposition 3.1.4* or *Exercise* 1, p. 124, there exists a basis $\mathbf{a}_1, \ldots, \mathbf{a}_r$ for Span $[H]$ such that $\mathbf{a}_j$ is in H for $j = 1, \ldots, r$. Let P be defined by (3.1). *Proposition 3.1.5* implies that $H \cap P$ is a finite set of generators for H and that every element in H can be expressed as a unique linear combination of $\mathbf{a}_1, \ldots, \mathbf{a}_r$ with rational coefficients.

Because $H \cap P$ is finite, expressing every element in $H \cap P$ as a unique linear combination of $\mathbf{a}_1, \ldots, \mathbf{a}_r$ with rational coefficients requires a only finite set of rational numbers and, in particular, a finite set of denominators. Let μ be the least common multiple of these denominators.

Let $\mathbf{x}$ be an arbitrary element of H. Then $\mathbf{x}$ can be written as $\mathbf{x} = \sum_{j=1}^{r} c_j \mathbf{a}_j$ with unique rational coefficients (*Proposition 3.1.5*). Set

$$\mathbf{x}_1 = \mathbf{x} - \sum_{j=1}^{r} [c_j] \mathbf{a}_j = \sum_{j=1}^{r} (c_j - [c_j]) \mathbf{a}_j.$$

As before, $\mathbf{x}_1$ is in $H \cap P$. There exist integers k_j such that

$$c_j - [c_j] = \frac{k_j}{\mu}$$

for $j = 1, \ldots, r$, and

$$c_j = \frac{k_j + [c_j]\mu}{\mu}.$$

Therefore, every element of H can be expressed as a unique linear combination of $\mathbf{a}_1, \ldots, \mathbf{a}_r$ with rational coefficients of the form k/μ, $k \in \mathbb{Z}$.

If $\mathbf{b}_1, \ldots, \mathbf{b}_r$ is any set of r linearly independent vectors in H, then $\mathbf{b}_1, \ldots, \mathbf{b}_r$ is another basis for Span $[H]$ and there exists a $r \times r$ matrix B such that

$$\mathbf{b}_j = \sum_{k=1}^{r} B(k, j)\mathbf{a}_k$$

and such that every $B(k, j)$ is a rational number of the form k/μ. Consequently, μB is a matrix of integers and $\text{Det}\,[\mu B] = \mu^r \text{Det}\,[B]$ is an integer. Applying *Proposition 2.2.7* to $V = \text{Span}\,[H]$, shows that B is invertible. Thus $|\text{Det}\,[\mu B]|$ is a positive integer, and $\mathbf{b}_1, \ldots, \mathbf{b}_r$ can be chosen to minimize $|\text{Det}\,[\mu B]|$. Therefore, there exists a set $\mathbf{b}_1, \ldots, \mathbf{b}_r$ of linearly independent vectors in H such that $|\text{Det}\,[\mu B]|$ is the minimal possible value among all sets of r linearly independent vectors in H. By replacing $\mathbf{b}_1$ with $-\mathbf{b}_1$ if necessary, we can assume that $\text{Det}\,[B] = k/\mu^r > 0$, $k \in \mathbb{Z}^+$.

To complete the proof, it suffices to show that

$$H = \langle \mathbf{b}_1, \ldots, \mathbf{b}_r \rangle = \left\{ \sum_{j=1}^{r} m_j \mathbf{b}_j : m_j \in \mathbb{Z} \right\}.$$

Let $\mathbf{x}$ be an arbitrary element of H. Then $\mathbf{x} = \sum_{j=1}^{r} c_j \mathbf{b}_j$. Set

$$\mathbf{y} = \mathbf{x} - \sum_{j=1}^{r} [c_j]\mathbf{b}_j = \sum_{j=1}^{r} (c_j - [c_j])\mathbf{b}_j = \sum_{j=1}^{r} d_j \mathbf{b}_j$$

with $0 \le d_j < 1$. Observe that $\mathbf{y}$ is in H, and it suffices to show that $d_j = 0$ for $j = 1, \ldots, r$.

Suppose some $0 < d_j < 1$. Without loss of generality $j = 1$ by reordering $\mathbf{b}_1, \ldots, \mathbf{b}_r$. Consider the vectors $\mathbf{y}, \mathbf{b}_2, \ldots, \mathbf{b}_r$. A routine calculation shows that they are linearly independent. Because they are in H, there exists a unique $r \times r$ invertible matrix A such that $\mathbf{y} = \sum_{k=1}^{r} A(k, 1)\mathbf{a}_k$ and $\mathbf{b}_j = \sum_{k=1}^{r} A(k, j)\mathbf{a}_k$ for $2 \le j \le r$. Set

$$D = \begin{bmatrix} d_1 & 0 & \cdots & 0 \\ d_2 & 1 & \cdots & 0 \\ \vdots & & & \vdots \\ d_r & 0 & \cdots & 1 \end{bmatrix},$$

and note that $\text{Det}\,[D] = \text{Det}\,[D^t] = d_1$ by *Exercise* 3, p. 113. Clearly, $\mathbf{y} = \sum_{j=1}^{r} D(j, 1)\mathbf{b}_j$ and $\mathbf{b}_i = \sum_{j=1}^{r} D(j, i)\mathbf{b}_j$. The uniqueness of A and another routine calculation shows that $A = BD$. Therefore, $\text{Det}\,[A] = \text{Det}\,[B]\text{Det}\,[D] = d_1 \text{Det}\,[B] < \text{Det}\,[B]$, contradicting choice of B and proving that $d_1 = 0$. □

Corollary 3.1.7 *If H is a discrete subgroup of $\mathbb{R}^n$ of rank r, then H is isomorphic to $\mathbb{Z}^r$.*

Proof. Let $\mathbf{b}_1, \ldots, \mathbf{b}_r$ be given by the theorem. Define $\varphi : \mathbb{Z}^r \to H$ by $\varphi(m_1, \ldots, m_r) = \sum_{j=1}^{r} m_j \mathbf{b}_j$. Then φ is obviously onto, and it is one-to-one because $\mathbf{b}_1, \ldots, \mathbf{b}_r$ are linearly independent. Thus φ is an algebraic isomorphism from $\mathbb{Z}^r$ onto H. It is continuous, because both topologies are discrete. □

Proposition 3.1.4 can now be refined and stated as the theorem that captures the essence of much of this section.

Theorem 3.1.8 *If H is a closed subgroup of $\mathbb{R}^n$ of rank $r > 0$ with $p = \operatorname{Dim}[L_H]$ and $q = r - p$, then there exists a basis $\mathbf{u}_1, \ldots, \mathbf{u}_n$ of $\mathbb{R}^n$ with the following properties:*

(a) The vectors $\mathbf{u}_k$ are in H for $1 \leq k \leq r$ and $\mathbf{u}_1, \ldots, \mathbf{u}_r$ is a basis for Span $[H]$.

(b) If $p > 0$, then $\mathbf{u}_1, \ldots, \mathbf{u}_p$ is a basis for L_H.

(c) If $q > 0$, then

$$H \cap \operatorname{Span}[\mathbf{u}_{p+1}, \ldots, \mathbf{u}_r] = \left\{ \sum_{j=p+1}^{r} m_j \mathbf{u}_j : m_j \in \mathbb{Z} \right\}.$$

Proof. Let $\mathbf{v}_1, \ldots, \mathbf{v}_n$ be a basis of $\mathbb{R}^n$ given by *Proposition 3.1.4*. It also follows from this result that $H' = H \cap \operatorname{Span}[\mathbf{v}_{p+1}, \ldots, \mathbf{v}_r]$ is a discrete subgroup of $\mathbb{R}^n$. The heart of the argument is applying *Theorem 3.1.6* to H'.

Assume $q > 0$. Obviously, $\operatorname{Span}[H'] \subset \operatorname{Span}[\mathbf{v}_{p+1}, \ldots, \mathbf{v}_r]$. Because the linearly independent vectors $\mathbf{v}_{p+1}, \ldots, \mathbf{v}_r$ are in H and hence in H', it follows that $\operatorname{Span}[\mathbf{v}_{p+1}, \ldots, \mathbf{v}_r] = \operatorname{Span}[H']$ and the rank of H' equals q. By *Theorem 3.1.6* there exist a set of linearly independent vectors $\mathbf{b}_1, \ldots, \mathbf{b}_q$ in H' such that

$$H' = \left\{ \sum_{j=1}^{r} m_j \mathbf{b}_j : m_j \in \mathbb{Z} \right\}.$$

Clearly,

$$\operatorname{Span}[\mathbf{b}_1, \ldots, \mathbf{b}_q] = \operatorname{Span}[H'] = \operatorname{Span}[\mathbf{v}_{p+1}, \ldots, \mathbf{v}_r].$$

and

$$H \cap \operatorname{Span}[\mathbf{b}_1, \ldots, \mathbf{b}_q] = H' = \left\{ \sum_{j=1}^{r} m_j \mathbf{b}_j : m_j \in \mathbb{Z} \right\}.$$

Set $\mathbf{u}_j = \mathbf{v}_j$ for $j = 1, \ldots, p$ and for $j = r+1, \ldots, n$. For $p + 1 \leq j \leq r$ set $\mathbf{u}_j = \mathbf{b}_{j-p}$. A routine calculation shows that $\mathbf{u}_1, \ldots, \mathbf{u}_n$ is a basis for $\mathbb{R}^n$ and the three properties follow from the construction of $\mathbf{u}_1, \ldots, \mathbf{u}_n$. □

Corollary 3.1.9 *Let H be a closed subgroup of $\mathbb{R}^n$ of rank $r > 0$ with $p = \operatorname{Dim}[L_H]$. There exists a basis $\mathbf{u}_1, \ldots, \mathbf{u}_n$ for $\mathbb{R}^n$ such that $\mathbf{x} = \sum_{j=1}^{n} c_j \mathbf{u}_j$ is in H if and only if c_j is in $\mathbb{Z}$ for $j = p+1, \ldots, r$ and $c_j = 0$ for $j > r$.*

Proof. . Let $\mathbf{u}_1, \ldots, \mathbf{u}_n$ be given by the theorem. If $\mathbf{x}$ is in $H \subset \mathrm{Span}\,[H]$, then $\mathbf{x} = \sum_{j=1}^{r} c_j \mathbf{u}_j$. Obviously, $\mathbf{y} = \sum_{j=1}^{p} c_j \mathbf{u}_j$ is in $L_H \subset H$. It follows that $\mathbf{x} - \mathbf{y} = \sum_{j=p+1}^{r} c_j \mathbf{u}_j$ is in $H \cap \mathrm{Span}\,[\mathbf{u}_{p+1}, \ldots, \mathbf{u}_r]$. Hence by this theorem c_j is an integer for $j = p+1, \ldots, r$ and $\mathbf{x}$ satisfies the required criteria.

Conversely, if $\mathbf{x} = \sum_{j=1}^{n} c_j \mathbf{u}_j$ with c_j is in $\mathbb{Z}$ for $j = p+1, \ldots, r$ and $c_j = 0$ for $j > r$, then $\mathbf{y} = \sum_{j=1}^{p} c_j \mathbf{u}_j$ is in $L_H \subset H$ and $\mathbf{w} = \sum_{j=p+1}^{r} c_j \mathbf{u}_j$ is in $H \cap \mathrm{Span}\,[\mathbf{u}_{p+1}, \ldots, \mathbf{u}_r]$ by part *(c)* of *Theorem 3.1.8* because $c_j \in \mathbb{Z}$ for $j = p+1, \ldots, r$. In particular, $\mathbf{y}$ and $\mathbf{w}$ are in H. Thus $\mathbf{x} = \mathbf{y} + \mathbf{w}$ is in H. □

Theorem 3.1.10 *If H is a closed subgroup of $\mathbb{R}^n$ of rank r, then H is isomorphic to $\mathbb{R}^p \times \mathbb{Z}^q$, where $p = \mathrm{Dim}\,[L_H]$ and $q = r - p$. In particular, H is isomorphic to $\mathbb{R}^p$ when $q = 0$, and H is isomorphic to $\mathbb{Z}^q$ when $p = 0$.*

Proof. Let $\mathbf{u}_1, \ldots, \mathbf{u}_n$ be a basis for $\mathbb{R}^n$ given by *Corollary 3.1.9*. Define $T : \mathbb{R}^r \to \mathrm{Span}\,[H]$ by $T(y_1, \ldots, y_r) = \sum_{j=1}^{r} y_j \mathbf{u}_j$. Clearly T is a one-to-one linear function of $\mathbb{R}^r$ onto $\mathrm{Span}\,[H]$. Let φ be the restriction of T to

$$\mathbb{R}^p \times \mathbb{Z}^q = \{(x_1, \ldots, x_p, n_1, \ldots, n_q) : x_j \in \mathbb{R} \text{ and } n_k \in \mathbb{Z}\}.$$

It follows from the properties of $\mathbf{u}_1, \ldots, \mathbf{u}_n$ and T that φ is an algebraic isomorphism of $\mathbb{R}^r \times \mathbb{Z}^q$ onto H.

Proposition 2.4.7 implies that T and its inverse are continuous. It follows that φ is continuous as is its inverse, which is the restriction of T^{-1} to H. Thus φ is the required isomorphism of the metric group $\mathbb{R}^r \times \mathbb{Z}^q$ onto H. □

Consequently, $\mathbb{R}^n$ has closed subgroups isomorphic to $\mathbb{R}^p \times \mathbb{Z}^q$ for all nonnegative integers such that $0 \le p + q \le n$. The extremes, $p = n$ and $p + q = 0$ correspond to the subgroups $\mathbb{R}^n$ and $\{\mathbf{0}\}$. When $p < n$ and $p + q > 0$, there are infinitely many distinct closed subgroups of $\mathbb{R}^n$ isomorphic to $\mathbb{R}^p \times \mathbb{Z}^q$.

Although all the results in this section are stated and proved for $\mathbb{R}^n$, they are valid for any finite-dimensional linear space over $\mathbb{R}$. To see why, let V be a linear space over $\mathbb{R}$ with $\mathrm{Dim}\,[V] = n$. Then there exists a linear isomorphism T of V onto $\mathbb{R}^n$ by *Theorem 2.3.9* and the following hold for a closed subgroup H of V:

(a) H is a closed subgroup of V if and only if $T(H)$ is a closed subgroup of $\mathbb{R}^n$.

(b) H is discrete if and only if $T(H)$ is discrete.

(c) $T(L_H) = L_{T(H)}$ and $\mathrm{Dim}\,[L_H] = \mathrm{Dim}\,[L_{T(H)}]$.

(d) $T(\mathrm{Span}\,[H]) = \mathrm{Span}\,[T(H)]$ and $\mathrm{Dim}\,[\mathrm{Span}\,[H]] = \mathrm{Dim}\,[\mathrm{Span}\,[T(H)]]$.

(e) The ranks of H and $T(H)$ are equal.

(f) If H is closed, then V/H and $\mathbb{R}^n/T(H)$ are isomorphic metric groups.

Consequently, all the results in this section can be applies to a closed subgroup H of V via $T(H)$.

EXERCISES

1. Let E be a subset of a finite-dimensional linear space V, and let $m = \text{Dim}\,[\text{Span}\,[E]]$. Show that there exists a basis $\mathbf{v}_1, \ldots, \mathbf{v}_m$ of Span $[E]$ such that $\mathbf{v}_j$ is an element of E for $j = 1, \ldots, m$.

2. Let G be an Abelian group. Given $g_1, \ldots, g_n$ in G, prove that there exists a unique algebraic homomorphism $\varphi : \mathbb{Z}^n \to G$ such that $\varphi(\mathbf{e}_j) = g_j$ for $j = 1, \ldots, n$.

3. Show that $\mathbb{R}$ contains a subgroup algebraically isomorphic to $\mathbb{Z}^2$.

4. Prove that any subgroup of $\mathbb{R}$ that is algebraically isomorphic to $\mathbb{Z}^2$ is dense in $\mathbb{R}$.

5. Let H be a closed subgroup of $\mathbb{R}^n$ and let $\mathbf{u}_1, \ldots, \mathbf{u}_n$ be a basis of $\mathbb{R}^n$ given by *Theorem 3.1.8.* Consider the norm defined in *Exercise* 6, p. 91. Show that if $\mathbf{u}$ is in L_H and $\mathbf{v}$ is in $H \setminus L_H$, then $\|\mathbf{u} - \mathbf{v}\|_1 > 1/2$.

6. Let H be a closed subgroup of $\mathbb{R}^n$. Prove that L_H is an open subgroup of H. (One approach is to use *Exercise 5* above.)

7. Given integers p and r such that $0 < p < r$ and linear independent vectors $\mathbf{u}_1, \ldots, \mathbf{u}_r$, set

$$H = \left\{ \sum_{j=1}^{r} c_j \mathbf{u}_j : c_j \in \mathbb{R} \text{ for } 1 \leq j \leq p \text{ and } c_j \in \mathbb{Z} \text{ for } p < j \leq r \right\}.$$

   Prove that H is a closed subgroup of $\mathbb{R}^n$. (One approach is to use *Exercise* 7, p. 91.)

8. Let $\varphi : \mathbb{Z}^n \to \mathbb{Z}^m$ be a homomorphism. Prove the following:

   (a) There exists a linear function $T : \mathbb{R}^n \to \mathbb{R}^m$ such that T restricted to $\mathbb{Z}^n$, denoted by $T|\mathbb{Z}^n$, equals φ.

   (b) If φ is onto, then $n \geq m$.

   (c) If φ is an isomorphism, then $n = m$.

9. Let H be a discrete subgroup of $\mathbb{R}^n$. Prove that H is isomorphic to $\mathbb{Z}^m$ if and only if the rank of H is m.

10. Let H be a discrete subgroup of $\mathbb{R}^n$, and let A be an element of $\mathcal{M}_r(\mathbb{Z})$, the $r \times r$ matrices with integer entries. Suppose $\mathbf{a}_1, \ldots, \mathbf{a}_r$ be linearly independent vectors in H such that $\langle \mathbf{a}_1, \ldots, \mathbf{a}_r \rangle = H$. Set $\mathbf{b}_j = \sum_{k=1}^{r} A(k,j)\mathbf{a}_k$ for $j = 1, \ldots, r$. Prove that $\langle \mathbf{b}_1, \ldots, \mathbf{b}_r \rangle = H$ if and only if $\text{Det}\,[A] = \pm 1$.

11. Let A be in $\mathcal{M}_n(\mathbb{Z})$. Prove the following are equivalent:

    (a) $\operatorname{Det}[A] = \pm 1$.

    (b) The columns of A are linearly independent generators for $\mathbb{Z}^n$.

    (c) The rows of A are linearly independent generators for $\mathbb{Z}^n$.

3.2 Quotient Groups

Since $\mathbb{R}^n$ is an Abelian group, all its closed subgroups are normal and their quotient spaces are Abelian metric groups. By using the results of Section 3.1, it is not difficult to determine the quotient metric groups of $\mathbb{R}^n/H$, when H is a closed subgroup of $\mathbb{R}^n$. Some of these quotients are compact metric groups of the form $\mathbb{K}^m$, but others are not compact. Analyzing the closed subgroups and quotients of $\mathbb{K}^m$ will be the more substantive part of this section and will depend heavily on the structure of the closed subgroups of $\mathbb{R}^n$.

The function $\pi_n : \mathbb{R}^n \to \mathbb{K}^n$ defined by

$$\pi_n(\mathbf{x}) = \left(e^{2\pi i x_1}, \ldots, e^{2\pi i x_n}\right).$$

is easily seen to be a homomorphism of metric groups. First, it is continuous because the coordinate functions $e^{2\pi i x_j}$ are all continuous. Second, it satisfies the homomorphism equation, $\pi_n(\mathbf{x}+\mathbf{y}) = \pi_n(\mathbf{x})\pi_n(\mathbf{y})$, because $e^{2\pi i(x+y)} = e^{2\pi i x}e^{2\pi i y}$ for all real numbers x and y. Note that π_n is onto and its kernel is $\mathbb{Z}^n$. The homomorphism π_n will play a central role in the rest of this section.

Recall that $\mathbb{R}^n$ is both locally compact and σ-compact. Thus whenever $\varphi : \mathbb{R}^n \to G$ is a homomorphism onto a locally compact metric group G with kernel K, *Corollary 1.5.19* applies and $\mathbb{R}^n/K$ is isomorphic to G. In particular, it follows that $\mathbb{R}^n/\mathbb{Z}^n$ is isomorphic to $\mathbb{K}^n$ because $\pi_n : \mathbb{R}^n \to \mathbb{K}^n$ is an onto homomorphism. To be specific, $\mathbf{x} + \mathbb{Z}^n \mapsto \pi_n(\mathbf{x})$ is the natural isomorphism from $\mathbb{R}^n/\mathbb{Z}^n$ onto $\mathbb{K}^n$ provided by *Corollary 1.5.19*.

Since $\mathbb{K}^n$ is compact, it is both locally compact and σ-compact. It follows that $\mathbb{K}^m \times \mathbb{R}^n$ is also locally compact and σ-compact by applying *Exercises 7 and 12*, p. 61.

Theorem 3.2.1 *If H is a closed subgroup of $\mathbb{R}^n$ of rank r and $p = \operatorname{Dim}[L_H]$, then $\mathbb{R}^n/H$ is isomorphic to $\mathbb{K}^q \times \mathbb{R}^{n-r}$, where $q = r - p$.*

Proof. It suffices by *Corollary 1.5.19* to construct a homomorphism $\varphi : \mathbb{R}^n \to \mathbb{K}^q \times \mathbb{R}^{n-r}$ with kernel H. Let $\mathbf{u}_1, \ldots, \mathbf{u}_n$ be a basis for $\mathbb{R}^n$ given by *Corollary 3.1.9.* Then,

$$\varphi\left(\sum_{j=1}^{n} c_j \mathbf{u}_j\right) = \left(e^{2\pi i c_{p+1}}, \ldots, e^{2\pi i c_r}, c_{r+1}, \ldots, c_n\right).$$

is the required homomorphism. □

If H is a closed subgroup of $\mathbb{K}$ and $H \neq \mathbb{K}$, then *Theorem 1.4.20* implies that *(a)* H is a discrete finite cyclic subgroup of $\mathbb{K}$ and *(b)* $\mathbb{K}/H$ is isomorphic to $\mathbb{K}$. The first part is not true even for $\mathbb{K}^2$. For example, $\{(z,1) : z \in \mathbb{K}\}$ is a closed subgroup of $\mathbb{K}^2$ that is not finite, and $\{(1,1),(i,1),(1,i),(i,i)\}$ is a finite discrete subgroup of $\mathbb{K}^2$ that is not cyclic. There is a generalization of the second part that can be obtained as a corollary of *Theorem 3.2.1*. Like $\mathbb{R}^n$, every closed subgroup H of $\mathbb{K}^n$ is normal and $\mathbb{K}^n/H$ is an Abelian metric group. Unlike $\mathbb{R}^n$, these quotient groups are always compact because $\mathbb{K}^n$ is compact. We use the convention that $\mathbb{K}^0$ is the trivial group consisting of one element.

Corollary 3.2.2 *If H is a closed subgroup of $\mathbb{K}^n$, then $\mathbb{K}^n/H$ is isomorphic to $\mathbb{K}^q$ for some integer q such that $0 \leq q \leq n$.*

Proof. Let $\pi : \mathbb{K}^n \to \mathbb{K}^n/H$ be the canonical homomorphism and consider the homomorphism $\varphi = \pi \circ \pi_n : \mathbb{R}^n \to \mathbb{K}/H$. The kernel of φ is $K = \pi_n^{-1}(H)$, and $\mathbb{R}^n/K$ is isomorphic to $\mathbb{K}^q/H$ by *Corollary 1.5.19*. Since the identity of $\mathbb{K}^n$ is in H, the group K contains the kernel of π_n, which is $\mathbb{Z}^n$. Thus the rank of K is n and $\mathbb{R}^n/K$ is isomorphic to $\mathbb{K}^q$ for $q = n - \mathrm{Dim}\,[L_K]$ by *Theorem 3.2.1*. □

The structure of the subgroups of $\mathbb{K}^n$ can be completely determined. *Theorem 3.2.3* provides an important first step, and the full result appears as *Corollary 8.3.5*.

Theorem 3.2.3 *If H is a closed subgroup of $\mathbb{K}^n$, then H contains an open-and-closed subgroup H' such that*

(a) H' is isomorphic to $\mathbb{K}^m$ for some integer m satisfying $0 \leq m \leq n$

(b) H/H' is a finite Abelian group.

Proof. Let $G = \pi_n^{-1}(H)$. So G is a closed subgroup of $\mathbb{R}^n$, and a locally compact σ-compact metric group in its own right (see *Exercises 5 and 10*, p. 61). Furthermore, L_G is an open subgroup of the metric group G by *Exercise* 6, p. 124.

Let $\pi = \pi_n|G$, the restriction of π_n to G. So π is a homomorphism of the locally compact σ-compact metric group G onto the compact metric group H. It follows that π is an open function (*Theorem 1.5.18*) and hence $H' = \pi(L_G)$ is an open subgroup of H. Consequently, H' is also a closed (*Proposition 1.4.6*) and compact subgroup of $\mathbb{K}^n$. It follows that H/H' is Abelian, compact and discrete. Therefore, H/H' is a finite Abelian group.

Let $\pi' = \pi|L_G = \pi_n|L_G$, and let $K = L_G \cap \mathbb{Z}^n$, the kernel of π'. So π' is a homomorphism of L_G onto H', and L_G/K is isomorphic to $\mathbb{K}^m \times \mathbb{R}^{m'}$ for suitable integers m and m' by *Theorem 3.2.1*, because L_G is isomorphic to $\mathbb{R}^p$ and the comments in the last paragraph of Section 3.1 apply. Since L_G is locally compact and σ-compact, L_G/K is also isomorphic to H', which is compact. Therefore, $m' = 0$ and H' is isomorphic to $\mathbb{K}^m$. □

Suppose $\varphi : \mathbb{K}^n \to G$ is a homomorphism of $\mathbb{K}^n$ onto a metric group G with kernel K. Then clearly G is compact and isomorphic to $\mathbb{K}^m/K$. Hence G is isomorphic to $\mathbb{K}^q$ for some q such that $0 \leq q \leq n$ by *Corollary 3.2.2.* This raises an interesting question. Can we determine all the homomorphisms of $\mathbb{K}^n$ onto $\mathbb{K}^q$? The next few results about homomorphisms lay the foundation for addressing this and similar questions.

Proposition 3.2.4 *Let φ and φ' be homomorphisms from $\mathbb{R}^n$ to the metric group G. If there exists an open set U of $\mathbb{R}^n$ containing* $\mathbf{0}$ *such that $\varphi(\mathbf{x}) = \varphi'(\mathbf{x})$ for all $\mathbf{x}$ in U, then $\varphi = \varphi'$.*

Proof. It can be assumed without loss of generality that $U = \{\mathbf{x} : \|\mathbf{x}\| < r\}$ for some $r > 0$. Then $kU = \{\mathbf{x} : \|\mathbf{x}\| < kr\}$ for any positive integer k. It follows that

$$\mathbb{R}^n = \bigcup_{k=1}^{\infty} kU$$

and hence for any $\mathbf{x} \in \mathbb{R}^n$ there exist a positive integer k such that $(1/k)\mathbf{x} \in U$.

The group G will be written using multiplicative notation so that $\varphi(k\mathbf{x}) = \varphi(\mathbf{x})^k$ for a positive integer k. Given an arbitrary $\mathbf{x}$ in $\mathbb{R}^n$, let k be a positive integer such that $(1/k)\mathbf{x}$ is in U. Then,

$$\varphi(\mathbf{x}) = \varphi\left(\frac{k}{k}\mathbf{x}\right) = \varphi\left(\frac{1}{k}\mathbf{x}\right)^k = \varphi'\left(\frac{1}{k}\mathbf{x}\right)^k = \varphi'\left(\frac{k}{k}\mathbf{x}\right) = \varphi'(\mathbf{x})$$

and $\varphi = \varphi'$ □

Not only is a homomorphism from $\mathbb{R}^n$ to a metric group G uniquely determined by its values on a neighborhood of $\mathbf{0}$, but it is also possible to construct a homomorphism from a suitable function defined only on a neighborhood of $\mathbf{0}$.

Theorem 3.2.5 *Let U be an open set in $\mathbb{R}^n$ containing* $\mathbf{0}$, *and let $f : U \to G$ be a continuous function of U into a metric group G such that $f(\mathbf{x}+\mathbf{y}) = f(\mathbf{x})f(\mathbf{y})$ whenever $\mathbf{x}$, $\mathbf{y}$, and $\mathbf{x}+\mathbf{y}$ are all in U. Then there exists a unique homomorphism $\varphi : \mathbb{R}^n \to G$ such that $\varphi(\mathbf{x}) = f(\mathbf{x})$ whenever $\mathbf{x} \in U$. Moreover, $\varphi(\mathbb{R}^n)^-$ is an Abelian subgroup of G.*

Proof. It can be assumed without loss of generality that $U = \{\mathbf{x} : \|\mathbf{x}\| < r\}$ for some $r > 0$, so that $c\mathbf{x}$ is in U for all $\mathbf{x}$ in U and c such that $|c| \leq 1$. As in the previous proof, given an arbitrary $\mathbf{x} \in \mathbb{R}^n$ there exists a positive integer k such that $(1/k)\mathbf{x}$ is in U. In fact, $(1/k)\mathbf{x}$ is in U for all k sufficiently large. This suggests setting

$$\varphi(\mathbf{x}) = f\left(\frac{1}{k}\mathbf{x}\right)^k$$

when $(1/k)\mathbf{x}$ is in U, but it must first be shown that $f\big((1/k)\mathbf{x}\big)^k = f\big((1/m)\mathbf{x}\big)^m$ when both $(1/k)\mathbf{x}$ and $(1/m)\mathbf{x}$ are in U.

Suppose both $(1/k)$ and $(1/m)\mathbf{x}$ are in U. In this case, $(1/km)\mathbf{x}$ is also in U and

$$f\left(\frac{1}{k}\mathbf{x}\right)^k = f\left(\frac{m}{km}\mathbf{x}\right)^k = f\left(\frac{1}{km}\mathbf{x}\right)^{km} = f\left(\frac{k}{km}\mathbf{x}\right)^m = f\left(\frac{1}{m}\mathbf{x}\right)^m.$$

Thus φ is unambiguously defined on all of $\mathbb{R}^n$ and agrees with f on U because $(1/1)\mathbf{x} = \mathbf{x}$ is in U, when $\mathbf{x}$ is in U.

Now, let $\mathbf{x}$ and $\mathbf{y}$ be in $\mathbb{R}^n$ and choose k sufficiently large so that $(1/k)\mathbf{x}$, $(1/k)\mathbf{y}$, and $(1/k)(\mathbf{x}+\mathbf{y})$ are all in U. Then by hypothesis

$$f\left(\frac{1}{k}\mathbf{x}\right) f\left(\frac{1}{k}\mathbf{y}\right) = f\left(\frac{1}{k}(\mathbf{x}+\mathbf{y})\right) = f\left(\frac{1}{k}\mathbf{y}\right) f\left(\frac{1}{k}\mathbf{x}\right).$$

In particular, $f((1/k)\mathbf{x})$ and $f((1/k)\mathbf{y})$ commute in G. It follows that

$$f\left(\frac{1}{k}\mathbf{x}\right)^k f\left(\frac{1}{k}\mathbf{y}\right)^k = \left(f\left(\frac{1}{k}\mathbf{x}\right) f\left(\frac{1}{k}\mathbf{y}\right)\right)^k = f\left(\frac{1}{k}(\mathbf{x}+\mathbf{y})\right)^k.$$

Therefore,

$$\varphi(\mathbf{x})\varphi(\mathbf{y}) = \varphi(\mathbf{x}+\mathbf{y}),$$

and φ is an algebraic homomorphism.

Because $\varphi(\mathbf{x}) = f(\mathbf{x})$ on the open set U and f is continuous at $\mathbf{0}$, the function φ is continuous at $\mathbf{0}$. It follows from *Exercise* 5, p. 39, that φ is continuous on $\mathbb{R}^n$and a homomorphism from $\mathbb{R}^n$ to the metric group G . The uniqueness of φ follows from *Proposition 3.2.4*.

Finally, $\varphi(\mathbb{R}^n)$ is clearly an Abelian subgroup of G, and hence so is $\varphi(\mathbb{R}^n)^-$ by *Proposition 1.4.3*. □

Theorem 3.2.6 *If $\varphi : \mathbb{R}^m \to \mathbb{K}^n$ is a homomorphism, then there exists a unique linear map $T : \mathbb{R}^m \to \mathbb{R}^n$ such that $\varphi = \pi_n \circ T$.*

Proof. Let $R = \{\mathbf{x} \in \mathbb{R}^n : |x_i| < 1/4 \text{ for } 1 \leq i \leq n\}$, which is an open subset of $\mathbb{R}^n$ containing $\mathbf{0}$. Note that π_n is one-to-one on R. Since π_n is an open function and R is an open set, $V = \pi_n(R)$ is an open subset of $\mathbb{K}^n$ and $\pi_n|R$ is a homeomorphism of R onto V. Set $\psi = (\pi_n|R)^{-1}$, which is a homeomorphism of V onto R such that $\pi_n(\psi(\mathbf{z})) = \mathbf{z}$ for all $\mathbf{z} \in V$.

If $\mathbf{z}$, $\mathbf{w}$, and $\mathbf{zw} = (z_1w_1, \ldots, z_nw_n)$ are all in V, then

$$\pi_n(\psi(\mathbf{z}) + \psi(\mathbf{w})) = \pi_n(\psi(\mathbf{z}))\pi_n(\psi(\mathbf{w})) = \mathbf{zw} = \pi_n(\psi(\mathbf{zw})).$$

Since $\psi(\mathbf{z})$, $\psi(\mathbf{w})$, and $\psi(\mathbf{zw})$ are all in R and π_n is one-to-one on R, it follows that

$$\psi(\mathbf{z}) + \psi(\mathbf{w}) = \psi(\mathbf{zw}).$$

Set $U = \varphi^{-1}(V)$ and consider the function defined by $f = \psi \circ (\varphi|U) : U \to \mathbb{R}^n$. Obviously, f is continuous. Since φ is a homomorphism and $\psi(\mathbf{z}) + \psi(\mathbf{w}) = \psi(\mathbf{zw})$ when $\mathbf{z}$, $\mathbf{w}$, and $\mathbf{zw}$ are in V, it follows that $f(\mathbf{x}+\mathbf{y}) = f(\mathbf{x}) + f(\mathbf{y})$

when $\mathbf{x}$, $\mathbf{y}$, and $\mathbf{x}+\mathbf{y}$ are in U. Now *Theorem 3.2.5* applies and there exists a homomorphism $T : \mathbb{R}^m \to \mathbb{R}^n$ that agrees with f on the open set U. If $\mathbf{x}$ is in U, then $\pi_n \circ T(\mathbf{x}) = \pi_n \circ \psi \circ \varphi(\mathbf{x}) = \varphi(\mathbf{x})$ and $\pi_n \circ T = \varphi$ by *Proposition 3.2.4*. Furthermore, *Proposition 2.4.1* implies that T is linear.

If $S : \mathbb{R}^m \to \mathbb{R}^n$ is another linear function such that $\pi_n(S(\mathbf{x})) = \pi_n(T(\mathbf{x}))$ for all $\mathbf{x} \in \mathbb{R}^m$, then $S(\mathbf{x}) - T(\mathbf{x})$ is in $\mathbb{Z}^n$, the kernel of π_n. Consequently, the range of the linear map $S - T$ is a subspace of $\mathbb{R}^n$ and a subset of the discrete group $\mathbb{Z}^n$. It follows that the range of $S - T$ is the trivial subspace $\{\mathbf{0}\}$ and $S = T$. □

Of course, if $T : \mathbb{R}^m \to \mathbb{R}^n$ is a linear function, then $\pi_n \circ T : \mathbb{R}^m \to \mathbb{K}^n$ is a homomorphism. It follows from the above theorem that function $T \to \pi_n \circ T$ is, in fact, a one-to-one function from $\mathcal{L}(\mathbb{R}^m, \mathbb{R}^n)$ onto the set of homomorphisms of $\mathbb{R}^m$ to $\mathbb{K}^n$. This certainly qualifies as a complete description of the homomorphisms from $\mathbb{R}^m$ to $\mathbb{K}^n$. We would like to obtain a similar description of the homomorphisms $\varphi : \mathbb{K}^m \to \mathbb{K}^n$.

Proposition 3.2.7 *If $\varphi : \mathbb{K}^m \to \mathbb{K}^n$ is a homomorphism, then there exists a unique linear function $T : \mathbb{R}^m \to \mathbb{R}^n$ such that $\varphi \circ \pi_m = \pi_n \circ T$ and $T(\mathbb{Z}^m) \subset \mathbb{Z}^n$. Furthermore, if $T : \mathbb{R}^m \to \mathbb{R}^n$ is a linear function such that $T(\mathbb{Z}^m) \subset \mathbb{Z}^n$, then there exists a homomorphism $\varphi : \mathbb{K}^m \to \mathbb{K}^n$ such that $\varphi \circ \pi_m = \pi_n \circ T$.*

Proof. Applying *Theorem 3.2.6* to $\varphi \circ \pi_m$, proves the existence of a unique linear function $T : \mathbb{R}^m \to \mathbb{R}^n$ such that $\varphi \circ \pi_m = \pi_n \circ T$. If $\mathbf{x}$ is in $\mathbb{Z}^m$, then $\mathbf{x}$ is in the kernel of π_m, and hence in the kernel of $\varphi \circ \pi_m$. Now $\varphi \circ \pi_m = \pi_n \circ T$ implies that $T(\mathbf{x})$ must be in $\mathbb{Z}^n$, the kernel of π_n, proving that $T(\mathbb{Z}^m) \subset \mathbb{Z}^n$.

To define $\varphi(\mathbf{z})$ for the second statement, let $\mathbf{x}$ be an element of $\mathbb{R}^m$ such that $\pi_m(\mathbf{x}) = \mathbf{z}$ and set $\varphi(\mathbf{z}) = \pi_n(T(\mathbf{x}))$. If $\pi_m(\mathbf{x}) = \pi_m(\mathbf{x}')$, then $\mathbf{x} - \mathbf{x}'$ is in $\mathbb{Z}^m$, the kernel of π_m, and $T(\mathbf{x}) - T(\mathbf{x}') = T(\mathbf{x} - \mathbf{x}')$ is in $\mathbb{Z}^n$ by hypothesis. Therefore, $\pi_n(T(\mathbf{x})) = \pi_n(T(\mathbf{x}'))$, and φ is a well-defined function. Obviously, $\varphi \circ \pi_m = \pi_n \circ T$.

Given $\mathbf{z}$ and $\mathbf{w}$ in $\mathbb{K}^m$, let $\mathbf{x}$ and $\mathbf{w}$ be elements of $\mathbb{R}^m$ such that $\pi_m(\mathbf{x}) = \mathbf{z}$ and $\pi_m(\mathbf{y}) = \mathbf{w}$. Then $\pi_m(\mathbf{x}+\mathbf{y}) = \mathbf{zw}$ and

$$\varphi(\mathbf{zw}) = \pi_n(T(\mathbf{x}+\mathbf{y})) = \pi_n(T(\mathbf{x}))\pi_n(T(\mathbf{y})) = \varphi(\mathbf{z})\varphi(\mathbf{w}),$$

proving that φ is an algebraic homomorphism.

Let U be an open subset of $\mathbb{K}^n$. To prove that $\varphi^{-1}(U)$ is open, it suffices to prove that $\pi_m^{-1}(\varphi^{-1}(U))$ is open because π_m is an open function. It follows from $\varphi \circ \pi_m = \pi_n \circ T$ that $\pi_m^{-1}(\varphi^{-1}(U)) = T^{-1}(\pi_n^{-1}(U))$, which is open because T and π_n are continuous. Thus φ is the required homomorphism. □

It follows from *Theorem 3.2.7* that there is a one-to-one correspondence between the homomorphisms $\varphi : \mathbb{K}^m \to \mathbb{K}^n$ and the linear functions $T : \mathbb{R}^m \to \mathbb{R}^n$ such that $T(\mathbb{Z}^m) \subset \mathbb{Z}^n$. This can be put in a nicer form by using the matrix B of T with respect to the standard bases of $\mathbb{R}^m$ and $\mathbb{R}^n$. Specifically, $T(\mathbb{Z}^m) \subset \mathbb{Z}^n$ holds if and only if $T(\mathbf{e}_j) \in \mathbb{Z}^n$ for $j = 1, \ldots, m$ if and only if the entries of the matrix of T with respect to the standard bases are all integers.

Hence, there is a one-to-one onto correspondence between the homomorphisms from $\mathbb{K}^m$ to $\mathbb{K}^n$ and the $n \times m$ integral matrices, which will be denoted by $\mathcal{M}_{n\times m}(\mathbb{Z})$. Now using $\varphi \circ \pi_m = \pi_n \circ T$ and equation (2.6) for the ith coordinate of $T(\mathbf{x})$, we obtain the following general formula for an arbitrary homomorphism $\varphi : \mathbb{K}^m \to \mathbb{K}^n$:

$$\varphi(\exp(2\pi i x_1), \ldots, \exp(2\pi i x_m)) = \left(\exp\left(2\pi i \sum_{k=1}^{m} B(1,k)x_k\right), \ldots, \exp\left(2\pi i \sum_{k=1}^{m} B(n,k)x_k\right)\right) \tag{3.2}$$

where B is an arbitrary element of $\mathcal{M}_{n\times m}(\mathbb{Z})$.

Following our earlier notation, it will be convenient to denote $\mathcal{M}_{n\times n}(\mathbb{Z})$ simply by $\mathcal{M}_n(\mathbb{Z})$, when $m = n$, and in this case more can be said.

Proposition 3.2.8 *The group of automorphisms of* $\mathbb{K}^n$ *is algebraically isomorphic to the discrete subgroup*

$$H = \{A \in \mathcal{M}_n(\mathbb{Z}) : |\text{Det}\,[A]| = 1\}$$

of $\text{GL}(n, \mathbb{R})$.

Proof. Given $A \in \mathcal{M}_n(\mathbb{Z})$, let $\varphi_A : \mathbb{K}^n \to \mathbb{K}^n$ denote the homomorphism determined by A using (3.2). From the discussion preceding this equation, it follows that the function $A \to \varphi_A$ is a one-to-one function of $\mathcal{M}_n(\mathbb{Z})$ onto the set of homomorphisms of $\mathbb{K}^n$ to itself, which we will denote by $\mathcal{S}$.

Observe that $\mathcal{M}_n(\mathbb{Z})$ is a semigroup with identity I under matrix multiplication and $\mathcal{S}$ is a semigroup with identity ι under composition of functions. So given A and B in $\mathcal{M}_n(\mathbb{Z})$, there exists a unique C in $\mathcal{M}_n(\mathbb{Z})$ such that $\varphi_A \circ \varphi_B = \varphi_C$. Equation (3.2) shows that

$$\exp\left(\sum_{j=1}^{n} B(k,j)x_j\right)$$

is the kth coordinate of $\varphi_B((\exp(2\pi i x_1), \ldots, \exp(2\pi i x_n))$. Applying (3.2) again the mth coordinate of $\varphi_A \circ \varphi_B(\exp(2\pi i x_1), \ldots, \exp(2\pi i x_m)$ is given by

$$\exp\left(\sum_{k=1}^{n} A(m,k)\left(\sum_{j=1}^{n} B(k,j)x_j\right)\right) = \exp\left(\sum_{j=1}^{n}\left(\sum_{k=1}^{n} A(m,k)B(k,j)\right)x_j\right)$$

and

$$C(m,j) = \sum_{k=1}^{n} A(m,k)B(k,j).$$

Thus $C = AB$ and $A \to \varphi_A$ is a semigroup isomorphism of $\mathcal{M}_n(\mathbb{Z})$ onto $\mathcal{S}$.

Obviously, $\varphi_I = \iota$. It follows that $A \to \varphi_A$ maps the group of units of $\mathcal{M}_n(\mathbb{Z})$ onto the group of units of $\mathcal{S}$. Since the group of automorphisms of $\mathbb{K}^n$

is the group of units of $\mathcal{S}$, showing that $H = \{A \in \mathcal{M}_n(\mathbb{Z}) : |\text{Det}[A]| = 1\}$ is the group of units of $\mathcal{M}_n(\mathbb{Z})$ will complete the proof.

An element A of $\mathcal{M}_n(\mathbb{Z})$ is invertible if and only if $\text{Det}[A] \neq 0$. Invertibility of A does not guarantee that A is unit in $\mathcal{M}_n(\mathbb{Z})$ unless A^{-1} is also in $\mathcal{M}_n(\mathbb{Z})$. Clearly, $\text{Det}[A]$ and all the cofactors of A are integers. It follows from equation (2.38), that the entries of A^{-1} all have the form $k/\text{Det}[A]$ for some $k \in \mathbb{Z}$. If $\text{Det}[A] = \pm 1$, then the matrix A is invertible and A^{-1} is in $\mathcal{M}_n(\mathbb{Z})$. Conversely, suppose A^{-1} is in $\mathcal{M}_n(\mathbb{Z})$. Then both $\text{Det}[A]$ and $\text{Det}[A^{-1}]$ are integers such that $\text{Det}[A]\text{Det}[A^{-1}] = \text{Det}[AA^{-1}] = \text{Det}[I] = 1$. Therefore, $\text{Det}[A] = \pm 1$. □

EXERCISES

1. Let k be a positive integer and let $\mathbf{z}$ be in $\mathbb{K}^n$. Show that there exists $\mathbf{w}$ in $\mathbb{K}^n$ such that $\mathbf{w}^k = \mathbf{z}$.

2. Show that every homomorphism $\varphi : \mathbb{K} \to \mathbb{K}$ has the form $\varphi(z) = z^k$ for some integer k. Show that the group of automorphisms of $\mathbb{K}$ is algebraically isomorphic to $\mathbb{Z}_2$.

3. Give an example of a closed subgroup H of $\mathbb{K}^2$ with a subgroup H' such that H' is isomorphic to $\mathbb{K}$ and H/H' is isomorphic to $\mathbb{Z}_4$.

4. Let H be a closed subgroup of $\mathbb{K}^n$ and let H' be given by *Theorem 3.2.3.* Prove that H is isomorphic to $\mathbb{K}^q \times \mathbb{Z}_m$, if H/H' is isomorphic to $\mathbb{Z}_m$.

5. Prove that $\mathbb{K}^m$ and $\mathbb{K}^n$ are isomorphic if and only if $m = n$. (One approach is to assume that $m > n$ and apply *Proposition 3.2.7.*

6. Consider the linear function $T : \mathbb{R} \to \mathbb{R}^2$ defined by $T(t) = (\sqrt{2}\,t, t)$ and let $\varphi = \pi_2 \circ T$, which is homomorphism from $\mathbb{R}$ into $\mathbb{K}^2$. Show that
$$\varphi(\mathbb{Z})^{-} = \{(w, 1) : w \in \mathbb{K}\}.$$
Then show that $\varphi(\mathbb{R})^{-} = \mathbb{K}^2$.

7. Show that if $\varphi : \mathbb{K}^m \to \mathbb{K}$ is a homomorphism, then there exists $\mathbf{k} = (k_1, \ldots, k_m))$ in $\mathbb{Z}^m$ such that
$$\varphi(z_1, \ldots z_m) = \prod_{j=1}^{m} z_j^{k_j}.$$

8. Let V be a subspace of $\mathbb{R}^n$ and let $H = \pi_n(V)$, which is a subgroup of $\mathbb{K}^n$. Prove that H is closed if and only if the dimension of V equals the rank of $V \cap \mathbb{Z}^n$.

3.3 Dense Subgroups

The groups, $\mathbb{K}^n$, also have interesting subgroups that are not closed. For example, if α is irrational then the infinite cyclic subgroup of $\mathbb{K}$ generated by $a = \exp(2\pi i\alpha)$ is dense in $\mathbb{K}$ by *Exercise* 13, p. 51. It follows that $\varphi : \mathbb{Z} \to \mathbb{K}$ defined by $\varphi(k) = a^k$ is a homomorphism of the integers onto a dense subgroup of the circle. Similarly, *Exercise* 6, p. 131, provides an example of a homomorphism of $\mathbb{R}$ onto a dense subgroup of $\mathbb{K}^2$.

The primary goal of this section is to obtain Kronecker's classical criterion for when a finitely generated subgroup of $\mathbb{K}^n$ is dense in $\mathbb{K}^n$. This will require that we first gain an understanding of when finitely generated subgroups of $\mathbb{R}^n$ containing $\mathbb{Z}^n$ are dense in $\mathbb{R}^n$. We begin by studying a set function for subsets of $\mathbb{R}^n$ that will be useful in the analysis.

Let E be a nonempty subset of $\mathbb{R}^n$ and set

$$E^{\#} = \{\mathbf{x} \in \mathbb{R}^n : \mathbf{x} \cdot \mathbf{u} \in \mathbb{Z} \text{ for all } \mathbf{u} \in E\}.$$

Obviously, $\mathbf{0}$ is in $E^{\#}$. So $E \to E^{\#}$ a function from the nonempty subsets of $\mathbb{R}^n$ to themselves. Obviously, $\{\mathbf{0}\}^{\#} = \mathbb{R}^n$ and $(\mathbb{R}^n)^{\#} = \{\mathbf{0}\}$. It is easy to see that $(\mathbb{Z}^n)^{\#} = \mathbb{Z}^n$. Obviously, $E \subset F$ implies that $F^{\#} \subset E^{\#}$. The next two propositions develop additional basic properties of the $E^{\#}$ construction.

Proposition 3.3.1 *Let E be a nonempty subset of $\mathbb{R}^n$. Then $E^{\#}$ has the following properties:*

(a) $E^{\#}$ is a closed subset of $\mathbb{R}^n$.

(b) $E^{\#}$ is a subgroup of $\mathbb{R}^n$.

(c) $\left(E^{-}\right)^{\#} = E^{\#}$.

Proof. The function $f_{\mathbf{u}}(\mathbf{x}) = \mathbf{x} \cdot \mathbf{u}$ is clearly a continuous real-valued function on $\mathbb{R}^n$. Observe that

$$E^{\#} = \bigcap_{\mathbf{u} \in E} f_{\mathbf{u}}^{-1}(\mathbb{Z}). \tag{3.3}$$

Since $\mathbb{Z}$ is a closed subset of $\mathbb{R}$, the set $f_{\mathbf{u}}^{-1}(\mathbb{Z})$ is a closed set. Part *(a)* now follows from equation (3.3) because the intersection of closed sets is a closed set.

Turning to part *(b)*, the function $f_{\mathbf{u}}$ is also a homomorphism from $\mathbb{R}^n$ to $\mathbb{R}$. Hence, $f_{\mathbf{u}}^{-1}(\mathbb{Z})$ is a subgroup of $\mathbb{R}^n$ because $\mathbb{Z}$ is a subgroup of $\mathbb{R}$. It follows using equation (3.3) again that $E^{\#}$ is a subgroup because the intersection of subgroups is a subgroup.

For part *(c)*, observe that $\mathbf{x}$ is in $E^{\#}$ if and only if $f_{\mathbf{x}}(\mathbf{u})$ is in $\mathbb{Z}$ for all $\mathbf{u} \in E$ if and only if $E \subset f_{\mathbf{x}}^{-1}(\mathbb{Z})$. Since $f_{\mathbf{x}}^{-1}(\mathbb{Z})$ is closed, $E \subset f_{\mathbf{x}}^{-1}(\mathbb{Z})$ if and only if $E^{-} \subset f_{\mathbf{x}}^{-1}(\mathbb{Z})$. Therefore, $\mathbf{x}$ is in $E^{\#}$ if and only if $\mathbf{x}$ is in $\left(E^{-}\right)^{\#}$ and $\left(E^{-}\right)^{\#} = E^{\#}$. $\square$

Proposition 3.3.2 *If H is a closed subgroup of $\mathbb{R}^n$, then $H^{\#\#} = H$.*

Proof. By *Corollary 3.1.9* there exists a basis $\mathbf{u}_1, \ldots, \mathbf{u}_n$ for $\mathbb{R}^n$ such that

$$H = \left\{ \sum_{j=1}^{p} c_j \mathbf{u}_j + \sum_{k=p+1}^{r} m_k \mathbf{u}_k : c_j \in \mathbb{R} \text{ and } m_k \in \mathbb{Z} \right\}.$$

By *Exercise* 4, p. 138, there exists a basis $\mathbf{v}_1, \ldots, \mathbf{v}_n$ such that

$$\mathbf{u}_j \cdot \mathbf{v}_k = \begin{cases} 1 & \text{when } j = k \\ 0 & \text{when } j \neq k \end{cases}.$$

It follows that

$$\left(\sum_{j=1}^{n} c_j \mathbf{u}_j \right) \cdot \left(\sum_{k=1}^{n} d_k \mathbf{v}_k \right) = \sum_{j=1}^{n} c_j d_j.$$

Let $\mathbf{x} = \sum_{j=1}^{n} d_j \mathbf{v}_j$ be an element of $H^{\#}$. Applying the above formula for the dot product of an arbitrary element of H with $\mathbf{x}$, it is easy to see that d_j must equal zero for $j = 1, \ldots, p$ and d_j must be an integer n_j for $j = p+1. \ldots, r$. It follows that

$$H^{\#} = \left\{ \sum_{k=p+1}^{r} n_k \mathbf{v}_k + \sum_{j=r+1}^{n} d_j \mathbf{v}_j : n_k \in \mathbb{Z} \text{ and } d_j \in \mathbb{R} \right\},$$

and repeating the same argument yields

$$H^{\#\#} = \left\{ \sum_{j=1}^{p} c_j \mathbf{u}_j + \sum_{k=p+1}^{r} m_k \mathbf{u}_k : c_j \in \mathbb{R} \text{ and } m_k \in \mathbb{Z} \right\} = H$$

to complete the proof. □

As a corollary we obtain a criterion for when a subgroup of $\mathbb{R}^n$ is dense in $\mathbb{R}^n$ that will be used to prove the main result in this section.

Corollary 3.3.3 *If H is a subgroup of $\mathbb{R}^n$, then $H^{-} = H^{\#\#}$. Furthermore, H is dense in $\mathbb{R}^n$ if and only if $H^{\#} = \{\mathbf{0}\}$.*

Proof. Since $(H^{-})^{\#} = H^{\#}$ by part *(c)* of *Proposition 3.3.1*, it follows that $(H^{-})^{\#\#} = H^{\#\#}$. Because H^{-} is a closed subgroup, the proposition implies that

$$H^{-} = (H^{-})^{\#\#} = H^{\#\#}$$

to complete the proof of the first statement.

To prove the second statement, first assume that $H^{-} = \mathbb{R}^n$. Then $H^{\#} = (H^{-})^{\#} = \{\mathbf{0}\}$. Now suppose that $H^{\#} = \{\mathbf{0}\}$. Then $H^{-} = H^{\#\#} = \{\mathbf{0}\}^{\#} = \mathbb{R}^n$ and H is dense in $\mathbb{R}^n$. □

Given $\mathbf{a}_1, \ldots, \mathbf{a}_m$ in $\mathbb{R}^n$, we know that

$$H = \langle \mathbf{a}_1, \ldots, \mathbf{a}_m \rangle = \left\{ \sum_{j=1}^{m} k_j \, \mathbf{a}_j : k_j \in \mathbb{Z} \text{ for } j = 1, \ldots, m \right\} \tag{3.4}$$

is the *subgroup generated by* $\mathbf{a}_1, \ldots, \mathbf{a}_m$. In general, the group generated by $\mathbf{a}_1, \ldots, \mathbf{a}_m$ need not be discrete and may even be dense in $\mathbb{R}^n$. *Corollary 3.3.3* and $H^{\#}$ provide a test for whether or not H is dense in $\mathbb{R}^n$.

One interesting case is the subgroup H of $\mathbb{R}^n$ generated by $\mathbf{e}_1, \ldots, \mathbf{e}_n$ and some additional generators $\mathbf{a}_1, \ldots, \mathbf{a}_m$, so

$$H = \left\{ \mathbf{b} + \sum_{j=1}^{m} k_j \, \mathbf{a}_j : \mathbf{b} \in \mathbb{Z}^n \text{ and } k_j \in \mathbb{Z} \text{ for } j = 1, \ldots, m \right\}. \tag{3.5}$$

Because $\mathbb{Z}^n \subset H$, the description of $H^{\#}$ can be refined to prove a classical result.

Theorem 3.3.4 (Kronecker) *Let $\mathbf{a}_1, \ldots, \mathbf{a}_m$ be elements of $\mathbb{R}^n$ and let H be the subgroup of $\mathbb{R}^n$ given by equation (3.5). Then H is dense in $\mathbb{R}^n$ if and only if there does not exist $\mathbf{r} \in \mathbb{Z}^n \setminus \{\mathbf{0}\}$ such that $\mathbf{r} \cdot \mathbf{a}_k \in \mathbb{Z}$ for $k = 1, \ldots, m$.*

Proof. The preparatory results and discussion make the proof almost a triviality. By *Corollary 3.3.3*, it suffices to show that $H^{\#} = \{\mathbf{0}\}$ if and only if there does not exist $\mathbf{r} \in \mathbb{Z}^n \setminus \{\mathbf{0}\}$ such that $\mathbf{r} \cdot \mathbf{a}_k \in \mathbb{Z}$ for $k = 1, \ldots, m$.

Because $\mathbb{Z}^n \subset H$, it follows that $H^{\#} \subset (\mathbb{Z}^n)^{\#} = \mathbb{Z}^n$. Then, by *Exercise* 2, p. 138,

$$H^{\#} = \{\mathbf{x} \in \mathbb{Z}^n : \mathbf{x} \cdot \mathbf{a}_k \in \mathbb{Z} \text{ for } j = 1, \ldots, m\}.$$

Consequently, $H^{\#} \neq \{\mathbf{0}\}$ if and only if there exist some $\mathbf{r} \in \mathbb{Z}^n \setminus \{\mathbf{0}\}$ such that $\mathbf{r} \cdot \mathbf{a}_k \in \mathbb{Z}$ for $k = 1, \ldots, m$. Of course, this is the same as saying that $H^{\#} = \{\mathbf{0}\}$ if and only if there does not exist $\mathbf{r} \in \mathbb{Z}^n \setminus \{\mathbf{0}\}$ such that $\mathbf{r} \cdot \mathbf{a}_k \in \mathbb{Z}$ for $k = 1, \ldots, m$. □

The classical statement of Kronecker's Theorem for $m = 1$ is

Corollary 3.3.5 *Let $a_1, \ldots, a_n$ be n real numbers. In order that given any n real numbers $x_1, \ldots, x_n$ and $\varepsilon > 0$ there exist integers $p_1, \ldots, p_n, q$ such that*

$$|qa_k - p_k - x_k| < \varepsilon$$

for $k = 1, \ldots, n$ it is necessary and sufficient that there exists no relation of the form

$$\sum_{j=1}^{n} r_j a_j = b$$

for integers $r_1, \ldots, r_n$ that are not all zero and an integer b.

Proof. Apply Kronecker's Theorem with $m = 1$ to $\mathbf{a} = (a_1, \ldots, a_n)$ using the norm $\|\mathbf{x}\|_s = \sup\{|x_1|, \ldots, |x_n|\}$ for density. □

Notice that when $n = 1$, the above criterion reduces to the statement that a_1 is irrational. When $n > 1$, the criterion is there do not exist integers $r_1, \ldots, r_n$ that are not all zero and satisfy $\sum_{j=1}^{n} r_j a_j \in \mathbb{Z}$. Proving the existence of real numbers $a_1, \ldots, a_n$ that satisfy this criterion, requires looking at this criterion from a different perspective.

Although the definition of linear spaces in Section 2.1 was limited to the scalar fields $\mathbb{R}$ and $\mathbb{C}$, the definition works just as well with $\mathbb{Q}$, the rational numbers, as the scalar field. In particular, subspace, span, and linear independence can be defined for $\mathbb{Q}$ in the same way. More importantly, we can now think of $\mathbb{R}$ as a linear space over the scalar field $\mathbb{Q}$ because the conditions on page 64 obviously hold for c and d rational and x and y real.

Continuing in this vein, real numbers $b_1, \ldots, b_k$ are said to be *rationally independent* if there do not exist rational numbers $r_1, \ldots, r_k$ not all zero such that $r_1 b_1 + \ldots + r_k b_k = 0$. Since we can multiply $r_1 b_1 + \ldots + r_k b_k = 0$ by the least common multiple of the denominators of the rational numbers $r_1, \ldots, r_k$, it follows that $b_1, \ldots, b_k$ are rationally independent if and only if there do not exist integers $r_1, \ldots, r_k$ not all zero such that $r_1 b_1 + \ldots + r_k b_k = 0$. Thus an alternate version of *Corollary 3.3.5* is

Corollary 3.3.6 *Given* $\mathbf{a}$ *in* $\mathbb{R}^n$, *the subgroup*

$$H = \langle \mathbf{e}_1, \ldots, \mathbf{e}_n, \mathbf{a} \rangle$$

of $\mathbb{R}^n$ *is dense in* $\mathbb{R}^n$ *if and only if* $1, a_1, \ldots, a_n$ *are rationally independent.*

The next theorem pulls these ideas together using the concept of a countable set. A set E is *countable* if there exists a one-to-one function from $\mathbb{Z}^+ = \{k \in \mathbb{Z} : k \geq 1\}$, the counting numbers, onto E. This is the same as saying there is a sequence in E that includes every element of E exactly once. A set is said to be *uncountable* if it is neither finite nor countable. For the reader not familiar with countable sets, several exercises at the send of this section provide a set of basic facts about them.

The set of real numbers $\mathbb{R}$ is the prime example of an uncountable set (*Exercise* 15, p. 61). The facts that $\mathbb{R}$ is uncountable and $\mathbb{Q}^n$ is countable (*Exercise* 11, p. 139) will play a critical role in the next proof.

Theorem 3.3.7 *For every positive integer* n *there exist real numbers* $a_1, \ldots, a_n$ *such that* $1, a_1, \ldots, a_n$ *are rationally independent and the dimension of* $\mathbb{R}$ *as a* $\mathbb{Q}$ *linear space is not finite.*

Proof. The proof will make use of the following observation: Suppose $b_1, \ldots, b_m$ are rationally independent real numbers and consider

$$\text{Span}\,[b_1, \ldots, b_m] = \{c_1 b_1 + \ldots + c_m b_m : c_j \in \mathbb{Q} \text{ for } j = 0, \ldots, n\}.$$

If b_{m+1} is not in Span $[b_1, \ldots, b_m]$, then the proof of *Proposition 2.1.4* works for $\mathbb{Q}$ as well as for $\mathbb{R}$ and $\mathbb{C}$, and it follows that $b_1, \ldots, b_m, b_{m+1}$ are rationally independent.

The proof now proceeds by induction. Clearly, the number 1 is rationally independent and Span $[1] = \mathbb{Q} \neq \mathbb{R}$. Hence there exists a real number a_1 not in $\mathbb{Q}$. It follows that $1, a_1$ are rationally independent real numbers.

Next, suppose that $1, a_1, \ldots, a_n$ are rationally independent real numbers. To prove that there exists a_{n+1} such that $1, a_1, \ldots, a_n, a_{n+1}$ are rationally independent real numbers, it suffices to show that Span $[1, a_1, \ldots, a_n] \neq \mathbb{R}$. Since $\mathbb{R}$ is uncountable (*Exercise* 15, p. 61), the proof is further reduced to showing that Span $[1, a_1, \ldots, a_n]$ is countable.

The function $f : \mathbb{Q}^{n+1} \to$ Span $[1, a_1, \ldots, a_n]$ defined by $f(c_0, c_1, \ldots, c_n) = c_0 + c_1a_1 + \ldots, +c_na_n$ is obviously an onto function. It is one-to-one because $1, a_1, \ldots, a_n$ are rationally independent real numbers. Because $\mathbb{Q}^{n+1}$ is countable (*Exercise* 11, p. 139), Span $[1, a_1, \ldots, a_n]$ is countable . □

Corollary 3.3.8 *There exists* $\mathbf{a} = (a_1, \ldots, a_n) \in \mathbb{R}^n$ *such that the subgroup*

$$H = \langle \mathbf{e}_1, \ldots, \mathbf{e}_n, \mathbf{a} \rangle$$

of $\mathbb{R}^n$ *is dense in* $\mathbb{R}^n$.

Proof. Apply the theorem and *Corollary 3.3.6.* □

Turning to the torus, let H' denote the subgroup of $\mathbb{K}^n$ generated by the elements $\mathbf{z}_1 = \pi_n(\mathbf{a}_1), \ldots, \mathbf{z}_m = \pi_n(\mathbf{a}_m)$, where $\mathbf{a}_1, \ldots, \mathbf{a}_m$ are in $\mathbb{R}^n$. Thus,

$$H' = \langle \mathbf{z}_1, \ldots, \mathbf{z}_m \rangle = \left\{ \prod_{j=1}^{m} \mathbf{z}_j^{k_j} : k_j \in \mathbb{Z} \text{ for } j = 1, \ldots, m \right\}, \qquad (3.6)$$

Then $H = \pi_n^{-1}(H')$ is given by equation (3.5). Because π_n is continuous and open, it is easy to verify that H' is dense in $\mathbb{K}^n$ if and only if H is dense in $\mathbb{R}^n$. (See *Exercise* 15, p. 139, for the more general case.) Therefore, *Theorem 3.3.4* can be applied in this situation. In particular, we have the following results:

Theorem 3.3.9 *Let* $\mathbf{a}_1, \ldots, \mathbf{a}_m$ *be elements of* $\mathbb{R}^n$, *and consider the group* $H' = \langle \pi_n(\mathbf{a}_1), \ldots, \pi_n(\mathbf{a}_m) \rangle$ *in* $\mathbb{K}^n$. *Then* H' *is dense in* $\mathbb{K}^n$ *if and only if there does not exist* $\mathbf{r} \in \mathbb{Z}^n \setminus \{\mathbf{0}\}$ *such that* $\mathbf{r} \cdot \mathbf{a}_k \in \mathbb{Z}$ *for* $k = 1, \ldots, m$.

Corollary 3.3.10 *Let* $a_1, \ldots, a_n$ *be* n *real numbers. The cyclic subgroup of* $\mathbb{K}^n$ *generated by* $\mathbf{w} = (\exp(2\pi i a_1), \ldots, \exp(2\pi i a_n))$ *is dense in* $\mathbb{K}^n$ *if and only if* $1, a_1, \ldots, a_n$ *are rationally independent.*

A topological group G is *monothetic* provided there exists a homomorphism $\varphi : \mathbb{Z} \to G$ such that the subgroup $\varphi(\mathbb{Z})$ of G is dense in G. Equivalently, a topological group is monothetic if and only if there exists a in G such that the cyclic group $\{a^k : k \in \mathbb{Z}\}$ is dense in G because $\varphi(k) = a^k$ always defines a homomorphism of $\mathbb{Z}$ to G such that $\varphi(\mathbb{Z}) = \{a^k : k \in \mathbb{Z}\}$. All cyclic groups are monothetic, $\mathbb{K}$ is monothetic, but $\mathbb{R}$ is not monothetic.

A topological group G is *solenoidal* if there exists a homomorphism $\varphi : \mathbb{R} \to G$ such that the subgroup $\varphi(\mathbb{R})$ of G is dense in G. Obviously, $\mathbb{R}$ and its homomorphic image $\mathbb{K}$ are solenoidal, but $\mathbb{R}^n$ is not solenoidal for $n > 1$. All monothetic and solenoidal groups are Abelian because they contain dense Abelian subgroups.

Theorem 3.3.11 *The metric group* $\mathbb{K}^n$ *is both monothetic and solenoidal.*

Proof. It follows from *Theorem 3.3.7* that there exist real numbers $a_1, \ldots, a_m$ such that $1, a_1, \ldots, a_m$ are rationally independent. Then *Corollary 3.3.10* implies that $\mathbb{K}^n$ is monothetic.

As mentioned above, $\mathbb{K}$ is solenoidal because $\mathbb{K} = \exp(\mathbb{R})$. It follows from *Exercise* 6, p. 131, that $\mathbb{K}^2$ is solenoidal, and the same ideas used to show that $\mathbb{K}^2$ is solenoidal can now be employed to prove that $\mathbb{K}^n$ is solenoidal for $n \geq 2$.

Consider $\mathbb{K}^{n+1}$ for $n \geq 1$. Because $\mathbb{K}^n$ is monothetic, there exist real numbers $a_1, \ldots, a_n$ such that the cyclic group

$$\begin{aligned} H' &= \left\{ \left(\exp(2\pi i a_1), \ldots, \exp(2\pi i a_n) \right)^k : k \in \mathbb{Z} \right\} \\ &= \left\{ \left(\exp(2\pi i k a_1), \ldots, \exp(2\pi i k a_n) \right) : k \in \mathbb{Z} \right\} \end{aligned}$$

is dense in $\mathbb{K}^n$.

Define $\varphi : \mathbb{R} \to \mathbb{K}^{n+1}$ by $\varphi(t) = \left(\exp(2\pi i t a_1), \ldots, \exp(2\pi i t a_n), \exp(2\pi t) \right)$ and note that φ is a homomorphism. Furthermore, observe that for $k \in \mathbb{Z}$

$$\varphi(k) = \left(\exp(2\pi i k a_1), \ldots, \exp(2\pi i k a_n), 1 \right).$$

Let $\mathbf{z} = \left(\exp(2\pi i b_1), \ldots, \exp(2\pi i b_{n+1}) \right)$ be an arbitrary point in $\mathbb{K}^{n+1}$. Set

$$s = b_{n+1} - [b_{n+1}],$$

and

$$\mathbf{w} = \left(\exp\left(2\pi i (b_1 - s a_1)\right), \ldots, \exp\left(2\pi i (b_n - s a_n)\right), 1 \right).$$

Clearly, $\mathbf{z} = \mathbf{w}\varphi(s)$ because

$$\varphi(s) = \left(\exp(2\pi i s a_1), \ldots, \exp(2\pi i s a_n), \exp(2\pi i s) \right)$$

and $\exp(2\pi i s) = \exp(2\pi i b_{n+1})$.

Because H' is dense in $\mathbb{K}^n$, there exists a sequence of integers k_j such that

$$\left(\exp(2\pi i a_1), \ldots, \exp(2\pi i a_n) \right)^{k_j} = \left(\exp(2\pi i k_j a_1), \ldots, \exp(2\pi i k_j a_n) \right)$$

converges to

$$\left(\exp\left(2\pi i (b_1 - s a_1)\right), \ldots, \exp\left(2\pi i (b_n - s a_n)\right) \right)$$

in $\mathbb{K}^n$. It follows that

$$\varphi(k_j) = \big(\exp(2\pi i k_j\, a_1), \ldots, \exp(2\pi i k_j\, a_n), 1\big).$$

converges to $\mathbf{w}$ in $\mathbb{K}^{n+1}$. Hence, $\varphi(k_j + s) = \varphi(k_j)\varphi(s)$ converges to $\mathbf{w}\varphi(s) = \mathbf{z}$, proving that $\varphi(\mathbb{R})$ is dense in $\mathbb{K}^{n+1}$ and $\mathbb{K}^{n+1}$ is solenoidal. □

EXERCISES

1. Given $\mathbf{u} \in \mathbb{R}^n$, let $f_{\mathbf{u}} : \mathbb{R}^n \to \mathbb{R}$ denote the function defined by $\mathbf{f_u}(\mathbf{x}) = \mathbf{x} \cdot \mathbf{u}$. Show that $\mathbf{u} \mapsto f_{\mathbf{u}}$ defines a linear isomorphism from $\mathbb{R}^n$ onto $\mathcal{L}(\mathbb{R}^n, \mathbb{R})$.

2. Let $\mathbf{c}_1, \ldots, \mathbf{c}_m$ be elements of $\mathbb{R}^n$, and let $H = \langle \mathbf{c}_1, \ldots, \mathbf{c}_m \rangle$. Prove that
$$H^{\#} = \{\mathbf{x} \in \mathbb{R}^n : \mathbf{x} \cdot \mathbf{c}_k \in \mathbb{Z} \text{ for } j = 1, \ldots, m\}.$$

3. Let E be a nonempty subset of $\mathbb{R}^n$, and set
$$E^{\perp} = \{\mathbf{x} \in \mathbb{R}^n : \mathbf{x} \cdot \mathbf{y} = 0 \text{ for all } \mathbf{y} \in E\}$$
Show that $E^{\perp}$ has the following properties:
   (a) $\{\mathbf{0}\}^{\perp} = \mathbb{R}^n$ and $\left(\mathbb{R}^n\right)^{\perp} = \{\mathbf{0}\}$.
   (b) $E \subset F$ implies that $F^{\perp} \subset E^{\perp}$.
   (c) $E^{\perp}$ is a subspace of $\mathbb{R}^n$ and thus also closed.
   (d) If V is a subspace of $\mathbb{R}^n$, then $V^{\#} = V^{\perp}$.
   (e) If E is a subspace of $\mathbb{R}^n$, then $E^{\perp\perp} = E$.

4. Let $\mathbf{u}_1, \ldots, \mathbf{u}_n$ be a basis of $\mathbb{R}^n$. Show that there exists a basis $\mathbf{v}_1, \ldots, \mathbf{v}_n$ for $\mathbb{R}^n$ such that
$$\mathbf{u}_j \cdot \mathbf{v}_k = \begin{cases} 1 & \text{when } j = k \\ 0 & \text{when } j \neq k \end{cases}.$$
(One approach is to use *Exercise 1* above and the basis for $\mathcal{L}(\mathbb{R}^n, \mathbb{R})$ given by equation (2.3), page 74.)

5. Let H be a closed subgroup of $\mathbb{R}^n$. Prove the following:
   (a) H is discrete if and only if the rank of $H^{\#}$ is n.
   (b) H is a subspace of $\mathbb{R}^n$ if and only if $H^{\#}$ is a subspace of $\mathbb{R}^n$.

6. Let H be the subgroup of $\mathbb{R}^n$ generated by $\mathbf{a}_1, \ldots, \mathbf{a}_m$. Prove that if $m \leq n$, then H is not dense in $\mathbb{R}^n$.

7. Show that $\mathbb{Z}$ is countable.

8. Let E_k, $k = 1, \ldots$ be a sequence of disjoint finite sets. Show that

$$E = \bigcup_{k=1}^{\infty} E_k$$

is countable. Show that if the sets E_k are not assumed to be disjoint, then $E = \bigcup_{k=1}^{\infty} E_k$ is either finite or countable.

9. Use *Exercise 8* above to prove that $(\mathbb{Z}^+)^n$, $\mathbb{Z}^n$, $\mathbb{Q}^+$, and $\mathbb{Q}$ are countable. For example, to prove that $\mathbb{Z}^+ \times \mathbb{Z}^+$ is countable, let $E_1 = \{(1,1)\}$, $E_2 = \{(1,2),(2,1)\}$, $E_3 = \{(1,3),(2,2),(3,1)\}$, and so on.

10. Let F_k, $k = 1, \ldots$ be a sequence of disjoint countable sets. Show that $F = \bigcup_{k=1}^{\infty} F_k$ is countable. (One approach is to again use *Exercise 8* above.)

11. Prove that $\mathbb{Q}^n$ is countable for all n in $\mathbb{Z}^+$. (One approach is to use *Exercise 10* above and induction.)

12. Show that an infinite subset of a countable set is countable.

13. Let $\theta : E \to F$ from the set E onto the set F. Prove that if E is countable, then F is either finite or countable.

14. An element a of a group G is of *finite order* provided that $a^k = e$ for some positive integer k. Show that the elements of finite order in $\mathbb{K}^n$ are a countable dense subgroup of $\mathbb{K}^n$.

15. Let G and G' be metric groups, and let $\varphi : G \to G'$ be an onto homomorphism. Show that if H is a dense subgroup of G, then $\varphi(H)$ is dense in G'. Assuming φ is an open function, prove that H' is a dense subgroup of G' if and only if $\varphi^{-1}(H')$ is a dense subgroup of G.

16. Let $\mathbf{z}$ and $\mathbf{w}$ be elements of $\mathbb{K}^2$, and let $H = \langle \mathbf{z}, \mathbf{w} \rangle$. Show that there exist $\mathbf{z}$ and $\mathbf{w}$ such that H is dense in $\mathbb{K}^2$, but the subgroups generated by $\mathbf{z}$ and by $\mathbf{w}$ individually are not dense in $\mathbb{K}^2$.

Chapter 4

Matrix Groups

The general linear groups $\mathrm{GL}(n, \mathbb{R})$ and $\mathrm{GL}(n, \mathbb{C})$ were introduced at the end of Chapter 2 and shown to be locally compact σ-compact metric groups (*Theorem 2.5.14*). This chapter is devoted to the closed subgroups of $\mathrm{GL}(n, \mathbb{R})$ and $\mathrm{GL}(n, \mathbb{C})$. Their subgroup structure is very rich, and the results presented here are nowhere near as definitive as the results about the subgroups of $\mathbb{R}^n$ in Chapter 3. In fact, most of this chapter focuses on a few special families of subgroups of the general linear groups.

Section 4.1 begins with a series of simple examples that demonstrate the wide variety of the closed subgroups of the general linear groups. In particular, $\mathrm{GL}(n, \mathbb{R})$ is naturally a closed subgroup of $\mathrm{GL}(n, \mathbb{C})$, and $\mathrm{GL}(n, \mathbb{C})$ is isomorphic to a closed subgroup of $\mathrm{GL}(2n, \mathbb{R})$. In addition, the special linear subgroups $\mathrm{SL}(n, \mathbb{R})$ and $\mathrm{SL}(n, \mathbb{C})$ of $\mathrm{GL}(n, \mathbb{R})$ and $\mathrm{GL}(n, \mathbb{C})$, respectively, are shown to be manifolds

The orthogonal and unitary groups studied in Section 4.2 are compact subgroups of $\mathrm{GL}(n, \mathbb{R})$ and $\mathrm{GL}(n, \mathbb{C})$, respectively. They are also shown to be manifolds. The elements of the orthogonal group $\mathrm{O}(n)$ can also be thought of as Euclidean isometries of $\mathbb{R}^n$. In Section 6.3, the orthogonal group will be used as part of our analysis of the metric group of rigid motions of $\mathbb{R}^n$.

Triangular groups of matrices, the subject of Section 4.3, provide important examples of nilpotent and solvable metric groups that are closed subgroups of the general linear groups. The nilpotent ones are homeomorphic to $\mathbb{R}^q$ for some q. Although they are not Abelian, they are in some sense the next best thing to Abelian. Analogous to $\mathbb{R}^n$ and $\mathbb{Z}^n$, these nilpotent groups contain discrete subgroups with compact quotient spaces.

In Section 4.4, the exponential of a matrix is defined using power series. This is done in the more general context of a Banach algebra. The exponential of a matrix is always an invertible matrix. For a fixed matrix A, the exponential function of e^{tA}, $t \in \mathbb{R}$ is a homomorphism of $\mathbb{R}$ into $\mathrm{GL}(n, \mathbb{S})$, called a *one-parameter subgroup*. Thus the general linear groups contain a wealth of one-parameter subgroups.

4.1 General Linear Groups

This section partially explores the rich subgroup structure of $\mathrm{GL}(n, \mathbb{S})$. In contrast with $\mathbb{R}^n$, $\mathrm{GL}(n, \mathbb{S})$ has an incredible variety of subgroups. Whereas the subgroups of $\mathbb{R}^n$ are all of the form $\mathbb{R}^p \times \mathbb{Z}^q$, the general linear groups contain closed subgroups isomorphic to the group of rigid motions, every finite group, $\mathbb{R}^n$ and $\mathbb{K}^n$, among others.

As before it will be understood that all finite groups are metric groups with the discrete topology. Let α be an element of S_n, the permutation group on n symbols, and let A be an element of $\mathcal{M}_n(\mathbb{S})$, the algebra of $n \times n$ matrices for the real or complex scalar field. Recall from Section 2.5 that ${}_\alpha A$ denotes the matrix A with its rows permuted by α, that is, ${}_\alpha A(j,k) = A(\alpha(j), k)$ for all j and k. Also recall that $\mathrm{Det}\,[{}_\alpha I] = \mathrm{sgn}(\alpha) \neq 0$, where I is the identity matrix. So ${}_\alpha I$ is in $\mathrm{GL}(n, \mathbb{S})$.

Since

$$ {}_\alpha I A(j,k) = \sum_{m=1}^{n} I(\alpha(j), m) A(m,k) = A(\alpha(j), k), $$

it follows that

$$ {}_\alpha I A = {}_\alpha A. $$

In particular,

$$ ({}_\alpha I)({}_\beta I) = {}_\alpha({}_\beta I) = {}_{\alpha \circ \beta} I, $$

and the function $\varphi : S_n \to \mathrm{GL}(n, \mathbb{S})$ defined by $\varphi(\alpha) = {}_\alpha I$ is an algebraic homomorphism. Of course, φ is continuous because S_n is discrete. It is one-to-one because ${}_\alpha I = I$ if and only if $\alpha = \iota$, the identity element of S_n. Therefore, $\{{}_\alpha I : \alpha \in S_n\}$ is a finite subgroup of $\mathrm{GL}(n, \mathbb{S})$ isomorphic to S_n.

Proposition 4.1.1 *If G is a finite group containing n elements, then* $\mathrm{GL}(n, \mathbb{S})$ *contains a subgroup isomorphic to G.*

Proof. By *Exercise* 14, p. 12, there exists a subgroup H of S_n isomorphic to G. With $\varphi(\alpha) = {}_\alpha I$ as above, $\varphi(H)$ is the required subgroup of $\mathrm{GL}(n, \mathbb{S})$ isomorphic to G. □

An $n \times n$ matrix A is a *diagonal matrix* provided $A(j,k) = 0$, when $j \neq k$. Given scalars $a_1, \ldots, a_n$, it is convenient to denote the diagonal matrix A with $A(j,j) = a_j$ by $A = \mathrm{Diag}\,(a_1, \ldots, a_n)$. Note that

$$ \mathrm{Det}\,[\mathrm{Diag}\,(a_1, \ldots, a_n)] = \prod_{j=1}^{n} a_j $$

If $B = \mathrm{Diag}\,(b_1, \ldots, b_n)$ is another diagonal matrix, it is easy to see that $AB = \mathrm{Diag}\,(a_1 b_1, \ldots, a_n b_n)$, and the diagonal matrices form a semigroup.

Proposition 4.1.2 *The set of $n \times n$ matrices*

$$ \mathrm{DL}(n, \mathbb{S}) = \{\mathrm{Diag}\,(a_1, \ldots, a_n) : 0 \neq a_j \in \mathbb{S} \text{ for } j = 1, \ldots, n\} \qquad (4.1) $$

is a closed subgroup of the metric group $\mathrm{GL}(n,\mathbb{S})$ *and is isomorphic to the multiplicative metric group* $(\mathbb{S}\setminus\{0\})^n$.

Proof. Obviously, $\mathrm{DL}(n,\mathbb{S})$ is contained in $\mathrm{GL}(n,\mathbb{S})$. If $A=\mathrm{Diag}\,(a_1,\ldots,a_n)$ is in $\mathrm{DL}(n,\mathbb{S})$, then $A^{-1}=\mathrm{Diag}\,(1/a_1,\ldots,1/a_n)$ is in $\mathrm{DL}(n,\mathbb{S})$. It follows that $\mathrm{DL}(n,\mathbb{S})$ is a subgroup of $\mathrm{GL}(n,\mathbb{S})$.

The set V of all diagonal matrices is a subspace of $\mathcal{M}_n(\mathbb{S})$ and a closed subset of $\mathcal{M}_n(\mathbb{S})$. Since $\mathrm{DL}(n,\mathbb{S})=V\cap\mathrm{GL}(n,\mathbb{S})$, it follows that $\mathrm{DL}(n,\mathbb{S})$ is a closed subset of $\mathrm{GL}(n,\mathbb{S})$.

Finally, $\varphi(a_1,\ldots,a_n)=\mathrm{Diag}\,(a_1,\ldots,a_n)$, defines an isomorphism from $(\mathbb{S}\setminus\{0\})^n$ onto $\mathrm{DL}(n,\mathbb{S})$. □

Corollary 4.1.3 *The set of* $n\times n$ *complex matrices*

$$H=\{\mathrm{Diag}\,(a_1,\ldots,a_n): a_j\in\mathbb{C}\text{ and }|a_j|=1\ for\ j=1,\ldots,n\}$$

is a closed subgroup of the metric group $\mathrm{GL}(n,\mathbb{C})$ *and is isomorphic to* $\mathbb{K}^n$.

Proof. Clearly, H is a closed subgroup of $\mathrm{GL}(n,\mathbb{C})$. Observe that $\mathbb{K}^n$ is a closed subgroup of $(\mathbb{C}\setminus\{0\})^n$ and $H=\varphi(\mathbb{K}^n)$, where φ is the isomorphism defined in the proof of the proposition. □

Let k and m be positive integers such that $k+m=n$ and let A and B be $k\times k$ and $m\times m$ matrices over the same scalar field. Consider the block diagonal matrix

$$C=\begin{bmatrix} A & \mathrm{O} \\ \mathrm{O} & B\end{bmatrix}, \tag{4.2}$$

where O denotes a $k\times m$ matrix of all zeros in the upper-right corner and a $m\times k$ matrix of all zeros in the lower-left corner. (In general, O will be used to denote a matrix of all zeros of the appropriate size for the context.) If

$$C'=\begin{bmatrix} A' & \mathrm{O} \\ \mathrm{O} & B'\end{bmatrix}$$

is another such matrix, then

$$CC'=\begin{bmatrix} AA' & \mathrm{O} \\ \mathrm{O} & BB'\end{bmatrix}.$$

Furthermore, $\mathrm{Det}\,[C]=\mathrm{Det}\,[A]\mathrm{Det}\,[B]$ (*Exercise* 2, p. 148). So C is in $\mathrm{GL}(n,\mathbb{S})$ if and only if A is in $\mathrm{GL}(k,\mathbb{S})$ and B is in $\mathrm{GL}(m,\mathbb{S})$. When C is in $\mathrm{GL}(n,\mathbb{S})$, then

$$C^{-1}=\begin{bmatrix} A^{-1} & \mathrm{O} \\ \mathrm{O} & B^{-1}\end{bmatrix}.$$

Proposition 4.1.4 *If* H_1 *and* H_2 *are closed subgroups of the metric groups* $\mathrm{GL}(k,\mathbb{S})$ *and* $\mathrm{GL}(m,\mathbb{S})$, *respectively, and* $n=k+m$, *then*

$$H=\left\{\begin{bmatrix} A & \mathrm{O} \\ \mathrm{O} & B\end{bmatrix}: A\in H_1\text{ and }B\in H_2\right\}$$

is a closed subgroup of the metric group $\mathrm{GL}(n,\mathbb{S})$ *isomorphic to* $H_1\times H_2$.

Proof. It follows from the discussion preceding the statement of the proposition that H is a subgroup of $\mathrm{GL}(n, \mathbb{S})$. If

$$C_n = \begin{bmatrix} A_n & \mathrm{O} \\ \mathrm{O} & B_n \end{bmatrix},$$

is a sequence of matrices in H converging to some C, then C is of the form

$$C = \begin{bmatrix} A & \mathrm{O} \\ \mathrm{O} & B \end{bmatrix},$$

where A_n and B_n converge to A and B, respectively. Since H_1 and H_2 are closed subgroups, A and B are in H_1 and H_2, respectively. Hence, C is in H, and H is a closed subgroup. Clearly,

$$(A, B) \to \begin{bmatrix} A & \mathrm{O} \\ \mathrm{O} & B \end{bmatrix}$$

defines an isomorphism of $H_1 \times H_2$ onto H. □

Corollary 4.1.5 *The metric group* $\mathrm{GL}(n, \mathbb{S})$ *contains closed subgroups isomorphic to* $\mathrm{GL}(k, \mathbb{S})$ *for* $0 < k < n$.

Proof. Let $H_1 = \mathrm{GL}(k, \mathbb{S})$ and let $H_2 = \{I\}$. □

Recall that for a complex number $z = x + iy$, the complex conjugate of z is $\overline{z} = x - iy$. The function $z \mapsto \overline{z}$ is obviously a continuous function from $\mathbb{C}$ onto itself. Define $\overline{A}$, the conjugate of a complex matrix A, by

$$\overline{A}(j, k) = \overline{A(j, k)}.$$

Clearly, $\overline{\overline{A}} = A$ and $A \mapsto \overline{A}$ is a continuous one-to-one function of $\mathcal{M}_n(\mathbb{C})$ onto itself. Because $\overline{z + w} = \overline{z} + \overline{w}$ and $\overline{zw} = \overline{z}\,\overline{w}$, it follows that

$$\overline{A + B} = \overline{A} + \overline{B} \text{ and } \overline{AB} = \overline{A}\,\overline{B}$$

for all complex $n \times n$ matrices A and B. Hence, $A \mapsto \overline{A}$ defines an automorphism of $\mathrm{GL}(n, \mathbb{C})$. Since a complex number z is a real number if and only if $\overline{z} = z$,

$$\mathrm{GL}(n, \mathbb{R}) = \left\{A \in \mathrm{GL}(n, \mathbb{C}) : \overline{A} = A\right\}. \tag{4.3}$$

Applying *Exercise* 1, p. 148, to the right side of equation (4.3) shows that $\mathrm{GL}(n, \mathbb{R})$ is the set of fixed points of an automorphism of $\mathrm{GL}(n, \mathbb{C})$, and hence a closed subgroup of $\mathrm{GL}(n, \mathbb{C})$. This idea will be used again in Section 4.2.

What is more interesting is that $\mathrm{GL}(n, \mathbb{C})$ is isomorphic to a closed subgroup of $\mathrm{GL}(2n, \mathbb{R})$. In fact, except for the complex linear structure, the whole algebraic structure of $\mathcal{M}_n(\mathbb{C})$ can be found in $\mathrm{GL}(2n, \mathbb{R})$.

Theorem 4.1.6 *There exists a continuous one-to-one function* $\varphi : \mathcal{M}_n(\mathbb{C}) \to \mathcal{M}_{2n}(\mathbb{R})$ *such that the following hold:*

(a) For all A and B in $\mathcal{M}_n(\mathbb{C})$ and c in $\mathbb{R}$,

$$\begin{aligned} \varphi(A+B) &= \varphi(A)+\varphi(B) \\ \varphi(cA) &= c\varphi(A) \\ \varphi(I) &= I \\ \varphi(AB) &= \varphi(A)\varphi(B). \end{aligned}$$

(b) A is in $\mathrm{GL}(n,\mathbb{C})$ if and only if $\varphi(A)$ is in $\mathrm{GL}(2n,\mathbb{R})$.

(c) φ maps $\mathrm{GL}(n,\mathbb{C})$ isomorphically onto a closed subgroup of $\mathrm{GL}(2n,\mathbb{R})$.

Proof. From the proof of *Theorem 2.2.6*, let $\Phi_1 : \mathcal{L}(\mathbb{C}^n, \mathbb{C}^n) \to \mathcal{M}_n(\mathbb{C})$ and $\Phi_2 : \mathcal{L}(\mathbb{R}^{2n}, \mathbb{R}^{2n}) \to \mathcal{M}_{2n}(\mathbb{R})$ be the one-to-one onto functions that assign to a linear function in $\mathcal{L}(\mathbb{C}^n, \mathbb{C}^n)$ or $\mathcal{L}(\mathbb{R}^{2n}, \mathbb{R}^{2n})$ its matrix with respect to the standard basis $\mathbf{e}_1', \ldots, \mathbf{e}_n'$ of $\mathbb{C}^n$ or $\mathbf{e}_1, \ldots, \mathbf{e}_{2n}$ of $\mathbb{R}^{2n}$. The functions Φ_i, $i = 1, 2$ have the following properties for all $S, T \in \mathcal{L}(\mathbb{S}^n, \mathbb{S}^n)$ and $c \in \mathbb{S}$:

$$\begin{aligned} \Phi_i(S+T) &= \Phi_i(S)+\Phi_i(T) \\ \Phi_i(cS) &= c\,\Phi_i(S) \\ \Phi_i(\iota) &= I \\ \Phi_i(S\circ T) &= \Phi_i(S)\Phi_1(T). \end{aligned}$$

Moreover, T is invertible if and only if $\Phi_i(T)$ is invertible.

Next define $f : \mathbb{R}^{2n} \to \mathbb{C}^n$ by

$$f(x_1, \ldots, x_{2n}) = (x_1 + ix_2, x_3 + ix_4, \ldots, x_{2n-1} + ix_{2n}). \tag{4.4}$$

Clearly, f is a one-to-one onto real linear function. (Here is where the complex linear structure is lost, but the real linear structure remains.) Then define $\psi : \mathcal{L}(\mathbb{C}^n, \mathbb{C}^n) \to \mathcal{L}(\mathbb{R}^{2n}, \mathbb{R}^{2n})$ by $\psi(T) = f^{-1} \circ T \circ f$. It is easily verified that

$$\begin{aligned} \psi(S+T) &= \psi(S)+\psi(T) \\ \psi(cS) &= c\psi(S) \\ \psi(\iota) &= \iota \\ \psi(S\circ T) &= \psi(S)\circ\psi(T). \end{aligned}$$

hold for all $S, T \in \mathcal{L}(\mathbb{C}^n, \mathbb{C}^n)$ and $c \in \mathbb{R}$. The function ψ is a real linear one-to-one function; it is not onto because the dimensions of $\mathcal{L}(\mathbb{C}^n, \mathbb{C}^n)$ and $\mathcal{L}(\mathbb{R}^{2n}, \mathbb{R}^{2n})$ over the scalar field $\mathbb{R}$ are $2n^2$ and $4n^2$, respectively. Moreover, T is invertible if and only if $\psi(T)$ is invertible.

Set $\varphi = \Phi_2 \circ \psi \circ \Phi_1^{-1}$. Since the dimension of $\mathcal{M}_n(\mathbb{C})$ as a real linear space is finite, the function φ is continuous. The properties of Φ_1, ψ, and Φ_2 imply that φ is one-to-one and satisfies *(a)* and *(b)*.

It follows from *(b)* that

$$\varphi\big(\mathcal{M}_n(\mathbb{C})\big) \cap \mathrm{GL}(n,\mathbb{R}) = \varphi\big(\mathrm{GL}(n,\mathbb{C})\big).$$

Since $\varphi(\mathcal{M}_n(\mathbb{C}))$ is a subspace of $\mathcal{M}_{2n}(\mathbb{R})$ and hence closed, $\varphi(\mathrm{GL}(n,\mathbb{C}))$ is a closed subgroup of $\mathrm{GL}(2n,\mathbb{R})$. Clearly, φ restricted to $\mathrm{GL}(n,\mathbb{C})$ is an isomorphism of $\mathrm{GL}(n,\mathbb{C})$ onto $\varphi(\mathrm{GL}(n,\mathbb{C}))$. □

Corollary 4.1.7 *The metric group H is isomorphic to a closed subgroup of* $\mathrm{GL}(n,\mathbb{R})$ *for some n if and only if it is isomorphic to a closed subgroup of* $\mathrm{GL}(m,\mathbb{C})$ *for some m.*

With φ constructed as in the proof of *Theorem 4.1.6*, it is also possible to describe more precisely the matrices in $\varphi(\mathcal{M}_n(\mathbb{C}))$. Let $\mathbf{e}_1,\ldots,\mathbf{e}_{2n}$ be the standard basis for $\mathbb{R}^{2n}$ and let $\mathbf{e}'_1,\ldots,\mathbf{e}'_n$ be the standard basis for $\mathbb{C}^n$. If A is the matrix of $T \in \mathcal{L}(\mathbb{C}^n,\mathbb{C}^n)$ with respect to $\mathbf{e}'_1\ldots,\mathbf{e}'_n$, then $\varphi(A)$ is the matrix of $f^{-1}\circ T\circ f \in \mathcal{L}(\mathbb{R}^n,\mathbb{R}^n)$, where f is defined by equation (4.4).

For convenience of notation, let $A(j,k) = z_{jk} = x_{jk} + iy_{jk}$. Observe that $f(\mathbf{e}_k) = \mathbf{e}'_{(k+1)/2}$ when k is odd and $f(\mathbf{e}_k) = i\mathbf{e}'_{k/2}$ when k is even. Furthermore,

$$f^{-1}(z_{1k},\ldots,z_{nk}) = (x_{1k},y_{1k},\ldots,x_{nk},y_{n,k})$$

and

$$f^{-1}(iz_{1k},\ldots,iz_{nk}) = (-y_{1k},x_{1k},\ldots,-y_{nk},x_{nk}).$$

Because the columns of $\varphi(A)$ are the coefficients of $f^{-1}\circ T\circ f(\mathbf{e}_k)$ with respect to the basis $\mathbf{e}_1,\ldots,\mathbf{e}_{2n}$, the typical $2k-1$ and $2k$ columns of $\varphi(A)$ are:

$$\begin{matrix} x_{1k} & -y_{1k} \\ y_{1k} & x_{1k} \\ \vdots & \vdots \\ x_{nk} & -y_{nk} \\ y_{nk} & x_{nk} \end{matrix}$$

for $k = 1,\ldots,n$,

For example, if $n = 2$, then the matrices in $\varphi(\mathcal{M}_n(\mathbb{C}))$ are precisely those 4×4 matrices of the form

$$\begin{bmatrix} x_{11} & -y_{11} & x_{12} & -y_{12} \\ y_{11} & x_{11} & y_{12} & x_{12} \\ x_{21} & -y_{21} & x_{22} & -y_{22} \\ y_{21} & x_{21} & y_{22} & x_{22} \end{bmatrix}$$

More generally, $\varphi(\mathcal{M}_n(\mathbb{C}))$ consists precisely of the real $2n\times 2n$ matrices satisfying the equations

$$\begin{aligned} A(2j-1,2k-1) &= A(2j,2k) \\ A(2j-1,2k) &= -A(2j,2k-1) \end{aligned}$$

for $j = 1,\ldots,n$ and $k = 1,\ldots,n$.

The $n = 1$ case warrants additional comment. Since 1×1 matrices are just scalars, $\mathbb{S} = \mathcal{M}_1(\mathbb{S})$. In this case, it is easy to verify directly that

$$\varphi(z) = \varphi(x + iy) = \begin{bmatrix} x & -y \\ y & x \end{bmatrix}$$

satisfies the conclusion of *Theorem 4.1.6.* It follows that the set matrices given by

$$\left\{ \begin{bmatrix} x & -y \\ y & x \end{bmatrix} : x, y \in \mathbb{R} \right\}$$

is a copy of $\mathbb{C}$ sitting in the algebra $\mathcal{M}_2(\mathbb{R})$. Also note that $\text{Det}\,[\varphi(A)] = |z|^2$. Hence the metric groups $\mathbb{C} \setminus \{0\}$ and $\mathbb{K}$ are isomorphic to the closed subgroups of $\text{GL}(2, \mathbb{R})$ given by

$$\left\{ \begin{bmatrix} x & -y \\ y & x \end{bmatrix} : x, y \in \mathbb{R} \text{ and } x^2 + y^2 > 0 \right\}. \tag{4.5}$$

and

$$\left\{ \begin{bmatrix} x & -y \\ y & x \end{bmatrix} : x, y \in \mathbb{R} \text{ and } x^2 + y^2 = 1 \right\} \tag{4.6}$$

respectively.

A metric space X is a *manifold of dimension n* provided that for each x in X there exists an open neighborhood U of x homeomorphic to an open subset of $\mathbb{R}^n$. It is easy to see that a metric space X is an n-dimensional manifold if and only if for each x in X there exists an open neighborhood U of x homeomorphic to an open ball $B_r(\mathbf{v})$ of $\mathbb{R}^n$. Obviously, $\mathbb{R}^n$, $\mathbb{C}^n$ and their open subsets are manifolds of dimension n and $2n$, respectively. Notice that the dimension of $\mathbb{R}^n$ as a manifold coincides with its dimension as a real linear space. It is convenient to let $\mathbb{R}^0 = \{\mathbf{0}\}$. Then discrete metric spaces are zero-dimensional manifolds. Manifolds are a particularly rich class of metric spaces. For an introduction to the topology of manifolds in which groups play an important role see [8].

It follows that the metric group $\text{GL}(n, \mathbb{R})$ is a manifold of dimension n and $\text{GL}(n, \mathbb{C})$ is a manifold of dimension $2n$ because $\text{GL}(n, \mathbb{R})$ and $\text{GL}(n, \mathbb{C})$ are open subsets of $\mathcal{M}_n(\mathbb{R})$ and $\mathcal{M}_n(\mathbb{C})$, respectively. To prove that a metric group is a manifold of dimension n, it suffices to show that there exists an open neighborhood U of the identity e homeomorphic to an open ball $B_r(\mathbf{v})$.

Theorem 4.1.8 *The metric group* $\text{SL}(n, \mathbb{R})$ *is a manifold of dimension* $n^2 - 1$.

Proof. Since $\text{SL}(1, \mathbb{R})$ is the discrete multiplicative group $\{1, -1\}$, it is a zero-dimensional manifold.

Assume for the rest of the proof that $n > 1$. The idea of the proof is simply to solve for $A(n, n)$ in terms of the other $n^2 - 1$ entries of A, when $A \in \text{SL}(n, \mathbb{R})$ is close to I.

The function $\text{Det}\,[A(n|n)]$ is a continuous function on $\mathcal{M}_n(\mathbb{R})$ such that $\text{Det}\,[I(n|n)] = 1$. It follows that there exists an open set U' containing I such that $|\mathrm{Det}\,[A(n|n)] - 1| < 1/2$ for all $A \in U'$. Set $U = U' \cap \mathrm{SL}(n, \mathbb{R})$, which is an open set in $\mathrm{SL}(n, \mathbb{R})$ containing I. It will be shown that U is homeomorphic to an open subset of $\mathbb{R}^{n^2-1}$.

Let $\widehat{\mathcal{M}}_n(\mathbb{R})$ be the set of arrays of the form

$$\hat{A} = \begin{pmatrix} \hat{A}(1,1) & \hat{A}(1,2) & \dots & \hat{A}(1,n-1) & \hat{A}(1,n) \\ \vdots & \vdots & & \vdots & \vdots \\ \hat{A}(n-1,1) & \hat{A}(n-1,2) & \dots & \hat{A}(n-1,n-1) & \hat{A}(n-1,n) \\ \hat{A}(n,1) & \hat{A}(n,2) & \dots & \hat{A}(n,n-1) & \end{pmatrix}$$

In other words, $\widehat{\mathcal{M}}_n(\mathbb{R})$ is the set of $n \times n$ "matrices" with no $A(n,n)$ entry. Clearly, $\widehat{\mathcal{M}}_n(\mathbb{R})$ is just another way of writing $\mathbb{R}^{n^2-1}$ that turns out to be convenient for this proof.

There is a natural projection $p : \mathcal{M}_n(\mathbb{R}) \to \widehat{\mathcal{M}}_n(\mathbb{R})$ defined by simply erasing the $A(n,n)$ entry A. It is a continuous and open function like all projections. Set $V = p(U')$, which is an open subset of $\widehat{\mathcal{M}}_n(\mathbb{R})$ or $\mathbb{R}^{n^2-1}$. It suffices to show that U is homeomorphic to V.

Recall that

$$\mathrm{Det}\,(A) = \sum_{\alpha \in S_n} \mathrm{sgn}(\alpha) \prod_{i=1}^{n} A(i, \alpha(i)).$$

Let $H = \{\alpha \in S_n : \alpha(n) = n\}$. The restriction of α in H to $1, \dots, n-1$ is a permutation of $n-1$ symbols. Conversely, every element β of S_{n-1} can be regarded as the restriction of an element α of H to $1, \dots, n-1$. Furthermore, the sign of α in H is the same as the sign of its restriction to $1, \dots, n-1$ because the α-orbit of n is just $\{n\}$ and contributes nothing to $N(\alpha)$. (See the discussion preceding equation (1.2), page 10.) Applying these observations to the formula for the determinant yields the following:

$$\begin{aligned} \mathrm{Det}\,(A) &= \sum_{\alpha \in H} \mathrm{sgn}(\alpha) \prod_{i=1}^{n} A(i, \alpha(i)) + \sum_{\alpha \notin H} \mathrm{sgn}(\alpha) \prod_{i=1}^{n} A(i, \alpha(i)) \\ &= \sum_{\alpha \in S_{n-1}} A(n,n)\mathrm{sgn}(\alpha) \prod_{i=1}^{n-1} A(i, \alpha(i)) + \sum_{\alpha \notin H} \mathrm{sgn}(\alpha) \prod_{i=1}^{n} A(i, \alpha(i)) \\ &= A(n,n)\mathrm{Det}\,[A(n|n)] + \sum_{\alpha \notin H} \mathrm{sgn}(\alpha) \prod_{i=1}^{n} A(i, \alpha(i)). \end{aligned}$$

If A is in U', then $\mathrm{Det}\,[A(n|n)] \neq 0$ and $\mathrm{Det}\,[A] = 1$ can be solved for $A(n,n)$ as

$$A(n,n) = \frac{1 - \sum_{\alpha \notin H} \mathrm{sgn}(\alpha) \prod_{i=1}^{n} A(i, \alpha(i))}{\mathrm{Det}\,[A(n|n)]}$$

Thus by changing only the $A(n,n)$ entry of A, we can assume that $Det[A] = 1$ and that A is in U. It follows that $p(U) = p(U') = V$

Define $f : V \to U$ by

$$f(\hat{A})(n,n) = \frac{1 - \sum_{\alpha \notin H} \operatorname{sgn}(\alpha) \prod_{i=1}^{n} A(i, \alpha(i))}{\operatorname{Det}[A(n|n)]}$$

and

$$f(\hat{A})(j,k) = \hat{A}(j,k)$$

for $(j,k) \neq (n,n)$. Since for all j and k the functions $f(\hat{A})(j,k)$ are continuous functions, the function f is continuous. Clearly, f is the inverse of $p\,|U$ and the open set U of $\mathrm{SL}(n,\mathbb{R})$ is homeomorphic to the open set V of $\mathbb{R}^{n^2-1}$. $\square$

The more common proof of the preceding theorem takes advantage of the fact that the determinant is a polynomial function of n^2 variables, and hence has continuous partial derivatives of all orders. It proceeds by also using the formula

$$\operatorname{Det}[A] = A(n,n)\operatorname{Det}[A(n|n)] + \sum_{\alpha \notin H} \operatorname{sgn}(\alpha) \prod_{i=1}^{n} A(i, \alpha(i)),$$

to show that the partial derivative of the determinant with respect to the $A(n,n)$ variable at I is 1, and then invoking the implicit function theorem. There are pluses and minuses to both approaches. Since our focus is on metrics and groups, we opted for the explicit approach that requires no derivatives.

EXERCISES

1. Let G be a Hausdorff topological group and let $\varphi : G \to G$ be a homomorphism. Show that $\{x \in G : \varphi(x) = x\}$ is a closed subgroup of G. Use this to show that if a is an element of a topological group G, then $H = \{x : xa = ax\}$ is a closed subgroup of G.

2. Let C be the block diagonal matrix given by equation (4.2). Show that $\operatorname{Det}[C] = \operatorname{Det}[A]\operatorname{Det}[B]$.

3. Show that
$$H = \left\{ \begin{pmatrix} 1 & c \\ 0 & 1 \end{pmatrix} : c \in \mathbb{S} \right\}$$
is a closed subgroup of $\mathrm{GL}(2,\mathbb{S})$ isomorphic to $\mathbb{S}$ and use this to construct closed subgroups of general linear groups isomorphic to $\mathbb{S}^n$.

4. Show that the sphere $S^n = \{\mathbf{x} \in \mathbb{R}^{n+1} : \|\mathbf{x}\| = 1\}$ is a manifold of dimension n.

5. Let H be a discrete subgroup of the metric group G. Show that if G is a manifold of dimension n then the quotient space G/H is also a manifold of dimension n.

6. Show that if X and Y are manifolds of dimensions m and n, respectively, then $X \times Y$ is a manifold of dimension $m + n$.
7. Show that $\mathbb{K}^n$ is an n-dimensional manifold.
8. Show that $\mathrm{SL}(n, \mathbb{C})$ is a manifold of dimension $2n^2 - 2$.
9. Let $\mathrm{GL}^+(n, \mathbb{R}) = \{A \in \mathrm{GL}(n, \mathbb{R}) : \mathrm{Det}\,[A] > 0\}$ and prove the following:
   (a) $\mathrm{GL}^+(n, \mathbb{R})$ is a closed normal subgroup of $\mathrm{GL}(n, \mathbb{R})$.
   (b) $\mathrm{GL}(n, \mathbb{R})/\mathrm{GL}^+(n, \mathbb{R})$ is isomorphic to Z_2.
   (c) $\mathrm{GL}^+(n, \mathbb{R})$ is isomorphic to the product metric group $\mathbb{R}^+ \times \mathrm{SL}(n, \mathbb{R})$.
10. Show that $H = \{A \in \mathrm{GL}(n, \mathbb{C}) : \mathrm{Det}\,[A] \in \mathbb{R}^+\}$ is a closed normal subgroup of $\mathrm{GL}(n, \mathbb{C})$ and $\mathrm{GL}(n, \mathbb{C})/H$ is isomorphic to $\mathbb{K}$.

4.2 Orthogonal and Unitary Groups

In this section, $\mathrm{GL}(n, \mathbb{R})$ will be viewed as the closed subgroup of $\mathrm{GL}(n, \mathbb{C})$ defined by equation (4.3). From this perspective, closed subgroups of $\mathrm{GL}(n, \mathbb{R})$ are automatically also closed subgroups of $\mathrm{GL}(n, \mathbb{C})$. *Exercise* 1, p. 148, will used in the next few paragraphs to define several important closed subgroups of $\mathrm{GL}(n, \mathbb{R})$ and $\mathrm{GL}(n, \mathbb{C})$.

The function $A \to (A^t)^{-1} = (A^{-1})^t$ defines an automorphism of $\mathrm{GL}(n, \mathbb{C})$ (*Exercise* 8, p. 114). Consequently,

$$\mathrm{O}(n, \mathbb{C}) = \{A \in \mathrm{GL}(n, \mathbb{C}) : (A^t)^{-1} = A\}$$

defines a closed subgroup of $\mathrm{GL}(n, \mathbb{C})$ called the *complex orthogonal group*, and

$$\mathrm{O}(n) = \{A \in \mathrm{GL}(n, \mathbb{C}) : (A^t)^{-1} = A = \overline{A}\} = \mathrm{O}(n, \mathbb{C}) \cap \mathrm{GL}(n, \mathbb{R})$$

defines a closed subgroup of $\mathrm{GL}(n, \mathbb{R})$ called the *orthogonal group*.

The function

$$A \to \left((\overline{A})^t\right)^{-1}$$

is the composition of the two automorphisms, $A \to \overline{A}$ and $A \to (A^t)^{-1}$, of $\mathrm{GL}(n, \mathbb{C})$, and hence an automorphism of $\mathrm{GL}(n, \mathbb{C})$. The *unitary group* is the closed subgroup defined by

$$\mathrm{U}(n) = \left\{A \in \mathrm{GL}(n, \mathbb{C}) : \left((\overline{A})^t\right)^{-1} = A\right\}.$$

The automorphisms $A \to \overline{A}$ and $A \to (A^t)^{-1}$ commute, and hence the equation $\left((\overline{A})^t\right)^{-1} = A$ can be written as $(A^t)^{-1} = \overline{A}$ or as $A^{-1} = \overline{A}^t$. Let $A^* = \overline{A}^t$.

Then A is a *unitary matrix* if and only if $A^{-1} = A^*$. Clearly, $\mathrm{O}(n) = \mathrm{U}(n) \cap \mathrm{GL}(n, \mathbb{R})$, and a real matrix A is an *orthogonal matrix* if and only if $A^{-1} = A^t$.

The *special orthogonal groups* and the *special unitary groups* are defined by

$$\mathrm{SO}(n) = \mathrm{O}(n) \cap \mathrm{SL}(n, \mathbb{R})$$

and

$$\mathrm{SU}(n) = \mathrm{U}(n) \cap \mathrm{SL}(n, \mathbb{C})$$

respectively. They are clearly closed subgroups of $\mathrm{GL}(n, \mathbb{R})$ and $\mathrm{GL}(n, \mathbb{C})$. This section is devoted to the basic properties of these five metric groups, $\mathrm{O}(n)$, $\mathrm{U}(n)$, $\mathrm{O}(n, \mathbb{C})$, $\mathrm{SO}(n)$, and $\mathrm{SU}(n)$.

We begin with the geometric significance of $\mathrm{O}(n)$.

Proposition 4.2.1 *Let A be an $n \times n$ real matrix. The following are equivalent:*

(a) A is an orthogonal matrix.

(b) $\|\mathbf{x}A\| = \|\mathbf{x}\|$ for all $\mathbf{x}$ in $\mathbb{R}^n$.

(c) $(\mathbf{x}A) \cdot (\mathbf{y}A) = \mathbf{x} \cdot \mathbf{y}$ for all $\mathbf{x}$ and $\mathbf{y}$ in $\mathbb{R}^n$.

Proof. Given a real $n \times n$ matrix A, recall that $(\mathbf{x}A) \cdot \mathbf{y} = \mathbf{x} \cdot (\mathbf{y}A^t)$ for all $\mathbf{x}$ and $\mathbf{y}$ in $\mathbb{R}^n$ (*Exercise* 7, p. 82). First, suppose A is an orthogonal matrix. Then,

$$\|\mathbf{x}A\|^2 = (\mathbf{x}A) \cdot (\mathbf{x}A) = \mathbf{x} \cdot (\mathbf{x}AA^t) = \mathbf{x} \cdot (\mathbf{x}AA^{-1}) = \mathbf{x} \cdot \mathbf{x} = \|\mathbf{x}\|^2$$

for all $\mathbf{x}$ in $\mathbb{R}^n$. Hence, $\|\mathbf{x}A\| = \|\mathbf{x}\|$ for all $\mathbf{x}$ and *(a)* implies *(b)*.

Next, suppose $\|\mathbf{x}A\| = \|\mathbf{x}\|$ for all $\mathbf{x}$ in $\mathbb{R}^n$. In particular, $\|(\mathbf{x}+\mathbf{y})A\| = \|\mathbf{x}+\mathbf{y}\|$ which implies that $(\mathbf{x}+\mathbf{y})A \cdot (\mathbf{x}+\mathbf{y})A = (\mathbf{x}+\mathbf{y}) \cdot (\mathbf{x}+\mathbf{y})$. Simultaneously simplifying both sides of this equation yields:

$$\begin{aligned}
(\mathbf{x}+\mathbf{y})A \cdot (\mathbf{x}+\mathbf{y})A &= (\mathbf{x}+\mathbf{y}) \cdot (\mathbf{x}+\mathbf{y}) \\
(\mathbf{x}A) \cdot (\mathbf{x}A) + 2(\mathbf{x}A) \cdot (\mathbf{y}A) + (\mathbf{y}A) \cdot (\mathbf{y}A) &= \mathbf{x} \cdot \mathbf{x} + 2\mathbf{x} \cdot \mathbf{y} + \mathbf{y} \cdot \mathbf{y} \\
\|\mathbf{x}A\|^2 + 2(\mathbf{x}A) \cdot (\mathbf{y}A) + \|\mathbf{y}A\|^2 &= \|\mathbf{x}\|^2 + 2\mathbf{x} \cdot \mathbf{y} + \|\mathbf{y}\|^2 \\
\|\mathbf{x}\|^2 + 2(\mathbf{x}A) \cdot (\mathbf{y}A) + \|\mathbf{y}\|^2 &= \|\mathbf{x}\|^2 + 2\mathbf{x} \cdot \mathbf{y} + \|\mathbf{y}\|^2 \\
(\mathbf{x}A) \cdot (\mathbf{y}A) &= \mathbf{x} \cdot \mathbf{y}
\end{aligned}$$

and proves that *(b)* implies *(c)*.

In preparation for proving that *(c)* implies *(a)*, note that if $\mathbf{x} \cdot \mathbf{u} = 0$ for all $\mathbf{x} \in \mathbb{R}^n$, then $\mathbf{u} = \mathbf{0}$. Now suppose that $(\mathbf{x}A) \cdot (\mathbf{y}A) = \mathbf{x} \cdot \mathbf{y}$ for all $\mathbf{x}$ and $\mathbf{y}$ in $\mathbb{R}^n$ and rewrite this equation as follows:

$$\begin{aligned}
(\mathbf{x}A) \cdot (\mathbf{y}A) &= \mathbf{x} \cdot \mathbf{y} \\
\mathbf{x} \cdot (\mathbf{y}AA^t) &= \mathbf{x} \cdot \mathbf{y} \\
0 &= \mathbf{x} \cdot (\mathbf{y} - \mathbf{y}AA^t).
\end{aligned}$$

Since the above equations hold for all $\mathbf{x}$ and $\mathbf{y}$, the initial observation implies that $\mathbf{y} - \mathbf{y}AA^t = \mathbf{0}$ or $\mathbf{y}(I - AA^t) = \mathbf{0}$ for all $\mathbf{y}$. It follows that $I - AA^t = O$ or $I = AA^t$. Therefore, $A^{-1} = A^t$ and A is orthogonal. □

A similar result holds for unitary matrices (*Exercise* 1, p. 158). Several other unitary group results that parallel those for the orthogonal group also appear as exercises.

Proposition 4.2.2 *If A is a unitary matrix, then $|Det[A]| = 1$. If A is an orthogonal matrix, then* $\text{Det}\,[A] = \pm 1$.

Proof. Recall that $\text{Det}\,[A^t] = \text{Det}\,[A]$ and note that $\text{Det}\,[\overline{A}] = \overline{\text{Det}\,[A]}$. It follows that

$$\begin{aligned} |\text{Det}\,[A]|^2 &= \text{Det}\,[A]\,\overline{\text{Det}\,[A]} \\ &= \text{Det}\,[A]\,\text{Det}\,[\overline{A}] \\ &= \text{Det}\,[A]\,\text{Det}\,[(\overline{A})^t] \\ &= \text{Det}\,[A]\,\text{Det}\,[A^{-1}] \\ &= \text{Det}\,[AA^{-1}] = \text{Det}\,[I] = 1. \end{aligned}$$

The second statement is an immediate consequence of the first because orthogonal matrices are real unitary matrices. □

For $\mathbb{C}^n$, we use the Hermitian product defined in equation (2.12) as

$$\mathbf{z} \cdot \mathbf{w} = \sum_{j=1}^{n} z_j \overline{w}_j$$

and the norm $\|\mathbf{z}\|^2 - \mathbf{z} \cdot \mathbf{z}$. The basic properties of the Hermitian product can be found in *Exercise* 8, p. 82. Following the definition for $\mathbb{R}^n$, two vectors $\mathbf{z}$ and $\mathbf{w}$ in $\mathbb{C}^n$ are *orthogonal vectors* if $\mathbf{z} \cdot \mathbf{w} = 0$. A set of vectors $\mathbf{v}_1, \ldots, \mathbf{v}_p$ in either $\mathbb{R}^n$ or $\mathbb{C}^n$ are said to be *orthonormal* if they are orthogonal vectors of norm 1, that is, $\mathbf{v}_j \cdot \mathbf{v}_k = 0$ when $j \neq k$ and $\mathbf{v}_j \cdot \mathbf{v}_j = 1$ for $j = 1, \ldots, p$.

The defining condition $A^{-1} = A^t$ for a matrix A to be orthogonal can also be written as $AA^t = I$. As in Section 2.5, denote the rows of a matrix A by $A_1, \ldots, A_n$. Then the condition $AA^t = I$ can be written as n^2 algebraic equations. Specifically, a real matrix A is orthogonal if and only if it satisfies the equations

$$A_j \cdot A_m = \sum_{k=1}^{n} A(j,k)A(m,k) = \sum_{k=1}^{n} A(j,k)A^t(k,m) = 0 \tag{4.7}$$

when $j \neq m$, and

$$A_j \cdot A_j = \sum_{k=1}^{n} A(j,k)^2 = \sum_{k=1}^{n} A(j,k)A^t(k,j) = 1 \tag{4.8}$$

for $j = 1, \ldots, n$. The corresponding equations for a unitary matrix are

$$A_j \cdot A_m = \sum_{k=1}^{n} A(j,k)\overline{A}(m,k) = 0 \tag{4.9}$$

when $j \neq m$ and

$$A_j \cdot A_j = \sum_{k=1}^{n} |A(j,k)|^2 = \sum_{k=1}^{n} A(j,k)\overline{A}(j,k) = 1 \tag{4.10}$$

for $j = 1, \ldots, n$. These equations will be used to prove *Proposition 4.2.3*, which is actually two propositions, one for real and one for complex matrices. The alternate words for the complex case appear in braces behind the expression they replace.

Proposition 4.2.3 *Let A be a real {complex} matrix. The following are equivalent:*

(a) A is orthogonal {unitary}.

(b) The rows of A are an orthonormal basis for $\mathbb{R}^n$ $\{\mathbb{C}^n\}$.

(c) The columns of A are an orthonormal basis for $\mathbb{R}^n$ $\{\mathbb{C}^n\}$.

Proof. From the discussion preceding the statement of the proposition, we know that a real {complex} matrix A is orthogonal {unitary} if and only if the entries of A satisfy equations (4.7) and (4.8) {(4.9) and (4.10)}. These equations are precisely the conditions that the rows are an orthonormal set of vectors. Furthermore, they are automatically linearly independent by *Exercise* 3, p. 158, and a basis because there are n of them. This proves that *(a)* and *(b)* are equivalent.

To complete the proof, it will be shown that *(a)* and *(c)* are equivalent using the fact that *(a)* and *(b)* are equivalent. Because $\mathrm{O}(n)$ $\{\mathrm{U(m)}\}$ is a group, the matrix A is orthogonal {unitary} if and only if $A^{-1} = A^t$ $\{A^{-1} = (\overline{A})^t\}$ is orthogonal {unitary}. Since the rows of A^t are the columns of A, it follows that A is orthogonal if and only if the columns of A are an orthonormal basis for $\mathbb{R}^n$. {Since the rows of $(\overline{A})^t$ are the columns of $\overline{A}$, it follows that A is unitary if and only if the columns of $\overline{A}$ are an orthonormal basis for $\mathbb{C}^n$. Then, *Exercise* 4, p. 158, implies that the columns of $\overline{A}$ are an orthonormal basis for $\mathbb{C}^n$ if and only if the columns of A are an orthonormal basis for $\mathbb{C}^n$.} □

Theorem 4.2.4 *The groups* $\mathrm{U}(n)$, $\mathrm{SU}(n)$, $\mathrm{O}(n)$, *and* $\mathrm{SO}(n)$ *are compact.*

Proof. Since $\mathrm{SU}(n)$, $\mathrm{O}(n)$, and $\mathrm{SO}(n)$ are closed subgroups of $\mathrm{U}(n)$, it suffices to show that $\mathrm{U}(n)$ is compact. Because $\mathcal{M}_n(\mathbb{C})$ is a finite-dimensional normed linear space over $\mathbb{C}$, to prove that $\mathrm{U}(n)$ is compact it suffices by *Theorem 2.3.7* to show that $\mathrm{U}(n)$ is a closed and bounded set of $\mathcal{M}_n(\mathbb{C})$.

Just because U(n) is a closed set in GL$(n,\mathbb{C})$ does not, however, guarantee that it is a closed set in $\mathcal{M}_n(\mathbb{C})$. If A_k is a sequence in U(n) converging to A in $\mathcal{M}_n(\mathbb{C})$, then $|\text{Det}\,[A_k]|$ converges to $|\text{Det}\,[A]|$ and $|\text{Det}\,[A]| = 1$ by *Proposition 4.2.2*, proving that A is in GL$(n,\mathbb{C})$. It follows that A is in U(n) because U(n) is a closed subgroup of GL$(n,\mathbb{C})$. Hence, U(n) is a closed set in $\mathcal{M}_n(\mathbb{C})$.

Finally, equation (4.10) implies that

$$\|A\|^2 = \sum_{j=1}^{n}\sum_{k=1}^{n} |A(j,k)|^2 = n$$

and U(n) is a bounded set of $\mathcal{M}_n(\mathbb{C})$ with the Hermitian norm. □

By *Proposition 4.1.4*, the matrices of the form

$$C = \begin{bmatrix} I & \mathrm{O} \\ \mathrm{O} & A \end{bmatrix}$$

with $A \in \mathrm{O}(n-k)$, $0 < k < n$ are a closed subgroup of GL$(n,\mathbb{R})$ isomorphic to O$(n-k)$. An easy calculation using *Proposition 4.2.3* shows that these matrices are also in O(n). To summarize

$$\left\{ \begin{bmatrix} I & \mathrm{O} \\ \mathrm{O} & A \end{bmatrix} : A \in \mathrm{O}(n-k) \right\}$$

is a closed subgroup of O(n) isomorphic to O$(n-k)$. Thus we can and will regard O$(n-k)$ as a closed subgroup of O(n). In particular, we can investigate the quotient space O(n)/O$(n-1)$. Doing so requires the following result:

Proposition 4.2.5 (Gram-Schmidt Process) *If V is a subspace of $\mathbb{R}^n$ or $\mathbb{C}^n$, then V has an orthonormal basis. If $\mathbf{v}_1, \ldots, \mathbf{v}_m$ is an orthonormal basis for a subspace of $\mathbb{R}^n$ or $\mathbb{C}^n$, then there exist vectors $\mathbf{v}_{m+1} \ldots, \mathbf{v}_n$ such that $\mathbf{v}_1, \ldots, \mathbf{v}_n$ is an orthonormal basis for $\mathbb{R}^n$ or $\mathbb{C}^n$, respectively.*

Proof. For the first statement, let $\mathbf{u}_1, \ldots, \mathbf{u}_m$ be a basis for V. Set $\mathbf{v}_1 = \mathbf{u}_1$. Trivially, Span $[\mathbf{u}_1]$ = Span $[\mathbf{v}_1]$. Suppose orthogonal vectors $\mathbf{v}_1, \ldots, \mathbf{v}_k$ have been constructed so that Span $[\mathbf{u}_1, \ldots, \mathbf{u}_k]$ = Span $[\mathbf{v}_1, \ldots, \mathbf{v}_k]$. Set

$$\mathbf{v}_{k+1} = \mathbf{u}_{k+1} - \sum_{j=1}^{k} \frac{\mathbf{u}_{k+1} \cdot \mathbf{v}_j}{\mathbf{v}_j \cdot \mathbf{v}_j} \mathbf{v}_j, \tag{4.11}$$

and verify that $\mathbf{v}_1, \ldots, \mathbf{v}_{k+1}$ are orthogonal and satisfy Span $[\mathbf{u}_1, \ldots, \mathbf{u}_{k+1}]$ = Span $[\mathbf{v}_1, \ldots, \mathbf{v}_{k+1}]$. This process stops when $k = m$ with an orthogonal basis for V. Simply multiply $\mathbf{v}_j$ by $1/\|\mathbf{v}_j\|$ to make the basis orthonormal. (This iterative process for constructing orthonormal vectors from linear independent vectors is known as the Gram-Schmidt Process.)

For the second part, apply *Proposition 2.1.7* to obtain $n - m$ additional vectors $\mathbf{u}_{m+1}, \ldots, \mathbf{u}_n$ such that $\mathbf{v}_1, \ldots, \mathbf{v}_m, \mathbf{u}_{m+1}, \ldots, \mathbf{u}_n$ is a basis for $\mathbb{R}^n$ or

$\mathbb{C}^n$, respectively. Applying the Gram-Schmidt Process to this basis does not alter the first m vectors and produces an orthonormal basis for $\mathbb{R}^n$ or $\mathbb{C}^n$. □

The n-dimensional sphere $S^n = \{\mathbf{x} \in \mathbb{R}^{n+1} : \|\mathbf{x}\| = 1\}$ is an n-dimensional manifold (*Exercise* 4, p. 148). Note that $S^0 = \{1, -1\} = \mathrm{O}(1)$ and $S^1 = \mathbb{K}$. Clearly, $\mathrm{SO}(1)$ is the trivial group and $\mathrm{SO}(2)$ is isomorphic to $\mathbb{K}$ by *Exercise* 7, p. 158. Hence $\mathrm{SO}(2)/\mathrm{SO}(1)$ is homeomorphic to S^1. If we define $\mathrm{O}(0)$ to be the trivial group containing one element, then $\mathrm{O}(1)/\mathrm{O}(0)$ is homeomorphic to S^0. These facts are a small part of the following more general result:

Theorem 4.2.6 *The quotient space $\mathrm{O}(n)/\mathrm{O}(n-1)$ with $n \geq 1$ and the quotient space $\mathrm{SO}(n)/\mathrm{SO}(n-1)$ with $n \geq 2$ are homeomorphic to S^{n-1}.*

Proof. Equation (4.8) implies that each row of an orthogonal matrix is a point on S^{n-1}. The proofs of the two cases will be carried out in parallel and are based on an analysis of the function $f : \mathrm{O}(n) \to S^{n-1}$ defined by $f(A) = A_1$, the first row of A. It is obviously a continuous function. Given $\mathbf{v}_1 \in S^{n-1}$, there exist vectors $\mathbf{v}_2, \ldots, \mathbf{v}_n$ such that $\mathbf{v}_1, \ldots, \mathbf{v}_n$ is an orthonormal basis for $\mathbb{R}^n$ by *Proposition 4.2.5*. Let A be the matrix whose rows are given by $A_j = \mathbf{v}_j$. Then A is in $\mathrm{O}(n)$ by *Proposition 4.2.3* and $f(A) = \mathbf{v}_1$. Thus f is onto.

For the $\mathrm{SO}(n)$ case, we can assume that $n \geq 3$ by the remarks before the statement of the theorem. If the matrix A constructed in the preceding paragraph is not in $\mathrm{SO}(n)$, that is , $\mathrm{Det}\,[A] = -1$, then switching the last two rows of A will produce $A' \in \mathrm{SO}(n)$ such that $f(A') = f(A)$. So f also maps $\mathrm{SO}(n)$ onto S^{n-1} when $n \geq 3$.

If $C \in \mathrm{O}(n)$ is in the subgroup $\mathrm{O}(n-1)$, then, obviously, $C(1,1) = 1$. What is more interesting is that the converse is also true. Suppose C is in $\mathrm{O}(n)$ and $C(1,1) = 1$. Then equation (4.8) implies that $C_1 = (1, 0, \ldots, 0)$. Since $C^{\mathfrak{t}} = C^{-1}$ is also an orthogonal matrix and $C^{\mathfrak{t}}(1,1) = 1$, it follows that $(C^{\mathfrak{t}})_1 = (1, 0, \ldots, 0)$. Hence, the first row and first column of C is all zeros except for the one in the upper left corner, implying that C is in $\mathrm{O}(n-1)$. An orthogonal matrix C is in $\mathrm{O}(n-1)$ if and only if $C(1,1) = 1$.

If C is in $\mathrm{SO}(n)$ and $C(1,1) = 1$, then C is in $\mathrm{O}(n-1)$. Now *Theorem 2.5.12* implies $\mathrm{Det}\,[C] = \mathrm{Det}\,[C(1|1)]$ and $C \in \mathrm{SO}(n)$. A special orthogonal matrix C is in $\mathrm{SO}(n-1)$ if and only if $C(1,1) = 1$.

For A and B in $\mathrm{O}(n)$, it follows using $B^{-1} = B^{\mathfrak{t}}$ that:

(a) $\mathrm{O}(n-1)A = \mathrm{O}(n-1)B$ if and only if $AB^{-1} = AB^{\mathfrak{t}} \in \mathrm{O}(n-1)$ if and only if $A_1 \cdot B_1 = AB^{\mathfrak{t}}(1,1) = 1$.

(b) $\mathrm{SO}(n-1)A = \mathrm{SO}(n-1)B$ if and only if $AB^{-1} = AB^{\mathfrak{t}} \in \mathrm{SO}(n-1)$ if and only if $A_1 \cdot B_1 = AB^{\mathfrak{t}}(1,1) = 1$.

The Cauchy-Schwarz Inequality (*Proposition 1.2.1*) applied to the first rows of two matrices A and B in $\mathrm{O}(n)$ implies that

$$|A_1 \cdot B_1| \leq \|A_1\|\,\|B_1\| = 1,$$

because $\|A_1\| = 1 = \|B_1\|$ by (4.8). Furthermore, equality holds if and only if

$$B_1 = \frac{A_1 \cdot B_1}{A_1 \cdot A_1} A_1 = (A_1 \cdot B_1)A_1.$$

Applying this to the conclusion of the previous paragraph yields:

(a) For A and B in $\mathrm{O}(n)$, we have $\mathrm{O}(n-1)A = \mathrm{O}(n-1)B$ if and only if $A_1 = B_1$ if and only if $f(A) = f(B)$.

(b) For A and B in $\mathrm{SO}(n)$, we have $\mathrm{SO}(n-1)A = \mathrm{SO}(n-1)B$ if and only if $A_1 = B_1$ if and only if $f(A) = f(B)$.

Although the final steps of the proof are written for $\mathrm{O}(n)$, they work verbatim for $\mathrm{SO}(n)$.

It is now possible to define $\tilde{f} : \mathrm{O}(n)/\mathrm{O}(n-1) \to S^{n-1}$ by $\tilde{f}(\mathrm{O}(n-1)A) = f(A) = A_1$. The function $\tilde{f}$ is well defined and one-to-one because $\mathrm{O}(n-1)A = \mathrm{O}(n-1)B$ if and only if $f(A) = f(B)$. Clearly, $\tilde{f}$ is onto because f is onto.

Let $\pi : \mathrm{O}(n) \to \mathrm{O}(n)/\mathrm{O}(n-1)$ by $\pi(A) = \mathrm{O}(n-1)A$ and note that $\tilde{f} \circ \pi = f$. If V is an open subset of S^{n-1}, then $f^{-1}(V) = \pi^{-1}(\tilde{f}^{-1}(V))$ is open because f is continuous. Thus $\tilde{f}^{-1}(V)$ is open (*Theorem 1.4.7*) and $\tilde{f}$ is continuous. Since $\mathrm{O}(n)$ is compact by *Proposition 4.2.4*, the quotient space $\pi(\mathrm{O}(n)) = \mathrm{O}(n)/\mathrm{O}(n-1)$ is compact, and *Exercise* 2, p. 61, implies that $\tilde{f}$ is a homeomorphism. □

Let H be a closed subgroup of a metric group G and, as usual, let $\pi : G \to G/H$ by $\pi(x) = Hx$. Let V be an open subset of G/H containing $\pi(e) = H$. A *local section at the identity* on V is a continuous function $\sigma : V \to G$ such that $\pi(\sigma(Hx)) = Hx$ for all $Hx \in V$. If $V = G/H$, then σ is called a *global section.* Local sections at the identity provide information about the local structure of the topology of G.

Proposition 4.2.7 *If H is a closed subgroup of a metric group G and $\sigma : V \to G$ is a local section at the identity on V, then $\pi^{-1}(V)$ is a open neighborhood of the identity of G homeomorphic to $H \times V$.*

Proof. Define $f : H \times V \to G$ by setting $f(y, Hx) = y\sigma(Hx)$. Clearly, f is continuous, and $\sigma(Hx)$ is in Hx because $\pi(\sigma(Hx)) = Hx$. Hence, $y\sigma(Hx)$ is also in Hx and

$$\pi(f(y, Hx)) = \pi(y\sigma(Hx)) = Hx.$$

It follows that $f(H \times V) \subset \pi^{-1}(V)$. A routine calculation shows that

$$f^{-1}(x) = \left(x\sigma(\pi(x))^{-1}, \pi(x)\right)$$

which is continuous because π, σ, and the inverse function are continuous. □

Corollary 4.2.8 *If $\sigma : G/H \to G$ is a global section, then G is homeomorphic to $H \times G/H$.*

Corollary 4.2.9 *Let H be a closed subgroup of a metric group G and let $\sigma : V \to G$ be a local section at the identity. If H and G/H are manifolds, then G is a manifold.*

Proof. Since an open subset of a manifold is obviously a manifold, $H \times V$ is a manifold by *Exercise* 6, p. 149. It now follows from the proposition that the open set $\pi^{-1}(V)$ of G containing the identity is also a manifold. Thus there exists an open neighborhood of the identity homeomorphic to an open subset of $\mathbb{R}^n$ for some n, and G is a manifold. □

Theorem 4.2.10 *The orthogonal group $\mathrm{O}(n)$ is a manifold of dimension*

$$n(n-1)/2.$$

Proof. The proof proceeds by induction on n. It was already pointed out that $\mathrm{O}(1)$ is a zero-dimensional manifold. Assume $\mathrm{O}(n-1)$ is a manifold of dimension $(n-1)(n-2)/2$. By *Theorem 4.2.6*, $\mathrm{O}(n)/\mathrm{O}(n-1)$ is homeomorphic to S^{n-1}, a $(n-1)$-dimensional manifold. To prove that $\mathrm{O}(n)$ is a manifold, it suffices by *Corollary 4.2.9* to construct a local section at I from $\mathrm{O}(n)/\mathrm{O}(n-1)$ to $\mathrm{O}(n)$. Using *Exercise* 6, p. 149, it follows that the dimension of $\mathrm{O}(n)$ is

$$\frac{(n-1)(n-2)}{2} + n - 1 = \frac{n(n-1)}{2}$$

Since $\mathrm{O}(n-1)A \to A_1$ defines a homeomorphism from $\mathrm{O}(n)/\mathrm{O}(n-1)$ onto to S^{n-1}, the local section at I can be constructed on S^{n-1} using $f(A) = A_1$ instead of π. Let $V = \{\mathbf{x} \in S^{n-1} : x_1 > 0\}$, which is an open subset of S^{n-1}. Obviously, $f(I)$ is in V. We need to construct a continuous function $\sigma : V \to \mathrm{O}(n)$ such that $\sigma(\mathbf{x})_1 = \mathbf{x}$.

The first step is to construct continuous functions $\mathbf{f}_j : V \to \mathbb{R}^n$ for $j = 2, \ldots, n$ such that $\mathbf{x}, \mathbf{f}_2(\mathbf{x}), \ldots, \mathbf{f}_n(\mathbf{x})$ is an orthogonal basis for $\mathbb{R}^n$. If $\mathbf{x}$ is in V, then $\mathbf{x}, \mathbf{e}_2, \ldots, \mathbf{e}_n$ is a basis for $\mathbb{R}^n$ because $x_1 > 0$. Using the Gram-Schmidt Process, equation (4.11), set

$$\mathbf{f}_2(\mathbf{x}) = \mathbf{e}_2 - \frac{\mathbf{e}_2 \cdot \mathbf{x}}{\mathbf{x} \cdot \mathbf{x}}\mathbf{x} = \mathbf{e}_2 - x_2\mathbf{x}$$

and proceeding inductively for $k > 2$ set

$$\mathbf{f}_k(\mathbf{x}) = \mathbf{e}_k - x_k\mathbf{x} - \sum_{j=2}^{k-1} \frac{\mathbf{e}_k \cdot \mathbf{f}_j(\mathbf{x})}{\mathbf{f}_j(\mathbf{x}) \cdot \mathbf{f}_j(\mathbf{x})} \mathbf{f}_j(\mathbf{x}).$$

to construct an orthogonal basis for each $\mathbf{x}$ in V. Because they are a basis, $\|\mathbf{f}_j(\mathbf{x})\| > 0$ for $2 \le j \le n$.

Clearly, the functions $\mathbf{e}_k - x_k\mathbf{x}$ for $k = 2, \ldots, n$ are continuous on V. So $\mathbf{f}_2(\mathbf{x})$ is a continuous function of $\mathbf{x}$. If $\mathbf{f}_2(\mathbf{x}), \ldots, \mathbf{f}_{k-1}(\mathbf{x})$ are continuous functions on V, then so is

$$\sum_{j=2}^{k-1} \frac{\mathbf{e}_k \cdot \mathbf{f}_j(\mathbf{x})}{\mathbf{f}_j(\mathbf{x}) \cdot \mathbf{f}_j(\mathbf{x})} \mathbf{f}_j(\mathbf{x})$$

and $\mathbf{f}_k$ is continuous on V for $k = 2, \ldots, n$. It follows that the functions $\|\mathbf{f}_j(\mathbf{x})\|$ are continuous from V to $\mathbb{R}^+$. Therefore,

$$\sigma_1(\mathbf{x}) = \mathbf{x},\ \sigma_2(\mathbf{x}) = \frac{\mathbf{f}_2(\mathbf{x})}{\|\mathbf{f}_2(\mathbf{x})\|}, \ldots, \sigma_n(\mathbf{x}) = \frac{\mathbf{f}_n(\mathbf{x})}{\|\mathbf{f}_n(\mathbf{x})\|}$$

is an orthonormal basis for $\mathbb{R}^n$ that depends continuously on $\mathbf{x} \in V$.

Finally, let $\sigma(\mathbf{x})$ be the orthogonal matrix whose rows are given by σ_j for $j = 1, \ldots, n$. □

There are parallel theorems to *Theorems 4.2.10 and 4.2.6* for the unitary group, $\mathrm{U}(n)$.

Theorem 4.2.11 *The quotient space* $\mathrm{U}(n)/\mathrm{U}(n-1)$ *with* $n \geq 1$ *is homeomorphic to* S^{2n-1}.

Proof. Notice that

$$\left\{ \mathbf{z} \in \mathbb{C}^n : \sum_{j=1}^{n} |z_j|^2 = 1 \right\} = S^{2n-1}.$$

Hence equation (4.10) implies that every row of a unitary matrix is a point on S^{2n-1}. As in the orthogonal case, the proof is based on an analysis of the continuous function $f : \mathrm{U}(n) \to S^{2n-1}$ defined by $f(A) = A_1$, the first row of A. In fact, with just a few clarifying comments the same proof works.

Since the Gram-Schmidt Process works in both $\mathbb{R}^n$ and $\mathbb{C}^n$, the function f is onto in both the $\mathrm{O}(n)$ and the $\mathrm{U}(n)$ case. It is also easy to see that a unitary matrix C is in $\mathrm{U}(n-1)$ if and only if $C(1,1) = 1$ using equation (4.10).

To prove that $\mathrm{U}(n-1)A = \mathrm{U}(n-1)B$ if and only if $A_1 = B_1$, use the same argument with the Hermitian product for $\mathbb{C}^n$ and the corresponding generalization of the Cauchy-Schwarz Inequality (*Exercises 8 and 9*, p. 82).

The final steps are a straightforward quotient space argument that works in both cases with only changing the names of the groups. □

Theorem 4.2.12 *The unitary group* $\mathrm{U}(n)$ *is manifold of dimension* n^2.

Proof. An element of $\mathrm{U}(1)$ is just a nonzero complex number z such that $z^{-1} = \bar{z}$, which is equivalent to saying that $z \in \mathbb{K} = S^1$. Thus $\mathrm{U}(1) = S^1$ and the result holds for $n = 1$.

Assume $\mathrm{U}(n-1)$ is a manifold of dimension $(n-1)^2$. Construct a local section at the identity from $\mathrm{U}(n)/\mathrm{U}(n-1)$ to $\mathrm{U}(n)$ as in the proof of *Theorem 4.2.10*, but using the Gram-Schmidt Process for $\mathbb{C}^n$ instead for $\mathbb{R}^n$. Then it follows that $\mathrm{U}(n)$ is a manifold of dimension $(n-1)^2 + 2n - 1 = n^2$ by applying *Theorem 4.2.11* and *Corollary 4.2.9*. □

EXERCISES

1. Let A be an $n \times n$ complex matrix. Using the Hermitian product $\mathbf{z} \cdot \mathbf{w} = \sum_{j=1}^{n} z_j \overline{w}_j$, show that the following are equivalent:

   (a) A is an unitary matrix.

   (b) $\|\mathbf{z}A\| = \|\mathbf{z}\|$ for all $\mathbf{z}$ in $\mathbb{C}^n$.

   (c) $(\mathbf{z}A) \cdot (\mathbf{w}A) = \mathbf{z} \cdot \mathbf{w}$ for all $\mathbf{z}$ and $\mathbf{w}$ in $\mathbb{C}^n$.

2. Show that $\mathrm{O}(n)/\mathrm{SO}(n)$ is isomorphic to Z_2 and $\mathrm{U}(n)/\mathrm{SU}(n)$ is isomorphic to $\mathbb{K}$.

3. Show that if $\mathbf{v}_1, \ldots, \mathbf{v}_m$ is a set of orthonormal vectors in $\mathbb{R}^n$ or $\mathbb{C}^n$, then they are linearly independent.

4. Let $\mathbf{v}_1, \ldots, \mathbf{v}_m$ be a set of vectors in $\mathbb{C}^n$. Show that $\mathbf{v}_1, \ldots, \mathbf{v}_m$ is orthonormal if and only if $\overline{\mathbf{v}}_1, \ldots, \overline{\mathbf{v}}_m$ is orthonormal.

5. Show that if $\mathbf{u}_1, \ldots, \mathbf{u}_m$ is a set of orthonormal vectors in $\mathbb{R}^n$ or $\mathbb{C}^n$ and $\mathbf{v}$ is in $\mathrm{Span}\,[\mathbf{u}_1, \ldots, \mathbf{u}_m]$, then

$$\mathbf{v} = \sum_{j=1}^{m} (\mathbf{v} \cdot \mathbf{u}_j)\mathbf{u}_j \tag{4.12}$$

6. Show that if $\mathbf{u}_1, \ldots, \mathbf{u}_m$ is a set of orthogonal vectors in $\mathbb{R}^n$ or $\mathbb{C}^n$, then

$$\left(\sum_{i=1}^{m} \mathbf{u}_j\right) \cdot \left(\sum_{j=1}^{m} \mathbf{u}_j\right) = \sum_{j=1}^{m} \|\mathbf{u}_j\|^2. \tag{4.13}$$

7. Show that

$$\mathrm{SO}(2) = \left\{ \begin{bmatrix} \cos(\theta) & \sin(\theta) \\ -\sin(\theta) & \cos(\theta) \end{bmatrix} : \theta \in \mathbb{R} \right\}$$

   and that $\mathrm{SO}(2)$ is isomorphic to $\mathbb{K}$.

8. Show that if A is in $\mathrm{O}(n, \mathbb{C})$, then $\mathrm{Det}\,[A] = \pm 1$.

9. Given real numbers a and b such that $a^2 - b^2 = 1$, show that there exist c and d in $\mathbb{C}$ such that the matrix

$$A = \begin{bmatrix} a & ib \\ c & d \end{bmatrix}$$

   is in $\mathrm{O}(n, \mathbb{C})$. Use this to prove that $\mathrm{O}(2, \mathbb{C})$ is not compact.

10. Show that $\mathrm{O}(n, \mathbb{C})$ is not compact for $n \geq 2$.

4.3 Triangular Groups

A matrix A in $\mathcal{M}_n(\mathbb{S})$ is *triangular* if $A(j,k) = 0$ when $k < j$. In other words, in a triangular matrix all the entries are equal to zero below the main diagonal where the indices are equal. To be more precise, these matrices are upper triangular and one can define lower triangular matrices by requiring that $A(j,k) = 0$ when $k > j$. But we will restrict our attention to upper triangular matrices and not use the adjective "upper".

In *Exercise* 3, p. 113, it was shown that the determinant of a triangular matrix is simply the product of the diagonal elements or

$$\mathrm{Det}\,[A] = \prod_{j=1}^{n} A(j,j). \tag{4.14}$$

So triangular matrices are generally not invertible and do not form a subgroup of $\mathrm{GL}(n,\mathbb{S})$. There are, however, interesting subgroups of $\mathrm{GL}(n,\mathbb{S})$ that consist entirely of triangular matrices. They will be constructed from more elementary pieces.

For each integer $p \geq 0$, set

$$T_p = \{A \in \mathcal{M}_n(\mathbb{S}) : A(j,k) = 0 \text{ when } k < j + p\}. \tag{4.15}$$

Note that T_0 is just the set of all triangular matrices, $T_p = \{\mathrm{O}\}$ when $p \geq n$, and

$$T_0 \supset T_1 \supset \ldots \supset T_n.$$

Consequently, the statement that A is in T_p means that A satisfies the conditions to be in T_p but does not rule out the possibility that it might also be in T_q for some $q > p$.

Each T_p is a subspace, and hence a closed subset of $\mathcal{M}_n(\mathbb{S})$. A routine calculation shows that

$$\mathrm{Dim}\,[T_p] = 1 + 2 + \ldots + n - p = \frac{(n-p)(n-p+1)}{2}$$

for $p = 0, \ldots, n$. Thus T_p is homeomorphic to $\mathbb{R}^{(n-p)(n-p+1)/2}$ or $\mathbb{R}^{(n-p)(n-p+1)}$ according as $\mathbb{S}$ is $\mathbb{R}$ or $\mathbb{C}$.

One way or another everything in this section depends on the following simple lemma.

Lemma 4.3.1 *For non-negative integers p and q,*

$$T_p T_q \subset T_{p+q}. \tag{4.16}$$

In particular, $T_0 T_0 \subset T_0$ and $T_1^n = \{\mathrm{O}\}$.

Proof. Let A and B be in T_p and T_q, respectively. If $\sum_{k=1}^{n} A(j,k)B(k,m) \neq 0$, then $A(j,k)B(k,m) \neq 0$ for some k and both $A(j,k)$ and $B(k,m)$ must be nonzero. It follows that $m \geq k + q$ and $k \geq j + p$. Hence, $m \geq j + p + q$, and $\sum_{k=1}^{n} A(j,k)B(k,m) = 0$ when $m < j + p + q$ and AB is in T_{p+q}. □

Proposition 4.3.2 *If $n \geq 2$ and A is in T_1, then $I + A$ is an invertible matrix and*

$$(I + A)^{-1} = I + \sum_{k=1}^{n-1} (-1)^k A^k. \tag{4.17}$$

If $n \geq 2$, then

$$H_p = \{I + A : A \in T_p\} \tag{4.18}$$

is a closed subgroup of $\mathrm{SL}(n, \mathbb{S})$ for $p \geq 1$.

Proof. Let A be an element of T_p. Then

$$\begin{aligned} (I + A)\left(I + \sum_{k=1}^{n-1}(-1)^k A^k\right) &= \\ I + \sum_{k=1}^{n-1}(-1)^k A^k + A + A\sum_{k=1}^{n-1}(-1)^k A^k &= \\ I + \sum_{k=1}^{n-1}(-1)^k A^k + A + \sum_{k=1}^{n-1}(-1)^k A^{k+1} &= \\ I + \sum_{k=1}^{n-1}(-1)^k A^k - \sum_{k=1}^{n}(-1)^{k-1} A^k &= \\ I + (-1)^n A^n &= I, \end{aligned}$$

and equation (4.17) holds because $A^n = O$ by the lemma.

Since T_p is a closed set in $\mathcal{M}_n(\mathbb{S})$, its translate $H_p = I + T_p$ is also a closed set in $\mathcal{M}_n(\mathbb{S})$. It follows from equation (4.14) that $H_p = I + T_p$ is contained in $\mathrm{SL}(n, \mathbb{S})$ and is a closed set in $\mathrm{SL}(n, \mathbb{S})$.

If A and B are elements of T_p, then $(I + A)(I + B) = I + A + B + AB$. Since AB is in $T_{2p} \subset T_p$ by the lemma, $A + B + AB$ is in T_p and $(I + A)(I + B)$ is in H_p. Similarly, $\sum_{k=1}^{n}(-1)^k A^k$ is contained in T_p and $(I + A)^{-1}$ is in H_p. Thus H_p is a subgroup of $\mathrm{SL}(n, \mathbb{S})$. □

Corollary 4.3.3 *The metric groups $H_p = I + T_p$ are manifolds homeomorphic to $\mathbb{R}^{(n-p)(n-p+1)/2}$ or $\mathbb{R}^{(n-p)(n-p+1)}$ according as $\mathbb{S}$ is $\mathbb{R}$ or $\mathbb{C}$, and hence they are both locally compact and σ-compact.*

Proof. The function $A \to I + A$ defines a homeomorphism of T_p onto H_p. □

Proposition 4.3.4 *If $1 \leq p \leq n - 1$, then H_{p+1} is a closed normal subgroup of the metric group H_p and H_p/H_{p+1} is isomorphic to the normed linear space $\mathbb{S}^{n-p}$.*

Proof. It is clear from the discussion thus far that H_{p+1} is a closed subgroup of H_p. The core of the proof is setting the stage to apply *Corollary 1.5.19.*

The projection functions $A \to A(j,k)$ are all continuous on $\mathcal{M}_n(\mathbb{S})$ and so the functions $\varphi_j : H_p \to \mathbb{S}$ for $j = 1, \ldots, n-p$ defined by $\varphi_j(I+A) = A(j, j+p)$ are continuous on H_p.

If A and B are in T_p, then AB is in T_{2p} by the lemma, and the $(j, j+p)$ entry of AB equals 0. It follows that the $(j, j+p)$ entry of $(I+A)(I+B) = I + A + B + AB$ is $A(j, j+p) + B(j, j+p)$. Therefore,

$$\varphi_j\big((I+A)(I+B)\big) = A(j,j+p) + B(j,j+p) = \varphi_j(I+A) + \varphi_j(I+B),$$

and φ_j is a homomorphism of H_p onto $\mathbb{S}$.

Now define $\varphi : H_p \to \mathbb{S}^{n-p}$ by

$$\varphi(I+A) = \big(\varphi_1(I+A), \ldots, \varphi_{n-p}(I+A)\big).$$

Obviously, φ is a homomorphism. It is also easy to see that it is onto and the kernel of φ is H_{p+1}. Because the metric groups H_p are all locally compact and σ-compact, *Corollary 1.5.19* now applies to complete the proof. □

Technically, our notation here should probably indicate the dependence of T_p and H_p on the size of the matrices and the choice of scalars, but that is rather cumbersome. Instead, we single out just H_1 for more complete notation. Specifically, we let $H_1 = \mathrm{NL}(n, \mathbb{S})$ and continue to use the less precise notation T_p and H_p with the understanding that the matrices are $n \times n$ with entries from $\mathbb{R}$ or $\mathbb{C}$. The letter "N" stands for *nilpotent*, which will be defined shortly.

Of course, there are invertible triangular matrices with scalars other than 1 on the diagonal. Because $T_0 T_0 \subset T_0$, it is a semigroup. The group of units of the semigroup T_0 will be denoted by $\mathrm{TL}(n, \mathbb{S})$ and is called the *triangular linear group*. It can be thought of in several ways: first, $\mathrm{TL}(n, \mathbb{S}) = \mathrm{GL}(n, \mathbb{S}) \cap T_0$, second

$$\begin{aligned} \mathrm{TL}(n,\mathbb{S}) &= \{A \in T_0 : \mathrm{Det}\,[A] \neq 0\} \\ &= \{A \in T_0 : A(j,j) \neq 0 \text{ for } j = 1, \ldots, n\} \end{aligned}$$

by equation (4.14), and third

$$\mathrm{TL}(n,\mathbb{S}) = \mathrm{DL}(n,\mathbb{S}) + T_1. \tag{4.19}$$

The matrix groups $\mathrm{TL}(n, \mathbb{S})$ and its closed subgroup $\mathrm{NL}(n, \mathbb{S})$ are the main subjects of this section.

Proposition 4.3.5 *The following hold for $n \geq 2$:*

(a) $\mathrm{TL}(n, \mathbb{S})$ *is a closed subgroup of* $\mathrm{GL}(n, \mathbb{S})$.

(b) $\mathrm{NL}(n, \mathbb{S})$ *is a normal subgroup of* $\mathrm{TL}(n, \mathbb{S})$, *and* $\mathrm{TL}(n, \mathbb{S})/\mathrm{NL}(n, \mathbb{S})$ *is isomorphic to* $\mathrm{DL}(n, \mathbb{S})$.

Proof. Since $\mathrm{TL}(n, \mathbb{S}) = \mathrm{GL}(n, \mathbb{S}) \cap T_0$ and T_0 is a closed subset of $\mathcal{M}_n(\mathbb{S})$, it follows that $\mathrm{TL}(n, \mathbb{S})$ is a closed subset of $\mathrm{GL}(n, \mathbb{S})$. If A and B are in $\mathrm{TL}(n, \mathbb{S})$, AB is in $\mathrm{TL}(n, \mathbb{S})$ because $\mathrm{Det}\, AB = \mathrm{Det}\,[A]\mathrm{Det}\,[B] \neq 0$ and AB is in T_0 by *Lemma 4.3.1*. Hence, $\mathrm{TL}(n, \mathbb{S})$ is closed under multiplication.

Because a typical element of $\mathrm{TL}(n, \mathbb{S})$ can be written as $A + B$ with $A \in \mathrm{DL}(n, \mathbb{S})$ and $B \in T_1$ by (4.19), it follows that $A + B = A(I + A^{-1}B)$. Clearly, $A^{-1}B$ is in T_1 by *Lemma 4.3.1*. *Proposition 4.3.2* implies that $(I + A^{-1}B)^{-1}$ is in H_1 and can be written as $(I + A^{-1}B)^{-1} = I + C$ with C in T_1. Thus

$$(A + B)^{-1} = (I + A^{-1}B)^{-1}A^{-1} = (I + C)A^{-1} = A^{-1} + CA^{-1}$$

is in $\mathrm{DL}(n, \mathbb{S}) + T_1 = \mathrm{TL}(n, \mathbb{S})$ because CA^{-1} is in T_1. Therefore, the inverse of an element of $\mathrm{TL}(n, \mathbb{S})$ is in $\mathrm{TL}(n, \mathbb{S})$, and it is a closed subgroup of $\mathrm{GL}(n, \mathbb{S})$. This completes the proof of the first part.

For the second part, it suffices to construct a homomorphism from the metric group $\mathrm{TL}(n, \mathbb{S})$ onto the metric group $\mathrm{DL}(n, \mathbb{S})$ with kernel H_1 and apply *Corollary 1.5.19* again. Define $\varphi : \mathrm{TL}(n, \mathbb{S}) \to \mathrm{DL}(n, \mathbb{S})$ by

$$\varphi(A)(j, k) = \begin{cases} A(j, k) & \text{when } j = k \\ 0 & \text{when } j \neq k \end{cases}.$$

The function φ is obviously continuous. If B is another element of $\mathrm{TL}(n, \mathbb{S})$, then $\varphi(AB)(j, j) = A(j, j)B(j, j)$ by *Exercise* 2, p. 168. It follows that $\varphi(AB) = \varphi(A)\varphi(B)$ and the kernel of φ is $\mathrm{NL}(n, \mathbb{S})$. □

A group G is *solvable* provided that there exist subgroups

$$G = G_0 \supset G_1 \supset \ldots \supset G_m = \{e\}$$

such that

(a) G_{j+1} is a normal subgroup of G_j for $j = 1, \ldots, m - 1$

(b) G_j / G_{j+1} is an Abelian group.

It follows from *Propositions 4.3.4 and 4.3.5* that $\mathrm{TL}(n, \mathbb{S})$ and H_p are solvable groups. In spite of this similarity, there are fundamental differences in the structures of these two classes of groups.

The *center of a group* G, denoted by $Z(G)$, is defined by

$$Z(G) = \{x \in G : xy = yx \text{ for all } y \in G\}.$$

It is easily verified that the center of a group is an Abelian normal subgroup. The center will be used to study the structure of the matrix groups $\mathrm{TL}(n, \mathbb{S})$ and $\mathrm{NL}(n, \mathbb{S})$.

Theorem 4.3.6 *The center of* $\mathrm{NL}(n, \mathbb{S})$ *is* H_{n-1}, *and the center of* $\mathrm{TL}(n, \mathbb{S})$ *is the subgroup* $\{cI : c \in \mathbb{S} \text{ and } c \neq 0\}$.

Proof. Let A and B be in T_1 and consider $I+A$ and $I+B$ in $\mathrm{NL}(n,\mathbb{S})$. Then

$$(I+A)(I+B)=(I+B)(I+A)$$

if and only if

$$I+A+B+AB=I+B+A+BA$$

if and only if $AB=BA$. If B is in T_{n-1}, then both AB and BA are in $T_n=\{\mathrm{O}\}$, and $AB=\mathrm{O}=BA$. It follows that H_{n-1} is contained in the center of $\mathrm{NL}(n,\mathbb{S})$.

Suppose B is not in T_{n-1}. So B must be in T_1 and have a nonzero entry other than the $(1,n)$ coordinate. Consequently, $B(m,m+p)\neq 0$ for some minimal p such that $1\leq p<n-1$. There are two cases.

If $m+p<n$, define A in T_1 by

$$A(j,k)=\begin{cases}1 & \text{if } j=m+p \text{ and } k=m+p+1\\ 0 & \text{otherwise}\end{cases}.$$

An easy calculation shows that the $(m,m+p+1)$ entries of AB and BA are 0 and $B(m,m+p)\neq 0$, respectively. Thus $I+B$ is not in the center of $\mathrm{NL}(n,\mathbb{S})$.

If $m+p=n$, then $1<m$ because $p<n-1$. Define A in T_1 by

$$A(j,k)=\begin{cases}1 & \text{if } j=m-1 \text{ and } k=m\\ 0 & \text{otherwise}\end{cases}$$

In this case $AB(m-1,m+p)=B(m,m+p)\neq 0$ and $BA(m-1,m+p)=0$. So again $I+B$ is not in the center of $\mathrm{NL}(n,\mathbb{S})$. Therefore, the center of $\mathrm{NL}(n,\mathbb{S})$ is H_{n-1}. (Both constructions of A will be used again in the proof of *Proposition 4.3.8.*)

The center of $\mathrm{TL}(n,\mathbb{S})$ obviously contains $\{cI : c\in\mathbb{S} \text{ and } c\neq 0\}$. To prove that the center is contained in $\{cI : c\in\mathbb{S} \text{ and } c\neq 0\}$, we need to examine when an arbitrary element D of $\mathrm{DL}(n,\mathbb{S})$ commutes with an arbitrary element A of $\mathrm{TL}(n,\mathbb{S})$. Since any element A of $\mathrm{TL}(n,\mathbb{S})$ can be written in the form $A=(I+B)C$ for some B in T_1 and C in $\mathrm{DL}(n,\mathbb{S})$, the equation to be studied is $(I+B)CD=D(I+B)C$. It can be rewritten as

$$(I+B)CD=(I+DBD^{-1})DC$$

which simplifies to $B=DBD^{-1}$ because $\mathrm{DL}(n,\mathbb{S})$ is an Abelian subgroup of $\mathrm{GL}(n,\mathbb{S})$. Therefore, $(I+B)CD=D(I+B)C$ if and only if $B=DBD^{-1}$.

Suppose $A=(I+B)C$ is in the center. Then, $B=DBD^{-1}$ for all D in $\mathrm{DL}(n,\mathbb{S})$. If B is not equal to O, then there exists $j<k$ such that $B(j,k)\neq 0$. Choose D in $\mathrm{DL}(n,\mathbb{S})$ such that $D(j,j)\neq D(k,k)$. Then the (j,k) entry of DBD^{-1} is $D(j,j)B(j,k)D(k,k)^{-1}\neq B(j,k)$ and $B\neq DBD^{-1}$. It follows that $B=\mathrm{O}$ and $A=(I+\mathrm{O})C=C\in\mathrm{DL}(n,\mathbb{S})$. Therefore, the center of $\mathrm{TL}(n,\mathbb{S})$ is contained in $\mathrm{DL}(n,\mathbb{S})$.

Now, suppose $D\in\mathrm{DL}(n,\mathbb{S})$ is in the center if $\mathrm{TL}(n,)$. If $D(j,j)\neq D(k,k)$, we can assume that $j<k$ and choose a B in T_1 such that $B(j,k)\neq 0$. Then, as in the previous paragraph, $B\neq DBD^{-1}$ and D is not in the center of $\mathrm{TL}(n,)$. Therefore, the center of $\mathrm{TL}(n,\mathbb{S})$ is contained in $\{cI : c\in\mathbb{S} \text{ and } c\neq 0\}$. □

Proposition 4.3.7 *The groups H_p for $1 < p < n$ are normal subgroups of* $\mathrm{NL}(n, \mathbb{S})$.

Proof. Let A and B be in T_1 and T_p, respectively. It suffices to show that $(I+A)(I+B)(I+A)^{-1} = I + B'$ for some B' in T_p. This requires a careful calculation. First, expanding the middle term yields

$$\begin{aligned}(I+A)(I+B)(I+A)^{-1} &= (I+A)I(I+A)^{-1} + (I+A)B(I+A)^{-1} \\ &= I + (B+AB)(I+A)^{-1} \\ &= I + B(I+A)^{-1} + AB(I+A)^{-1}.\end{aligned}$$

Using equation (4.17) to simplify the two terms containing $(I+A)^{-1}$, produces

$$\begin{aligned}B(I+A)^{-1} &= B\left(I + \sum_{k=1}^{n-1}(-1)^k A^k\right) \\ &= B + \sum_{k=1}^{n-1}(-1)^k BA^k \\ &= B - BA + \sum_{k=2}^{n-1}(-1)^k BA^k\end{aligned}$$

and

$$\begin{aligned}AB(I+A)^{-1} &= AB\left(I + \sum_{k=1}^{n-1}(-1)^k A^k\right) \\ &= AB + \sum_{k=1}^{n-1}(-1)^k ABA^k.\end{aligned}$$

Set

$$B_1 = \sum_{k=2}^{n-1}(-1)^k BA^k + \sum_{k=1}^{n-1}(-1)^k ABA^k.$$

By *Lemma 4.3.1* each BA^k for $k = 2, \ldots, n-1$ and each ABA^k $k = 1, \ldots, n-1$ is in T_{p+2}. Since T_{p+2} is a subspace of $\mathcal{M}_n(\mathbb{S})$, it follows that B_1 is in T_{p+2}. Similarly, $AB - BA$ is in T_{p+1}. Finally, putting the pieces together

$$(I+A)(I+B)(I+A)^{-1} = I + B + AB - BA + B_1 \tag{4.20}$$

and $B + AB - BA + B_1$ is in T_p. □

Proposition 4.3.8 *If A is in T_1 and B is in T_p for $p \geq 1$, then*

$$(I+A)(I+B)(I+A)^{-1}(I+B)^{-1}$$

is in H_{p+1}. If B is in T_p and B is not in T_{p+1} for some p such that $1 \leq p \leq n-2$, then there exists A in T_1 such that $(I+A)(I+B)(I+A)^{-1}(I+B)^{-1}$ is not in T_{p+2}

Proof. Using (4.17), write

$$(I+B)^{-1} = I + \sum_{k=1}^{n-1}(-1)^k B^k$$

and multiply it times the righthand side of equation (4.20). Thus

$$\begin{aligned}
(I+A)(I+B)(I+A)^{-1}(I+B)^{-1} &= \\
(I+B+AB-BA+B_1)\left(I+\sum_{k=1}^{n-1}(-1)^k B^k\right) &= \\
I+\sum_{k=1}^{n-1}(-1)^k B^k + B + B\sum_{k=1}^{n-1}(-1)^k B^k &+ \\
+AB-BA+(AB-BA)\left(\sum_{k=1}^{n-1}(-1)^k B^k\right) &+ \\
B_1\left(I+\sum_{k=1}^{n-1}(-1)^k B^k\right) &= \\
I+AB-BA &+ \\
\sum_{k=2}^{n-1}(-1)^k B^k + \sum_{k=1}^{n-1}(-1)^k B^{k+1} &+ \\
(AB-BA)\left(\sum_{k=1}^{n-1}(-1)^k B^k\right) &+ \\
B_1\left(I+\sum_{k=1}^{n-1}(-1)^k B^k\right).
\end{aligned}$$

Furthermore,

$$\begin{aligned}
\sum_{k=2}^{n-1}(-1)^k B^k + \sum_{k=1}^{n-1}(-1)^k B^{k+1} &= \\
\sum_{k=2}^{n-1}(-1)^k B^k + \sum_{k=2}^{n}(-1)^{k-1} B^k &= \mathrm{O}
\end{aligned}$$

because $B^n = \mathrm{O}$. Consequently,

$$\begin{aligned}
(I+A)(I+B)(I+A)^{-1}(I+B)^{-1} &= \\
I+AB-BA &+ \\
(AB-BA)\left(\sum_{k=1}^{n-1}(-1)^k B^k\right) &+ \\
B_1\left(I+\sum_{k=1}^{n-1}(-1)^k B^k\right).
\end{aligned}$$

Clearly, $AB - BA$ is in T_{p+1}, and every other term is in T_{p+2} except I. Therefore, $(I + A)(I + B)(I + A)^{-1}(I + B)^{-1}$ is in H_{p+1} and

$$(I + A)(I + B)(I + A)^{-1}(I + B)^{-1} = I + AB - BA + B_2$$

for some $B_2 \in T_{p+2}$.

For the second statement, if B is in $T_p \backslash T_{p+1}$, then $B(m, m+p) \neq 0$ for some m. Construct A in T_1 as in the proof of *Theorem 4.3.6* for both $m + p < n$ and $m + p = n$. In both cases the calculations show that either the $(m, m + p + 1)$ or the $(m - 1, n)$ entry of $AB - BA$ is nonzero. It follows that $AB - BA$ is in $T_{p+1} \setminus T_{p+2}$. □

Proposition 4.3.9 *The center of* $\mathrm{NL}(n, \mathbb{S})/H_{p+1}$ *is* H_p/H_{p+1}.

Proof. Two left cosets $H_{p+1}(I + A)$ and $H_{p+1}(I + B)$ in $\mathrm{NL}(n, \mathbb{S})/H_{p+1}$ commute provided $H_{p+1}(I + A)(I + B) = H_{p+1}(I + B)(I + A)$ or equivalently $(I + A)(I + B)(I + A)^{-1}(I + B)^{-1}$ is in H_{p+1}. It follows from *Proposition 4.3.8* that $H_{p+1}(I + A)$ and $H_{p+1}(I + B)$ commute, if $I + B$ is in H_p. Therefore, H_p/H_{p+1} is contained in the center of $\mathrm{NL}(n, \mathbb{S})/H_{p+1}$.

Consider $H_{p+1}(I + B)$ such that $I + B$ is not in H_p. So B is in some T_q such that $1 \leq q < p$. By the second statement of *Proposition 4.3.8*, there exists $A \in T_1$ such that $(I + A)(I + B)(I + A)^{-1}(I + B)^{-1}$ is not in H_{q+2}. Since $q + 2 \leq p + 1$, it follows that $H_{q+2} \supset H_{p+1}$ and $(I + A)(I + B)(I + A)^{-1}(I + B)^{-1}$ is not in H_{p+1}. In particular, $H_p(I + B)$ is not in the center of $\mathrm{NL}(n, \mathbb{S})/H_{p+1}$ when $I + B$ is not in H_p. Therefore, the center of $\mathrm{NL}(n, \mathbb{S})/H_{p+1}$ is H_p/H_{p+1}. □

Let G be a group and recall that $Z(G)$ is an Abelian normal subgroup of G. Set $G^0 = \{e\}$ and $G^1 = Z(G)$. Then $G^1/G^0 = Z(G/G^0)$. If π is the canonical algebraic homomorphism of G onto G/G^1, then $G^2 = \pi^{-1}(Z(G/G^1))$ is a subgroup of G such that $G/G^2 = Z(G/G^1)$. Moreover, G^2 is a normal subgroup of G because $Z(G/G^1)$ is a normal subgroup of G/G^1. This process can be repeated. There exists a normal subgroup G^3 of G containing G^2 such that $G^3/G^2 = Z(G/G^2)$. In general, G^{p+1} is a normal subgroup of G containing G^p such that $G^{p+1}/G^p = Z(G/G^p)$. The sequence

$$\{e\} = G^0 \subset G^1 \subset \ldots \subset G^p \subset \ldots$$

of normal subgroups of G is called the *upper central series* of G.

If the center of G is trivial, that is, $Z(G) = \{e\}$, then $G^p = \{e\}$ for all p. If G is Abelian, then $G^p = G$ for all $p > 0$. In fact, if $G^p = G$ for some $p > 0$, then G/G^{p-1} is Abelian and $G^q = G$ for all $q \geq p$. This leads to a definition. A group G is *nilpotent* provided $G^p = G$ for some $p > 0$. Abelian groups are, of course, nilpotent. Nilpotent groups are solvable groups because $G^{p+1}/G^p = Z(G/G^p)$ is an Abelian group.

If G is a Hausdorff topological group, then the center is a closed subgroup by *Exercise* 4, p. 168, and consequently, all the groups in the upper central series for G are closed subgroups of G.

Theorem 4.3.10 *The groups* $\mathrm{NL}(n, \mathbb{S})$ *are nilpotent but not Abelian for* $n > 2$.

Proof. *Proposition 4.3.9* shows that the upper central series of $\mathrm{NL}(n, \mathbb{S})$ is just

$$\{O\} = H_n \subset H_{n-1} \subset \ldots \subset H_1 = \mathrm{NL}(n, \mathbb{S}),$$

that is, $G^0 = H_n$, $G^1 = H_{n-1}$, and $G^p = H_{n-p}$ for $0 \leq p \leq n-1$. Therefore, $\mathrm{NL}(n, \mathbb{S})$ is a nilpotent metric group. Furthermore, $\mathrm{NL}(n, \mathbb{S})$ is Abelian if and only if $\mathrm{NL}(n, \mathbb{S}) = H_1$ if and only if $n = 2$. (See *Exercise* 1, p. 168.) □

Theorem 4.3.11 *The triangular linear group* $\mathrm{TL}(n, \mathbb{S})$ *is solvable but not nilpotent for* $n > 1$.

Proof. It has already been pointed out that the solvability of $\mathrm{TL}(n, \mathbb{S})$ follows from *Propositions 4.3.4 and 4.3.5*.

The first step is a calculation. Consider B an element of T_p for some $p \geq 1$ and two elements of $\mathrm{DL}(n, \mathbb{S})$ denoted by C and D. Then,

$$\begin{aligned}
[(I+B)C]\,D\,[(I+B)C]^{-1}\,D^{-1} &= \\
(I+B)CDC^{-1}(I+B)^{-1}D^{-1} &= \\
(I+B)D\left(I + \sum_{k=1}^{n}(-1)^k B^k\right)D^{-1} &= \\
(I+B)\left(I + \sum_{k=1}^{n}(-1)^k \left(DBD^{-1}\right)^k\right) &= \\
I + B - DBD^{-1} + \sum_{k=2}^{n}(-1)^k \left(DBD^{-1}\right)^k &+ \\
B\left(\sum_{k=1}^{n}(-1)^k \left(DBD^{-1}\right)^k\right)&.
\end{aligned}$$

Set

$$B_1 = \sum_{k=2}^{n}(-1)^k \left(DBD^{-1}\right)^k + B\left(\sum_{k=1}^{n}(-1)^k \left(DBD^{-1}\right)^k\right)$$

and note that B_1 is in T_{2p}. Therefore,

$$[(I+B)C]\,D\,[(I+B)C]^{-1}\,D^{-1} = I + B - DBD^{-1} + B_1 \tag{4.21}$$

with B_1 in T_{2p} by*Lemma 4.3.1* again.

It follows from *Theorem 4.3.6* that $G^1 = \{cI : c \in \mathbb{S} \text{ and } c \neq 0\}$, the center of $\mathrm{TL}(n, \mathbb{S})$. Suppose A is in G^2, that is, $G^1 A$ is in the center of $\mathrm{TL}(n, \mathbb{S})/G^1$. Then for all D in $\mathrm{DL}(n, \mathbb{S})$, we have $G^1 AD = G^1 DA$ or equivalently $ADA^{-1}D^{-1} \in G^1$. As before, $A = (I+B)C$ for some B in T_1 and C in $\mathrm{DL}(n, \mathbb{S})$.

If $B \neq O$, then B is in $T_p \setminus T_{p+1}$ with $1 \leq p \leq n-1$ Now equation (4.21) applies and $ADA^{-1}D^{-1} = I + B - DBD^{-1} + B_1$ with B_1 in T_{2p}. There exists j

such that $B(j, j+p) \neq 0$ for some j. As in the proof of *Theorem 4.3.6* choose D in $\mathrm{DL}(n, \mathbb{S})$ such that $D(j,j) \neq D(j+p, j+p)$ and it follows that the $(j, j+p)$ entry of DBD^{-1} is $D(j,j)B(j,j+p)D(j+p,j+p)^{-1} \neq B(j,j+p)$. Thus $(B - DBD^{-1})(j, j+p) \neq 0$. Since B_2 is in T_{2p} and $p \geq 1$, the $(j, j+p)$ entry of $I + B - DBD^{-1} + B_1$ is $(B - DBD^{-1})(j, j+p) \neq 0$ and $ADA^{-1}D^{-1}$ is not in $G^1 = \{cI : c \in \mathbb{S} \text{ and } c \neq 0\}$. Therefore, $B = O$ and $G^2 \subset \mathrm{DL}(n, \mathbb{S})$.

Now, suppose $D \in \mathrm{DL}(n, \mathbb{S})$ is in G^2. Then, as before $ADA^{-1}D^{-1}$ is in G^1 for all $A \in \mathrm{TL}(n, \mathbb{S})$. Again writing $A = (I + B)C$ with $B \in T_p$ for $p \geq 1$ and $C \in \mathrm{DL}(n, \mathbb{S})$ and applying (4.21), we obtain $ADA^{-1}D^{-1} = I + B - DBD^{-1} + B_1$ with $B_1 \in T_{2p}$. If $D(k,k) \neq D(m,m)$, we can assume that $k < m$, set $p = m - k$, and define B in T_p by

$$B(i,j) = \begin{cases} 1 & \text{if } i = k \text{ and } j = m \\ 0 & \text{otherwise} \end{cases}$$

As above the $(k, k+p)$ entry of $I + B - DBD^{-1} + B_1$ is nonzero and $ADA^{-1}D^{-1}$ is not in G^1. Therefore, D is in $\{cI : c \in \mathbb{S} \text{ and } c \neq 0\} = G^1$ and $G^2 \subset G^1$. It follows that $G^2 = G^1$ and $G^m = G^1$ for all m, in other words, the metric group $\mathrm{TL}(n, \mathbb{S})/\{cI : c \in \mathbb{S} \text{ and } c \neq 0\}$ has a trivial center. □

EXERCISES

1. Show that H_p is not Abelian for $n \geq 3$ and $1 \leq p \leq n-2$.

2. Let A and B be elements of T_0. Show that the (j,j) entry of AB is just $A(j,j)B(j,j)$ and $A^k(j,j) = (A(j,j))^k$.

3. Let G be a group and let H be a normal subgroup of G. Prove that G/H is Abelian if and only if H contains $\{xyx^{-1}y^{-1} : x, y \in G\}$. (Expressions of the form $xyx^{-1}y^{-1}$ are called *commutators*.)

4. Prove that the center of a Hausdorff topological group is a closed subgroup by showing that its complement is an open set.

5. Let H and K be subgroups of a group G. Prove that if K is a normal subgroup of G, then $HK = \{xy : x \in H \text{ and } y \in K\}$ is a subgroup of G.

6. Show that if A is in $\mathrm{TL}(n, \mathbb{S})$, then λ is a characteristic value (page 107) for A if and only if $\lambda = A(j,j)$ for some j.

7. Prove that if H is a subgroup of a solvable group G, then H is solvable.

8. Let $\mathrm{NL}(n, \mathbb{Z}) = \mathrm{NL}(n, \mathbb{R}) \cap \mathcal{M}_n(\mathbb{Z})$, so $\mathrm{NL}(n, \mathbb{Z})$ consists of the elements of $\mathrm{NL}(n, \mathbb{R})$ with integer coefficients. Prove the following for $n > 1$:

   (a) $\mathrm{NL}(n, \mathbb{Z})$ is a discrete subgroup of $\mathrm{NL}(n, \mathbb{R})$.

   (b) $\mathrm{NL}(n, \mathbb{Z})$ is not a normal subgroup of $\mathrm{NL}(n, \mathbb{R})$.

   (c) $\mathrm{NL}(n, \mathbb{R})/\mathrm{NL}(n, \mathbb{Z})$ is a compact manifold.

9. This exercise provides a description, using the groups $\mathbb{K}^m$, of the structure of the manifold $\mathrm{NL}(n, \mathbb{R})/\mathrm{NL}(n, \mathbb{Z})$ from *Exercise 8* above. Let $T_1(\mathbb{Z}) = T_1 \cap \mathcal{M}_n(\mathbb{Z})$ and prove the following:

   (a) $K_p = I + T_1(\mathbb{Z}) + T_p$ is a closed subgroup of $\mathrm{NL}(n, \mathbb{R})$ for $p = 1, \ldots, n$.

   (b) $K_1 = \mathrm{NL}(n, \mathbb{R}) \supset K_2 \supset \ldots \supset K_{n-1} \supset K_n = \mathrm{NL}(n, \mathbb{Z})$.

   (c) There exists a homomorphism φ of K_p onto $\mathbb{K}^{n-p}$ with kernel K_{p+1}.

   (d) K_{p+1} is a normal subgroup of K_p such that K_p/K_{p+1} is isomorphic to $\mathbb{K}^{n-p}$.

   (e) Letting $\pi_p : \mathrm{NL}(n, \mathbb{R}) \to \mathrm{NL}(n, \mathbb{R})/K_p$ be the canonical continuous open function $\pi_p(A) = K_p A$, there exists a continuous open function $\theta_p : \mathrm{NL}(n, \mathbb{R})/K_{p+1} \to \mathrm{NL}(n, \mathbb{R})/K_p$ such that $\pi_p = \theta_p \circ \pi_{p+1}$.

   (f) $\theta_p^{-1}(K_p) = K_p/K_{p+1}$.

   (g) $\theta_p^{-1}(K_p A) = \pi_{p+1}(K_p A)$, where on the left $K_p A$ is an element of $\mathrm{NL}(n, \mathbb{R})/K_p$ and on the right it is a coset in $\mathrm{NL}(n, \mathbb{R})$.

   (h) $\theta_p^{-1}(K_p A)$ is homeomorphic to $\mathbb{K}^{n-p}$ for all $K_p A \in \mathrm{NL}(n, \mathbb{R})/K_p$.

4.4 One-Parameter Subgroups

A *one-parameter subgroup* of a topological group G is a homomorphism $\varphi : \mathbb{R} \to G$. Although $\varphi(\mathbb{R})$ is a subgroup of G, it need not be closed. The solenoidal groups defined on page 137 are just the metric groups with dense one-parameter subgroups like the tori (*Theorem 3.3.11*). A torus has many one-parameter subgroups - some are dense, some are compact, and some are neither, but they all have the form $\varphi : \mathbb{R} \to \mathbb{K}^n$ by $\varphi(t) = \left(\exp(2\pi i t a_1), \ldots, \exp(2\pi i t a_n)\right)$ by *Theorem 3.2.6*.

The general linear groups also have a wealth of one-parameter subgroups, but none of them are dense because the general linear groups are not Abelian. The exponential function will play a leading role but in a more general setting than the complex plane. The key ideas occur naturally in the context of a normed algebra with identity and will be developed in this more general context before being applied to the normed algebra $\mathcal{M}_n(\mathbb{S})$ with identity I.

Let $\mathcal{A}$ be a normed algebra over $\mathbb{S}$ with identity $\mathbf{e}$ as defined on page 96. If $\mathbf{v}$ is in $\mathcal{A}$, then $\mathbf{v}^k$ is defined for all $k \in \mathbb{Z}^+$, and as usual $\mathbf{v}^0 = \mathbf{e}$ by convention. Given an infinite sequence $c_0, c_1, \ldots$ is in $\mathbb{S}$ and $\mathbf{v}$ in W, a *power series* in $\mathcal{A}$ is defined to be an expression of the form

$$\sum_{k=0}^{\infty} c_k \mathbf{v}^k = c_0 \mathbf{e} + c_1 \mathbf{v} + c_2 \mathbf{v}^2 + \ldots, \tag{4.22}$$

and to converge at $\mathbf{v}$ in W when the sequence of *partial sums*

$$s_m(\mathbf{v}) = \sum_{k=0}^{m} c_k \mathbf{v}^k = c_0\mathbf{e} + c_1\mathbf{v} + \ldots + c_m\mathbf{v}^m \tag{4.23}$$

is a convergent sequence in $\mathcal{A}$. If (4.22) converges to $\mathbf{u}$, then both $s_m(\mathbf{v})$ and $s_{m-1}(\mathbf{v})$ converge to $\mathbf{u}$. Hence, $c_m\mathbf{v}^m = s_m(\mathbf{v}) - s_{m-1}(\mathbf{v})$ converges to $\mathbf{u} - \mathbf{u} = \mathbf{0}$.

The fact that $\mathbb{R}$ and $\mathbb{C}$ are complete metric spaces with their usual metrics (*Exercise* 18, p. 62) plays a critical, but not always visible, role in the study of real and complex power series. Here completeness must be assumed because not every normed algebra with identity is complete. This leads to a well-studied class of spaces.

A *Banach algebra with identity* is a normed algebra with identity that is a complete metric space with respect to its usual metric $d(\mathbf{x}, \mathbf{y}) = \|\mathbf{x} - \mathbf{y}\|_a$. (Some, but not all authors, require a Banach algebra to have a complex linear structure.) For example, $\mathcal{M}_n(\mathbb{S})$ with norm $\|\cdot\|_\mathcal{B}$ is a Banach algebra with identity because $\mathbb{R}^m$ with its usual metric $d(\mathbf{x}, \mathbf{y}) = \|\mathbf{x} - \mathbf{y}\|$ is complete for all $m \geq 1$ and all the norms on $\mathbb{R}^m$ are equivalent (see *Exercise* 8, p. 179). So whatever we learn about power series in Banach algebras applies to it. Other examples of Banach algebras with identity can be found in *Exercise* 1, p. 178, and *Exercise* 7, p. 179.

Because a Banach algebra with identity is a complete metric space, a power series in it converges if and only if the sequence of partial sums is a Cauchy sequence, that is, given $\varepsilon > 0$ there exists N such that

$$\left|\sum_{k=0}^{q} c_k \mathbf{v}^k - \sum_{k=0}^{p} c_k \mathbf{v}^k\right| = \left|\sum_{k=p+1}^{q} c_k \mathbf{v}^k\right| < \varepsilon$$

when $q > p > N$.

Note that the partial sum depends on the choice of $\mathbf{v}$, and hence it was written as a function of $\mathbf{v}$ in (4.23). Of course, when (4.22) converges it converges to a function $f(\mathbf{v})$. Our initial goal is to obtain some basic information about the domain $f(\mathbf{v})$.

Both $\mathbb{R}$ and $\mathbb{C}$ are Banach algebras and can be used to study power series in an arbitrary Banach algebra. For any infinite sequence $c_0, c_1, \ldots$ in $\mathbb{R}$ or $\mathbb{C}$, there is a complex power series denoted by

$$\sum_{k=0}^{\infty} c_k z^k \tag{4.24}$$

with z in $\mathbb{C}$. We will review a few ideas about complex power series that are essential for understanding power series in a Banach algebras.

Let $d_0, d_1, \ldots$ be another sequence of real or complex numbers and w a complex number. If $|c_k z^k| \leq |d_k w^k|$ for all k , then

$$\left|\sum_{k=p}^{q} c_k z^k\right| \leq \sum_{k=p}^{q} |c_k z^k| \leq \sum_{k=p}^{q} |d_k w^k| \tag{4.25}$$

for $q > p > 0$. If $\sum_{k=0}^{\infty} |d_k w^k|$ converges, then its partial sums are a Cauchy sequence. It follows from (4.25) that the partial sums of $\sum_{k=0}^{\infty} |c_k z^k|$ and $\sum_{k=0}^{\infty} c_k z^k$ are also Cauchy sequences and both series converge because $\mathbb{C}$ is a complete metric space. This observation is just a convenient form of the *comparison test.* In particular, note that the convergence of $\sum_{k=0}^{\infty} |c_k z^k|$, called *absolute convergence*, implies the convergence of $\sum_{k=0}^{\infty} c_k z^k$.

Given a complex power series $\sum_{k=0}^{\infty} c_k z^k$, set

$$r_c = \sup\left\{ |z| : \sum_{k=0}^{\infty} c_k z^k \text{ converges} \right\} \tag{4.26}$$

The possible values for r_c are zero, any positive real number, and infinity. It is aptly called the *radius of convergence* because of the following fact:

Proposition 4.4.1 *If $\sum_{k=0}^{\infty} c_k \zeta^k$ converges and $|z| < |\zeta|$, then $\sum_{k=0}^{\infty} c_k z^k$ converges absolutely. In particular, $\sum_{k=0}^{\infty} c_k z^k$ converges and defines a function $f(z) = \sum_{k=0}^{\infty} c_k z^k$ on the open disk $\{z \in \mathbb{C} : |z| < r_c\}$.*

Proof. The geometric series $\sum_{k=0}^{\infty} z^k$ will be used to apply the comparison test to $\sum_{k=0}^{\infty} c_k z^k$. Recall (or prove by induction) that

$$\sum_{k=0}^{m} z^k = \frac{1 - z^{m+1}}{1 - z} \tag{4.27}$$

when $z \neq 1$. It follows that the geometric series converges to $1/(1-z)$ when $|z| < 1$ and diverges when $|z| > 1$. So the radius of convergence of the geometric series is 1.

Since $\sum_{k=0}^{\infty} c_k \zeta^k$ converges by hypothesis, the sequence $|c_k \zeta^k|$ converges to zero. Consequently, there exists a constant $\alpha > 0$ such that $|c_k \zeta^k| < \alpha$ for all k. If $|z| < |\zeta|$, then

$$|c_k z^k| = \left| c_k \frac{z^k}{\zeta^k} \zeta^k \right| = |c_k \zeta^k| \left| \frac{z}{\zeta} \right|^k \leq \alpha \left| \frac{z}{\zeta} \right|^k$$

and by setting $w = z/\zeta$

$$|c_k z^k| \leq |\alpha w^k| = \alpha |w^k|.$$

Because $|w| < 1$, the power series $\alpha \sum_{k=0}^{\infty} w^k$ converges. It follows from (4.25) that the series $\sum_{k=0}^{\infty} |c_k z^k|$ converges. The second statement is now clear. □

Proposition 4.4.2 *Let $\mathcal{A}$ be a Banach algebra over $\mathbb{S}$ with identity $\mathbf{e}$ and norm $\|\cdot\|_a$. Let $c_0, c_1, \ldots$ be a sequence in $\mathbb{S}$. If $r_c > 0$ for the complex power series $\sum_{k=0}^{\infty} c_k z^k$, then the power series $\sum_{k=0}^{\infty} c_k \mathbf{v}^k$ converges to a function on $B_{r_c}(\mathbf{0})$ in $\mathcal{A}$.*

Proof. Consider any positive real number r such that $r < r_c$. Then $\sum_{k=0}^{\infty} |c_k r^k|$ converges by *Proposition 4.4.1*. Thus its partial sums form a Cauchy sequence, and given $\varepsilon > 0$ there exists N such that

$$\sum_{k=p+1}^{q} |c_k| r^k = \sum_{k=p+1}^{q} |c_k r^k| < \varepsilon$$

for $q > p > N$.

If $\mathbf{v}$ in $\mathcal{A}$ satisfies $\|\mathbf{v}\|_a < r$, then

$$\left\| \sum_{k=0}^{q} c_k \mathbf{v}^k - \sum_{k=0}^{p} c_k \mathbf{v}^k \right\|_a = \left\| \sum_{k=p+1}^{q} c_k \mathbf{v}^k \right\|_a \leq \sum_{k=p+1}^{q} |c_k| \, \|\mathbf{v}\|_a^k < \sum_{k=p+1}^{q} |c_k| r^k < \varepsilon$$

for $q > p > N$. To summarize

$$\left\| \sum_{k=0}^{q} c_k \mathbf{v}^k - \sum_{k=1}^{p} c_k \mathbf{v}^k \right\|_a < \varepsilon \tag{4.28}$$

for $\|\mathbf{v}\|_a < r$ and $q > p > N$. Hence the partial sums of $\sum_{k=0}^{\mathbf{m}} c_k \mathbf{v}^k$ are a Cauchy sequence and converge because $\mathcal{A}$ is a complete metric space. It follows that $f(\mathbf{v}) = \sum_{k=0}^{\infty} c_k \mathbf{v}^k$ defines a function on the open ball $B_{r_c}(\mathbf{0})$ in $\mathcal{A}$. □

To prove that the function $f(\mathbf{v}) = \sum_{k=0}^{\infty} c_k \mathbf{v}^k$ is continuous on $B_{r_c}(\mathbf{0})$ in $\mathcal{A}$, we must introduce *uniform convergence* and prove a basic result about the uniform convergence. Let X and Y be metric spaces with metrics d_X and d_Y, respectively. A sequence of functions $f_k : X \to Y$ *converges uniformly* on X to a function $f : X \to Y$ provided that given $\varepsilon > 0$ there exists a positive integer N such that $d_Y(f_k(x), f(x)) < \varepsilon$ for all x in X and all $k \geq N$.

Proposition 4.4.3 *Let X and Y be metric spaces with metrics d_X and d_Y, respectively. If a sequence of continuous functions $f_k : X \to Y$ converges uniformly on X to a function $f : X \to Y$, then f is continuous on X.*

Proof. Let z be any point in X and let $\varepsilon > 0$. By uniform convergence, there exists a positive integer k such that $d_Y(f_k(x), f(x)) < \varepsilon/3$ for all $x \in X$. Then there exists $\delta > 0$ such that $d_Y(f_k(x), f_k(z)) < \varepsilon/3$ when $d_X(x, z) < \delta$. Thus

$$\begin{aligned} d_Y(f(x), f(z)) &\leq d_Y(f(x), f_k(x)) + d_Y(f_k(x), f_k(z)) + d_Y(f_k(z), f(z)) \\ &< \frac{\varepsilon}{3} + \frac{\varepsilon}{3} + \frac{\varepsilon}{3} \\ &= \varepsilon \end{aligned}$$

when $d_X(x, z) < \delta$. Therefore, f is continuous at every z in X. □

Theorem 4.4.4 *Let $\mathcal{A}$ be a Banach algebra over $\mathbb{S}$ with identity* $\mathbf{e}$ *and norm* $\|\cdot\|_a$. *Let $c_0, c_1, \ldots$ be a sequence in $\mathbb{S}$. If $r_c > 0$ for the complex power series $\sum_{k=0}^{\infty} c_k z^k$, then the power series $\sum_{k=0}^{\infty} c_k \mathbf{v}^k$ converges to a continuous function on $B_{r_c}(\mathbf{0})$ in $\mathcal{A}$.*

Proof. Consider any positive r such that $r < r_c$. From equation (4.28), we know there exists N such that

$$\left| \sum_{k=0}^{q} c_k \mathbf{v}^k - \sum_{k=1}^{p} c_k \mathbf{v}^k \right| < \varepsilon$$

for $q > p > N$ and $\mathbf{v} \in B_r(\mathbf{0})$. Taking the limit as q goes to infinity, yields

$$\left| f(\mathbf{v}) - \sum_{k=1}^{p} c_k \mathbf{v}^k \right| \leq \varepsilon$$

for $p > N$ and $\mathbf{v} \in B_r(\mathbf{0})$. Therefore, the sequence of functions $s_p(\mathbf{v}) = \sum_{k=0}^{p} c_k \mathbf{v}^k$ converges uniformly on $B_r(\mathbf{0})$ to $f(\mathbf{v})$ and $f(\mathbf{v})$ is continuous on the metric space $B_r(\mathbf{0})$ by *Proposition 4.4.3*. Because the sets $B_r(\mathbf{0})$ such that $0 < r < r_c$ form an open cover of $B_{r_c}(\mathbf{0})$, the function f is continuous on $B_{r_c}(\mathbf{0})$ by *Exercise* 2, p. 179. □

The radius of convergence for the familiar exponential series, $\sum_{k=0}^{\infty} z^k/k!$, is infinity. Hence

$$e^{\mathbf{v}} = \sum_{k=0}^{\infty} \frac{1}{k!} \mathbf{v}^k \tag{4.29}$$

defines an exponential function on a Banach algebra $\mathcal{A}$ to itself. It is not true, however, that $e^{\mathbf{u}+\mathbf{v}} = e^{\mathbf{u}} e^{\mathbf{v}}$ for all $\mathbf{u}$ and $\mathbf{v}$ in $\mathcal{A}$.

Theorem 4.4.5 *Let $\mathcal{A}$ be a Banach algebra with identity and define the exponential function on $\mathcal{A}$ by (4.29). If* $\mathbf{u}$ *and* $\mathbf{v}$ *are elements of $\mathcal{A}$ such that* $\mathbf{uv} = \mathbf{vu}$, *then*

$$e^{\mathbf{u}+\mathbf{v}} = e^{\mathbf{u}}\, e^{\mathbf{v}}. \tag{4.30}$$

Proof. It suffices to show that

$$\left\| \sum_{k=0}^{2m} \frac{1}{k!} (\mathbf{u}+\mathbf{v})^k - \left(\sum_{j=0}^{m} \frac{1}{j!} \mathbf{u}^j \right) \left(\sum_{k=0}^{m} \frac{1}{k!} \mathbf{v}^k \right) \right\|$$

goes to zero as m goes to infinity. This requires rewriting both pieces to eliminate the common terms and then analyzing the remaining terms.

Since $\mathbf{uv} = \mathbf{vu}$, the expression $(\mathbf{u}+\mathbf{v})^k$ can be expanded by the binomial formula (*Exercises 3 and 4*, p. 179) and

$$\frac{1}{k!}(\mathbf{u}+\mathbf{v})^k = \frac{1}{k!} \sum_{j=0}^{k} \frac{k!}{j!(k-j)!} \mathbf{u}^j \mathbf{v}^{k-j}$$

$$= \sum_{j=0}^{k} \frac{1}{j!(k-j)!}\mathbf{u}^j\mathbf{v}^{k-j}$$
$$= \sum_{j=0}^{k} \frac{1}{j!}\mathbf{u}^j \frac{1}{(k-j)!}\mathbf{v}^{k-j}.$$

It follows that

$$\sum_{k=0}^{2m} \frac{1}{k!}(\mathbf{u}+\mathbf{v})^k = \sum_{k=0}^{2m}\sum_{j=0}^{k} \frac{1}{j!}\mathbf{u}^j \frac{1}{(k-j)!}\mathbf{v}^{k-j}.$$

Set

$$\Lambda_1 = \left\{(j,k) \in \mathbb{Z}^2 : 0 \le j,\ 0 \le k \text{ and } j+k \le 2m\right\}.$$

A moments reflection shows that

$$\sum_{k=0}^{2m}\sum_{j=0}^{k} \frac{1}{j!}\mathbf{u}^j \frac{1}{(k-j)!}\mathbf{v}^{k-j} = \sum_{(j,k)\in\Lambda_1} \frac{1}{j!k!}\mathbf{u}^j\mathbf{v}^k.$$

Notice that Λ_1 is a triangular array in $\mathbb{Z}^2$ and it contains $1+2+\ldots+(2m+1) = (2m+1)(2m+2)/2 = (2m+1)(m+1)$ elements.

Turning to the second piece,

$$\left(\sum_{j=0}^{m} \frac{1}{j!}\mathbf{u}^j\right)\left(\sum_{k=0}^{m} \frac{1}{k!}\mathbf{v}^k\right) = \sum_{k=0}^{m}\sum_{j=0}^{m}\left(\frac{1}{j!}\mathbf{u}^j\right)\left(\frac{1}{k!}\mathbf{v}^k\right)$$
$$= \sum_{(j,k)\in\Lambda_2} \frac{1}{j!k!}\mathbf{u}^j\mathbf{v}^k.$$

where

$$\Lambda_2 = \left\{(j,k) \in \mathbb{Z}^2 : 0 \le j \le m \text{ and } 0 \le k \le m\right\}.$$

Clearly, Λ_2 is a square array containing $(m+1)^2$ elements and $\Lambda_2 \subset \Lambda_1$. Set $\Lambda_3 = \Lambda_1 \setminus \Lambda_2$. It contains $(2m+1)(m+1) - (m+1)^2 = m(m+1)$ elements. Furthermore,

$$\Lambda_3 = \left\{(j,k) \in \mathbb{Z}^2 : \max\{j,k\} \ge m+1; \text{and } j+k \le 2m\right\}.$$

Putting the pieces together,

$$\left\|\sum_{k=0}^{2m} \frac{1}{k!}(\mathbf{u}+\mathbf{v})^k - \left(\sum_{j=0}^{m} \frac{1}{j!}\mathbf{u}^j\right)\left(\sum_{k=0}^{m} \frac{1}{k!}\mathbf{v}^k\right)\right\| =$$

$$\left\|\sum_{(j,k)\in\Lambda_1} \frac{1}{j!k!}\mathbf{u}^j\mathbf{v}^k - \sum_{(j,k)\in\Lambda_2} \frac{1}{j!k!}\mathbf{u}^j\mathbf{v}^k\right\| =$$

$$\left\| \sum_{(j,k)\in\Lambda_3} \frac{1}{j!k!}\mathbf{u}^j\mathbf{v}^k \right\| \leq$$

$$\sum_{(j,k)\in\Lambda_3} \frac{1}{j!k!}\|\mathbf{u}^j\mathbf{v}^k\| \leq$$

$$\sum_{(j,k)\in\Lambda_3} \frac{1}{j!k!}\|\mathbf{u}\|^j\,\|\mathbf{v}\|^k.$$

Set $c = \max\{\|\mathbf{u}\|, \|\mathbf{v}\|, 1\}$. From the description of Λ_3 it follows that $j!k! \geq (m+1)!$ for $(j,k) \in \Lambda_3$ and $\|\mathbf{u}\|^j\,\|\mathbf{v}\|^k \leq c^{2m}$. Consequently,

$$\frac{1}{j!k!}\|\mathbf{u}\|^j\,\|\mathbf{v}\|^k \leq \frac{c^{2m}}{(m+1)!}$$

It follows that

$$\sum_{(j,k)\in\Lambda_3} \frac{1}{j!k!}\|\mathbf{u}\|^j\,\|\mathbf{v}\|^k \leq m(m+1)\frac{c^{2m}}{(m+1)!} = c^2\frac{(c^2)^{m-1}}{(m-1)!}$$

which goes to zero as m goes to infinity because $\sum_{k=0}^{\infty}(c^2)^k/k!$ converges. Therefore,

$$\left\| \sum_{k=0}^{2m} \frac{1}{k!}(\mathbf{u}+\mathbf{v})^k - \left(\sum_{j=0}^{m} \frac{1}{j!}\mathbf{u}^j\right)\left(\sum_{k=0}^{m} \frac{1}{k!}\mathbf{v}^k\right) \right\|$$

goes to zero as m goes to infinity. □

The remainder of this section will be devoted to the Banach algebra $\mathcal{M}_n(\mathbb{S})$ with identity I and norm $\|\cdot\|_{\mathcal{B}}$. By *Theorem 4.4.1*

$$e^A = \sum_{k=0}^{\infty} \frac{1}{k!}A^k$$

defines a continuous function on $\mathcal{M}_n(\mathbb{S})$. Clearly, $e^{\mathrm{O}} = I$. *Theorem 4.4.5* applied to A and $-A$ shows that

$$e^A\, e^{-A} = e^{A-A} = e^{\mathrm{O}} = I = e^{-A}\, e^A.$$

Hence, $(e^A)^{-1} = e^{-A}$ and e^A is in $\mathrm{GL}(n,\mathbb{S})$, and $A \to e^A$ is a continuous function from $\mathcal{M}_n(\mathbb{S})$ to $\mathrm{GL}(n,\mathbb{S})$.

Given a matrix A in $\mathcal{M}_n(\mathbb{S})$, *Theorem 4.4.5* also applies to sA and tA for $s, t \in \mathbb{R}$. So

$$e^{sA}\, e^{tA} = e^{(s+t)A},$$

and $\varphi_A(t) = e^{tA}$ is a one-parameter subgroup of $\mathrm{GL}(n,\mathbb{S})$. It is instructive to look at a number of examples of these one-parameter subgroups.

If $A = \mathrm{Diag}(a_1, \ldots, a_n)$, then $\varphi_A(t) = \mathrm{Diag}(e^{a_1 t}, \ldots, e^{a_n t})$. Next, consider $p \neq q$, $1 \leq p, q \leq n$ and define A by

$$A(j,k) = \begin{cases} a \neq 0 & \text{if } (j,k) = (p,q) \\ 0 & \text{otherwise} \end{cases}$$

Then $\varphi_A(t) = I + tA$. Slightly more general, if A is in T_1, then

$$\varphi_A(t) = I + tA + t^2A^2 + \ldots + t^{n-1}A^{n-1},$$

and $\varphi_A(t)$ is a one-parameter subgroup of $NL(n, \mathbb{S})$. Thus some of the one-parameter subgroups $\varphi_A(t) = e^{tA}$ are in fact polynomial functions in t.

Proposition 4.4.6 *Let A and B be elements of $\mathcal{M}_n(\mathbb{S})$. If B is invertible, then*

$$e^{BAB^{-1}} = Be^AB^{-1}.$$

Proof. Because the function $A \to BAB^{-1}$ is continuous and $BA^kB^{-1} = (BAB^{-1})^k$, we have

$$\begin{aligned} Be^AB^{-1} &= B\left(\lim_{q\to\infty} \sum_{k=0}^{p} \frac{1}{k!}A^k\right)B^{-1} \\ &= \lim_{q\to\infty} B\left(\sum_{k=0}^{p} \frac{1}{k!}A^k\right)B^{-1} \\ &= \lim_{q\to\infty} \sum_{k=0}^{p} \frac{1}{k!}(BAB^{-1})^k \\ &= e^{BAB^{-1}} \end{aligned}$$

as required. □

Similarly, using the continuity of $A \to A^{\mathfrak{t}}$ and $A \to \overline{A}$, we have:

Proposition 4.4.7 *If A is an element of $\mathcal{M}_n(\mathbb{S})$, then*

$$e^{A^{\mathfrak{t}}} = \left(e^A\right)^{\mathfrak{t}}$$

and

$$e^{\overline{A}} = \overline{\left(e^A\right)}$$

Corollary 4.4.8 *If A is an element of $\mathcal{M}_n(\mathbb{R})$ such that $A + A^{\mathfrak{t}} = \mathrm{O}$, then $\varphi(t) = e^{tA}$ is a one-parameter subgroup of $\mathrm{O}(n)$. If A is an element of $\mathcal{M}_n(\mathbb{C})$ such that $\overline{A} + A^{\mathfrak{t}} = \mathrm{O}$, then $\varphi(t) = e^{tA}$ is a one-parameter subgroup of $\mathrm{U}(n)$.*

We will need the following result about matrices for the last theorem and in Chapter 5. It makes use of the set T_0 of triangular matrices from Section 4.3.

Lemma 4.4.9 *For every A in $\mathcal{M}_n(\mathbb{C})$, there exists B in $\mathrm{GL}(n,\mathbb{C})$ such that A is in BT_0B^{-1} and*

$$\mathcal{M}_n(\mathbb{C}) = \bigcup_{B \in \mathrm{GL}(n,\mathbb{C})} BT_0B^{-1}. \tag{4.31}$$

Proof. Given A in $\mathcal{M}_n(\mathbb{C})$, consider the linear function T_A in $\mathcal{L}(\mathbb{C}^n, \mathbb{C}^n)$ defined by $T_A(\mathbf{z}) = \mathbf{z}A^{\mathfrak{t}}$. Recall that the matrix of T_A with respect to the standard basis is A (*Exercise* 5, p. 81, and *Exercise* 6, p. 82). By *Theorem 2.5.11*, there exists a basis $\mathbf{z}_1, \ldots, \mathbf{z}_n$ for $\mathbb{C}^n$ such that $T(\mathbf{z}_k) \in \mathrm{Span}\,[\mathbf{z}_1, \ldots, \mathbf{z}_k]$ for $k = 1, \ldots, n$. (This is where using $\mathbb{C}$ becomes essential because *Theorem 2.5.11* is false for linear spaces over $\mathbb{R}$.) Observe that the matrix A' of T_A with respect to the basis $\mathbf{z}_1, \ldots, \mathbf{z}_n$ is in T_0 (see the comments on page 76 about the content of the rows and columns of the matrix of a linear function). It follows from *Proposition 2.2.7* that there exists B in $\mathrm{GL}(n,\mathbb{C})$ such that $A = BA'B^{-1}$. Thus A is in BT_0B^{-1} and (4.31) holds. □

The final result shows how to calculate the determinant of e^A from the matrix A and provides one-parameter subgroups of $\mathrm{SL}(n,\mathbb{S})$. The *trace of a matrix A* is defines by $\mathrm{Tr}\,[A] = \sum_{j=1}^{n} A(j,j)$.

Theorem 4.4.10 *If A is in $\mathcal{M}_n(\mathbb{S})$, then*

$$\mathrm{Det}\,\left[e^A\right] = e^{\mathrm{Tr}\,[A]} \tag{4.32}$$

Proof. First observe that it suffices to prove the result for complex matrices because they contain the real matrices.

Consider A in T_0. Then A^k is in T_0 for all k and

$$e^A = \lim_{q \to \infty} \sum_{k=0}^{q} \frac{1}{k!} A^k$$

is in T_0 because it is closed. Furthermore, $A^k(j,j) = A(j,j)^k$ by *Exercise 2*, p. 168, and hence $e^A(j,j) = e^{A(j,j)}$. It follows from *Exercise* 3, p. 113, that

$$\begin{aligned}
\mathrm{Det}\,\left[e^A\right] &= \prod_{j=1}^{n} e^A(j,j) \\
&= \prod_{j=1}^{n} e^{A(j,j)} \\
&= e^{\sum_{j=1}^{n} A(j,j)} \\
&= e^{\mathrm{Tr}\,[A]}.
\end{aligned}$$

Hence, the result holds for A in T_0. The remainder of the proof reduces the general case to this special case.

If A is any matrix in $\mathcal{M}_n(\mathbb{C})$, then by *Lemma 4.4.9* there exists B in $\mathrm{GL}(n, \mathbb{C})$ and A' in T_0 such that $A = BA'B^{-1}$. Then using *Proposition 4.4.6*, *Exercise* 9, p. 114, and *Exercise* 11, p. 180, we obtain

$$\begin{aligned}
\mathrm{Det}\left[e^A\right] &= \mathrm{Det}\left[e^{BA'B^{-1}}\right] \\
&= \mathrm{Det}\left[Be^{A'}B^{-1}\right] \\
&= \mathrm{Det}\left[e^{A'}\right] \\
&= e^{\mathrm{Tr}\,[A']} \\
&= e^{\mathrm{Tr}\,[B^{-1}AB]} \\
&= e^{\mathrm{Tr}\,[A]}
\end{aligned}$$

to complete the proof. □

Corollary 4.4.11 *If* $\mathrm{Tr}\,[A] = 0$, *then* $\varphi(t) = e^{tA}$ *is a one-parameter subgroup of* $\mathrm{SL}(n, \mathbb{S})$.

The function $A \to e^A$ is also a differentiable function with continuous partial derivatives of all orders. The inverse function theorem applies at O and there is a differentiable function $A \to \ln(A)$ such that $e^{\ln A} = A$ in a neighborhood of I. The differentiable properties of the function e^A along with Lie algebras are powerful tools for studying the closed subgroups of $\mathrm{GL}(n, \mathbb{C})$. For example, it can shown that the closed subgroups of $\mathrm{GL}(n, \mathbb{C})$ are all manifolds and are called *matrix Lie groups.* Hall's book [3] is a modern comprehensive introduction to matrix Lie groups. More generally, Lie groups are metric groups that are smooth manifolds with smooth (infinitely differentiable) multiplication and inverse functions. For an extensive treatment of smooth manifolds including an introduction to general Lie groups see [9].

The one-parameter subgroups e^{At} also play an important role in the study of linear differential equations with constant (real or complex) coefficients. These equations are usually written as $\dot{\mathbf{x}} = A\mathbf{x}$, where $\mathbf{x}$ is a $n \times 1$ matrix. In this context, e^{At} is a fundamental matrix solution. Setting $\theta(t) = e^{A(t-s)}\boldsymbol{\xi}$, it follows from *Exercise* 14, p. 180, that $\dot{\theta}(t) = Ae^{A(t-s)}\boldsymbol{\xi} = A\theta(t)$. Since $\theta(s) = \boldsymbol{\xi}$, the function $\theta(t)$ is a solution of the general initial value problem for initial value problem $\dot{\mathbf{x}} = A\mathbf{x}$. For a full discussion of linear differential equations with constant coefficients, see Chapter 4 of [10].

EXERCISES

1. Let X be a compact metric space and consider the commutative normed algebra with identity from $\mathcal{C}(X)$ with the norm $\|f\|_s = \sup\{|f(x)| : x \in X\}$. (See *Exercise* 4, p. 99 and *Exercise* 3, p. 69 for background information.) Prove the following:

(a) A sequence f_k converges to f in $\mathcal{C}(X)$ if and only if f_k converges uniformly to f on X.

(b) $\mathcal{C}(X)$ is a Banach algebra with identity.

2. Let X and Y be metric spaces and let $f : X \to Y$ be a function. Prove that if there exists an open cover U_α, $\alpha \in \Lambda$ such that $f|U_\alpha$, the restriction of f to U_α, is continuous for all $\alpha \in \Lambda$, then f is continuous on X.

3. Let

$$\binom{n}{k} = \frac{n!}{k!(n-k)!}$$

where by definition $0! = 1$, and show that

$$\binom{n}{k} + \binom{n}{k-1} = \binom{n+1}{k}$$

for $1 \leq k \leq n$.

4. Let $\mathbf{u}$ and $\mathbf{v}$ be elements of an algebra $\mathcal{A}$. Prove by induction that if $\mathbf{uv} = \mathbf{vu}$, then

$$(\mathbf{u}+\mathbf{v})^n = \sum_{k=0}^{n} \binom{n}{k} \mathbf{u}^k \mathbf{v}^{n-k}$$

for $n \in \mathbb{Z}^+$. This formula is called the *binomial formula* and the integers $\binom{n}{k}$ are called *binomial coefficients.*.

5. Let X be a compact metric space and consider the Banach algebra $\mathcal{C}(X)$ in *Exercise 1* above. For $f \in \mathcal{C}(X)$, let e^f be given by *Theorem 4.4.4*. Show that $e^f(x) = e^{f(x)}$.

6. Let V and V' be normed linear spaces over $\mathbb{R}$ with norms $\|\cdot\|_a$ and $\|\cdot\|_b$. Consider the normed linear space $\mathcal{B}(V, V')$ with the norm $\|T\|_{\mathcal{B}} = \sup\{\|T(\mathbf{x})\|_b : \|\mathbf{x}\|_a \leq 1\}$. Show that a sequence T_k in $\mathcal{B}(V, V')$ converges to T in $\mathcal{B}(V, V')$ if and only if T_k converges uniformly to T on $\{\mathbf{x} \in V : \|\mathbf{x}\| \leq r\}$ for every $r > 0$.

7. Let V be a normed linear space that is complete with respect to the metric $d(\mathbf{x}, \mathbf{y}) = \|\mathbf{x} - \mathbf{y}\|_a$. (Such spaces are called *Banach spaces.*) Show that $\mathcal{B}(V, V)$ with the norm $\|\cdot\|_{\mathcal{B}}$ is a Banach algebra with identity.

8. Let V be a normed linear space with two equivalent norms $\|\cdot\|_a$ and $\|\cdot\|_b$. Prove the following:

(a) A sequence $\mathbf{x}_k$ is a Cauchy sequence for the metric $d_a(\mathbf{x}, \mathbf{y}) = \|\mathbf{x}-\mathbf{y}\|_a$ if and only if it is a Cauchy sequence for the metric $d_b(\mathbf{x}, \mathbf{y}) = \|\mathbf{x} - \mathbf{y}\|_a$.

(b) The linear space V is a Banach space (see *Exercise 7* above for definition) with the norm $\|\cdot\|_a$ if and only if it is a Banach space with respect to the norm $\|\cdot\|_b$.

9. Calculate e^A for

$$A = \begin{bmatrix} c & 1 & 0 & 0 & 0 \\ 0 & c & 1 & 0 & 0 \\ 0 & 0 & c & 1 & 0 \\ 0 & 0 & 0 & c & 1 \\ 0 & 0 & 0 & 0 & c \end{bmatrix}.$$

10. Given a real number θ, set

$$A = \begin{bmatrix} 0 & \theta \\ -\theta & 0 \end{bmatrix}$$

and show that

$$e^A = \begin{bmatrix} \cos\theta & \sin\theta \\ -\sin\theta & \cos\theta \end{bmatrix}.$$

11. Show that if A is in $\mathcal{M}_n(\mathbb{S})$ and B is in $\mathrm{GL}(n, \mathbb{S})$, then

$$\mathrm{Tr}\left[BAB^{-1}\right] = \mathrm{Tr}\,[A].$$

12. Let $\mathcal{A}$ be a Banach algebra with norm $\|\cdot\|_a$ and consider the power series $\sum_{k=0}^{\infty}(-1)^k \mathbf{v}^k$. Prove that it converges uniformly when $\|\mathbf{v}\|_a < 1$ to a continuous function f such that $f(\mathbf{v})(\mathbf{e}+\mathbf{v}) = \mathbf{e} = (\mathbf{e}+\mathbf{v})f(\mathbf{v})$ when $\|\mathbf{v}\|_a < 1$.

13. Let $\mathcal{A}$ be a Banach algebra. Prove that the group of units of $\mathcal{A}$ is an open subset of $\mathcal{A}$ and a metric group.

14. Prove that for A in $\mathcal{M}_n(\mathbb{C})$

$$\lim_{h\to 0} \frac{1}{h}\left(e^{hA} - I\right) = A$$

and use it to show that for t in $\mathbb{R}$

$$\lim_{h\to 0} \frac{1}{h}\left(e^{(t+h)A} - e^{tA}\right) = Ae^{tA}.$$

15. Calculate the dimensions of the subspaces $A + A^t = O$ and $\mathrm{Tr}\,[A] = 0$ of $\mathcal{M}_n(\mathbb{R})$ and show that they are invariant under the binary operation $(A, B) \to AB - BA$.

16. Consider $\mathcal{C}([0,1])$ with $\|f\|_s = \sup\{|f(x)| : 0 \le x \le 1\}$. Let W be the subspace of the polynomial functions $f(x) = c_0 + c_1x + \ldots + c_nx^n$ restricted to $[0,1]$. Show that the exponential function restricted to $[0,1]$ is in W^-, but not in W and there by proving that *Proposition 2.4.10* is a finite-dimensional result.

17. Show that $\mathcal{C}([0,1])$ with $\|f\|_s = \sup\{|f(x)| : 0 \le x \le 1\}$ is not σ-compact. (One approach is to use *Exercises 19, p. 62 and 9, p. 91.*)

Chapter 5

Connectedness of Topological Groups

The metric spaces $\mathbb{R}$ and $\mathbb{R} \setminus \{0\}$ are very different one-dimensional manifolds because $\mathbb{R} \setminus \{0\}$ consists of two disjoint pieces $\mathbb{R}^+$ and $-\mathbb{R}^+$. Moreover, both $\mathbb{R}^+$ and $-\mathbb{R}^+$ are open subsets of $\mathbb{R} \setminus \{0\}$. The first theorem in this chapter shows that $\mathbb{R}$ cannot be written as the disjoint union of two open subsets. Whether or not a metric space can be written as the union of two disjoint open sets is the distinction between a disconnected and connected topological space, which is the subject of this chapter. Thus $\mathbb{R}$ is connected and $\mathbb{R} \setminus \{0\}$ is disconnected.

Section 5.1 is devoted to the basic properties of connected topological spaces and their connected subsets. With a few general propositions, we can easily show that familiar metric groups like $\mathbb{C} \setminus \{0\}$, $\mathbb{R}^n$ and $\mathbb{K}^n$ are connected, and that $\mathrm{GL}(n, \mathbb{R})$ and $\mathrm{O}(n)$ are disconnected. Every topological group does, however, contains a largest connected subgroup, which is closed.

In Section 5.2, the deeper connectedness properties of the matrix groups are pursued. Specifically, we prove that $\mathrm{SO}(n)$, $\mathrm{U}(n)$, $\mathrm{GL}(n, \mathbb{C})$, and $\mathrm{GL}^+(n, \mathbb{R})$ are connected. The later is accomplished by obtaining a complete description of $\mathrm{GL}^+(n, \mathbb{R})/\mathrm{SO}(n)$ using positive matrices. As a bonus we obtain two important results about the structure of $\mathrm{GL}^+(n, \mathbb{R})$: one algebraic and one topological.

This chapter is also the junction at which we want to begin using products of a sequences of metric spaces. The product topology for the product of a sequence of metric spaces is introduced in the Section 5.1, and it is shown that the product of connected metric spaces is connected. Then, in Section 5.3 we prove that the product of a sequence of compact metric spaces is again compact. This necessitates a discussion of countable compactness, which is equivalent to compactness for metric spaces. Consequently, there are infinite products of metric groups like $\mathbb{K}^\infty$ that are compact and connected.

There are also compact metric groups like $\mathbb{Z}_2^\infty$ that are badly disconnected. Finally, Section 5.4 is devoted to totally disconnected groups. The a-adic groups are interesting examples of compact totally disconnected monothetic groups.

5.1 Connected Topological Spaces

In a topological space X, the sets X and ϕ are both open-and-closed subsets. The situation when they are the only subsets that are both open and closed is an important class of topological spaces. Specifically, a topological space X is *connected* provided that the only subsets of X that are both open and closed are X itself and the empty set ϕ. A topological space that is not connected is said to be *disconnected*.

Our goal in this section is to prove a variety of basic results about connected topological spaces. Along the way, we will determine the connectivity of a variety of familiar topological groups. When the underlying topological space of a topological group G is connected, G is called a *connected topological group*.

Obviously, the trivial group consisting of just an identity element is connected. But it is the only connected discrete group because in a discrete topological space every subset is both open and closed. So $\mathbb{Z}^n$ and all finite groups containing more than one element are disconnected.

A few frequently used observations about open-and-closed sets are worth mentioning at the outset. First, a subset U of a topological space X is both open and closed if and only if $X \setminus U$ is both open and closed. Similarly, U is both open and closed if and only if both U and $X \setminus U$ are open if and only if both U and $X \setminus U$ are closed. In particular, a topological space is connected if it is not the disjoint union of two nonempty open, alternatively closed, sets.

Theorem 5.1.1 *The metric group* $\mathbb{R}$ *is connected.*

Proof. If $\mathbb{R}$ is disconnected, then $\mathbb{R}$ is the disjoint union of two nonempty open sets U and V. When α and β are real numbers such that $\alpha < \beta$, it will be convenient to use the notation (α, β) for the open interval between α and β and $[\alpha, \beta]$ for the closed interval beginning with α and ending with β.

Let a and b be elements of U and V, respectively, and without loss of generality assume that $a < b$. Let

$$c = \sup \{\beta \in \mathbb{R} : a < \beta \text{ and } [a, \beta] \subset U\}.$$

Clearly, $a < c$ because U is open and $c \leq b$ by construction. If $a < \alpha < c$, then there exists β such that $\alpha < \beta < c$ and $[a, \beta] \subset U$ by the definition of c. Hence, α is in U and $(a, c) \subset U$.

Since $\mathbb{R}$ is the disjoint union of U and V, the number c is either in U or V. It suffices to show that each possibility leads to a contradiction.

If c is in U, there exists $\varepsilon > 0$ such that $(c - \varepsilon, c + \varepsilon) \subset U$ because U is open. It follows that $[a, c + \varepsilon/2] \subset U$, contradicting the definition of c.

If c is in V, there exists $\varepsilon > 0$ such that $(c - \varepsilon, c + \varepsilon) \subset V$ because V is open. Without loss of generality we can assume that $\varepsilon < c - a$. Consequently, $a < c - \varepsilon/2$ is in V contradicting $(a, c) \subset U$. □

With slight modifications this proof shows that every interval of $\mathbb{R}$ is a connected metric space. In particular, $\mathbb{R}^+$ is connected, but $\mathbb{R} \setminus \{0\}$ is disconnected because it is the disjoint union of the open sets $\mathbb{R}^+$ and $-\mathbb{R}^+$.

Proposition 5.1.2 *Let X and Y be topological spaces, and let $f : X \to Y$ be a continuous onto function. If X is connected, then Y is connected.*

Proof. If Y is the disjoint union of the nonempty open sets U and V, then X is the disjoint union of the nonempty open sets $f^{-1}(U)$ and $f^{-1}(V)$. □

Corollary 5.1.3 *The metric group $\mathbb{K}$ is connected.*

Proof. The function $f(x) = e^{2\pi ix}$ is a continuous function from $\mathbb{R}$ onto $\mathbb{K}$. Hence, $\mathbb{K}$ is connected because $\mathbb{R}$ is connected. □

Corollary 5.1.4 *The metric groups $\mathrm{GL}(n, \mathbb{R})$ and $\mathrm{O}(n)$ are disconnected.*

Proof. The function $\mathrm{Det}\,[A]$ is a continuous function of $\mathrm{GL}(n, \mathbb{R})$ onto the disconnected metric space $\mathbb{R} \setminus \{0\}$. So $\mathrm{GL}(n, \mathbb{R})$ cannot be connected. Similarly, $\mathrm{Det}\,[\mathrm{O}(n)]$ is the disconnected discrete space $\{-1, 1\}$. □

A nonempty subset Y of a topological space X is a *connected subset* of X provided that Y with the relative topology is a connected topological space. If U is an open-and-closed subset of X, then a connected subset of X is either contained in U or in $X \setminus U$. Moreover, if Y is a connected subset of X and $Y \subset Z \subset X$, then Y is a connected subset of Z because the relative topologies on Y from X and Z are equal. We will use this last observation in the next two proofs.

Proposition 5.1.5 *If a topological space X contains a dense connected subset, then X is connected. In particular, if Y is a connected subset of the topological space X, then Y^- is also a connected subset of X.*

Proof. Suppose D is a connected dense subset of X. If X is the disjoint union of two nonempty open sets U and V, then D is the disjoint union of the relatively open sets $D \cap U$ and $D \cap V$ of D. They are nonempty because D is dense in X, contradicting the connectivity of D. This proves the first statement.

If Y is a connected subset of X, then it is also a connected subset of Y^-. Since Y is dense in Y^-, the first part of the proposition applies. □

Proposition 5.1.6 *Suppose Y_α, $\alpha \in \Lambda$ is a family of connected subsets of a topological space X. If $Y_\alpha \cap Y_\beta \neq \phi$ for all α and β in Λ, then*

$$Y = \bigcup_{\alpha \in \Lambda} Y_\alpha$$

is a connected subset of X.

Proof. Because each Y_α is also a connected subset of Y, we can assume without loss of generality that

$$X = \bigcup_{\alpha \in \Lambda} Y_\alpha.$$

Suppose X is the disjoint union of open sets U and V. It suffices to show that either U or V is empty.

For each $\alpha \in \Lambda$, the connected subset Y_α is the disjoint union of the open sets $Y_\alpha \cap U$ and $Y_\alpha \cap V$. It follows that either $Y_\alpha \subset U$ or $Y_\alpha \subset V$.

Suppose that $Y_\beta \subset U$ for some $\beta \in \Lambda$ and let α be an arbitrary element of Λ. If $Y_\alpha \subset V$, then $Y_\beta \cap Y_\alpha \subset U \cap V = \phi$, contradicting the hypothesis that $Y_\beta \cap Y_\alpha \neq \phi$. Consequently, Y_α is contained in U for all $\alpha \in \Lambda$ and $V = \phi$. Therefore, $X = \bigcup_{\alpha \in \Lambda} Y_\alpha$ is either contained in U and V is empty or visa a versa. □

Corollary 5.1.7 *The metric groups $\mathbb{R}^n$ and $\mathbb{K}^n$ are connected.*

Proof. For each $\mathbf{x}$ in $\mathbb{R}^n$ such that $\|\mathbf{x}\| = 1$, let $Y_\mathbf{x} = \{t\mathbf{x} : t \in \mathbb{R}\}$. Since $t \to t\mathbf{x}$ is a continuous function from $\mathbb{R}$ onto $Y_\mathbf{x}$, the subsets $Y_\mathbf{x}$ of $\mathbb{R}^n$ are connected. The proposition applies because

$$\mathbb{R}^n = \bigcup_{\|\mathbf{x}\|=1} Y_\mathbf{x}$$

and

$$\mathbf{0} \in \bigcap_{\|\mathbf{x}\|=1} Y_\mathbf{x},$$

showing that $\mathbb{R}^n$ is connected.

Then $\mathbb{K}^n$ is connected because it is a continuous image of $\mathbb{R}^n$. □

Let X be a topological space and let x be a point in X. Set

$$C(x) = \bigcup \{Y \subset X : Y \text{ is connected and } x \in Y\}$$

Then $C(x)$ is a connected subset of X by *Proposition 5.1.6* and is called the *component* of x. Obviously, $C(x)$ is the largest connected set containing x. *Proposition 5.1.5* implies that $C(x)$ is always a closed set. Note that $y \in C(x)$ if and only if $C(y) = C(x)$. Consequently, the components of a topological space decompose it into disjoint closed connected sets.

Theorem 5.1.8 *If G is a topological group with identity element e, then $C(e)$ is a closed connected normal subgroup of G.*

Proof. It has already been shown that $C(e)$ is closed and connected. Let a be an element of $C(e)$. Since $x \to xa^{-1}$ is a continuous function, $C(e)a^{-1}$ is a connected subset of G, and e is in $C(e)a^{-1}$ because a is in $C(e)$. Hence, $C(e)a^{-1} \subset C(e)$. It follows that $C(e)C(e)^{-1} \subset C(e)$, and $C(e)$ is a subgroup.

Let b be an arbitrary element of G. Since the function $x \to bxb^{-1}$ is continuous on G and maps e to itself, $b\,C(e)b^{-1}$ is a connected subset of G containing e. Hence, $b\,C(e)b^{-1} \subset C(e)$, and $C(e)$ is a normal subgroup of G. □

Theorem 5.1.9 *Let G be a metric group with a closed subgroup H. If H and the quotient space G/H are connected, then G is connected.*

Proof. Suppose G is not connected, and hence the union of disjoint nonempty open sets U and V. Since the canonical homomorphism π is an open function, G/H is the union of the nonempty open sets $\pi(U)$ and $\pi(V)$. Because G/H is connected, $\pi(U) \cap \pi(V) \neq \phi$ and $Ha \in \pi(U) \cap \pi(V)$ for some coset Ha. Consequently, $U \cap Ha \neq \phi$ and $V \cap Ha \neq \phi$, It follows that Ha is the disjoint union of the nonempty relative open sets $U \cap Ha$ and $V \cap Ha$. This is impossible because the function $x \mapsto xa$ is continuous and maps the connected set H onto Ha, proving that the latter is also connected. Therefore, G cannot be the union of disjoint nonempty open sets and must be connected. □

Theorem 5.1.9 was stated for metric groups because the only quotient topological groups that we have studied are for closed subgroups of metric groups. *Exercises 18, 19, and 20* lead to a proof that the quotient of a Hausdorff topological group by a closed subgroup is again a Hausdorff topological group. Consequently, "metric group" can be replaced by "Hausdorff topological group" in *Theorem 5.1.9* without altering the proof.

A topological space X is *totally disconnected* provided that $C(x) = \{x\}$ for each x in X.

Theorem 5.1.10 *If G is a metric group, then $G/C(e)$ is a totally disconnected metric group.*

Proof. It follows from *Theorem 5.1.8* that $G/C(e)$ is a metric group and the component of the identity of $G/C(e)$, which for simplicity will be denoted by C, is a closed subgroup of $G/C(e)$. Consequently, $G_1 = \pi^{-1}(C)$ is a closed subgroup of G containing $C(e)$ such that $G_1/C(e) = C$. Now *Theorem 5.1.9* implies that G_1 is connected because $C(e)$ and C are connected. Hence, $G_1 \subset C(e)$ and $G_1 = C(e)$. Therefore, $C = \{C(e)\}$ is the trivial subgroup of $G/C(e)$. Finally, *Exercise* 12, p. 192, implies that $G/C(e)$ is totally disconnected. □

The metric group $\mathbb{R} \setminus \{0\}$ with identity 1 provides a simple example of this result. Here $C(1) = \mathbb{R}^+$ and $\mathbb{R} \setminus \{0\}/C(1)$ is isomorphic to $\mathbb{Z}_2$. It is also worth mentioning that *Theorem 5.1.10* is another result that can easily be extended to Hausdorff topological groups.

The remainder of this section is devoted to the connected properties of product spaces. We will start by extending the notion of the Cartesian product of sets, product metrics and product topologies to a more general situation.

Let $X_1, \ldots, X_j, \ldots$ be an infinite sequence of nonempty sets. These sets need not be distinct. The *Cartesian product* product of $X_1, \ldots, X_j, \ldots$ is defined by

$$\prod_{j=1}^{\infty} X_j = \{\mathbf{x} = (x_1, \ldots, x_j, \ldots) : x_j \in X_j \text{ for } j \geq 1\},$$

and for $m \in \mathbb{Z}^+$ the projections

$$p_m : \prod_{j=1}^{\infty} X_j \to X_j$$

are defined by $p_m(x_1, \ldots, x_j, \ldots) = x_m$. In other words, $\prod_{j=1}^{\infty} X_j$ is the set of all sequences $\mathbf{x} = (x_1, \ldots, x_j, \ldots)$ such that each x_j is chosen from X_j. (We continue to use boldface to denote points that have coordinates, in this case infinitely many.) When $X_j = X$ for all j, the Cartesian product will be written more simply as X^{∞} and is the set of all infinite sequences in X or functions $j \to x_j$ from $\mathbb{Z}^+$ to X.

Now, suppose $X_1, \ldots, X_j, \ldots$ is an infinite sequence of metric spaces with metrics $d_1, \ldots, d_j, \ldots$. By *Theorem 1.2.7*, it can be assumed for all j that $d_j(x_j, y_j) < 1$ for all x_j and y_j in X_j. Then for every $\mathbf{x}$ and $\mathbf{y}$ in $\prod_{j=1}^{\infty} X_j$, set

$$d(\mathbf{x}, \mathbf{y}) = \sum_{j=1}^{\infty} \frac{d_j(x_j, y_j)}{2^j} < \sum_{j=1}^{\infty} \frac{1}{2^j} = 1 \tag{5.1}$$

It is easily verified that $d(\mathbf{x}, \mathbf{y})$ defines a metric on $\prod_{j=1}^{\infty} X_j$.

The first result shows that the topology coming from this metric is a generalization of the product topology for a finite number of metric spaces. In particular, the topology on $\prod_{j=1}^{\infty} X_j$ can be described easily in terms of the topologies of the metric spaces in the sequence.

Theorem 5.1.11 *Let $X_1, \ldots, X_j, \ldots$ be an infinite sequence of metric spaces with metrics $d_1, \ldots, d_j, \ldots$ such that $d_j(x_j, y_j) < 1$ for all j and all x_j and y_j in X_j. Let d be the metric on $\prod_{j=1}^{\infty} X_j$ defined by (5.1), and let U be a subset of $\prod_{j=1}^{\infty} X_j$. Then the following are equivalent:*

(a) U is an open set for the metric d.

(b) For every point $\mathbf{x} \in U$ there exists a positive integer m and open sets U_j of X_j such that $U_j = X_j$ for $j > m$ and

$$\mathbf{x} \in \prod_{j=1}^{\infty} U_j \subset U.$$

Proof. Let U be an open subset of $\prod_{j=1}^{\infty} X_j$ and let $\mathbf{x}$ be in U. There exists $r > 0$ such that $B_r(\mathbf{x}) \subset U$. (It will be clear from the context which metric is being used to define the open balls used in the proof.) Choose a positive integer m such that

$$\sum_{j=m+1}^{\infty} \frac{1}{2^j} < \frac{r}{2}.$$

Set $U_j = B_{r/2}(x_j)$ for $1 \leq j \leq m$ and $U_j = X_j$ for $j > m$. Let $\mathbf{y}$ be an element of $\prod_{j=1}^{\infty} U_j$. It will be shown that $d(\mathbf{y}, \mathbf{x}) < r$. Specifically,

$$\begin{aligned} d(\mathbf{y}, \mathbf{x}) &< \frac{r}{2} \sum_{j=1}^{m} \frac{1}{2^j} + \sum_{j=m+1}^{\infty} \frac{1}{2^j} \\ &< \frac{r}{2} + \frac{r}{2} = r. \end{aligned}$$

It follows that

$$\mathbf{x} \in \prod_{j=1}^{\infty} U_j \subset B_r(\mathbf{x}) \subset U.$$

and *(a)* implies *(b)*.

Now, suppose that U satisfies condition *(b)* and $\mathbf{x}$ is any element of U. So there exists m and a sequence of open sets U_j of X_j such that $U_j = X_j$ for $j > m$, and $\mathbf{x} \in \prod_{j=1}^{\infty} U_j \subset U$. For $1 \leq j \leq m$, there exists $r_j > 0$ such that $B_{r_j}(x_j) \subset U_j$. Set

$$r = \min\left\{\frac{r_j}{2^j} : 1 \leq j \leq m\right\}.$$

It suffices to show that $B_r(\mathbf{x}) \subset U$.

Suppose $d(\mathbf{y}, \mathbf{x}) < r$. So $d(y_j, x_j)/2^j < r$ for all j, It follows that

$$d(y_j, x_j) < 2^j r \leq r_j$$

for $1 \leq j \leq m$. Thus $y_j \in B_{r_j}(x_j) \subset U_j$ for $1 \leq j \leq m$ and $\mathbf{y} \in \prod_{j=1}^{\infty} U_j \subset U$, proving that $B_r(\mathbf{x}) \subset U$. □

Corollary 5.1.12 *Let $X_1, \ldots, X_j, \ldots$ be an infinite sequence of metric spaces. If m is a positive integer and U_j are open subsets of X_j such that $U_j = X_j$ for $j > m$, then $\prod_{j=1}^{\infty} U_j$ is an open subset of $\prod_{j=1}^{\infty} X_j$.*

Proof. Condition *(b)* is satisfied for every point in $\mathbf{x} \in \prod_{j=1}^{\infty} U_j$. □

As in the case of finite products of topological spaces, the characterization of open sets for the product of metric spaces provides the definition of the product topology for a sequence of topological spaces. Let $X_1, \ldots, X_j, \ldots$ be an infinite sequence of topological spaces, and let $\mathcal{T}$ be the collection of subsets U such that for every point $\mathbf{x} \in U$ there exists a positive integer m and open sets U_j of X_j such that $U_j = X_j$ for $j > m$ and

$$\mathbf{x} \in \prod_{j=1}^{\infty} U_j \subset U.$$

Proving that $\mathcal{T}$ is a topology quickly reduces to showing that the intersection of two sets in $\mathcal{T}$ is again in $\mathcal{T}$. The proofs of the corresponding fact for a finite number of topological spaces (page 24) can be easily modified for the infinite case. This topology is also called a *product topology.*

With minor changes, the proofs of *Propositions 1.2.16, 1.2.17, and 1.3.7* can be used to prove the following propositions:

Proposition 5.1.13 *If $\prod_{j=1}^{\infty} X_j$ is the Cartesian product of a sequence of topological spaces with the product topology, then the projections are continuous open functions.*

Proposition 5.1.14 *Let $\prod_{j=1}^{\infty} X_j$ be the Cartesian product of a sequence of topological spaces with the product topology, and let Y be a topological space. A function $f : Y \to \prod_{j=1}^{\infty} X_j$ is continuous if and only if the functions $f_j = p_j \circ f$ are continuous for all $j \in \mathbb{Z}^+$.*

The definition of the direct product of a finite sequence of groups extends in the obvious way to an infinite sequence of groups $G_1, \ldots, G_j, \ldots$. This natural group structure on $\prod_{j=1}^{\infty} G_j$ is also called the *direct product.*

Proposition 5.1.15 *If $G_1, \ldots, G_m, \ldots$ is an infinite sequence of topological groups, then the direct product $\prod_{j=1}^{\infty} G_j$ with the product topology is a topological group.*

Corollary 5.1.16 *The groups $\mathbb{R}^\infty$, $\mathbb{Z}^\infty$, $\mathbb{K}^\infty$, $(\mathbb{Z}_n)^\infty$ with the product topology are metric groups.*

The following proposition will be useful in showing that the product of connected spaces is connected.

Proposition 5.1.17 *Let $X_1, \ldots, X_j, \ldots$ be a sequence of topological spaces. Given $\mathbf{y} \in \prod_{j=1}^{\infty} X_j$, the set*

$$D_{\mathbf{y}} = \bigcup_{m=1}^{\infty} \{\mathbf{x} : x_j = y_j \text{ for } j > m\}$$

is dense in $\prod_{j=1}^{\infty} X_j$ with the product topology.

Proof. The set $D_{\mathbf{y}}$ intersects every set of the form $\prod_{j=1}^{\infty} U_j$ such that U_j is open for all j and $U_j = X_j$ for $j > m$. Hence, $D_{\mathbf{y}}$ intersects every open set in $\prod_{j=1}^{\infty} X_j$ by *Theorem 5.1.11.* □

We are now in a position to study the connectedness of product spaces for both finite and infinite sequences of topological spaces, and we can treat both cases simultaneously.

Theorem 5.1.18 *Let X_j be a finite or infinite sequence of topological spaces. Then $\prod X_j$ with the product topology is connected if and only if X_j is connected for every j.*

Proof. If the product space is connected, then every X_j is connected by *Proposition 5.1.2* because the projections are continuous and onto.

Now suppose that X_j is connected for every j and fix $\mathbf{y}$ in $\prod X_j$. The key step in showing that the product space is connected is to show that if $x_j = y_j$ for $j > N$, then $\mathbf{x}$ and $\mathbf{y}$ are in a connected subset of $\prod X_j$. In the finite case it will be understood that N is less than or equal to the number of factors in the product space. The argument proceeds by induction.

When $N = 1$, the subset $\{\mathbf{x} : x_j = y_j \text{ for } j > 1\}$ is the continuous image of X_1 by mapping x_1 to $(x_1, y_2, \ldots, y_j, \ldots)$, and it is connected.

Suppose it is true that if $x_j = y_j$ for $j > N$, then $\mathbf{x}$ and $\mathbf{y}$ are in a connected subset of $\prod X_j$. Consider $\mathbf{x}$ such that $x_j = y_j$ for $j > N + 1$. Define $\mathbf{u}$ by

$$u_j = \begin{cases} x_j & \text{for } 1 \leq j \leq N \\ y_j & \text{for } j > N, \end{cases}$$

Now the induction hypothesis applies to $\mathbf{u}$, and there exists a connected set C containing both $\mathbf{u}$ and $\mathbf{y}$. As in the $N = 1$ case, the subset $D = \{\mathbf{v} : v_j = u_j \text{ for } j \neq N + 1\}$ is a continuous image of X_{N+1} and connected. Since $\mathbf{u} \in C \cap D$, the subset $C \cup D$ is connected. Clearly, $\mathbf{x}$ is in D. So $\mathbf{x}$ and $\mathbf{y}$ are in the connected set $C \cup D$, completing the induction.

Therefore, given $\mathbf{y}$ in $\prod X_j$ and $N \geq 1$, the set $\{\mathbf{x} : x_j = y_j \text{ for } j > N\}$ is contained in the component $C(\mathbf{y})$. In the case of a finite sequence of m spaces,

$$\{\mathbf{x} : x_j = y_j \text{ for } j > m\} = \prod_{j=1}^{m} X_j,$$

proving that the product space is a single component and connected.

In the cases of an infinite sequence of topological spaces,

$$\bigcup_{N=1}^{\infty} \{\mathbf{x} : x_j = y_j \text{ for } j > N\} \subset C(\mathbf{y}) \subset \prod_{j=1}^{\infty} X_j.$$

Because $C(\mathbf{y})$ is closed, *Proposition 5.1.17* implies

$$\prod_{j=1}^{\infty} X_j = \overline{\left(\bigcup_{N=1}^{\infty} \{\mathbf{x} : x_j = y_j \text{ for } j > N\} \right)} \subset C(\mathbf{y}) \subset \prod_{j=1}^{\infty} X_j.$$

Thus $C(\mathbf{y}) = \prod_{j=1}^{\infty} X_j$. □

Corollary 5.1.19 *The metric groups $\mathbb{R}^{\infty}$ and $\mathbb{K}^{\infty}$ are connected.*

Theorem 5.1.20 *Let X_j be a finite or infinite sequence of topological spaces. If $\mathbf{x}$ is an element of $\prod X_j$ with the product topology, then the component $C(\mathbf{x})$ is the $\prod C(x_j)$ with the product topology.*

Proof. By *Theorem 5.1.18* the product space $\prod C(x_j)$ is a connected topological space. And by *Exercise* 16, p. 192, it is a connected subset of $\prod X_j$.

It follows that $\prod C(x_j) \subset C(\mathbf{x})$. To prove the reverse inclusion, note that $\pi_j(C(\mathbf{x}))$ is connected and contains x_j. Consequently, $\pi_j(C(\mathbf{x})) \subset C(x_j)$ for all j and $C(\mathbf{x}) \subset \prod C(x_j)$. □

Corollary 5.1.21 *If X_j is a finite or infinite sequence of totally disconnected topological spaces, then $\prod_j X_j$ with the product topology is totally disconnected.*

The corollary implies that the metric group $\mathbb{Z}^\infty$ is totally disconnected. It is not discrete, however, because the set consisting of just the identity element $\mathbf{0} = (0, \ldots, 0 \ldots)$ is not open by *Theorem 5.1.11*. Similarly, if G_j is any infinite sequence of nontrivial finite groups (with the discrete topology), then the product $\prod_{j=1}^{\infty} G_j$ is a totally disconnected, but not a discrete metric group.

We restricted our attention to products of sequences of topological spaces because we wanted the product spaces to be metric spaces when the spaces in the sequence were metric. Product spaces like $\mathbb{K}^{\mathbb{R}}$, where the coordinates of a point are indexed by $\mathbb{R}$ not $\mathbb{Z}^+$, will be limited to a cameo appearance.

It is easy to see that condition *(b)* in *Theorem 5.1.11* is equivalent to the following condition:

(c) For every point $\mathbf{x} \in U$, there exists a finite subset F of $\mathbb{Z}^+$ and open sets U_j of X_j such that $U_j = X_j$ for $j \notin F$ and

$$\mathbf{x} \in \prod_{j=1}^{\infty} U_j \subset U.$$

Condition *(c)* naturally generalizes to any indexing set for the spaces X_j.

Suppose X_α is a family of topological spaces indexed by any set Λ, in particular Λ need not be finite or countable. To avoid degenerate situations, we will also assume every X_α contains at least two points. Then each $\mathbf{x}$ in $\prod_{\alpha \in \Lambda} X_\alpha$ has an α coordinate denoted by x_α in X_α for each α in Λ.

In this context, let $\mathcal{T}$ be the collection of subsets U of $\prod_{\alpha \in \Lambda} X_\alpha$ such that for every point $\mathbf{x} \in U$ there exists a finite subset F of Λ and open sets U_α of X_α such that $U_\alpha = X_\alpha$ for $\alpha \notin F$ and

$$\mathbf{x} \in \prod_{\alpha \in \Lambda} U_\alpha \subset U.$$

The collection of sets $\mathcal{T}$ is a topology for $\prod_{\alpha \in \Lambda} X_\alpha$, and is the most general form of the *product topology*. Note that it includes our two earlier versions of product topologies. Furthermore, extending *Propositions 5.1.13, 5.1.14, and 5.1.15* is primarily an exercise in change of notation.

The proof of *Theorem 5.1.18* can be easily modified to prove that $\prod_{\alpha \in \Lambda} X_\alpha$ is connected if and only if every X_α is connected. The key step is to show that if F is a finite subset of Λ and $x_\alpha = y_\alpha$ for α not in F, then $\mathbf{x}$ and $\mathbf{y}$ are in a connected subset of $\prod_{\alpha \in \Lambda} X_\alpha$, using induction on the number of elements in F.

Proposition 5.1.22 *Let X_α be a collection of metric spaces indexed by Λ with each space containing at least two points, and let*

$$X = \prod_{\alpha \in \Lambda} X_\alpha$$

with the product topology. If Λ is uncountable, then X is not a metric space.

Proof. If X is metric, then it is first countable. Given $\mathbf{x}$ in X, there exists a sequence of open neighborhoods V_k of $\mathbf{x}$ such that

$$\bigcap_{k=1}^{\infty} V_k = \{\mathbf{x}\}.$$

It follows from the definition of the topology on X that without loss of generality we can assume each V_k is an open set in the form

$$V_k = \prod_{\alpha \in \Lambda} U_{k\,\alpha}$$

such that every $U_{k\,\alpha}$ is an open set in X_α and $U_{k\,\alpha} = X_\alpha$ for $\alpha \notin F_k$, a finite subset of Λ.

From by *Exercise* 8, p. 139, we know that

$$\bigcup_{k=1}^{\infty} F_k$$

is a countable subset of Λ. Because Λ is neither finite nor countable, there exists $\beta \in \Lambda$ such that β is not in any F_k. Since we are assuming the topological spaces X_α are all nontrivial, there exists a $y \in X_\beta$ such that $y \neq x_\beta$. Define $\mathbf{y}$ in X by

$$\mathbf{y}_\alpha = \begin{cases} x_\alpha & \text{when } \alpha \neq \beta \\ y & \text{when } \alpha = \beta \end{cases}$$

It follows that $\mathbf{y} \neq \mathbf{x}$ and $\mathbf{y} \in \bigcap_{k=1}^{\infty} V_k$. Therefore, X is not first countable and cannot be a metric. □

If G is a metric group, then $G^{\mathbb{R}}$, which can be thought of as all functions from $\mathbb{R}$ to G, is a topological group but never a metric group because $\mathbb{R}$ is uncountable. Thus $\mathbb{R}^{\mathbb{R}}$, $\mathbb{K}^{\mathbb{R}}$, and $\mathrm{GL}(n,\mathbb{C})^{\mathbb{R}}$ are topological groups. In fact they are Hausdorff and connected by *Exercise* 21, p. 193, but for the most part lie outside the theory developed in this book. One of the tools needed to study them is nets because sequential arguments do not suffice in nonmetric spaces.

EXERCISES

1. Show that

$$\{(x, \sin(1/x)) : x > 0\} \cup \{(0, y) : -1 \leq y \leq 1\}$$

is a connected subset of $\mathbb{R}^2$.

2. Show that the metric group $\mathbb{C} \setminus \{0\}$ is connected.

3. Show that every normed linear space is connected.

4. Show that every solenoidal group is connected.

5. Use *Proposition 5.1.2* to prove the Intermediate Value Theorem: Let f be a continuous real-valued function on the closed interval $[a, b]$. If $f(a) < d < f(b)$, then there exists $c \in [a, b]$ such that $f(c) = d$.

6. Prove that if C is a connected subset of a topological space X and C is both open and closed in X, then C is a component of X.

7. Show that the components of a manifold are open sets.

8. Prove that every closed connected subgroup of $\mathbb{K}^n$ is isomorphic to $\mathbb{K}^m$ for some m.

9. Show that if E is a *symmetric* ($E^{-1} = E$) subset of a group G, then $\langle E \rangle = \bigcup_{k=1}^{\infty} E^k$, where $\langle E \rangle$ (page 118) denotes the smallest subgroup of G containing E.

10. Let G be a topological group with identity e. Prove that if U is a symmetric neighborhood of e, then
$$\langle U \rangle = \bigcup_{k=1}^{\infty} U^k$$
and $\langle U \rangle$ is an open-and-closed subgroup of G.

11. Prove that if G is a connected topological group and V is a neighborhood of e, then $G = \langle V \rangle = \bigcup_{k=1}^{\infty} V^k$.

12. Show that the components of a topological group G are the cosets of $C(e)$, the component if the identity of G, and that a topological group is totally disconnected if and only if $C(e) = \{e\}$.

13. Let H be a closed subgroup of $\mathbb{K}^n$ that is isomorphic to $\mathbb{K}^m$ with $0 < m < n$. Prove that $\mathbb{K}^n/H$ is isomorphic to $\mathbb{K}^{n-m}$.

14. Prove that $\mathbb{R}^m \times \mathbb{Z}^n$ is isomorphic to $\mathbb{R}^p \times \mathbb{Z}^q$ if and only if $m = p$ and $n = q$. (The $m = p = 0$ case is *Exercise* 8, p. 124.)

15. Let $G = \prod_{k=1}^{\infty} G_k$ be the direct product of a sequence of topological groups with the product topology, and let e_k denote the identity element of G_k. Prove that
$$\bigcup_{m=1}^{\infty} \{\mathbf{x} : x_k = e_k \text{ for } k > m\}$$
is a dense subgroup of G called the *weak direct product.*

16. Let $X_1, \ldots, X_j, \ldots$ be an infinite sequence of topological spaces. If Y_j is a sequence of nonempty subsets of X_j with the relative topology, then $\prod_{j=1}^{\infty} Y_j \subset \prod_{j=1}^{\infty} X_j$ and the product topology on $\prod_{j=1}^{\infty} Y_j$ equals the relative product topology from $\prod_{j=1}^{\infty} X_j$.

17. Let G be a topological group. Prove that the following are equivalent:
    (a) Given distinct points x and y in G, there exits an open set U containing either x or y but not both.
    (b) $\{a\}$ is a closed set in G for every a in G.
    (c) G is a Hausdorff topological group.

18. Let H be a subgroup of a Hausdorff topological group G, and define $\pi : G \to G/H$ by $\pi(x) = Hx$. Prove the following:
    (a) $\mathcal{T} = \{U \subset G/H : \pi^{-1}(U) \text{ is open}\}$ is a topology on G/H called the *quotient topology.*
    (b) The quotient topology $\mathcal{T}$ is a Hausdorff topology on G/H if and only if the set $\{(x, y) \in G \times G : xy^{-1} \notin H\}$ is an open subset of $G \times G$.
    (c) The quotient topology $\mathcal{T}$ is a Hausdorff topology on G/H if and only if H is a closed subgroup of G.

19. Let $f : X \to Y$ be a continuous open function from a topological space X onto a topological space Y. Suppose Z is another topological space, and $g : Y \to Z$ is a function. Prove that g is continuous if and only if $g \circ f$ is continuous.

20. Let H be a closed normal subgroup of a Hausdorff topological group G. Use *Exercise 19* above to prove that G/H is a Hausdorff topological group with the topology $\mathcal{T}$ from *Exercise 18* above. (For the sharpest results about quotient topologies see [4] Section 5.)

21. Suppose X_α is a family of topological spaces indexed by a set Λ, and let $X = \prod_{\alpha\in\Lambda} X_\alpha$ with the product topology, as defined on page 190. Prove the following:
    (a) If every X_α is Hausdorff, then X is Hausdorff.
    (b) If every X_α is regular, then X is regular.
    (c) If every X_α is connected, then X is connected.

5.2 Connected Matrix Groups

It was easy to show that the general linear group $\mathrm{GL}(n, \mathbb{R})$ and the orthogonal group $O(n)$ were disconnected using the determinant function. Showing that some of the other important matrix groups in Chapter 4 are connected is less simple and is the subject of this section. Recall that $\mathbb{R}\setminus\{0\}$ is disconnected and $\mathbb{C}\setminus\{0\}$ is connected. This distinction between the metric groups of nonzero real and complex numbers plays a major role in the study of connectivity properties of real and complex matrix groups. The first proposition is a simple illustration of this point

Proposition 5.2.1 *The complex diagonal linear groups* $\mathrm{DL}(n, \mathbb{C})$ *and the complex triangular linear groups* $\mathrm{TL}(n, \mathbb{C})$ *are connected. The real diagonal linear groups* $\mathrm{DL}(n, \mathbb{R})$ *and the real triangular linear groups* $\mathrm{TL}(n, \mathbb{R})$ *are not connected.*

Proof. The metric groups $\mathrm{DL}(n, \mathbb{C})$ and $\mathrm{TL}(n, \mathbb{C})$ are homeomorphic to $(\mathbb{C} \setminus \{0\})^n$ and

$$(\mathbb{C} \setminus \{0\})^n \times \mathbb{C}^{(n-1)n/2}$$

respectively. Therefore, they are connected by *Theorem 5.1.18*. Similarly, the metric groups $\mathrm{DL}(n, \mathbb{R})$ and $\mathrm{TL}(n, \mathbb{R})$ are homeomorphic to $(\mathbb{R} \setminus \{0\})^n$ and

$$(\mathbb{R} \setminus \{0\})^n \times \mathbb{R}^{(n-1)n/2}$$

respectively. They are not connected by applying *Theorem 5.1.18* in the other direction. □

Exercises 1 and 2, p. 206, explore the components of $\mathrm{DL}(n, \mathbb{R})$ and $\mathrm{TL}(n, \mathbb{R})$ in more detail.

Theorem 5.2.2 *The complex general linear groups* $\mathrm{GL}(n, \mathbb{C})$ *are connected.*

Proof. Let B be an element of $\mathrm{GL}(n, \mathbb{C})$. Given A in $\mathcal{M}_n(\mathbb{C})$, notice that A is in $\mathrm{GL}(n, \mathbb{C})$ if and only if BAB^{-1} is in $\mathrm{GL}(n, \mathbb{C})$. Recall that $\mathrm{TL}(n, \mathbb{S}) = \mathrm{GL}(n, \mathbb{S}) \cap T_0$. It now follows from *Lemma 4.4.9* that

$$\mathrm{GL}(n, \mathbb{C}) = \bigcup_{B \in \mathrm{GL}(n, \mathbb{C})} B\,[\mathrm{TL}(n, \mathbb{C})]\,B^{-1}.$$

The function $A \to BAB^{-1}$ is an automorphism of $\mathrm{GL}(n, \mathbb{C})$ onto itself and $B\,[\mathrm{TL}(n, \mathbb{C})]B^{-1}$ is a connected subset of $\mathrm{GL}(n, \mathbb{C})$ because $\mathrm{TL}(n, \mathbb{C})$ is connected. Since I is in $B\,[\mathrm{TL}(n, \mathbb{C})]B^{-1}$ for all B, *Theorem 5.1.6* and the previous equation imply that $\mathrm{GL}(n, \mathbb{C})$ is connected. □

Corollary 5.2.3 *The complex special linear groups* $\mathrm{SL}(n, \mathbb{C})$ *are connected.*

Proof. Define a function $f : \mathrm{GL}(n, \mathbb{C}) \to \mathrm{SL}(n, \mathbb{C})$ by

$$f(A) = \mathrm{Diag}\left[\frac{1}{\mathrm{Det}\,[A]}, 1, \ldots, 1\right] A.$$

Observe that f is continuous and onto. Hence, $\mathrm{SL}(n, \mathbb{C})$ is connected. □

Theorem 5.2.4 *The special orthogonal groups* $\mathrm{SO}(n)$ *and the unitary groups* $\mathrm{U}(n)$ *are connected.*

Proof. Use induction on n. To start $\mathrm{SO}(1) = \{1\}$. Assume $\mathrm{SO}(n)$ is connected and $n > 1$. Consider $\mathrm{SO}(n+1)$. Then $\mathrm{SO}(n+1)/\mathrm{SO}(n)$ is homeomorphic to S^n by *Theorem 4.2.6*, and S^n is connected by *Exercise* 7, p. 207, because $n \geq 1$. Thus $\mathrm{SO}(n)$ is a closed connected subgroup of $\mathrm{SO}(n+1)$ with a connected quotient. Therefore, $\mathrm{SO}(n+1)$ is connected by *Theorem 5.1.9*.

For the unitary case use the above argument starting with $U(1) = \mathbb{K}$ and *Theorem 4.2.11*. □

Corollary 5.2.5 *The component of the identity of* $\mathrm{O}(n)$ *is* $\mathrm{SO}(n)$, *and* $\mathrm{O}(n)$ *has two components.*

Proof. The determinant restricted to $\mathrm{O}(n)$ is a homomorphism of $\mathrm{O}(n)$ onto the discrete multiplicative group $\{1, -1\}$. Since its kernel is $\mathrm{SO}(n)$, it follows that $\mathrm{SO}(n)$ is an open-and-closed subset of $\mathrm{O}(n)$. Because $\mathrm{SO}(n)$ is connected, it is the component of $\mathrm{O}(n)$ containing I by *Exercise* 6, p. 192. It follows from *Exercise* 12, p. 192, that $\mathrm{O}(n)$ has two components because $\mathrm{O}(n)/\mathrm{SO}(n)$ is isomorphic to $\mathbb{Z}_2$. □

Similarly, we want to show that $\mathrm{GL}^+(n, \mathbb{R})$ is the component of the identity of $\mathrm{GL}(n, \mathbb{R})$. It is an open-and-closed subset of $\mathrm{GL}(n, \mathbb{R})$ because $\mathrm{GL}^+(n, \mathbb{R}) = \mathrm{Det}^{-1}[\mathbb{R}^+]$ and $\mathbb{R}^+$ is an open-and-closed subset $\mathbb{R} \setminus \{0\}$. But it does not follow that $\mathrm{GL}^+(n, \mathbb{R})$ is connected. Moreover, the proof that $\mathrm{GL}(n, \mathbb{C})$ is connected cannot be adapted to $\mathrm{GL}^+(n, \mathbb{R})$ because there is no result comparable to *Theorem 2.5.11* for real matrices. It will suffice by *Theorem 5.1.9* to show that $\mathrm{GL}^+(n, \mathbb{R})/\mathrm{SO}(n)$ is connected. Although this approach requires us to study the positive definite matrices in some detail, it has the added benefit of revealing some basic structural properties of general linear groups.

Recall that the Hermitian product for $\mathbb{C}^n$ is defined by $\mathbf{z} \cdot \mathbf{w} = \sum_{j=1}^{n} z_j \overline{w}_j$. Its basic properties including the complex Cauchy-Schwarz Inequality appear in *Exercises 8 and 9*, p. 82. It will be used consistently when we are working with vectors in $\mathbb{C}^n$. The restriction of the Hermitian product to vectors in $\mathbb{R}^n$ is, of course, the usual dot product.

As before, it is convenient to set $A^* = (\overline{A})^t$ for $A \in \mathcal{M}_n(\mathbb{C})$. Then $\mathbf{z}A \cdot \mathbf{w} = \mathbf{z} \cdot \mathbf{w}A^*$ (*Exercise* 10, p. 82), and A is unitary if and only if $A^{-1} = A^*$. Also note that $(AB)^* = B^*A^*$ and $(B^*)^* = B$. These basic facts will be frequently used.

A matrix A in $\mathcal{M}_n(\mathbb{C})$ is called a *Hermitian matrix* or a *self-adjoint matrix* if $A = A^*$.

Proposition 5.2.6 *The characteristic values of a Hermitian matrix are all real numbers.*

Proof. If λ is a characteristic value (page 107) of A, then $A - \lambda I$ is not invertible and there exists $\mathbf{v} \neq \mathbf{0}$ in $\mathbb{C}^n$ such that $\mathbf{v}(A - \lambda I) = \mathbf{0}$ (*Theorem 2.5.9*) or simply $\mathbf{v}A = \lambda\mathbf{v}$. It follows that

$$\begin{aligned}
\lambda(\mathbf{v} \cdot \mathbf{v}) &= (\lambda\mathbf{v} \cdot \mathbf{v}) \\
&= \mathbf{v}A \cdot \mathbf{v} \\
&= \mathbf{v} \cdot \mathbf{v}A^* \\
&= \mathbf{v} \cdot \mathbf{v}A \\
&= \mathbf{v} \cdot \lambda\mathbf{v} \\
&= \overline{\lambda}(\mathbf{v} \cdot \mathbf{v}).
\end{aligned}$$

Therefore, $(\lambda - \overline{\lambda})(\mathbf{v} \cdot \mathbf{v}) = 0$ and $\lambda - \overline{\lambda} = 0$ because $\mathbf{v} \cdot \mathbf{v} > 0$. □

A matrix A with real entries is Hermitian if and only if $A = A^{\mathfrak{t}}$. A matrix satisfying $A = A^{\mathfrak{t}}$ is said to be a *symmetric matrix.* So a real matrix is Hermitian if and only if it is symmetric.

Corollary 5.2.7 *If A is a real symmetric matrix, then every characteristic value of A is real.*

Proposition 5.2.8 *If A is a Hermitian matrix and B is a unitary matrix, then BAB^{-1} is Hermitian.*

Proof. Since $B^{-1} = B^*$, we have

$$(BAB^{-1})^* = (BAB^*)^* = (B^*)^* A^* B^* = BAB^{-1}$$

and BAB^{-1} is Hermitian. □

Corollary 5.2.9 *If A is a real symmetric matrix and B is an orthogonal matrix, the BAB^{-1} is a real symmetric matrix.*

Proof. A real matrix is a Hermitian matrix if and only if it is symmetric and an orthogonal matrix is a real unitary matrix. Hence, BAB^{-1} is a real symmetric matrix. □

Let $f(z)$ be the polynomial of degree n in the complex variable z. The Fundamental Theorem of Algebra states that there exists a complex number λ such that $f(\lambda) = 0$ when $n \geq 1$. One simple proof of this result that appears in a number of complex analysis books, such as [1], is to apply Liouville's theorem to $1/f(z)$, which is differentiable and bounded on $\mathbb{C}$, if $f(z) \neq 0$ for all z. It then follows by induction from the Fundamental Theorem of Algebra and division of polynomials that

$$f(z) = \beta(z - \lambda_1)^{n_1}(z - \lambda_2)^{n_2} \ldots (z - \lambda_p)^{n_p}$$

where β is a complex number, the roots λ_j for $j = 1, \ldots, p$ are all distinct, and $n_1 + \ldots + n_p = n$.

Let A be an element of $\mathcal{M}_n(\mathbb{C})$. If $f(z) = \text{Det}\,[A - zI]$ is the characteristic polynomial of A, then $\beta = (-1)^n$ and the λ_j are the distinct characteristic values for A. The exponent n_j is called the *algebraic multiplicity* of λ_j. So the sum of the algebraic multiplicities of the characteristic values equals n.

For each λ_j, set

$$N(A - \lambda_j I) = \{\mathbf{v} \in \mathbb{C}^n : \mathbf{v}(A - \lambda_j I) = \mathbf{0}\}. \tag{5.2}$$

Note that $N(A - \lambda_j I)$ is the kernel of the linear map $\mathbf{v} \to \mathbf{v}(A - \lambda_j I)$ and a subspace of $\mathbb{C}^n$. Since λ_j is a characteristic value of A, the matrix $A - \lambda_j I$ is not invertible and there exists a nonzero vector $\mathbf{v} \in \mathbb{C}^n$ such that $\mathbf{v}(A - \lambda_j I) = \mathbf{0}$. Thus the dimension of $N(A - \lambda_j I)$ at least one.

When A is in $\mathcal{M}_n(\mathbb{R})$, a characteristic value λ_j can still be complex, making $(A-\lambda_j I)$ complex. So we can either think about A as a complex matrix or only consider

$$N(A - \lambda_j I) = \{\mathbf{v} \in \mathbb{R}^n : \mathbf{v}(A - \lambda_j I) = \mathbf{0}\}, \tag{5.3}$$

when λ_j is real. When λ_j is real, there always exists a nonzero vector $\mathbf{v} \in \mathbb{R}^n$ such that $\mathbf{v}(A - \lambda_j I) = \mathbf{0}$ and the dimension of the subspace $N(A - \lambda_j I)$ of $\mathbb{R}^n$ is at least one. Fortunately, for real symmetric matrices every characteristic value is real.

In either case, the nonzero vectors in $N(A - \lambda_j)$ are called *characteristic vectors*. In particular, $\mathbf{v} \neq \mathbf{0}$ is a characteristic vector for λ_j if and only if $\mathbf{v}A = \lambda_j \mathbf{v}$. The dimension of $N(A - \lambda_j)$ is often referred to as the *geometric multiplicity* of λ_j.

For the rest of this section, there will be a parallelism between real symmetric and Hermitian matrices. We will present the real symmetric proofs because our primary goal is to prove that $\mathrm{GL}^+(n, \mathbb{R})$ is connected. When essentially the same result and proof work, we will state the result with the alternate Hermitian version wording in braces. In the real symmetric case, $N(A - \lambda_j I)$ will always refer to (5.3) and in the Hermitian case to (5.2). In the Hermitian, case A^* replaces A^t and the Hermitian product replaces the dot product.

As usual $\mathbf{e}_1, \ldots, \mathbf{e}_n$ will denote the standard basis of $\mathbb{S}^n$. Observe that if A is an $n \times n$ matrix, then $\mathbf{e}_j A$ is just the jth row of A. This simple fact will be very useful in what follows.

Theorem 5.2.10 *Let A be a real symmetric {Hermitian} matrix. If λ is a characteristic value of A, then the algebraic multiplicity of λ equals the geometric multiplicity or dimension of $N(A - \lambda I)$.*

Proof. Let $f(z)$ be the characteristic polynomial of A and let m denote the algebraic multiplicity of λ. So m is the highest power $(z - \lambda)$ that divides $f(z)$. Denote the dimension of $N(A - \lambda I)$ by p. So we must show that $p = m$.

Let $\mathbf{v}_1, \ldots, \mathbf{v}_p$ be an orthonormal basis for $N(A - \lambda I)$ and extend it to an orthonormal basis of $\mathbb{R}^n$, using *Proposition 4.2.5*. Define B to be the $n \times n$ real matrix whose j^{th} row is $\mathbf{v}_j$. Then B is an orthogonal matrix by *Proposition 4.2.3* and BAB^{-1} is a real symmetric matrix by *Corollary 5.2.9*. Furthermore, A and BAB^{-1} have the same characteristic polynomial, characteristic values, and algebraic multiplicities by *Exercise* 9, p. 114.

Observe that for $1 \leq j \leq p$

$$\mathbf{e}_j BAB^{-1} = \mathbf{v}_j AB^{-1} = \lambda \mathbf{v}_j B^{-1} = \lambda \mathbf{e}_j BB^{-1} = \lambda \mathbf{e}_j.$$

Consequently,

$$BAB^{-1} = \begin{bmatrix} \lambda I & \mathrm{O} \\ C & D \end{bmatrix}$$

where I is the $p \times p$ identity matrix, O is the $p \times n - p$ zero matrix, C is a $n - p \times p$ matrix, and D is $n - p \times n - p$ matrix. Because BAB^{-1} is symmetric,

C is also a zero matrix and D is symmetric. Therefore,

$$BAB^{-1} = \begin{bmatrix} \lambda I & O \\ O & D \end{bmatrix}.$$

It follows from *Exercise* 2, p. 148, that

$$f(z) = (-1)^p(z-\lambda)^p g(z)$$

where $g(z)$ is the characteristic polynomial of D. Hence, $p \leq m$ because m is the highest power $(z-\lambda)$ that divides $f(z)$.

It is also obvious from the form of BAB^{-1} that Span $\{\mathbf{e}_1, \ldots, \mathbf{e}_p\}$ is contained in $N(BAB^{-1} - \lambda I)$. Suppose $\mathbf{v}$ is in $N(BAB^{-1} - \lambda I)$, so $\mathbf{v}BAB^{-1} = \lambda\mathbf{v}$ or $(\mathbf{v}B)A = \lambda(\mathbf{v}B)$. Thus $\mathbf{v}B$ is in $N(A - \lambda I)$ and

$$\mathbf{v}B = \sum_{j=1}^{p} \alpha_j \mathbf{v}_j = \sum_{j=1}^{p} \alpha_j \mathbf{e}_j B.$$

Multiplying on the right by B^{-1} shows that $\mathbf{v}$ is in Span $\{\mathbf{e}_1, \ldots, \mathbf{e}_p\}$. Therefore,

$$\text{Span}\,\{\mathbf{e}_1, \ldots, \mathbf{e}_p\} = N(BAB^{-1} - \lambda I).$$

If $p < m$, then $(z-\lambda)$ divides $g(z)$ and λ is a characteristic value for D. So there exists $\mathbf{u} = (u_{p+1}, \ldots, u_n)$ in $\mathbb{R}^{n-p}$ such that $\mathbf{u}D = \lambda\mathbf{u}$. Set $\mathbf{v} = (0, \ldots, 0, u_{p+1}, \ldots, u_n)$ in $\mathbb{R}^n$. Clearly, $\mathbf{v}BAB^{-1} = \lambda\mathbf{v}$ and

$$\mathbf{v} \in N(BAB^{-1} - \lambda I) = \text{Span}\,\{\mathbf{e}_1, \ldots, \mathbf{e}_p\}$$

which contradicts the construction of $\mathbf{v}$. Therefore, $p = m$. □

If generalized eigenspaces are used to define the geometric multiplicity, then the algebraic and geometric multiplicities are equal in general. This is an important first step in the construction of canonical forms for complex $n \times n$ matrices. One source for this material is Chapter 4 of [10], where it is used to study linear differential equations with constant coefficients.

Proposition 5.2.11 *Let A be a real symmetric {Hermitian} matrix . If $\mathbf{v}_1$ and $\mathbf{v}_2$ are characteristic vectors in $\mathbb{R}^n$ $\{\mathbb{C}^n\}$ for distinct characteristic values λ_1 and λ_2, then $\mathbf{v}_1$ and $\mathbf{v}_2$ are orthogonal.*

Proof. Observe that

$$\begin{aligned} (\lambda_1\mathbf{v}_1)\cdot\mathbf{v}_2 &= (\mathbf{v}_1 A)\cdot\mathbf{v}_2 \\ &= \mathbf{v}_1\cdot\mathbf{v}_2 A^t \\ &= \mathbf{v}_1\cdot\mathbf{v}_2 A \\ &= \mathbf{v}_1\cdot(\lambda_2\mathbf{v}_2) \end{aligned}$$

because A is symmetric. It follows that

$$\lambda_1(\mathbf{v}_1\cdot\mathbf{v}_2) = \lambda_2(\mathbf{v}_1\cdot\mathbf{v}_2)$$

or

$$(\lambda_1 - \lambda_2)(\mathbf{v}_1 \cdot \mathbf{v}_2) = 0.$$

Because $\lambda_1 \neq \lambda_2$, it follows that $\lambda_1 - \lambda_2 \neq 0$ and $\mathbf{v}_1 \cdot \mathbf{v}_2 = 0$. □

Theorem 5.2.12 *If A is a real symmetric {Hermitian} matrix, then there exists an orthonormal basis of $\mathbb{R}^n$ {$\mathbb{C}^n$} consisting of characteristic vectors and an orthogonal matrix {unitary matrix} B such that BAB^{-1} is a diagonal matrix.*

Proof. Let $\lambda_1, \ldots, \lambda_p$ be the distinct characteristic values of A with algebraic multiplicities $n_1, \ldots, n_p$. By *Theorem 5.2.10*, the dimension of $N(A - \lambda_j I)$ is n_j.

Let $\mathbf{v}_1, \ldots, \mathbf{v}_{n_1}$ and $\mathbf{v}_{n_1+1}, \ldots, \mathbf{v}_{n_1+n_2}$ be an orthonormal bases for $N(A - \lambda_1 I)$ and $N(A - \lambda_2 I)$, respectively. Now it follows from *Proposition 5.2.11* that $\mathbf{v}_1, \ldots, \mathbf{v}_{n_1+n_2}$ is an orthonormal set of vectors in $\mathbb{R}^n$. Repeating this construction for $j = 3, \ldots, p$ with $N(A - \lambda_j I)$ produces an orthonormal basis $\mathbf{v}_1, \ldots, \mathbf{v}_n$ for $\mathbb{R}^n$ because $n = n_1 + \ldots + n_p$. This proves the first part.

Set B equal to the $n \times n$ real matrix whose j^{th} row is $\mathbf{v}_j$. Clearly, B is an orthogonal matrix. If $\mathbf{v}_j A = \lambda_i \mathbf{v}_j$, then

$$\mathbf{e}_j BAB^{-1} = \mathbf{v}_j AB^{-1} = \lambda_i \mathbf{v}_j B^{-1} = \lambda_i \mathbf{e}_j BB^{-1} = \lambda_i \mathbf{e}_j$$

proving that BAB^{-1} is a diagonal matrix. □

Corollary 5.2.13 *If A is a real symmetric {Hermitian} matrix with distinct characteristic values $\lambda_1, \ldots, \lambda_p$, then given $\mathbf{x}$ in $\mathbb{R}^n$ {$\mathbb{C}^n$}, there exist orthogonal vectors $\mathbf{u}_1, \ldots . \mathbf{u}_p$ such that $\mathbf{u}_i \in N(A - \lambda_i I)$ and $\mathbf{x} = \sum_{i=1}^{p} \mathbf{u}_i$.*

Proof. By the theorem there exists an orthonormal basis of $\mathbb{R}^n$ consisting of characteristic vectors $\mathbf{v}_1, \ldots, \mathbf{v}_n$. So there exist scalars $\alpha_1, \ldots, \alpha_n$ such that $\mathbf{x} = \alpha_1 \mathbf{v}_1 + \ldots + \alpha_n \mathbf{v}_n$, and each $\mathbf{v}_j$ belongs to exactly one of the subspaces $N(A - \lambda_i I)$, $i = 1, \ldots, p$. Let $\Lambda_i = \{j : \mathbf{v}_j \in N(A - \lambda_i I)\}$. Note that the Λ_i partition the set $\{1, \ldots, n\}$ into nonempty disjoint sets. Set

$$\mathbf{u}_i = \sum_{j \in \Lambda_i} \alpha_j \mathbf{v}_j$$

It follows that each $\mathbf{u}_i$ is in $N(A - \lambda_i I)$ and $\mathbf{x} = \sum_{i=1}^{p} \mathbf{u}_i$. The vectors $\mathbf{u}_i$ are orthogonal by *Proposition 5.2.11.* □

The usual norm on $\mathcal{M}_n(\mathbb{S})$ can also be expressed as

$$\|A\|_{\mathcal{B}} = \sup\{\|\mathbf{v}A\| : \|\ \mathbf{v}\| = 1\}$$

because $\|c\mathbf{v}A\| \leq \|\mathbf{v}A\|$, when $|c| \leq 1$. If $\mathbf{v}$ is a characteristic vector for the characteristic value λ, then $\mathbf{v}A = \lambda \mathbf{v}$. We can assume that $\|\mathbf{v}\| = 1$. It follows that $|\lambda| \leq \|A\|_{\mathcal{B}}$. When A is a real symmetric matrix or a Hermitian matrix, and consequently has only real characteristic values, we can do even better.

Proposition 5.2.14 *If A is a real symmetric {Hermitian} matrix with distinct characteristic values $\lambda_1, \ldots, \lambda_p$, then*

$$\|A\|_{\mathcal{B}} = \max\{|\lambda_j| : j = 1, \ldots, p\}.$$

Proof. It follows from the comments preceding the statement of the proposition that $\mu = \max\{|\lambda_j| : j = 1, \ldots, p\} \leq \|A\|_{\mathcal{B}}$.

Given $\mathbf{x} \in \mathbb{R}^n$ with $\|\mathbf{x}\| = 1$, there exist orthogonal vectors $\mathbf{u}_1, \ldots, \mathbf{u}_p$ such that $\mathbf{u}_i \in N(A - \lambda_i I)$ for $i = 1, \ldots, p$ and $\mathbf{x} = \mathbf{u}_1 + \ldots + \mathbf{u}_p$ by *Corollary 5.2.13*. It follows from (4.13) that $1 = \|\mathbf{x}\|^2 = \sum_{i=1}^{p} \|\mathbf{u}_i\|^2$. Then,

$$\begin{aligned}
\mathbf{x}A \cdot \mathbf{x}A &= \left(\sum_{i=1}^{p} \mathbf{u}_i A\right) \cdot \left(\sum_{j=1}^{p} \mathbf{u}_j A\right) \\
&= \left(\sum_{i=1}^{p} \lambda_i \mathbf{u}_i\right) \cdot \left(\sum_{j=1}^{p} \lambda_j \mathbf{u}_i\right) \\
&= \sum_{i=1}^{p} \lambda_i^2 \|\mathbf{u}_i\|^2 \\
&\leq \mu^2 \sum_{i=1}^{p} \|\mathbf{u}_i\|^2 \\
&= \mu^2.
\end{aligned}$$

Therefore,

$$\|\mathbf{x}A\| = (\mathbf{x}A \cdot \mathbf{x}A)^{1/2} \leq \mu$$

and $\|A\|_{\mathcal{B}} \leq \mu = \max\{|\lambda_j| : j = 1, \ldots, p\}$. $\square$

A real symmetric matrix A is a *positive definite matrix* provided $\mathbf{x}A \cdot \mathbf{x} > 0$ for all nonzero $\mathbf{x} \in \mathbb{R}^n$. A Hermitian matrix is positive definite provided that $\mathbf{z}A \cdot \mathbf{z} > 0$ for all nonzero $\mathbf{z} \in \mathbb{C}^n$. *By definition, a positive definite matrix is always either real symmetric or Hermitian.* The proofs will continue to focus on the real symmetric case.

Proposition 5.2.15 *If A is in* $\mathrm{GL}(n, \mathbb{R})$ $\{\mathrm{GL}(n, \mathbb{C})\}$, *then $A^{\mathrm{t}}A$ $\{A^*A\}$ is positive definite.*

Proof. The matrix $A^{\mathrm{t}}A$ is symmetric because $(A^{\mathrm{t}}A)^{\mathrm{t}} = A^{\mathrm{t}}(A^{\mathrm{t}})^{\mathrm{t}} = A^{\mathrm{t}}A$. Since A is invertible, A^{t} is invertible and $\mathbf{x}A^{\mathrm{t}} \neq \mathbf{0}$ when $\mathbf{x} \neq \mathbf{0}$. Now $\mathbf{x}A^{\mathrm{t}}A \cdot \mathbf{x} = \mathbf{x}A^{\mathrm{t}} \cdot \mathbf{x}A^{\mathrm{t}} > 0$ for all $\mathbf{x} \neq \mathbf{0}$. $\square$

Proposition 5.2.16 *If A is a positive definite $n \times n$ matrix and B is in* $\mathrm{O}(n)$ $\{\mathrm{U}(n)\}$, *then BAB^{-1} is positive definite.*

Proof. If $\mathbf{x} \neq \mathbf{0}$, then $\mathbf{x}B \neq 0$ because B is invertible and $\mathbf{x}BAB^{-1} \cdot \mathbf{x} = \mathbf{x}BAB^{\mathrm{t}} \cdot \mathbf{x} = (\mathbf{x}B)A \cdot (\mathbf{x}B) > 0$ because A is positive definite. $\square$

Proposition 5.2.17 *Let A be a real symmetric {Hermitian} matrix. Then A is positive definite if and only if the characteristic values of A are all positive.*

Proof. Suppose A is positive definite. Let λ be any characteristic value for A, and let $\mathbf{v}$ be a characteristic vector for λ. Then $\lambda(\mathbf{v}\cdot\mathbf{v}) = (\lambda\mathbf{v})\cdot\mathbf{v} = \mathbf{v}A\cdot\mathbf{v} > 0$ because A is positive definite. It follows that $\lambda > 0$ because $\mathbf{v}\cdot\mathbf{v} > 0$.

Now suppose the distinct characteristic values $\lambda_1, \ldots, \lambda_p$ of A are all positive. Given $\mathbf{x}$ in $\mathbb{R}^n$, there exist orthogonal a vectors $\mathbf{u}_1, \ldots. \mathbf{u}_p$ such that $\mathbf{u}_i \in N(A - \lambda_i I)$ and $\mathbf{x} = \sum_{i=1}^{p} \mathbf{u}_i$. Then,

$$\begin{aligned}
\mathbf{x}A\cdot\mathbf{x} &= \left(\sum_{i=1}^{p} \mathbf{u}_i\right) A \cdot \left(\sum_{j=1}^{p} \mathbf{u}_j\right) \\
&= \left(\sum_{i=1}^{p} \mathbf{u}_i A\right) \cdot \left(\sum_{j=1}^{p} \mathbf{u}_j\right) \\
&= \left(\sum_{i=1}^{p} \lambda_i \mathbf{u}_i\right) \cdot \left(\sum_{j=1}^{p} \mathbf{u}_j\right) \\
&= \sum_{i=1}^{p} \lambda_i \mathbf{u}_i \cdot \mathbf{u}_i \\
&= \sum_{i=1}^{p} \lambda_i \|\mathbf{u}_i\|^2
\end{aligned}$$

Note that there are no negative terms in the last expression because the λ_i are all positive. Consequently, $\mathbf{x}A\cdot\mathbf{x} \geq \lambda_i\|\mathbf{u}_i\|^2$ for all i. If $\mathbf{x} \neq \mathbf{0}$, then some $\mathbf{u}_i \neq \mathbf{0}$ and $\mathbf{x}A\cdot\mathbf{x} > 0$. □

Corollary 5.2.18 *If A is a positive definite matrix, then* $\operatorname{Det}[A] > 0$ *and A is invertible.*

Proof. By *Theorem 5.2.12*, there exists an orthogonal matrix B such that $BAB^{-1} = D$ is a diagonal matrix. Then D is positive definite by *Proposition 5.2.16*. The characteristic values of a diagonal matrix are just the diagonal entries of D. Because D and A have the same characteristic values, every entry on the diagonal of D is positive and $\operatorname{Det}[D] > 0$. So $\operatorname{Det}[A] = \operatorname{Det}[BAB^{-1}] = \operatorname{Det}[D] > 0$ and A is invertible. □

Proposition 5.2.19 *The set $\mathcal{P}_n(\mathbb{R})$ $\{\mathcal{P}_n(\mathbb{C})\}$ of positive definite matrices is a closed connected subset of* $\mathrm{GL}^+(n,\mathbb{R})$ $\{\mathrm{GL}(n,\mathbb{C})\}$.

Proof. Consider a sequence A_k of positive definite matrices converging to A in $\mathrm{GL}^+(n,\mathbb{R})$. To show that $\mathcal{P}_n(\mathbb{R})$ is closed, it suffices to show that A is positive definite. Since the symmetric matrices are obviously a closed subset of $\mathrm{GL}(n,\mathbb{R})$, the matrix A is symmetric and has only real characteristic values that are nonzero because A is invertible. Clearly, $\mathbf{x}A_k$ converges to $\mathbf{x}A$, and $\mathbf{x}A_k \cdot \mathbf{x}$ converges to $\mathbf{x}A \cdot \mathbf{x}$. So $\mathbf{x} \neq \mathbf{0}$ implies that $\mathbf{x}A \cdot \mathbf{x} \geq 0$.

If A has a negative characteristic value λ and $\mathbf{u}$ is a characteristic vector for λ, then $\mathbf{u}A \cdot \mathbf{u} = \lambda(\mathbf{u} \cdot \mathbf{u}) < 0$. Therefore, the characteristic values of A are all positive, and A is positive definite by *Proposition 5.2.17* proving that $\mathcal{P}_n(\mathbb{R})$ is closed in $\mathrm{GL}^+(n, \mathbb{R})$.

The component of the identity of the diagonal linear group $\mathrm{DL}(n, \mathbb{R})$ is the set of diagonal matrices with positive entries on the diagonal (see *Exercise* 1, p. 206) and will be denoted by $\mathcal{D}_n$. *Proposition 5.2.17* implies that $\mathcal{D}_n = \mathcal{P}_n(\mathbb{R}) \cap \mathrm{DL}(n, \mathbb{R})$. It follows from *Proposition 5.2.16* that for every orthogonal matrix B, the set $B^{-1}\mathcal{D}_n B$ is a connected subset of $\mathcal{P}_n(\mathbb{R})$ containing I. By *Theorem 5.2.12*,

$$\mathcal{P}_n(\mathbb{R}) = \bigcup_{B \in \mathrm{O}(n)} B^{-1}\mathcal{D}_n B$$

and $\mathcal{P}_n(\mathbb{R})$ is connected by *Proposition 5.1.6*. □

Theorem 5.2.20 *If A is in $\mathcal{P}_n(\mathbb{S})$, then there exists a unique matrix $\rho(A)$ in $\mathcal{P}_n(\mathbb{S})$ such that $\rho(A)^2 = A$. The function $A \mapsto \rho(A)$ is continuous from $\mathcal{P}_n(\mathbb{S})$ onto itself.*

Proof. The first step for $\mathbb{S} = \mathbb{R}$ is to prove that for every A in $\mathcal{P}_n(\mathbb{R})$ there exists a matrix C in $\mathcal{P}_n(\mathbb{R})$ such that $C^2 = A$. There exists B in $\mathrm{O}(n)$ such that

$$BAB^{-1} = D = \mathrm{Diag}\,[\alpha_1, \ldots, \alpha_n]$$

with every $\alpha_j > 0$. Each α_j has a unique positive square root $\sqrt{\alpha_j}$. It follows that $E = \mathrm{Diag}\,[\sqrt{\alpha_1}, \ldots, \sqrt{\alpha_n}]$ is in $\mathcal{D}_n$ and $E^2 = D$. Furthermore,

$$(B^{-1}EB)^2 = B^{-1}EBB^{-1}EB = B^{-1}E^2B = B^{-1}DB = A.$$

and $B^{-1}EB$ is positive definite by *Proposition 5.2.16*. Thus A has at least one square root in $\mathcal{P}_n(\mathbb{R})$.

The second step is to prove that there is exactly one positive definite square root of $A \in \mathcal{P}_n(\mathbb{R})$. Suppose C is in $\mathcal{P}_n(\mathbb{R})$ and $C^2 = A$. With B as above, it is easy to see that $(BCB^{-1})^2 = D = \mathrm{Diag}\,[\alpha_1, \ldots, \alpha_n]$. So it suffices to show that a diagonal matrix with positive entries on the diagonal has exactly one positive square root in $\mathcal{P}_n(\mathbb{R})$.

Suppose $C^2 = D = \mathrm{Diag}\,[\alpha_1, \ldots, \alpha_n]$ and C is in $\mathcal{P}_n(\mathbb{R})$. It must be shown that $C = E = \mathrm{Diag}\,[\sqrt{\alpha_1}, \ldots, \sqrt{\alpha_n}\,]$. Let $\lambda_1, \ldots, \lambda_p$ be the distinct characteristic values of C with algebraic multiplicities $n_1, \ldots, n_p$. Because $C \in \mathcal{P}_n(\mathbb{R})$, $\lambda_i > 0$ for $1 \leq i \leq p$ and $\lambda_1^2, \ldots, \lambda_p^2$ are distinct positive real numbers.

If $\mathbf{v} \in N(C - \lambda_i I)$, then $\mathbf{v}C = \lambda_i \mathbf{v}$ and $\mathbf{v}D = \mathbf{v}C^2 = \lambda_i^2 \mathbf{v}$. Hence, λ_i^2 is a characteristic value of D for each i. Denote the algebraic multiplicity of λ_i^2 by m_i. It follows that

$$N(C - \lambda_i I) \subset N(D - \lambda_i^2 I)$$

and then, using *Theorem 5.2.10*

$$n_i = \text{Dim}\,[N(C - \lambda_i I)] \leq \text{Dim}\,[N(D - \lambda_i^2 I)] = m_i.$$

Consequently,

$$n = n_1 + \ldots + n_p \leq m_1 + \ldots + m_p \leq n$$

because $\lambda_1^2, \ldots, \lambda_p^2$ are distinct characteristic values of D, but might not be all of them. It is now obvious that $n_i = m_i$ for $1 \leq i \leq p$ and $m_1 + \ldots + m_p = n$. Therefore, $\lambda_1^2, \ldots, \lambda_p^2$ is actually the full list of distinct characteristic values of D, and $N(C - \lambda_i I) = N(D - \lambda_i^2 I)$ for $i = 1, \ldots, p$. Since every $\mathbf{e}_j$ is a characteristic vector for D, every $\mathbf{e}_j$ is a characteristic vector for C, and C is in $\mathcal{D}$. It is now obvious that $C = \text{Diag}\,[\sqrt{\alpha_1}, \ldots, \sqrt{\alpha_n}\,]$ because $C^2 = \text{Diag}\,[\alpha_1, \ldots, \alpha_n]$.

Thus far, we have shown that for each A in $\mathcal{P}_n(\mathbb{R})$ there exist a unique $\rho(A)$ in $\mathcal{P}_n(\mathbb{R})$ such that $\rho(A)^2 = A$. So $\rho : \mathcal{P}_n(\mathbb{R}) \to \mathcal{P}_n(\mathbb{R})$ is a well-defined function. Since A in $\mathcal{P}_n(\mathbb{R})$ implies A^2 is also in $\mathcal{P}_n(\mathbb{R})$ (*Exercise* 5, p. 206), $\rho(A^2) = A$ and ρ is onto. Furthermore, the characteristic values of $\rho(A)$ are precisely the positive square roots of the characteristic values of A. Consequently, *Proposition 5.2.14* implies that

$$\|\rho(A)\|_{\mathcal{B}} = \sqrt{\|A\|_{\mathcal{B}}}.$$

It remains to show that ρ is continuous. Consider a sequence A_k in $\mathcal{P}_n(\mathbb{R})$ converging to A in $\mathcal{P}_n(\mathbb{R})$. It suffices to show that $\rho(A_k)$ converges to $\rho(A)$. By *Exercise* 5, p. 91, there exists a constant $c > 0$ such that $\|A_k\|_{\mathcal{B}} \leq c$ and $\|\rho(A_k)\|_{\mathcal{B}} \leq \sqrt{c}$ for all k..

If $\rho(A_k)$ does not converges to $\rho(A)$, there exists $\varepsilon > 0$ and a subsequence A_{k_j} such that $\|\rho(A_{k_j}) - \rho(A)\| \geq \varepsilon$. By passing to a subsequence if necessary, we can assume by compactness of $\{A \in \mathcal{M}_n : \|A\|_{\mathcal{B}} \leq \sqrt{c}\}$ that $\rho(A_{k_j})$ converges to some B not equal to $\rho(A)$. Since matrix multiplication is continuous, $A_{k_j} = \rho(A_{k_j})^2$ converges to B^2. Of course, A_{k_j} converges to A because A_k does, and so $A = B^2$. It follows that $\text{Det}\,[B] \neq 0$ and B is in $\text{GL}(n, \mathbb{R})$. Thus the matrix B is in $\mathcal{P}_n(\mathbb{R})$ because it is closed in $\text{GL}(n, \mathbb{R})$ by *Proposition 5.2.19.* Therefore, $B = \rho(A)$ because $\rho(A)$ is the unique square root of A in $\mathcal{P}_n(\mathbb{R})$, but by construction $B \neq \rho(A)$. This contradiction completes the proof that the function ρ is continuous on $\mathcal{P}_n(\mathbb{R})$. □

Next, we prove a general proposition about the elements of the quotient spaces $\text{GL}(n, \mathbb{R})/\text{O}(n)$ and $\text{GL}^+(n.\mathbb{R})/\text{SO}(n)$.

Proposition 5.2.21 *The following hold:*

(a) $\text{O}(n)A = \text{O}(n)B$ $\{\text{U}(n)A = \text{U}(n)B\}$ *if and only if* $A^{\mathsf{t}}A = B^{\mathsf{t}}B$ $\{A^*A = B^*B\}$ *for* A *and* B *in* $\text{GL}(n, \mathbb{R})$.

(b) $\text{SO}(n)A = \text{SO}(n)B$ *if and only if* $A^{\mathsf{t}}A = B^{\mathsf{t}}B$ *for* A *and* B *in* $\text{GL}^+(n, \mathbb{R})$.

(c) If A *is in* $\text{GL}(n, \mathbb{R})$ $\{\text{GL}(n, \mathbb{C})\}$, *then* $\rho(A^{\mathsf{t}}A)$ $\{\rho(A^*A)\}$ *is the unique positive definite matrix in* $\text{O}(n)A$ $\{\text{U}(n)A\}$.

(d) If A in $\mathrm{GL}^+(n,\mathbb{R})$, *then* $\rho(A^t A)$ *is the unique positive definite matrix in* $\mathrm{SO}(n)A$.

Proof. For part *(a)*, $\mathrm{O}(n)A = \mathrm{O}(n)B$ if and only if AB^{-1} is in $\mathrm{O}(n)$ if and only if $(AB^{-1})^t = (AB^{-1})^{-1}$. The last equation simplifies to $A^t A = B^t B$. Therefore, $\mathrm{O}(n)A = \mathrm{O}(n)B$ if and only if $A^t A = B^t B$.

The proof of part *(b)* is almost the same as part *(a)*. Specifically, $\mathrm{SO}(n)A = \mathrm{SO}(n)B$ if and only if AB^{-1} is in $\mathrm{SO}(n)$ if and only if AB^{-1} is in $\mathrm{O}(n)$ because $\mathrm{Det}\,[AB^{-1}] = \mathrm{Det}\,[A]/\mathrm{Det}\,[B] > 0$ by the hypothesis that A and B are in $\mathrm{GL}^+(n,\mathbb{R})$.

For part *(c)*, the matrix $A^t A$ is positive definite by *Proposition 5.2.15*, and $\rho(A^t A)^t \rho(A^t A) = \rho(A^t A)^2 = A^t A$ because the elements of $\mathcal{P}_n(\mathbb{R})$, like $\rho(A^t A)$, are symmetric. Thus $\mathrm{O}(n)\rho(A^t A) = \mathrm{O}(n)A$ by part *(a)*, and $\rho(A^t A)$ is a positive definite matrix in $\mathrm{O}(n)A$.

If P is a positive definite matrix in O(n)A, then $\mathrm{O}(n)A = \mathrm{O}(n)P$ and $A^t A = P^t P = P^2$. By the uniqueness of the square root $P = \rho(A^t A)$.

The proof of *(d)* is essentially the same as *(c)*. □

Theorem 5.2.22 *If A is in* $\mathrm{GL}(n,\mathbb{R})$ $\{\mathrm{GL}(n,\mathbb{C})\}$, *then there exist unique matrices P in* $\mathcal{P}_n(\mathbb{R})$ $\{\mathcal{P}_n(\mathbb{C})\}$ *and R in* $\mathrm{O}(n)$ $\{\mathrm{U}(n)\}$ *such that* $A = PR$.

Proof. Given A in $\mathrm{GL}(n,\mathbb{R})$, the unique positive definite matrix in $\mathrm{O}(n)A^t$ is $P = \rho(AA^t)$, and there exists a unique B in $\mathrm{O}(n)$ such that $A^t = BP$. Set $R = B^{-1}$. Then $A = (BP)^t = PB^t = PB^{-1} = PR$.

If $A = P_1 R_1$ with P_1 in $\mathcal{P}_n(\mathbb{R})$ and R_1 in $\mathrm{O}(n)$, then $A^t = R_1^{-1} P_1$ and P_1 is a positive definite matrix in $\mathrm{O}(n)A^t$. It follows that $P_1 = P$ and $R_1 = R$. □

It is easy to see that $\mathcal{P}_1(\mathbb{C}) = \mathbb{R}^+$. Since we know that $\mathrm{U}(1) = \mathbb{K}$ and $\mathrm{GL}(1,\mathbb{C}) = \mathbb{C}\setminus\{0\}$, it follows from *Theorem 5.2.22* that every nonzero complex number z can be written as $z = re^{i\theta}$. So *Theorem 5.2.22* can be viewed as a generalization of polar coordinates for nonzero complex numbers to $\mathrm{GL}(n,\mathbb{C})$.

Theorem 5.2.23 *The function* $\theta(A) = \mathrm{SO}(n)\rho(A)$ $\{\theta(A) = \mathrm{U}(n)\rho(A)\}$ *is a homeomorphism of* $\mathcal{P}_n(\mathbb{R})$ $\{\mathcal{P}_n(\mathbb{C})\}$ *onto* $\mathrm{GL}^+(n,\mathbb{R})/\mathrm{SO}(n)$ $\{\mathrm{GL}^+(n,\mathbb{C})/\mathrm{U}(n)\}$.

Proof. The function θ is the composition of two continuous functions $A \to \rho(A)$ and $B \to \mathrm{SO}(n)B$, and hence continuous. Now $\mathrm{SO}(n)\rho(A) = \mathrm{SO}(n)\rho(B)$ if and only if

$$\rho(A)^t \rho(A) = \rho(B)^t \rho(B)$$

or equivalently

$$A = \rho(A)^2 = \rho(B)^2 = B.$$

Thus θ is one-to-one.

Given $\mathrm{SO}(n)A$ with A in $\mathrm{GL}^+(n,\mathbb{R})$, the matrix $A^t A$ is positive definite and $\rho(A^t A)$ is the unique positive definite matrix in $\mathrm{SO}(n)A$ by part *(d)* of *Proposition 5.2.21*. Hence, $\theta(A^t A) = \mathrm{SO}(n)A$ and θ is onto.

It remains to show that θ^{-1} is continuous. Since $\mathrm{SO}(n)A = \mathrm{SO}(n)B$ if and only if $A^t A = B^t B$, the function $\psi(\mathrm{SO}(n)A) = A^t A$ is a well-defined function of

$\mathrm{GL}^+(n,\mathbb{R})/\mathrm{SO}(n)$ to $\mathcal{P}_n(\mathbb{R})$. The function $\gamma(A) = A^t A$ is obviously a continuous function of $\mathrm{GL}^+(n,\mathbb{R})$ onto $\mathcal{P}_n(\mathbb{R})$ such that $\psi \circ \pi = \gamma$, where $\pi(A) = \mathrm{SO}(n)A$. Since π is an open continuous function (*Corollary 1.4.8*), ψ is continuous by *Exercise* 19, p. 193.

Finally, it is easy to check that $\theta^{-1} = \psi$, completing the proof that θ is a homeomorphism. □

Corollary 5.2.24 *The following hold:*

(a) *The metric group* $\mathrm{GL}^+(n,\mathbb{R})$ *is connected.*

(b) $\mathrm{GL}^+(n,\mathbb{R})$ *is the component of the identity for* $\mathrm{GL}(n,\mathbb{R})$.

(c) *The metric group* $\mathrm{SL}(n,\mathbb{R})$ *is connected.*

Proof. For part *(a)*, *Theorem 5.1.9* applies because $\mathrm{SO}(n)$ is connected by *Theorem 5.2.4* and $\mathcal{P}_n(\mathbb{R})$ is connected by *Proposition 5.2.19*.

Part *(b)* now follows from part *(a)* and *Exercise* 6, p. 192.

For part *(c)*, observe that $\mathrm{SL}(n,\mathbb{R})$ is the continuous image of $\mathrm{GL}^+(n,\mathbb{R})$ using the function in the proof of *Corollary 5.2.3*. □

Not only is $\mathrm{GL}^+(n,\mathbb{R})$ connected, it is homeomorphic to the product of two connected metric spaces.

Corollary 5.2.25 *The metric group* $\mathrm{GL}^+(n,\mathbb{R})$ $\{\mathrm{GL}(n,\mathbb{C})\}$ *is homeomorphic to* $\mathrm{SO}(n) \times \mathcal{P}_n(\mathbb{R})$ $\{\mathrm{U}(n) \times \mathcal{P}_n(\mathbb{C})\}$.

Proof. Observe that the function $\mathrm{SO}(n)A \mapsto \rho(A^t A)$ is a global section for $\mathrm{GL}^+(n,\mathbb{R})/\mathrm{SO}(n)$. Now apply *Corollary 4.2.8* and the theorem. □

Finally, by pushing the study of positive matrices a little further, we can show that they are homeomorphic to a Euclidean space and obtain the following theorem.

Theorem 5.2.26 *The metric group* $\mathrm{GL}^+(n,\mathbb{R})$ *is homeomorphic to the metric space* $\mathrm{SO}(n) \times \mathbb{R}^{n(n+1)/2}$

Proof. Let $\mathcal{S}_n(\mathbb{R})$ denote the real symmetric matrices, which are a $n(n+1)/2$ dimensional linear subspace of $\mathcal{M}_n(\mathbb{R})$. So $\mathcal{S}_n(\mathbb{R})$ is homeomorphic to $\mathbb{R}^{n(n+1)/2}$.

If C is in $\mathcal{S}_n(\mathbb{R})$, then by *Theorem 5.2.12* there exists an orthogonal matrix B such that $BCB^{-1} = D = \mathrm{Diag}\,[\alpha_1, \ldots, \alpha_n]$. Consequently,

$$\exp(C) = \exp(B^{-1}DB) = B^{-1}\exp(D)B = B^{-1}\mathrm{Diag}\,[e^{\alpha_1}, \ldots, e^{\alpha_n}]B$$

and $\exp(C)$ is in $\mathcal{P}_n(\mathbb{R})$. Thus $C \to \exp(C)$ maps $\mathcal{S}_n(\mathbb{R})$ continuously into $\mathcal{P}_n(\mathbb{R})$. It now suffices to show that exp restricted to $\mathcal{S}_n(\mathbb{R})$ is a homeomorphism onto $\mathcal{P}_n(\mathbb{R})$.

If A is in $\mathcal{P}_n(\mathbb{R})$, then there exists B in $\mathrm{O}(n)$ and $\beta_j > 0$ for $j = 1, \ldots, n$ such that $BAB^{-1} = \mathrm{Diag}\,[\beta_1, \ldots, \beta_n]$. Set $D = \mathrm{Diag}\,[\ln(\beta_1), \ldots, \ln(\beta_n)]$ and $C = B^{-1}DB$. Clearly, $\exp(C) = A$ and the map is onto.

Suppose C is in $\mathcal{S}_n(\mathbb{R})$ and $\exp(C) = A$. Since $\exp\left((1/2)C\right)^2 = \exp(C) = A$, it follows that $\rho(A) = \exp\left((1/2)C\right)$ and by iteration

$$\rho^k(A) = \exp\left(\frac{1}{2^k}C\right)$$

Exercise 14, p. 180 with $h = 1/2^k$ implies that

$$\lim_{k\to\infty} \frac{\rho^k(A) - I}{1/2^k} = \lim_{k\to\infty} \frac{\exp\left(\frac{1}{2^k}C\right) - I}{1/2^k} = C$$

Because the left side of the above equation depends only on A, it follows that the exponential function is one-to-one on $\mathcal{S}_n(\mathbb{R})$.

Having shown that exp restricted to $\mathcal{S}_n(\mathbb{R})$ is one-to-one and onto $\mathcal{P}_n(\mathbb{R})$, there exists an inverse $\ln : \mathcal{P}_n(\mathbb{R}) \to \mathcal{S}_n(\mathbb{R})$, and it remains to show that this inverse is continuous.

From the onto part of the proof, we know that if the characteristic values of A are $\lambda_1, \ldots, \lambda_p$, then the characteristic values of $\ln(A)$ are $\ln(\lambda_1), \ldots, \ln(\lambda_p)$. Using *Proposition 5.2.14* and the properties of $\ln : \mathbb{R}^+ \to \mathbb{R}$, it follows that

$$\| \ln(A)\|_{\mathcal{B}} = \max\left\{\ln(\|A\|_{\mathcal{B}}), \ln(\|A^{-1}\|_{\mathcal{B}})\right\}$$

Since $\|A\|_{\mathcal{B}}$ and $\|A^{-1}\|_{\mathcal{B}}$ are continuous functions, $\| \ln(A)\|_{\mathcal{B}}$ is a continuous function on $\mathcal{P}_n(\mathbb{R})$.

Suppose the sequence A_k in $\mathcal{P}_n(\mathbb{R})$ converges to A. The continuity of $\| \ln(A)\|_{\mathcal{B}}$ implies there exists a constant $c > 0$ such that $\| \ln(A_k)\|_{\mathcal{B}} \leq c$ for all k. If $\ln(A_k)$ does not converges to $\ln(A)$, there exists $\varepsilon > 0$ and a subsequence A_{k_j} such that $\| \ln(A_{k_j}) - \ln(A)\|_{\mathcal{B}} \geq \varepsilon$. By passing to a subsequence of the subsequence, we can assume by the compactness of $\{C \in \mathcal{M}_n(\mathbb{R}) : \|C\|_{\mathcal{B}} \leq c\}$ that $\ln(A_{k_j})$ converges to some $B \in \mathcal{S}_n$ not equal to $\ln(A)$. Consequently, $\exp(\ln(A_{k_j}))$ converges to $\exp(B) \neq A$. But $\exp(\ln(A_{k_j})) = A_{k_j}$ converges to A. This contradiction completes the proof. □

EXERCISES

1. Show that the components of $\mathrm{DL}(n, \mathbb{R})$ are in one-to-one correspondence with the subsets of $\{1, 2, \ldots, n\}$. Determine the group $\mathrm{DL}(n, \mathbb{R})/C(I)$.

2. Determine the components of $\mathrm{TL}(n, \mathbb{R})$ and the group $\mathrm{TL}(n, \mathbb{R})/C(I)$.

3. Prove that the nilpotent linear groups $\mathrm{NL}(n, \mathbb{S})$ are connected.

4. Prove that the matrix groups $\mathrm{SU}(n)$ are connected.

5. Show that if A is positive definite, then A^n is positive definite for all $n \in \mathbb{Z}$.

6. Show that $\mathcal{P}_n(\mathbb{R})$ is not a subgroup of $\mathrm{GL}(n, \mathbb{R})$

7. Show that the sphere $S^n = \{\mathbf{x} \in \mathbb{R}^{n+1} : \|\mathbf{x}\| = 1\}$ is connected for $n \geq 1$.

8. Prove that a real $n \times n$ matrix A is symmetric if and only if $\mathbf{x}A \cdot \mathbf{y} = \mathbf{x} \cdot \mathbf{y}A$ for all $\mathbf{x}$ and $\mathbf{y}$ in $\mathbb{R}^n$. State and prove the analogous fact for Hermitian matrices.

9. Show that a 2×2 matrix is positive definite if and only if both the determinant and the trace are positive.

10. Prove that the components of a manifold are open sets.

11. Let X be a connected manifold, and let y be a point in X. Prove that for all x in X there exists a continuous function $f : [0, 1] \to X$ such that $f(0) = y$ and $f(1) = x$.

12. Show that $\mathrm{SL}(n, \mathbb{R})$ is homeomorphic to $\mathrm{SO}(n) \times \mathbb{R}^{(n+2)(n-1)/2}$.

13. Prove that A is in $\mathrm{SL}(n, \mathbb{C})$ if and only if there exists P in $\mathcal{P}_n(\mathbb{C}) \cap \mathrm{SL}(n, \mathbb{C})$ and R in $\mathrm{SU}(n)$ such that $A = PR$.

14. State and prove the analogue of *Theorem 5.2.26* for $GL(n, \mathbb{C})$.

5.3 Compact Product Spaces

This section continues the discussion of infinite products of metric spaces begun in Section 5.1. The goal is to prove that the product of a sequence of compact metric spaces is also compact. Although the material in this section is not about connectivity, which is the main subject of this chapter, it is essential for the study of an important class of totally disconnected metric groups in Section 5.4, and will extend out knowledge about compact metric groups in general.

A metric space is *countably compact* provided that every sequence contains a convergent subsequence. *Proposition 1.5.9* implies that every compact metric space is countably compact. One of the main results of this section will be to prove the converse, namely, if a metric space is countably compact, then it is compact.

A metric space is *separable* provided that it contains a countable dense subset. (Recall from page 135 that a set E is *countable* if there exists a one-to-one function from $\mathbb{Z}^+ = \{k \in \mathbb{Z} : k \geq 1\}$ onto E.) For example, $\mathbb{R}$ is separable because the rational numbers $\mathbb{Q}$ are obviously dense and $\mathbb{Q}$ is countable by *Exercise* 9, p. 139.

Proposition 5.3.1 *If X is a countably compact metric space, then X is separable.*

Proof. The first step is to prove that for any positive integer n there exists a finite subset F_n of X such that

$$\bigcup_{x \in F_n} B_{1/n}(x) = X. \tag{5.4}$$

Suppose this is false. Let x_1 be in X. Then $B_{1/n}(x_1) \neq X$ and there exists $x_2 \notin B_{1/n}(x_1)$. It follows that $d(x_1, x_2) \geq 1/n$ and that $B_{1/n}(x_1) \cup B_{1/n}(x_2) \neq X$. It is now apparent that by induction we can construct a sequence x_k in X such that for all k

$$\bigcup_{j=1}^{k} B_{1/n}(x_j) \neq X,$$

and

$$d(x_i, x_j) \geq 1/n$$

when $i \neq j$.

Because X is countably compact, there exists a subsequence x_{k_j} converging to some x in X. So $d(x_{k_j}, x) < 1/2n$ for large j. Therefore, $d(x_{k_i}, x_{k_j}) < 1/n$ for large i and j, contradicting the construction of the sequence x_k. Consequently, there exists a finite subset F_n of X satisfying equation (5.4).

Set

$$D = \bigcup_{n=1}^{\infty} F_n.$$

Then D is dense by equation (5.4) and countable because every F_n is finite. □

A topological space X is said to be *second countable* when there exists a sequence of nonempty open sets V_j such that for every open set U in X and every x in U there exists V_j satisfying $x \in V_j \subset U$. Therefore, in a second countable topological space X there exist open sets V_j indexed by $\mathbb{Z}^+$ such that for any open set U there exists a subset E of $\mathbb{Z}^+$ satisfying

$$U = \bigcup_{j \in E} V_j.$$

Although every metric space is first countable, not every metric space or metric group is second countable. For example, $\mathbb{R}_d$, the real numbers with the discrete topology, is not second countable because points are open sets and the set of real numbers is not a countable set.

Proposition 5.3.2 *A metric space X is separable if and only if it is second countable.*

Proof. If X is separable there exists a dense sequence x_k in X. Given an open set U of X and x in U, it is easy to see that there exists positive integers k and n such that $x \in B_{1/n}(x_k) \subset U$. The collection of open sets $B_{1/n}(x_k)$ are indexed by the set $\mathbb{Z}^+ \times \mathbb{Z}^+$, which is countable. Thus X is second countable.

For the second half of the proof, suppose X is second countable. So there exists a sequence of open sets V_j such that for every open set U in X and x in U there exists V_j satisfying $x \in V_j \subset U$. For each j, let x_j be an element of V_j. It follows that the sequence x_j is dense because every nonempty open set contains at least one V_j. □

Corollary 5.3.3 *Every compact or countably compact metric space is both separable and second countable.*

Proof. Since a compact metric space is countably compact, all compact and countably compact spaces are separable by *Proposition 5.3.1.* Now the proposition applies. □

Theorem 5.3.4 (Lindelof) *Let X be a second countable topological space. If U_α, $\alpha \in \Lambda$, is a covering of X by open sets, then there exists a finite or countable subset Γ of Λ such that U_α, $\alpha \in \Gamma$, covers X.*

Proof. By hypothesis, there exists a sequence of nonempty open sets V_j such that for every open set U in X and x in U there exists V_j satisfying $x \in V_j \subset U$. Let U_α, $\alpha \in \Lambda$, be a collection of open sets indexed by Λ that cover X. Set

$$E = \{j \in \mathbb{Z}^+ : V_j \subset U_\alpha \text{ for some } \alpha \in \Lambda\}.$$

Certainly, E is not empty and is either finite or countable by *Exercise* 12, p. 139. For each $j \in E$ the subset $A_j = \{\alpha \in \Lambda : V_j \subset U_\alpha\}$ of Λ is not empty. Choose $\alpha_j \in A_j$. Thus $V_j \subset U_{\alpha_j}$ for all $j \in E$. It follows from *Exercise* 13, p. 139, that the subset $\Gamma = \{\alpha_j \in \Lambda : j \in E\}$ is either finite or countable. It remains to be shown that U_{α_j} with $\alpha_j \in \Gamma$ covers X.

Consider x in X. Because U_α, $\alpha \in \Lambda$ covers X, there exists $\alpha \in \Lambda$ such that x is in U_α. So there exists $j \in \mathbb{Z}^+$ such that $x \in V_j \subset U_\alpha$. It follows that j is in E and $x \in V_j \subset U_{\alpha_j}$. Thus

$$X \subset \bigcup_{\alpha_j \in \Gamma} U_{\alpha_j}$$

and the proof is finished. □

Theorem 5.3.5 *A metric space X is compact if and only if it is countably compact.*

Proof. It has been shown already that compact implies countably compact in *Proposition 1.5.9.*

Suppose X is countably compact, and let U_α, $\alpha \in \Lambda$, be an open cover of X. By *Corollary 5.3.3* and the Lindelof Theorem, there exists a finite or countable subset Γ of Λ such that U_α, $\alpha \in \Gamma$, covers X. If Γ is finite the proof is finished.

If Γ is infinite, there is a one-to-one onto function $\theta : \mathbb{Z}^+ \to \Gamma$. It suffices to show that there exists a positive integer k such that

$$X \subset \bigcup_{i=1}^{k} U_{\theta(i)}.$$

If this not the case, then for each k there exists x_k in X such that x_k not in $\bigcup_{j=1}^k U_{\theta(j)}$. Because X is countably compact, the sequence x_k has a convergent subsequence x_{k_j}. The limit x of the subsequence x_{k_j} must lie in some $U_{\theta(m)}$. There exists N such that x_{k_j} is in $U_{\theta(m)}$ for $j \geq N$, and there must exist $k_j > m$ such that

$$x_{k_j} \in U_{\theta(m)} \subset \bigcup_{i=1}^{k_j} U_{\theta(i)},$$

contradicting the construction of the sequence x_k. □

Theorem 5.3.5 will be used to prove that the product of a sequence of compact metric spaces is compact. This will be accomplished by using the Cantor diagonal process to construct a convergent subsequence. This is a clever process that requires a thorough understanding of subsequences. To rigorously address the subtle issues in the proof, we will invest a little extra ink and paper to present the argument in more detail than usually seen.

It will also be a notational challenge. To prepare for this proof, observe that a sequence in a set X is just a function from $\mathbb{Z}^+$ to X with x_k denoting the image of the positive integer k.

What is a subsequence? The notation x_{k_j} means that for each integer j there is an integer k_j such that the jth term of the subsequence is the $k_j{}^{\text{th}}$ term of the original sequence with an important condition. The next term of the subsequence, $x_{k_{j+1}}$, must come after x_{k_j} in the original sequence x_k, that is, k_{j+1} must be bigger than k_j, and the function $j \to k_j$ is increasing. (This condition was just used in the last sentence of the proof of *Theorem 5.3.5*.) Conversely, an increasing function $\theta : \mathbb{Z}^+ \to \mathbb{Z}^+$ determines a subsequence by setting $k_j = \theta(j)$.

An easy induction argument shows that every increasing function $\theta : \mathbb{Z}^+ \to \mathbb{Z}^+$ satisfies $\theta(j) \geq j$. If θ_1 and θ_2 are two increasing functions from $\mathbb{Z}^+$ into itself, then $\theta_1 \circ \theta_2$ is increasing. So the subsequence of a subsequence is a subsequence of the original sequence.

For the remainder of this section, it will be convenient to use functional notation for a sequence in a set X by letting $x_k = \xi(k)$ with $\xi : \mathbb{Z}^+ \to X$. This will avoid notation like $x_{k_{j_i}}$ and worse for subsequences of subsequences ad nauseam. In this notation, the subsequences of ξ are all the functions $\xi \circ \theta$ where $\theta : \mathbb{Z}^+ \to \mathbb{Z}^+$ is an increasing function. If $\psi : \mathbb{Z}^+ \to \mathbb{Z}^+$ is another increasing function, then $\xi \circ \theta \circ \psi$ is a subsequence of a subsequence. With this notation we will be able to manage constructing a sequence of subsequences one after the other in X using $\xi \circ \theta_1 \circ \ldots \circ \theta_n$, a blend of the two notations for sequences.

Theorem 5.3.6 *Let X_j be a sequence of metric spaces. The product space $\prod_{j=1}^{\infty} X_j$ is compact if and only if X_j is compact for every j.*

Proof. If $\prod_{j=1}^{\infty} X_j$ is compact, then each X_j is compact because the projections are continuous and onto.

Now suppose that X_j is compact for every j. By *Theorem 5.3.5*, it suffices to prove that $\prod_{j=1}^{\infty} X_j$ is countably compact. Let $\boldsymbol{\xi}(k)$ be a sequence in $\prod_{j=1}^{\infty} X_j$. Then the jth coordinate function of $\boldsymbol{\xi}$ is a sequence in X_j denoted by $\xi_j(k)$ and $\boldsymbol{\xi}(k) = (\xi_1(k), \xi_2(k), \ldots)$. Because each X_j is compact, any sequence in X_j contains a convergent subsequence.

Thus there exists an increasing function $\theta_1 : \mathbb{Z}^+ \to \mathbb{Z}^+$ such that the subsequence $\xi_1 \circ \theta_1$ converges in X_1. Then the sequence $\xi_2 \circ \theta_1$ in X_2 has a convergent subsequence $\xi_2 \circ \theta_1 \circ \theta_2$. Of course, $\xi_1 \circ \theta_1 \circ \theta_2$ converges because the a subsequence of a convergent sequence converges to the same limit. It is now clear by induction that there exists a sequence of increasing functions $\theta_j : \mathbb{Z}^+ \to \mathbb{Z}^+$ such that

$$\xi_i \circ \theta_1 \circ \ldots \circ \theta_j$$

is a convergent sequence in X_i for $1 \leq i \leq j$. We now have an infinite collection of sequences indexed by i and j. The trick is to use the diagonal terms. Define a function $\psi : \mathbb{Z}^+ \to \mathbb{Z}^+$ by setting

$$\psi(k) = \theta_1 \circ \ldots \circ \theta_k(k). \tag{5.5}$$

The following computational lemma will be useful:

Lemma 5.3.7 *For $j \geq 0$, the function $\varphi_j : \{n \in \mathbb{Z}^+ : n \geq j+1\} \to \mathbb{Z}^+$ defined by $\varphi_j(k) = \theta_{j+1} \circ \ldots \circ \theta_k(k)$ is increasing and satisfies $\varphi_j(j+1) > j$.*

Proof. For $k \geq j+1$, the function $\theta_{j+1} \circ \ldots \circ \theta_k$ is increasing because the composition of increasing functions is increasing. Since θ_k is increasing, $\theta_{k+1}(k+1) \geq k+1$ and

$$\begin{aligned} \varphi_j(k+1) = \theta_{j+1} \circ \ldots \circ \theta_k \circ \theta_{k+1}(k+1) &\geq \theta_{j+1} \circ \ldots \circ \theta_k(k+1) \\ &> \theta_{j+1} \circ \ldots \circ \theta_k(k) \\ &= \varphi_j(k). \end{aligned}$$

Hence φ_j is increasing and $\varphi_j(j+1) = \theta_{j+1}(j+1) \geq j+1 > j$. $\square$

Returning to the function ψ defined by (5.5), note that $\varphi_0 = \psi$ and ψ is increasing by the lemma. So $\boldsymbol{\xi} \circ \psi$ is a subsequence of $\boldsymbol{\xi}$. It suffices to show that $\boldsymbol{\xi} \circ \psi$ converges in $\prod_{j=1}^{\infty} X_j$. This is equivalent to showing that $\xi_j \circ \psi$ converges for every j by *Exercise* 6, p. 213.

For a given j, the terms of the sequence $\xi_j \circ \psi(k)$ such that $k \leq j$ do not effect the convergence of the sequence. To take advantage of this fact, extend the definition of φ_j from $\{n \in \mathbb{Z}^+ : n \geq j+1\}$ to $\mathbb{Z}^+$ by setting $\varphi_j(k) = k$ for $1 \leq k \leq j$, so

$$\varphi_j(k) = \begin{cases} k & \text{for } 1 \leq k \leq j \\ \theta_{j+1} \circ \ldots \circ \theta_k(k) & \text{for } k > j \end{cases}$$

It follows from the lemma that for each j this extension is now an increasing function from $\mathbb{Z}^+$ to $\mathbb{Z}^+$. Therefore, $\xi_j \circ \theta_1 \circ \ldots \circ \theta_j \circ \varphi_j$ converges in X_j because it is a subsequence of the convergent sequence $\xi_j \circ \theta_1 \circ \ldots \circ \theta_j$ in X_j. Finally, observe that the two subsequences $\xi_j \circ \psi$ and $\xi_j \circ \theta_1 \circ \ldots \circ \theta_j \circ \varphi_j$ of ξ_j agree for $k > j$. Thus $\xi_j \circ \psi$ converges for all j and the sequence $\boldsymbol{\xi} \circ \psi$ converges. Therefore, X is countably compact and compact by *Theorem 5.3.5*. □

Theorem 5.3.8 *Let X_j be a sequence of metric spaces. The product space $\prod_{j=1}^{\infty} X_j$ is locally compact if and only if X_j is locally compact for all j and there exists a positive integer m such that X_j is compact for $j > m$.*

Proof. Suppose $\prod_{k=1}^{\infty} X_k$ is locally compact, and let $\mathbf{x}$ be a point in it. There exists a compact neighborhood U of $\mathbf{x}$. By *Theorem 5.1.11* there exists a positive integer m and open sets U_k of X_k such that $U_k = X_k$ for $k > m$, and

$$\mathbf{x} \in \prod_{k=1}^{\infty} U_k \subset U.$$

Since the projections $\pi_j : \prod_{k=1}^{\infty} X_k \to X_j$ are continuous open functions, $\pi_j(U)$ is a compact neighborhood of x_j in X_j. Thus X_j is locally compact for every j because $\mathbf{x}$ was an arbitrary point in the product space. Furthermore, X_j is compact for $j > m$ because $X_j = U_j = \pi_j(U)$ for $j > m$. This completes the first one-half of the proof.

Now suppose that X_k is locally compact for all k and there exists a positive integer m such that X_k is compact for $k > m$. Let $\mathbf{x}$ be in the product space. For $1 \leq j \leq m$, there exist open neighborhoods U_j of x_j such that U_j^- is compact. For $j > m$, set $U_j = X_j$. Then $\prod_{j=1}^{\infty} U_j$ is an open set containing $\mathbf{x}$.

Setting $C_j = U_j^-$, the sets C_j are compact for all j and

$$\prod_{j=1}^{\infty} U_j \subset \prod_{j=1}^{\infty} C_j.$$

Theorem 5.3.6 implies that $\prod_{j=1}^{\infty} C_j$ is a compact metric space with the product topology. It then follows from *Exercise* 16, p. 192, that $C = \prod_{j=1}^{\infty} C_j$ is a compact subset of $\prod_{j=1}^{\infty} X_j$. Furthermore, C is a neighborhood of $\mathbf{x}$ because it contains the open set $\prod_{j=1}^{\infty} U_j$, proving that $\prod_{k=1}^{\infty} X_k$ is locally compact. □

It follows from *Theorem 5.3.8* that a countable product of locally compact spaces that are not compact is not locally compact. In particular, $\mathbb{R}^{\infty}$ is a metric group that is not locally compact, while $\mathbb{R}^n \times \mathbb{K}^{\infty}$ is a locally compact metric group for all positive integers n.

Theorem 5.3.6 holds in much greater generality than presented here. The most general version is known as the Tychonoff Theorem and is arguably the most important theorem in general topology. Because of our focus on metric groups the Tychonoff Theorem will not be used, but because of its importance it merits a discussion that places it in its proper context visa vie metric groups.

The Tychonoff Theorem states that the product of any collection of compact topological spaces with the product topology is compact (see Chapter 5 of [7] or Chapter 6 of [16] for a proof). In particular, the Tychonoff Theorem implies, for example, that the topological groups $\mathbb{K}^{\mathbb{R}}$, $U(n)^{\mathbb{R}}$, and $\mathbb{Z}_p^{\mathbb{R}}$ are compact. If you dig deep enough somewhere along the way every proof of the Tychonoff Theorem, no matter how slick, uses a version of the following axiom:

> **Axiom of Choice**: If X_α is a nonempty set for each α in an index set Λ, then there exists a function f on Λ such that $f(\alpha)$ is in X_α.

The Axiom of Choice is logically equivalent to several other important mathematical principles including The Maximal Principle, Tukey's Lemma, the Well-ordering Principle, and Zorn's Lemma (see Chapter 0 of [7] for a full discussion of the equivalent forms of the Axiom of Choice). In fact, the Tychonoff Theorem is equivalent to the Axiom Choice [6].

There is also a countable version of the axiom of choice:

> **Axiom of Countable Choice**: If X_α is a nonempty set for each α in a countable index set Λ, then there exists a function f on Λ such that $f(\alpha)$ is in X_α.

We have been using the Axiom of Countable Choice without mention and will continue to do so because it is intuitively natural. For example, it was used in the proof of the Lindelof Theorem (*Theorem 5.3.4*), where "Choose $\alpha_j \in A_j$" appears in the proof. At that point all we knew is that E is a nonempty subset of positive integers indexing a family of nonempty unknown subsets of Λ. The Axiom of Countable Choice guarantees that we can choose the α_j as required. Consequently, it has also been used to prove *Theorem 5.3.6*.

EXERCISES

1. Let X_j be a sequence of metric spaces, and let F_j be a closed subset of X_j for all j. Show that $\prod F_j$ is a closed subset of $\prod X_j$.
2. Let X be a metric space and let Y be a subset of X with the relative topology. Prove the following:
   (a) If X is second countable, then Y is second countable.
   (b) If X is separable, then Y is separable.
3. Prove that a second countable topological space is first countable.
4. Let $X_1, \ldots, X_m$ be second countable or separable metric spaces. Prove that the product space $X_1 \times \ldots \times X_m$ is second countable and separable.
5. Let X_j be a sequence of second countable or separable metric spaces. Prove that the product space $\prod_{j=1}^{\infty} X_j$ is second countable and separable.
6. Let X_j be a sequence of metric spaces. As in the proof of *Theorem 5.3.6*, let $\boldsymbol{\xi} : \mathbb{Z}^+ \to \prod_{j=1}^{\infty} X_j$ be a sequence in the product space with coordinates $\boldsymbol{\xi}(k) = (\xi_1(k), \xi_2(k), \ldots)$. Show that $\boldsymbol{\xi}$ converges in $\prod_{j=1}^{\infty} X_j$ with the product topology if and only if ξ_j converges in X_j for all j.

7. Show that a metric space is countably compact if and only if every countable cover has a finite subcover.

8. Let H be an open subgroup of a second countable metric group G. Show that the quotient space G/H is finite or countable.

9. Show that the metric group $\mathbb{R}^\infty$ is a real linear space with scalar multiplication $(c, \mathbf{x}) = (cx_1, cx_2, \ldots)$. Show that this scaler multiplication is continuous from $\mathbb{R} \times \mathbb{R}^\infty$ to $\mathbb{R}^\infty$. Prove that $\mathbb{R}^\infty$ is not a normed linear space.

5.4 Totally Disconnected Groups

At the end of Section 5.1, it was pointed out that if G_j is any sequence of finite groups with the discrete topology, then the product $\prod_{j=1}^{\infty} G_k$ is a totally disconnected, but is not a discrete metric group. We now know that $\prod_{k=1}^{\infty} G_k$ is compact. So it is easy to produce a variety of totally disconnected compact metric groups. There are also interesting compact monothetic metric groups called **a**-adic groups whose underlying metric space is a countable products of finite discrete spaces. In this section, we take a closer look at totally disconnected groups. The first step is to gain a better understanding of the topology of such spaces.

Let X_j be a sequence of finite sets with the discrete metric. Hence, the space $\prod_{j=1}^{\infty} X_j$ is a compact totally disconnected metric space by *Theorem 5.3.6* and *Corollary 5.1.21*. The sets X_j can be thought of as finite lists of symbols and the points in $\prod_{j=1}^{\infty} X_j$ as a sequence of symbols, where the jth symbol is one of the symbols in X_j. For example, if $X_j = \{0,1\}$ for all $j \geq 1$, then $\prod_{j=1}^{\infty} X_j = \{0,1\}^\infty$ is all sequences of the symbols 0 and 1. If, continuing our earlier notation, $\boldsymbol{\xi}(k) = (\xi_1(k), \xi_2(k), \ldots)$ is a sequence in $\prod_{j=1}^{\infty} X_j$, what does it mean when $\boldsymbol{\xi}(k)$ converges to $\mathbf{x}$?

Proposition 5.4.1 *Let X_j be a sequence of finite discrete metric spaces and*

$$X = \prod_{j=1}^{\infty} X_j$$

with the product topology. Then the following hold:

(a) If $\mathbf{y} = (y_1, \ldots, y_j, \ldots)$ is a point in X and m is in $\mathbb{Z}^+$, then

$$U_m = \{\mathbf{x} \in X : x_j = y_j \text{ for } 1 \leq j \leq m\} \tag{5.6}$$

for $m \in \mathbb{Z}^+$ is a sequence of open-and-closed sets and a countable neighborhood base at $\mathbf{y}$.

(b) A sequence $\boldsymbol{\xi}(k) = (\xi_1(k), \xi_2(k), \ldots)$ converges to $\mathbf{y}$ in X if and only if for every $m \in \mathbb{Z}^+$ there exists $N > 0$ such that $(\xi_1(k), \ldots, \xi_m(k)) = (y_1, \ldots, y_m)$ when $k > N$.

Proof. Set $V_j = \{y_j\} \subset X_j$ for $1 \leq j \leq m$ and $V_j = X_j$ for $j > m$. Then

$$U_m = \prod_{j=1}^{\infty} V_j.$$

Because points are open sets in discrete metric spaces, each U_m is an open set by *Corollary 5.1.12*. Similarly, *Exercise* 1, p. 213, implies U_m is closed because points are closed sets in a discrete space. It follows from *Theorem 5.1.11* that the sequence of sets U_m is a countable neighborhood base at $\mathbf{y}$. This proves part *(a)*. Using the definition of convergence of a sequence, part *(b)* is an immediate consequence of part *(a)*. □

A metric space is *zero-dimensional* provided that at every point there is a countable neighborhood base consisting of open-and-closed sets. Obviously, discrete metric spaces are zero-dimensional. It follows from the above proposition that the product space of a sequence of finite discrete metric spaces is zero-dimensional.

Recall that if U is an open-and-closed subset of a metric space X, then a connected subset of X is either contained in U or in $X \setminus U$. If U_m is a countable neighborhood base of open-and-closed subsets at x an element of the metric space X, then $\bigcap_{m=1}^{\infty} U_m = \{x\}$ because for every $k \in \mathbb{Z}^+$ there exists m_k such that $U_{m_k} \subset B_{1/k}(x)$. It follows that

$$x \in C(x) \subset \bigcap_{m=1}^{\infty} U_m = \{x\}$$

and $C(x) = \{x\}$. *Hence, every zero-dimensional metric space is totally disconnected.* The converse is not generally true (see the example of Knaster and Kuratowski in [16] on page 210), but it is true for locally compact metric spaces. To prove this result, we will use the following structural property of the components of a compact metric space.

Theorem 5.4.2 *If x is a point in a compact metric space X, then there exists a sequence of open-and-closed sets U_m such that*

$$C(x) = \bigcap_{m=1}^{\infty} U_m$$

and $U_{m+1} \subset U_m$ for all m.

Proof. Set

$$Y = \bigcap \{U \subset X : x \in U \text{ and } U \text{ is both open and closed}\}.$$

Clearly, Y is a closed subset of X, and Y contains $C(x)$. The proof will proceed by first showing that there exists a sequence of open-and-closed sets U_m such that

$$Y = \bigcap_{m=1}^{\infty} U_m$$

and then proving that $Y = C(x)$.

If U is an open-and-closed subset of X containing x, then $X \setminus U \subset X \setminus Y$ because $Y \subset U$. Moreover, if y is not in Y, there exists U an open-and-closed subset of X containing x such that y is not in U and so y is in $X \setminus U$. It follows that the collection sets of the form $X \setminus U$ such that U is an open-and-closed subset of X containing x is an open cover of the metric space $X \setminus Y$.

Every compact metric space is second countable by *Corollary 5.3.3* and $X \setminus Y$ is second countable by *Exercise* 2, p. 213. Now the Lindelof Theorem (*Theorem 5.3.4*) implies that there exist a sequence U_m of open-and-closed subsets of X containing x such that the sets $X \setminus U_m$ cover $X \setminus Y$. It follows that

$$X \setminus Y = \bigcup_{m=1}^{\infty} X \setminus U_m$$

Consequently,

$$Y = X \setminus (X \setminus Y) = X \setminus \left(\bigcup_{m=1}^{\infty} X \setminus U_m \right) = \bigcap_{m=1}^{\infty} U_m$$

by de Morgan formulas. Because a finite intersection of sets that are both open and closed is both open and closed, it can be assumed that $U_{m+1} \subset U_m$ for all m.

To prove that $Y = C(x)$ it suffices to show that Y is connected because $C(x) \subset Y$. If Y is not connected, there exist nonempty disjoint closed subsets Y_1 and Y_2 of Y such that $Y = Y_1 \cup Y_2$. Since Y is closed in X, the sets Y_1 and Y_2 are also closed subsets of X. It will be assumed that x is in Y_1.

By *Exercise* 3, p. 222, there exists $r > 0$ such that the sets $V_i = \{w \in X : d(w, Y_i) \leq r\}$ are disjoint open sets such that $Y \subset V_1 \cup V_2$. If for all m, U_m is not contained in $V_1 \cup V_2$, then there exists w_m in U_m such w_m is not in $V_1 \cup V_2$. Because X is compact, there exists a subsequence w_{m_k} converging to w. Since $U_{m+1} \subset U_m$ for all m, the sequence w_{m_k} is in U_m for large k, implying that w is in the closed set U_m for all m. Hence, w is in Y. But w is also in the intersection of the closed sets $\{w \in X : d(w, Y_i) \geq r\}$ for $i = 1, 2$, and therefore, not in Y. This contradiction proves there exists m such that $U_m \subset V_1 \cup V_2$.

Consider m such that $U_m \subset V_1 \cup V_2$. Consequently, $U_m = (U_m \cap V_1) \cup (U_m \cap V_2)$ and the sets are $U_m \cap V_i$ are disjoint. Since V_1 and V_2 are open subsets of X, the sets $U_m \cap V_1$ and $U_m \cap V_2$ are open subsets of X and of U_m with the relative topology. Because U_m is their disjoint union, they are also closed subsets of U_m. Since U_m is also a closed subset of X, the sets $U_m \cap V_1$ and $U_m \cap V_2$ are closed subsets of X.

In particular, $U_m \cap V_1$ is an open-and-closed subset of X containing x. Therefore, $Y \subset U_m \cap V_1$ and $Y \cap U_m \cap V_2 = \phi$. Clearly, $Y_2 \subset Y \cap U_m \cap V_2$. Thus $Y_2 = \phi$, contradicting the assumption that Y was disconnected. Therefore, Y is connected and must equal $C(x)$. □

Corollary 5.4.3 *If X is a compact totally disconnected metric space, then X is zero-dimensional.*

Proof. If X is totally disconnected, then by the theorem for every x in X there exists a sequence of open-and-closed sets U_m such that $\{x\} = C(x) = \bigcap_{m=1}^{\infty} U_m$ and $U_{m+1} \subset U_m$ for all m.

To prove that U_m is a countable neighborhood base at x, it suffices to show that given an open neighborhood V of x there exists m such that $U_m \subset V$. If for all $m \in \mathbb{Z}^+$, the neighborhood V does not contain U_m, then there exists a sequence $x_m \in U_m \setminus V$. Since X is compact, there exists a subsequence x_{m_j} converging to y. Clearly, y is in the closed set $X \setminus V$. Because $U_k \subset U_m$ for all $k \geq m$, it follows that x_{m_j} is in U_m for all $j \geq m$ and that y is in the closed set U_m for all $m \in \mathbb{Z}^+$. So y is in $\bigcap_{m=1}^{\infty} U_m = \{x\}$ and $y = x \in V$, contradicting $y \in X \setminus V$. Therefore, U_m is a countable neighborhood base at x and X is zero-dimensional by definition. □

Proposition 5.4.4 *If X is a totally disconnected locally compact metric space, then X is zero-dimensional.*

Proof. Let d be a metric for X, and let x be an arbitrary point in X. Because X is locally compact, there exists $r > 0$ such that $B_r(x)^-$ is compact. Then, $B_r(x)^-$ with the relative topology is a totally disconnected metric space by *Exercise* 4, p. 222, and zero-dimensional by *Corollary 5.4.3.*

For $0 < \varepsilon \leq r$, the open sets $B_\varepsilon(x)$ of X are also open sets in the relative topology on $B_r(x)^-$. It follows from the definition of zero-dimensional that for $0 < \varepsilon \leq r$, there exists an open-and-closed set U_ε in the relative topology on $B_r(x)^-$ such that $x \in U_\varepsilon \subset B_\varepsilon(x)$. There exists a closed set F of X such that $U_\varepsilon = F \cap B_r(x)^-$. Thus U_ε is a closed subset of X because $B_r(x)^-$ is a closed set in X. And there exists an open set U of X such that $U_\varepsilon = U \cap B_r(x)^-$. Then $U_\varepsilon = U \cap B_\varepsilon(x)$ because $U_\varepsilon \subset B_\varepsilon(x)$ for $0 < \varepsilon \leq r$, and U_ε is an open subset of X. It is now obvious that $U_{r/m}$, $m \in \mathbb{Z}^+$ is a countable neighborhood base of open-and-closed sets of X at x, and X is zero-dimensional. □

Corollary 5.4.5 *A locally compact metric group is zero-dimensional if and only if it is totally disconnected.*

Proposition 5.4.6 *If U is an open-and-compact neighborhood of the identity in a metric group G, then there exists an open-and-compact subgroup H of G contained in U.*

Proof. Because U is an open neighborhood of the identity and U is a compact set contained in U, there exists V an open neighborhood of e such that $VU \subset U$

by *Exercise* 5, p. 222. Set $W = V \cap V^{-1}$, so that W is a symmetric open neighborhood of e such that $WU \subset U$. It follows from *Exercise* 10, p. 192, that $H = \bigcup_{k=1}^{\infty} W^k$ is an open-and-closed subgroup of G. So it remains to prove that $H \subset U$, and thus compact.

Note that $W \subset U$ because $WU \subset U$ and $e \in U$. If $W^k \subset U$, then $W^{k+1} = WW^k \subset WU \subset U$ and by induction $W^k \subset U$ for all $k \geq 1$. So $H \subset U$. □

Having done the hard work, we can now easily obtain two substantive results about totally disconnected groups.

Theorem 5.4.7 *If G is a totally disconnected locally compact metric group, then there exists a countable neighborhood base at the identity e consisting of subgroups that are both open and compact.*

Proof. Because G is locally compact, *Proposition 5.4.4* implies that there exists a countable neighborhood base at e consisting of open-and-compact neighborhoods U_m of e. By *Proposition 5.4.6,* for each m there exists an open-and-compact subgroup H_m contained in U_m. It follows that H_m is also a countable neighborhood base at e. □

Theorem 5.4.8 *If G is an infinite Abelian compact totally disconnected metric group, then there exists a sequence G_j finite groups such that G is isomorphic to a closed subgroup of the product metric group $\prod_{j=1}^{\infty} G_j$.*

Proof. By *Theorem 5.4.7,* there exists a countable neighborhood base at e consisting of open-and-compact subgroups H_m. Each H_m is a normal subgroup because G is Abelian and $G_m = G/H_m$ is finite because G is compact and H_m is open. Let $\pi_m : G \to G_m$ be the canonical homomorphism of G onto G_m. Define $\varphi : G \to \prod_{m=1}^{\infty} G_m$ by $\varphi(x) = (\pi_1(x), \ldots, \pi_m(x), \ldots)$. Then, φ is continuous by *Proposition 5.1.14* and a homomorphism. It follows that $\varphi(G)$ is a compact and hence closed subgroup of $\prod_{m=1}^{\infty} G_m$. If $x \neq e$ in G, then there exists m such that x is not in H_m and $\pi_m(x)$ is not the identity element of G_m. It follows that x is not in the kernel of φ and φ is one-to-one. Finally, φ^{-1} is continuous on $\varphi(G)$ by *Exercise* 2, p. 61. □

The remainder of the section is devoted to the construction of the **a**-*adic groups.* Let a be a positive integer. If b is a non-negative integer, the Division Algorithm states that there exists unique integers d and r such that $0 \leq r < a$ and $b = ad + r$. In other words, d is the largest integral divisor of a and r is the remainder. In the group $\mathbb{Z}_a = \{0, \ldots, a-1\}$, the sum of x plus y is defined by using the Division Algorithm to obtain integers d and r such that $0 \leq r < a$ and $x + y = ad + r$ in $\mathbb{Z}$, and then setting $x + y = r$ in $\mathbb{Z}_a$. This process makes no use of the divisor d. In the **a**-adic groups, the d carries into the next calculation and so on, ad infinitum.

Let $\mathbf{a} = (a_0, a_1, \ldots, a_k, \ldots)$ be an infinite sequence of integers such that $a_k \geq 2$ for $k \geq 0$. Set

$$\Gamma_{\mathbf{a}} = \prod_{k=0}^{\infty} \{0, \ldots, a_k - 1\}. \tag{5.7}$$

Thus $\mathbf{x} = (x_0, \ldots, x_k, \ldots)$ is a point in $\Gamma_{\mathbf{a}}$ if and only if every x_k is a non-negative integer such that $0 \le x_k < a_k$. This is the same metric space as the underlying metric space for $\prod_{j=1}^{\infty} \mathbb{Z}_{a_j}$, but the two group structures are very different.

Define a binary operation $\mathbf{x} + \mathbf{y} = \mathbf{z}$ on $\Gamma_{\mathbf{a}}$ inductively using the Division Algorithm as follows:

$$\begin{aligned} x_0 + y_0 &= a_0 d_0 + z_0 \text{ and } 0 \le z_0 < a_0 && (5.8)\\ x_1 + y_1 + d_0 &= a_1 d_1 + z_1 \text{ and } 0 \le z_1 < a_1 \\ &\vdots \\ x_k + y_k + d_{k-1} &= a_k d_k + z_k \text{ and } 0 \le z_k < ak && (5.9) \end{aligned}$$

In this binary operation, one coordinate in a summand can effect infinitely many in the sum. Suppose, for example, that $a_k = 2$ for all k and denote $\Gamma_{\mathbf{a}}$ by simply Γ_2. Let $\mathbf{x} = (1, \ldots, 1, \ldots)$ and $\mathbf{y} = (1, 0, \ldots, 0, \ldots)$. Then $\mathbf{x} + \mathbf{y} = (0, \ldots, 0, \ldots)$. So the only 1 in $\mathbf{y}$, changes every coordinate in $\mathbf{x}$. Every $\Gamma_{\mathbf{a}}$ exhibits similar behavior.

Set

$$F_m = \{\mathbf{x} \in \Gamma_{\mathbf{a}} : x_k = 0 \text{ for } k > m\}$$

and

$$F = \bigcup_{k=0}^{\infty} F_k.$$

If $\mathbf{x}$ and $\mathbf{y}$ are in F_m, then it is easy to see that $\mathbf{x} + \mathbf{y}$ is in F_n for some $n \ge m$, although it need not be in F_m.

Define φ_m mapping F_m into the non-negative integers by

$$\varphi_m(\mathbf{x}) = x_0 + \sum_{k=1}^{m} x_k \left(\prod_{j=0}^{k-1} a_j \right) \tag{5.10}$$

If $m < n$, then $F_m \subset F_n$ and $\varphi_n(\mathbf{x}) = \varphi_m(\mathbf{x})$ for $\mathbf{x} \in F_m$. So we can define φ mapping F to the non-negative integers by setting $\varphi(\mathbf{x}) = \varphi_m(\mathbf{x})$, if $\mathbf{x} \in F_m$. The next lemma will be the key to understanding the binary operation on $\Gamma_{\mathbf{a}}$ defined by equations (5.8) and (5.9)

Lemma 5.4.9 *If* $\mathbf{x}$ *and* $\mathbf{y}$ *are in* F, *then* $\varphi(\mathbf{x}+\mathbf{y}) = \varphi(\mathbf{x}) + \varphi(\mathbf{y})$. *Furthermore,* φ *is a one-to-one function of* F *onto the non-negative integers.*

Proof. Given $\mathbf{x}$ and $\mathbf{y}$ are in F, we can assume that $\mathbf{x}$, $\mathbf{y}$, and $\mathbf{z} = \mathbf{x} + \mathbf{y}$ are in F_m for some m. It follows that $d_m = 0$. Solving for z_k in the definition of $\mathbf{x} + \mathbf{y}$ and substituting into the formula for $\varphi(\mathbf{z})$, produces

$$\varphi(\mathbf{z}) = x_0 + y_0 - a_0 d_0 + \sum_{k=1}^{m} (x_k + y_k + d_{k-1} - a_k d_k) \left(\prod_{j=0}^{k-1} a_j \right)$$

$$
\begin{aligned}
&= \varphi(\mathbf{x}) + \varphi(\mathbf{y}) - a_0 d_0 + \sum_{k=1}^{m} (d_{k-1} - a_k d_k) \left(\prod_{j=0}^{k-1} a_j \right) \\
&= \varphi(\mathbf{x}) + \varphi(\mathbf{y}) + \sum_{k=1}^{m} d_{k-1} \left(\prod_{j=0}^{k-1} a_j \right) - \sum_{k=0}^{m} d_k \left(\prod_{j=0}^{k} a_j \right) \\
&= \varphi(\mathbf{x}) + \varphi(\mathbf{y}) + \sum_{i=0}^{m-1} d_i \left(\prod_{j=0}^{i} a_j \right) - \sum_{k=0}^{m} d_k \left(\prod_{j=0}^{k} a_j \right) \\
&= \varphi(\mathbf{x}) + \varphi(\mathbf{y}) - d_m \left(\prod_{j=0}^{m} a_j \right) \\
&= \varphi(\mathbf{x}) + \varphi(\mathbf{y})
\end{aligned}
$$

which proves the first part of the proposition.

Suppose $\varphi(\mathbf{x}) = \varphi(\mathbf{y})$, or

$$
x_0 + \sum_{k=1}^{m} x_k \left(\prod_{j=0}^{k-1} a_j \right) = y_0 + \sum_{k=1}^{m} y_k \left(\prod_{j=0}^{k-1} a_j \right).
$$

Both sides are in the form $b = da_0 + r$ with $0 \le r < a_0$, and hence $x_0 = y_0$. Then it follows that

$$
x_1 + \sum_{k=2}^{m} x_k \left(\prod_{j=1}^{k-1} a_j \right) = y_1 + \sum_{k=2}^{m} y_k \left(\prod_{j=1}^{k-1} a_j \right),
$$

and using the same argument with a_1 implies that $x_1 = y_1$. Induction now obviously works to complete the proof that $\mathbf{x} = \mathbf{y}$.

Let $\mathbf{0} = (0, \ldots, 0, \ldots)$ and $\mathbf{g} = (1, 0, \ldots, 0, \ldots)$. Clearly $\varphi(\mathbf{0}) = 0$ and $\varphi(\mathbf{g}) = 1$. Arguing by induction suppose there exists $\mathbf{x} \in F$ such that $\varphi(\mathbf{x}) = n$, then $\varphi(\mathbf{x} + \mathbf{g}) = \varphi(\mathbf{x}) + \varphi(\mathbf{g}) = n + 1$. $\square$

Theorem 5.4.10 *If Let $\mathbf{a} = (a_0, a_1, \ldots, a_k, \ldots)$ is an infinite sequence of integers such that $a_k \ge 2$, then $\Gamma_{\mathbf{a}}$ defined (5.7) is a totally disconnected compact monothetic metric group under the binary operation defined by equations (5.8) and (5.9).*

Proof. The first task is to show that $\Gamma_{\mathbf{a}}$ is a group. To that end, let φ be given by *Proposition 5.4.9.* If $\mathbf{x}$, $\mathbf{y}$, and $\mathbf{z}$ are in F, then using the associativity of addition in $\mathbb{Z}$

$$
\begin{aligned}
\varphi[(\mathbf{x} + \mathbf{y}) + \mathbf{z}] &= \varphi(\mathbf{x} + \mathbf{y}) + \varphi(\mathbf{z}) \\
&= \varphi(\mathbf{x}) + \varphi(\mathbf{y}) + \varphi(\mathbf{z}) \\
&= \varphi(\mathbf{x}) + \varphi(\mathbf{y} + \mathbf{z}) \\
&= \varphi[\mathbf{x} + (\mathbf{y} + \mathbf{z})].
\end{aligned}
$$

Because φ is one-to-one, $(\mathbf{x}+\mathbf{y})+\mathbf{z} = \mathbf{x}+(\mathbf{y}+\mathbf{z})$, and addition in $\Gamma_{\mathbf{a}}$ is associative on F.

If $\mathbf{x}$, $\mathbf{y}$,and $\mathbf{z}$ are in $\Gamma_{\mathbf{a}}$ and $m > 0$, then there exist $\mathbf{x}'$, $\mathbf{y}'$, and $\mathbf{z}'$ in F with the same k^{th} coordinates as $\mathbf{x}$, $\mathbf{y}$ and $\mathbf{z}$ for $k \le m$. Since the kth coordinates of $(\mathbf{x}+\mathbf{y})+\mathbf{z}$ and $\mathbf{x}+(\mathbf{y}+\mathbf{z})$ depend only on the jth coordinates of $\mathbf{x}$, $\mathbf{y}$ and $\mathbf{z}$ for $0 \le j \le k$, It follows from the first paragraph that

$$[(\mathbf{x}+\mathbf{y})+\mathbf{z}]_k = [(\mathbf{x}'+\mathbf{y}')+\mathbf{z}']_k = [\mathbf{x}'+(\mathbf{y}'+\mathbf{z}')]_k = [\mathbf{x}+(\mathbf{y}+\mathbf{z})]_k$$

for $0 \le k \le m$. Consequently, $[(\mathbf{x}+\mathbf{y})+\mathbf{z}]_k = [\mathbf{x}+(\mathbf{y}+\mathbf{z})]_k$ for all $k \ge 0$ because m was an arbitrary positive integer. Therefore, addition is associative on $\Gamma_{\mathbf{a}}$.

Clearly, $\mathbf{0} = (0,\ldots,0,\ldots)$ is an identity element, and addition is Abelian in $\Gamma_{\mathbf{a}}$. Of course, $-\mathbf{0} = \mathbf{0}$. Given $\mathbf{x} \ne \mathbf{0}$ in $\Gamma_{\mathbf{a}}$, define $-\mathbf{x}$ as follows:

$$m = \min\{k : x_k \ne 0\} \tag{5.11}$$

and

$$(-\mathbf{x})_k = \begin{cases} 0 & \text{for } 0 \le k < m \\ a_k - x_k & \text{fot } k = m \\ a_k - x_k - 1 & \text{for } k > m \end{cases} \tag{5.12}$$

Then $-\mathbf{x}+\mathbf{x} = \mathbf{0}$ by *Exercise* 6, p. 222. Thus $\Gamma_{\mathbf{a}}$ is a group.

By *Theorems 5.1.20 and 5.3.6*, the space $\Gamma_{\mathbf{a}}$ is compact and totally disconnected . The continuity of multiplication and the inverse function is a consequence of *Proposition 5.4.1* part *(b)*.

The set F is dense in $\Gamma_{\mathbf{a}}$ by *Proposition 5.4.1.* Using the *Lemma 5.4.9* twice, $\varphi(n\mathbf{g}) = n$ for all $n \ge 0$ by induction and then $F = \{n\mathbf{g} : n \ge 0\}$. Thus $\Gamma_{\mathbf{a}} \supset \langle \mathbf{g}\rangle^{-} \supset F^{-} = \Gamma_{\mathbf{a}}$ and $\Gamma_{\mathbf{a}}$ is monothetic. □

Proposition 5.4.11 *If $\Gamma_{\mathbf{a}}$ is an* **a**-*adic group, then $\Lambda_m = \{\mathbf{x} \in \Gamma_{\mathbf{a}} : x_j = 0 \text{ for } j = 0,\ldots,m\}$ is a open-and-closed subgroup of $\Gamma_{\mathbf{a}}$ and $\Gamma_{\mathbf{a}}/\Lambda_m$ is isomorphic to $\mathbb{Z}_{p_m}$, where $p_m = a_0 a_1 \cdots a_m$.*

Proof. Obviously, Λ_m is an open-and-closed subgroup. Observe that $\Lambda_m + \mathbf{x} = \Lambda_m + \mathbf{z}$ if and only if $\mathbf{z} = \mathbf{x}+\mathbf{y}$ for some $\mathbf{y} \in \Lambda_m$ if and only if $x_i = z_i$ for $0 \le i \le m$. Hence, $\Gamma_{\mathbf{a}}/\Lambda_m$ contains p_m elements. Every element of $\Gamma_{\mathbf{a}}/\Lambda_m$ has the form $\Lambda_m + k\mathbf{g}$. Thus $\Gamma_{\mathbf{a}}/\Lambda_m$ is a cyclic group containing p_m elements, and all such groups are isomorphic to $\mathbb{Z}_{p_m}$. □

EXERCISES

1. Prove that $\{0,1\}^{\infty}$ is uncountable

2. Let H be a closed subgroup of a compact metric group G. Show that G/H is finite if and only if H is open

3. Let X be a metric space with metric d. Recall from *Exercise* 12, p. 28, that the distance from a point x to a closed set F is defined by $d(x, F) = \inf\{d(x,y) : y \in F\}$ and $\{x \in X : d(x, F) < r\}$ is open. Let C be a compact subset of X, and let F be a closed subset of X. Prove the following:

   (a) If $C \cap F = \phi$, then $\inf\{d(x,y) : x \in C \text{ and } y \in F\} > 0$.

   (b) If $C \cap F = \phi$, then there exists $r > 0$ such that $\{x : d(x, C) < r\}$ and $\{y : d(y, F) < r\}$ are disjoint open sets.

4. Let Y be a subset of a totally disconnected metric space X. Show that Y (with the relative topology of course) is also totally disconnected.

5. Let G be a metric group, and let U be an open neighborhood of the identity e. Prove that if C is a compact subset of G contained in U, then there exists an open neighborhood V of e such that $VC \subset U$.

6. Given $\mathbf{x} \neq \mathbf{0}$ in $\Gamma_{\mathbf{a}}$, let $-\mathbf{x}$ be defined by equations (5.11) and (5.12). Prove that $-\mathbf{x} + \mathbf{x} = \mathbf{0}$.

7. Show that if p is a prime, then Γ_p is isomorphic to the subgroup of $\prod_{j=1}^{\infty} \mathbb{Z}_{p^j}$ generated by $(1, \ldots, 1, \ldots)$.

8. Let X_j be a sequence of zero-dimensional metric spaces. Prove that the product space $\prod_{j=1}^{\infty} X_j$ is zero-dimensional.

9. Show that the only open subgroups of Γ_p with p prime are the Λ_m from *Proposition 5.4.11*.

10. Let $\Gamma_{\mathbf{a}}$ be an $\mathbf{a}$-adic group and let $\mathbf{g}$ be an element of $\Gamma_{\mathbf{a}}$ such that $\langle \mathbf{g} \rangle^{-} = \Gamma_{\mathbf{a}}$. Show that $H = \{(-k\mathbf{g}, k) : k \in \mathbb{Z}\}$ is a discrete subgroup of the metric group $\Gamma_{\mathbf{a}} \times \mathbb{R}$ and that $(\Gamma_{\mathbf{a}} \times \mathbb{R})/H$ is a compact solenoidal group called the $\mathbf{a}$-*adic solenoid*.

11. Let G be a compact metric group. Show that there exists a sequence of open-and-closed subgroups H_m of G such that

$$C(e) = \bigcap_{m=1}^{\infty} H_m.$$

Chapter 6

Metric Groups of Functions

Functions are a natural source of groups. The permutation groups have already played an important role in the study of the metric groups. Many groups of functions are also interesting topological groups. In this chapter, we construct a metric on the set of continuous function from one metric space to another, and then use it to obtain a number of important metric groups of functions. This does, however, require some restrictions on the metric spaces. Specifically, locally compact second countable metric spaces will play a key role throughout this chapter.

We begin by revisiting the normed linear space $\mathcal{C}(X)$ of continuous function from a compact metric space X to $\mathbb{R}$. Recall that it has the additional structure of a commutative Banach algebra with identity. In Section 6.1 we prove the important Stone-Weierstrass Theorem about dense subalgebras of $\mathcal{C}(X)$ and extend it to continuous complex-valued functions. Convergence of a sequence of functions in $\mathcal{C}(X)$ is just uniform convergence on X. As we move to generalize the topology on $\mathcal{C}(X)$ to the set of continuous functions from one metric space to another, the algebraic structure disappears, but uniform convergence on compact sets will be a constant through out the chapter.

The primary construction (Section 6.2) is a metric on the set of continuous functions from a locally compact second countable metric space to a metric space. This topology can be completely described in terms of the compact subsets of the domain and the open subsets of the range. For this reason, it is called the compact-open topology. The remainder of the Section 6.2 is devoted to establishing several key theorems about the compact-open topology.

In Section 6.3, the group of isometries of a locally compact metric space onto itself provides a good example of a metric group with the compact-open topology. To illustrate how the compact-open topology ties in with earlier topics, the group of isometries of $\mathbb{R}^n$ is studied in more detail.

The more challenging problem of defining a suitable metric for the group of homeomorphisms of a locally compact second countable metric space onto itself is discussed in Section 6.4. When X is a compact metric space, the group of homeomorphisms of X onto itself is still a metric group with the compact-open topology. If X is just a locally compact second countable metric space, however, the inverse function is not always continuous when the compact-open topology is used for the group of homeomorphisms of X onto itself. Resolving this difficulty requires the construction of the one-point compactification of a locally compact second countable metric space, a construction that is of interest in its own right.

Homomorphisms of an Abelian metric group to an Abelian metric group can be multiplied by multiplying their values. This produces a group that is the subject of Section 6.5. When the range is locally compact and second countable, this group of homomorphisms is a metric group with the compact-open topology. The final result in the section and the most important result in the chapter shows that the metric group of homomorphisms of an Abelian locally compact second countable metric group to the circle group $\mathbb{K}$ is again an Abelian locally compact second countable metric group.

6.1 Real-Valued Functions

Let X be a compact metric space and as usual let $\mathcal{C}(X)$ denote the linear space of continuous real-valued functions on X. Then $\mathcal{C}(X)$ is a Banach algebra with the norm $\|f\|_s = \sup\{|f(x)| : x \in X\}$ by *Exercise* 1, p. 178. Every normed linear space is an Abelian metric group with additional algebraic structure and so $\mathcal{C}(X)$ is an example of a metric group of functions. Because it is a Banach algebra, it is also complete.

If $h : X \to X'$ is a homeomorphism of X onto X', then it is easy to see that $\|f \circ h\|_s = \|f\|_s$ for all $f \in \mathcal{C}(X')$. It follows that $H(f) = f \circ h$ defines an isometry of $\mathcal{C}(X')$ onto $\mathcal{C}(X)$. Furthermore, H is clearly an isomorphism of algebras with identity, so $\mathcal{C}(X)$ and $\mathcal{C}(X')$ are isometrically isomorphic Banach algebras, when X and X' are homeomorphic compact metric spaces.

Let f_k be a sequence in $\mathcal{C}(X)$. Then f_k converges to f in $\mathcal{C}(X)$ if and only if given $\varepsilon > 0$ there exists $N > 0$ such that

$$d(f_k, f) = \|f_k - f\|_s = \sup\left\{|f_k(x) - f(x)| : x \in X\right\} < \varepsilon$$

when $k > N$. Thus f_k converges to f in $\mathcal{C}(X)$ if and only if f_k converges uniformly to f on X. Uniform convergence on compact sets will play an key role through out this chapter.

The theme of this section is approximation. When can every function in $\mathcal{C}(X)$ be uniformly approximated by functions in a particular subset of $\mathcal{C}(X)$? The right subsets to consider are subalgebras, that is, subsets $\mathcal{A}$ of $\mathcal{C}(X)$ that are closed under addition, scaler multiplication, and multiplication in $\mathcal{C}(X)$. Uniform approximation, of course, means given f in $\mathcal{C}(X)$ and $\varepsilon > 0$, prove there exists g in $\mathcal{A}$ such that $d(f, g) = \sup\{|f(x) - g(x)| : x \in X\} < \varepsilon$. The basic question can now be rephrased as: Given a subalgebra $\mathcal{A}$ of $\mathcal{C}(X)$, when is $\mathcal{A}^- = \mathcal{C}(X)$? We begin with the classical Weierstrass Theorem for the subalgebra of polynomial functions in $\mathcal{C}([a, b])$.

We will need the binomial formula (*Exercise* 4, p. 179) again. For $n \geq 0$, recall that the binomial coefficients are defined by

$$\binom{n}{k} = \frac{n!}{k!(n-k)!}$$

where by definition $0! = 1$. If x and y are real or complex numbers and n is a positive integer, then

$$(x+y)^n = \sum_{k=0}^{n} \binom{n}{k} x^k y^{n-k}. \tag{6.1}$$

Setting $y = 1 - x$ in equation (6.1) produces the useful identity

$$\sum_{k=0}^{n} \binom{n}{k} x^k (1-x)^{n-k} = 1. \tag{6.2}$$

It will be convenient to set

$$\tau_k(x) = \binom{n}{k} x^k (1-x)^{n-k} \tag{6.3}$$

for $0 \leq k \leq n$. It follows from equation (6.2) that

$$\sum_{k=0}^{n} \tau'(x) = 0 \tag{6.4}$$

where, as you would expect, $\tau'(x)$ denotes the derivative of $\tau(x)$. If you multiply the left side of (6.4) by $x(1-x)$, then the result can be rewritten using the functions $\tau_k(x)$ instead of their derivatives. This straightforward but slightly tedious calculation is the substance of *Exercise* 1, p. 232, and equation (6.10) from that exercise will play an essential role in the proof of the classical Weierstrass Theorem about polynomial approximation.

Let $\mathcal{P}([a,b])$ denote the polynomial functions restricted to the closed interval $[a,b] = \{x \in \mathbb{R} : a \leq x \leq b\}$, where $a \leq b$. Obviously, $\mathcal{P}([a,b])$ is a subalgebra of $\mathcal{C}([a,b])$. In proving the Weierstrass Theorem, we will make use of the sequence of Bernstein polynomials $B_n(f)$ associated with a function f in $\mathcal{C}([0,1])$. The *Bernstein polynomials of* f are defined for $n \in \mathbb{Z}^+$ by

$$B_n(f)(x) = \sum_{k=0}^{n} f\left(\frac{k}{n}\right) \binom{n}{k} x^k (1-x)^{n-k} = \sum_{k=0}^{n} f\left(\frac{k}{n}\right) \tau_k(x). \tag{6.5}$$

Theorem 6.1.1 (Weierstrass) *If a and b are real numbers with $a \leq b$, then*

$$\mathcal{P}([a,b])^- = \mathcal{C}([a,b])$$

Proof. When $a = b$, note that $\mathcal{C}([a,b]) = \mathcal{C}(\{a\})$ is isomorphic to $\mathbb{R}$ and $\mathcal{C}(\{a\}) = \mathcal{P}(\{a\})$. So the interesting case is when $a < b$.

Since $h(x) = (b-a)x+a$ is a homeomorphism of $[0,1]$ onto $[a,b]$, $H(f) = f \circ h$ is an isometric isomorphism of $\mathcal{C}([a,b])$ onto $\mathcal{C}([0,1])$. Note that $h^{-1}(x) = (x-a)/(b-a)$ and $H^{-1}(g) = g \circ h^{-1}$. Because h and h^{-1} are linear, it follows that $H(f)$ is a polynomial if and only if f is a polynomial. Thus $H(\mathcal{P}([a,b])) = \mathcal{P}([0,1])$ and it suffices to prove the result for $[0,1]$.

The next step is to establish a key formula that will be used to show that given f in $\mathcal{C}([0,1])$, the Bernstein polynomials $B_n(f)$ converge uniformly to f on $[0,1]$. Using (6.10) from *Exercise* 1, p. 232, with $q = 0$, equation (6.4) times $x(1-x)$ can be written as

$$\sum_{k=0}^{n} (k - nx)\tau_k(x) = 0.$$

Differentiating this formula with respect to x produces

$$\sum_{k=0}^{n} (-n)\tau_k(x) + \sum_{k=0}^{n} (k - nx)\tau_k'(x) = 0$$

and simplifies to

$$-n + \sum_{k=0}^{n} (k - nx)\tau_k'(x) = 0$$

using (6.2). Therefore,

$$\sum_{k=0}^{n} (k - nx)\tau_k'(x) = n$$

and

$$x(1-x)\sum_{k=0}^{n} (k - nx)\tau_k'(x) = nx(1-x).$$

Now, apply (6.10) with $q = 1$ to the left-hand side of the above equation to obtain

$$\sum_{k=0}^{n} (k - nx)^2\tau_k(x) = nx(1-x)$$

and then divide by n^2 to get the formula

$$\sum_{k=0}^{n} \left(\frac{k}{n} - x\right)^2 \tau_k(x) = \frac{x(1-x)}{n}. \tag{6.6}$$

Let f be in $\mathcal{C}([0,1])$ and let $B_n(x) = B_n(f)(x)$ be its sequence of Bernstein polynomials. It follows from (6.2) and (6.3) that

$$f(x) - B_n(x) = \sum_{k=0}^{n} \tau_k(x) \left[f(x) - f\left(\frac{k}{n}\right) \right]$$

and, noting that $\tau_k(x) \geq 0$ on $[0,1]$,

$$|f(x) - B_n(x)| \leq \sum_{k=0}^{n} \tau_k(x) \left| f(x) - f\left(\frac{k}{n}\right) \right|.$$

Let $\varepsilon > 0$. To show that there exists N such that the right-hand side is less than ε for all x in $[0,1]$ and $n > N$, the sum will be split into two parts depending on x and different arguments used to estimate each piece.

Since f is uniformly continuous on $[0,1]$ by *Exercise* 3, p. 232, there exists $\delta > 0$ such that $|f(x) - f(y)| < \varepsilon/2$ when $|x - y| < \delta$. Let $\Lambda = \{k : |x - k/n| < \delta\}$ and note that Λ depends on both x and n. For fixed x and n,

$$\sum_{k \in \Lambda} \tau_k(x) \left| f(x) - f\left(\frac{k}{n}\right) \right| < \sum_{k \in \Lambda} \tau_k(x) \frac{\varepsilon}{2} \leq \frac{\varepsilon}{2} \sum_{k=0}^{n} \tau_k(x) = \frac{\varepsilon}{2}$$

Thus the first key estimate is

$$\sum_{k \in \Lambda} \tau_k(x) \left| f(x) - f\left(\frac{k}{n}\right) \right| < \frac{\varepsilon}{2}, \tag{6.7}$$

and the right-hand side of this estimate is independent of both x and n.

Since f is continuous on a closed interval, it is bounded and there exists $C > 0$ such that $|f(x)| \leq C$ for all $x \in [0,1]$. The maximum value of $x(1-x)$ on $[0,1]$ is clearly $1/4$. Using $|x - k/n| \geq \delta$ for $k \notin \Lambda$ and (6.6), we obtain

$$\begin{aligned}
\sum_{k \notin \Lambda} \tau_k(x) \left| f(x) - f\left(\frac{k}{n}\right) \right| &\leq 2C \sum_{k \notin \Lambda} \tau_k(x) = \\
\frac{2C}{\delta^2} \sum_{k \notin \Lambda} \delta^2 \tau_k(x) &\leq \frac{2C}{\delta^2} \sum_{k \notin \Lambda} \left(\frac{k}{n} - x\right)^2 \tau_k(x) \leq \\
\frac{2C}{\delta^2} \sum_{k=0}^{n} \left(\frac{k}{n} - x\right)^2 \tau_k(x) &= \frac{2C}{\delta^2} \frac{x(1-x)}{n} = \\
\frac{2Cx(1-x)}{\delta^2 n} &\leq \frac{2C}{4\delta^2 n}
\end{aligned}$$

The second key estimate is

$$\sum_{k \notin \Lambda} \tau_k(x) \left| f(x) - f\left(\frac{k}{n}\right) \right| \leq \frac{C}{2\delta^2 n} \tag{6.8}$$

Here the right-hand side is independent of x and goes to 0 as n goes to infinity. In particular,

$$\frac{C}{2\delta^2 n} < \frac{\varepsilon}{2}$$

when $n > C/(\varepsilon\delta^2)$.

Set $N = C/(\varepsilon\delta^2)$. Combining equations (6.7) and (6.8), shows that

$$|f(x) - B_n(x)| < \frac{\varepsilon}{2} + \frac{C}{2\delta^2 n} < \varepsilon$$

for all x in $[0,1]$ and $n > N$. Therefore, for all ε, there exists N such that $\|f - B_n(f)\|_s < \varepsilon$ when $n > N$. Consequently, f is in $\mathcal{P}([a,b])^-$ and $\mathcal{P}([a,b])^- = \mathcal{C}(X)$ to complete the proof. □

There are no polynomials in $\mathcal{C}(X)$ in general, but polynomials do play a role in understanding when a subalgebra is dense in $\mathcal{C}(X)$ because functions in $\mathcal{C}(X)$ can be composed with polynomials.

Proposition 6.1.2 *Let X be a compact metric space, and let $\mathcal{A}$ be subalgebra of $C(X)$ containing a nonzero constant function. If f is in $\mathcal{A}$ and g is in $\mathcal{C}([-\|f\|_s, \|f\|_s])$, then $g \circ f$ is in $\mathcal{A}^-$.*

Proof. If a subalgebra of $C(X)$ contains a nonzero constant function, then it contains all the constant functions. Let $p(y) = \alpha_0 + \alpha_1 y + \ldots + \alpha_n y^n$ be a polynomial. Then,

$$p \circ f = \alpha_0 + \alpha_1 f + \alpha_2 f^2 + \ldots + \alpha_n f^n$$

and $p \circ f$ is in $\mathcal{A}$.

Given $\varepsilon > 0$, the polynomial $p(x)$ can be chosen so that

$$\sup\left\{|g(y) - p(y)| : -\|f\|_s \leq y \leq \|f\|_s\right\} < \varepsilon$$

by *Theorem 6.1.1.* It follows that:

$$\sup\left\{|g(f(x)) - p(f(x))| : x \in X\right\} < \varepsilon$$

and

$$\|g \circ f - p \circ f\|_s < \varepsilon.$$

Therefore, $g \circ f$ is in $\mathcal{A}^-$ because $p \circ f$ is in $\mathcal{A}$. □

The following obvious corollary of this proposition will play a key role in determining when subalgebras of $\mathcal{C}(X)$ are dense:

Corollary 6.1.3 *Let X be a compact metric space, and let $\mathcal{A}$ be subalgebra of $C(X)$ containing a nonzero constant function. If $f \in \mathcal{A}$, then $|f(x)|$ is in $\mathcal{A}^-$.*

Proof. Let $g(y) = |y|$. □

Let S be a set and let $f_j : S \to \mathbb{R}$ for $j = 1, \ldots, k$ be a finite collection of real-valued functions on S. Define the maximum and minimum functions of $f_1, \ldots, f_k$ by setting

$$(f_1 \vee \ldots \vee f_k)(x) = \max\{f_1(x), \ldots, f_k(x)\}$$

and

$$(f_1 \wedge \ldots \wedge f_k)(x) = \min\{f_1(x), \ldots, f_k(x)\}.$$

Proposition 6.1.4 *Let X be a compact metric space, and let $\mathcal{A}$ be a closed subalgebra of $C(X)$ containing a nonzero constant function. If $f_1, \ldots, f_k$ are functions in $\mathcal{A}$, then $f_1 \vee \ldots \vee f_k$ and $f_1 \wedge \ldots \wedge f_k$ are in $\mathcal{A}$*

Proof. First, consider two functions f_1 and f_2 in $C(X)$. An easy calculation shows that

$$f_1 \vee f_2 = \frac{f_1 + f_2 + |f_1 - f_2|}{2}$$

and

$$f_1 \wedge f_2 = \frac{f_1 + f_2 - |f_1 - f_2|}{2}$$

Since $\mathcal{A}$ is closed, it follows from *Corollary 6.1.3* that $f_1 \vee f_2$ and $f_1 \wedge f_2$ are in $\mathcal{A}$. The general case follows by induction because $f_1 \vee \ldots \vee f_k = (f_1 \vee \ldots \vee f_{k-1}) \vee f_k$ and $f_1 \wedge \ldots \wedge f_k = (f_1 \wedge \ldots \wedge f_{k-1}) \wedge f_k$. □

A subalgebra $\mathcal{A}$ of $C(X)$ *separates points* of X provided that for every pair of distinct points x and y in X there exists f in $C(X)$ such that $f(x) \neq f(y)$. If $f(x) \neq f(y)$ for some $f \in C(X)$, set

$$g(w) = a\,\frac{f(w) - f(y)}{f(x) - f(y)} + b\,\frac{f(w) - f(x)}{f(y) - f(x)}$$

and observe that $g(x) = a$ and $g(y) = b$. This leads to a definition and a simple observation. A subalgebra $\mathcal{A}$ of $C(X)$ has the *two-point property* provided that for every pair of distinct points x and y in X and real numbers a and b there exists f in $\mathcal{A}$ such that $f(x) = a$ and $f(y) = b$. It follows from the above equation that a subalgebra $\mathcal{A}$ of $C(X)$ separates points if and only if it has the two-point property. Now we are ready to assemble the pieces.

Theorem 6.1.5 (Stone-Weierstrass) *Let X be a compact metric space. If $\mathcal{A}$ is a subalgebra of $C(X)$ that separates points and contains a nonzero constant function, then $\mathcal{A}^- = C(X)$.*

Proof. When X contains only one point, every function is a constant function and the only subalgebra containing a nonzero constant function is $C(X)$. So, without loss of generality we can assume that X contains more than one point.

Let f be an element of $C(X)$. It suffices to show that for every $\varepsilon > 0$ there exists h in $\mathcal{A}^-$ such that $\|f - h\|_s < \varepsilon$. Because $\mathcal{A}$ separates points of X, it has the two-point property. *Proposition 6.1.4* can be applied to $\mathcal{A}^-$ by *Exercise* 2, p. 232, and can be used to construct the required h.

Fix w in X. For each $y \neq w$, there exists g_y in $\mathcal{A}$ such that $g_y(w) = f(w)$ and $g_y(y) = f(y)$. Set $U_y = \{x \in X : g_y(x) < f(x) + \varepsilon\}$. Clearly, U_y is an open set containing w and y. Since y is arbitrary, the collection of sets $\{U_y : y \in X\}$ is an open cover of X, and there exists a finite subcover $U_{y_1}, \ldots, U_{y_k}$ by compactness. Set $h_w = g_{y_1} \wedge \ldots \wedge g_{y_k}$. Let x be in X. Then x is in U_{y_j} for some j and $g_{y_j}(x) < f(x) + \varepsilon$. It follows that $h_w(x) < f(x) + \varepsilon$ for all $x \in X$. In addition, $h_w(w) = f(w)$ and h_w is in $\mathcal{A}^-$ by *Proposition 6.1.4*.

Therefore, for each $w \in X$ there exists h_w in $\mathcal{A}^-$ such that $h_w(w) = f(w)$ and $h_w(x) < f(x) + \varepsilon$ for all $x \in X$. Set $V_w = \{x \in X : f(x) - \varepsilon < h_w(x)\}$. Then, w is in V_w and $\{V_w : w \in X\}$ is an open covering of X. Again there exists a finite subcover $V_{w_1}, \ldots, V_{w_m}$. Set $h = h_{w_1} \vee \ldots \vee h_{w_m}$. By *Proposition 6.1.4*, h is in $\mathcal{A}^-$. Let x be in X. Then x is in some V_{w_j} and $f(x) - \varepsilon < h_{w_j}(x)$ Thus $f(x) - \varepsilon < h(x)$ for all x. Clearly, $h(x) < f(x) + \varepsilon$ for all x, because $h_w(x) < f(x) + \varepsilon$ for all x and w in X. Therefore,

$$f(x) - \varepsilon < h(x) < f(x) + \varepsilon$$

for all $x \in X$, and there exists h in $\mathcal{A}^-$ such that $\|f - h\|_s < \varepsilon$. □

Let $\mathcal{C}(X, \mathbb{C})$ denote the set of continuous complex-valued functions on X, a compact metric space. Then $\mathcal{C}(X, \mathbb{C})$ is a complex normed algebra with $|f(x)|$ equal to the complex absolute value and $\|f\|_s = \sup\{|f(x)| : x \in X\}$. As one would expect there is a theorem similar to *Theorem 6.1.5* for $\mathcal{C}(X, \mathbb{C})$. It does, however, require an additional hypothesis. Showing why an additional hypothesis is necessary requires complex analysis and is beyond the scope of this book, but we can at least indicate where the difficulty lies.

Let $X = \{z \in \mathbb{C} : |z| \leq 1\}$, the closed unit disc and consider the function $f(z) = \overline{z}$ in $\mathcal{C}(X, \mathbb{C})$. On the one hand, it is not differentiable with respect to the complex variable z because it does not satisfy the Cauchy-Riemann conditions. On the other hand, using Morera's Theorem it can be shown that the uniform limit of a sequence of polynomials on the unit disc can be differentiated with respect to the complex variable z on $\{z \in \mathbb{C} : |z| < 1\}$. Hence, complex polynomials cannot uniformly approximate $f(z) = \overline{z}$ on the closed unit disk X. The results mentioned can be found in [1].

The way around this difficulty is to require that the subalgebra of $\mathcal{C}(X, \mathbb{C})$ is closed under conjugation. Every function f in $\mathcal{C}(X, \mathbb{C})$ can be written as $f(x) = u(x) + iv(x)$, where $u(x)$ and $v(x)$ are in $\mathcal{C}(X) \subset \mathcal{C}(X, \mathbb{C})$. The conjugate of f is defined by

$$\overline{f}(x) = \overline{f(x)} = u(x) - iv(x).$$

Theorem 6.1.6 (Complex Stone-Weierstrass) *Let X be a compact metric space and let $\mathcal{A}$ be a complex subalgebra. If $\mathcal{A}$ contains a nonzero constant function, separates points, and contains the conjugate of every function in it, then $\mathcal{A}^- = \mathcal{C}(X, \mathbb{C})$.*

Proof. Set $\mathcal{B} = \mathcal{A} \cap \mathcal{C}(X)$ and observe that $\mathcal{B}$ is a real subalgebra of $\mathcal{C}(X, \mathbb{C})$, that is, $\mathcal{B}$ is closed under addition and multiplication of functions and scalar multiplication of a functions by *real* numbers.

If $f(x) = u(x) + iv(x)$ is in $\mathcal{A}$, then u and v are in $\mathcal{B}$ because $u = (f + \overline{f})/2$ and $v = (f - \overline{f})/(2i)$. Consequently, $\mathcal{A} = \{u + iv : u \in \mathcal{B} \text{ and } v \in \mathcal{B}\}$ and it suffices to show that $\mathcal{B}^- = \mathcal{C}(X)$.

Obviously, $\mathcal{B}$ contains a nonzero constant function because $\mathcal{A}$ does. Given x and y in X such that $x \neq y$ there exists $f = u + iv$ in $\mathcal{A}$ such that $f(x) \neq f(y)$. Then $u(x) \neq u(y)$ or $v(x) \neq v(y)$. It follows that $\mathcal{B}$ separates points of X. Now $\mathcal{B}^- = \mathcal{C}(X)$ by the Stone-Weierstrass Theorem. □

When X is a metric space with metric d, it is trivial to show that continuous real-valued functions separate points; just use functions of the form $f(x) = d(x, y)$ with y fixed. We have not, however, proved a comparable result, such as Urysohn's Lemma (see Chapter 4 of [7] or Chapter 5 of [16]), that applies to compact Hausdorff spaces. In fact, from our perspective $C(X)$ could consist of just constant functions when X is compact Hausdorff, but not metric. For this reason, we have included the metric hypothesis in the real and complex Stone-Weierstrass Theorems, even though the proofs presented can be used when X is a compact Hausdorff space.

Recall from (*Theorem 1.2.7* on page 20) that

$$d(x, y) = \frac{|x - y|}{1 + |x - y|}$$

is a bounded metric on $\mathbb{R}$ equivalent to the usual metric $|x - y|$. Using this fact, it is not difficult to show that the formula

$$d(f, g) = \sum_{k=1}^{\infty} \frac{1}{2^k} \sup \left\{ \frac{|f(x) - g(x)|}{1 + |f(x) - g(x)|} : |x| \leq k \right\} \tag{6.9}$$

defines a metric on $C(\mathbb{R})$, set of continuous real-valued functions defined on $\mathbb{R}$. The best way to understand this metric is in terms of the convergence of sequences of functions (see *Exercise* 9, p. 233).

Although $C(\mathbb{R})$ is an algebra, we will show that there does not exist a norm $\|\cdot\|_a$ on $C(\mathbb{R})$ such that the metrics $d(f, g)$ given by (6.9) and $\|f - g\|_a$ are equivalent. Suppose $\|\cdot\|_a$ is a norm on $C(\mathbb{R})$ such that the metric $\|f - g\|_a$ is equivalent to $d(f, g)$.

Consider the sequence of functions defined by

$$f_k(x) = \begin{cases} 0 & \text{for } x \leq k \\ x - k & \text{for } x \geq k \end{cases}.$$

Then f_k converges to the zero function with respect to the metric. In fact, $\alpha_k f_k$ converges to the zero function with respect to the metric for any sequence of real numbers α_k. It follows that $\|\alpha_k f_k\|_a \to 0$ as $k \to \infty$. Since f_k is not the zero function, $\|f_k\|_a \neq 0$. Let $\alpha_k = 1/\|f_k\|_a$. Then, we have the following contradiction:

$$1 = \|\alpha_k f_k\|_a \to 0.$$

The metric space $C(\mathbb{R})$ is, however, a metric group under addition of functions by *Exercise* 10, p. 233. Furthermore, the definition of the metric given by (6.9) can be extended to more general spaces of functions and produces interesting metric groups. Section 6.2 will study the basic properties of these metric spaces of functions.

EXERCISES

1. Show that
$$x(1-x)\tau_k'(x) = (k-nx)\tau(x)$$
for $0 \le k \le n$, and use it to show that
$$x(1-x)\left(\sum_{k=0}^{n}(k-nx)^q\tau_k'(x)\right) = \sum_{k=0}^{n}(k-nx)^{q+1}\tau_k(x), \tag{6.10}$$
for $q \ge 0$.

2. Let X be a compact metric space. Prove that if $\mathcal{A}$ is a subalgebra of $\mathcal{C}(X)$, then $\mathcal{A}^{-}$ is a subalgebra of $\mathcal{C}(X)$.

3. Let X and Y be metric spaces with metrics d_X and d_Y, respectively. A function $f : X \to Y$ is *uniformly continuous* on X provided that given $\varepsilon > 0$ there exists $\delta > 0$ such that $d_Y(f(x), f(y)) < \varepsilon$ when $d_X(x,y) < \delta$. Prove that if X is compact, then every continuous function $f : X \to Y$ is uniformly continuous on X.

4. Let $\mathcal{E}$ be the even polynomials, that is $p(-x) = p(x)$ for all $x \in \mathbb{R}$, and let $\mathcal{E}([a,b])$ denote the even polynomial functions restricted to the closed interval $[a,b]$. Prove the following:
   (a) $\mathcal{E}([a,b])$ is a subalgebra of $\mathcal{C}([a,b])$.
   (b) If $0 \le a < b$, then $\mathcal{E}([a,b])^{-} = \mathcal{C}([a,b])$.
   (c) If $a > 0$, then $\mathcal{E}([-a,a])^{-}$ is the subalgebra of even functions in $\mathcal{C}([-a,a])$.

5. Let $f_1, \ldots, f_k$ be continuous real-valued functions on a metric space X. Show that $f_1 \vee \ldots \vee f_k$ and $f_1 \wedge \ldots \wedge f_k$ are also continuous on X.

6. Show that
$$d(f,g) = \sup\left\{\frac{|f(x)-g(x)|}{1+|f(x)-g(x)|} : a \le x \le b\right\}$$
defines a metric on $\mathcal{C}([a,b])$ equivalent to the usual metric $\|f-g\|_s$.

7. Show that $f \to \max\{f(x) : x \in X\}$ and $f \to \min\{f(x) : x \in X\}$ are continuous real-valued functions on $\mathcal{C}(X)$.

8. Let X and Y be compact metric spaces and let
$$\mathcal{A} = \left\{\sum_{i=1}^{m} f_i(x)g_i(y) : f_i \in \mathcal{C}(X), g_i \in \mathcal{C}(Y) \text{ for } 1 \le i \le m \text{ and } m \in \mathbb{Z}^+\right\}$$
Prove that $\mathcal{A}$ is a subalgebra of $\mathcal{C}(X \times Y)$ and $\mathcal{A}^{-} = \mathcal{C}(X \times Y)$.

9. Let the sequence f_k and f be in $\mathcal{C}(\mathbb{R})$. Prove that the following are equivalent:

   (a) The sequence f_k converges to f with respect to the metric defined by equation (6.9).

   (b) The sequence f_k converges uniformly on $[-m, m]$ to f for every m in $\mathbb{Z}^+$.

   (c) The sequence of the restrictions $f_k|[-m, m]$ converges to $f|[-m, m]$ in the normed linear space $\mathcal{C}([-m, m])$.

10. Show that $\mathcal{C}(\mathbb{R})$ with the metric (6.9) is a metric group, and show that the polynomial functions are dense in it.

6.2 The Compact-Open Topology

Let X and Y be metric spaces, and let $\mathcal{C}(X, Y)$ denote the set of continuous functions from X to Y. There are a number of ways the set $\mathcal{C}(X, Y)$ can be given a topological structure. One of them is the compact-open topology. It has the property that the evaluation function $(f, x) \to f(x)$ is a continuous function from $\mathcal{C}(X, Y) \times X$ with the product topology to Y. Like the product topology, the compact-open topology is not always a metric topology. It is a metric space when X is locally compact and second countable and in this context produces a variety of metric groups.

We begin with a discussion of locally compact second countable metric spaces. This includes compact metric spaces because they are second countable by *Corollary 5.3.3.* The analysis and results should be viewed as generalizations from compact metric spaces to the much larger class of locally compact second countable metric spaces.

Proposition 6.2.1 *If X is a locally compact metric space, then the following are equivalent:*

(a) X is second countable.

(b) X is separable.

(c) X is σ-compact.

Proof. In metric spaces second countable and separable are equivalent by *Proposition 5.3.2.* The locally compact hypothesis will be used to prove the equivalence of separable and σ-compact.

Suppose x is separable. Because X is locally compact, for every x in X there exist $r > 0$ such that $B_r(x)^-$ is compact, and then $B_\rho(x)^-$ is compact for all $0 < \rho < r$. So there are plenty of compact sets available, but to construct a sequence of compact sets C_j such that $X = \bigcup_{j=1}^{\infty} C_j$ it is necessary to avoid using smaller and smaller compact sets. To this end set

$$\nu(x) = \min \{ n \in \mathbb{Z}^+ : B_{1/n}(x)^{-} \text{ is compact} \}.$$

There exists a countable dense subset $D = \{x_1, \ldots, x_j \ldots\}$ in X, because X is separable. Let x be an arbitrary point in X and $r > 0$ such that $B_r(x)^{-}$ is compact. There exists n in $\mathbb{Z}^+$ such that $1/n < r/2$ and x_j in D such that x_j is in $B_{1/n}(x)$. It follows that

$$x \in B_{1/n}(x_j) \subset B_r(x).$$

Hence, $B_{1/n}(x_j)^{-} \subset B_r(x)^{-}$ and $B_{1/n}(x_j)^{-}$ is compact. Thus $\nu(x_j) \leq n$, and

$$x \in B_{1/n}(x_j) \subset B_{1/\nu(x_j)}(x_j) \subset B_{1/\nu(x_j)}(x_j)^{-}.$$

It follows that

$$X = \bigcup_{j=1}^{\infty} B_{1/\nu(x_j)}(x_j)^{-}$$

and *(b)* implies (c).

Now suppose X is σ-compact and C_j is a sequence of compact sets such that $X = \bigcup_{j=1}^{\infty} C_j$. Since the finite union of compact sets is compact, we can assume without loss of generality that $C_j \subset C_{j+1}$ for all j by replacing C_j with $\bigcup_{i=1}^{j} C_i$.

By compactness, for each j there exist $D_j = \{x_1, \ldots, x_{p_j}\} \subset C_j$ such that

$$C_j \subset \bigcup_{k=1}^{p_j} B_{1/j}(x_k).$$

Then $D = \bigcup_{j=1}^{\infty} D_j$ is a countable union of finite sets and finite or countable.

To show that D is dense, let x be a point in X and $\varepsilon > 0$. The point x is in some C_i, and hence in C_j for all $j \geq i$. Choose $j \geq i$ such that $1/j < \varepsilon$. Then x is in $B_{1/j}(x_k)$ for some x_k in D_j and $d(x, x_k) < \varepsilon$. □

If a locally compact metric space is σ-compact, and hence the union of a sequence of compact sets, the Baire Category Theorem implies that some of these compact sets contain open sets (see *Corollary 1.5.17*). Rather than just knowing that some of the compact sets contain open sets, it is useful to build a sequence of compact sets that contain increasing large open sets. The next proposition shows how this can always be arranged.

Proposition 6.2.2 *If X is a locally compact second countable metric space, then there exists a sequence of open sets W_j with the following properties:*

(a) $X = \bigcup_{j=1}^{\infty} W_j = \bigcup_{j=1}^{\infty} W_j^{-}$.

(b) W_j^{-} *is compact.*

(c) $W_j^- \subset W_{j+1}$.

Proof. If X is compact, simply set $W_j = X$ for all j. For the rest of the proof, we assume X is not compact.

By the previous proposition, X is σ-compact and there exists a sequence of compact sets C_j such that $X = \bigcup_{j=1}^{\infty} C_j$. As before, we can assume that $C_j \subset C_{j+1}$ for all j. Open sets W_j will be constructed inductively such that

1. $C_j \subset W_j$
2. W_j^- is compact
3. $W_{j-1}^- \subset W_j$

By the local compactness of X and the compactness of C_1, there exist $y_1, \ldots, y_{p(1)}$ in C_1 and positive numbers $r_1, \ldots, r_{p(1)}$ such that $B_{r_i}(y_i)^-$ is compact for $1, \ldots, p(1)$ and

$$C_1 \subset \bigcup_{i=1}^{p(1)} B_{r_i}(y_i) = W_1.$$

Clearly, W_1^- is compact.

Suppose W_i have been constructed for $i = 1, \ldots, j$ satisfying conditions 1, 2, and 3. Then $C_{j+1} \cup W_j^-$ is compact, and there exists $y_1, \ldots, y_{p(j+1)}$ in $C_{j+1} \cup W_j^-$ and positive numbers $r_1, \ldots, r_{p(j+1)}$ such that $B_{r_i}(y_i)^-$ is compact for $1, \ldots, p(j+1)$ and

$$C_{j+1} \cup W_j^- \subset \bigcup_{i=1}^{p(j+1)} B_{r_i}(y_i) = W_{j+1}.$$

Clearly, W_{j+1} satisfies conditions 1, 2, and 3.

Finally, condition 1 implies that $X = \bigcup_{j=1}^{\infty} W_j$. □

The primary advantage of using *Proposition 6.2.2* to construct compact sets of the form W_j^- is that every compact subset of X is contained in some W_j^- because the sets W_j are an open cover of X such that $W_j \subset W_{j+1}$.

Let X be a locally compact second countable metric space, and let W_j be a sequence of open sets in X given by *Proposition 6.2.2*. Set $K_j = W_j^-$. Given a metric space Y, we want to construct a metric on $\mathcal{C}(X, Y)$, the set of continuous functions from X to Y. By *Theorem 1.2.7*, there exists a metric for Y such that $d_Y(y, y') < 1$ for all y and y' in Y. Now define $d(f, g)$ on $\mathcal{C}(X, Y)$ by

$$d(f,g) = \sum_{j=1}^{\infty} \frac{1}{2^j} \sup \{d_Y(f(x), g(x)) : x \in K_j\}. \tag{6.11}$$

Note that the infinite series converges because

$$0 \leq \sup \{d_Y(f(x), g(x)) : x \in K_j\} \leq 1$$

for all $j \in \mathbb{Z}^+$. Equation (6.11) is clearly a generalization of (6.9) from the end of Section 6.1. When X is compact, $W_j = X$ for all j and (6.11) simplifies to $d(f,g) = \sup\{d_Y(f(x), g(x)) : x \in X\}$.

It is easy to verify that equation (6.11) defines a metric on $\mathcal{C}(X,Y)$ when X is locally compact and second countable. Although the metric defined by (6.11) depends on the choice of the bounded metric d_Y and the open sets W_j in X satisfying the conclusion of *Proposition 6.2.2*, we will show that they all produce the same topology on $\mathcal{C}(X,Y)$, that is, they are all equivalent.

Because $K_j \subset K_{j+1}$ for all j and $\sum_{j=1}^{\infty} 1/2^j = 1$, for a given positive integer m we have

$$\sum_{j=1}^{m} \frac{1}{2^j} \sup\left\{d_Y(f(x), g(x)) : x \in K_j\right\} \le \sup\left\{d_Y(f(x), g(x)) : x \in K_m\right\}.$$

It follows that

$$d(f,g) \le \sup\{d_Y(f(x), g(x)) : x \in K_m\} + \frac{1}{2^m} \tag{6.12}$$

because $\sum_{j=m+1}^{\infty} 1/2^j = 1/2^m$ and $\sup\{d_Y(f(x), g(x)) : x \in K_j\} \le 1$ for all j. This estimate will be useful in developing the basic properties of this metric and the resulting topology on $\mathcal{C}(X,Y)$.

Given a compact subset C of X and an open set V of Y, set

$$N(C,V) = \{f \in \mathcal{C}(X,Y) : f(C) \subset V\}.$$

Notice that if $C \subset C'$, then $N(C,V) \supset N(C',V)$; and if $V \subset V'$, then $N(C,V) \subset N(C,V')$. Also $N(C,V)$ always contains the constant functions with values in V, so $N(C,V)$ is not empty when C and V are not empty.

Proposition 6.2.3 *Let X and Y be metric spaces with X locally compact and second countable. If C is a compact subset of X and V is an open subset of Y, then $N(C,V)$ is an open subset of $\mathcal{C}(X,Y)$ with respect to the metric given by equation (6.11)*

Proof. Let f be in $N(C,V)$. Since $f(C)$ is compact and contained in V, it follows from *Exercise* 3, p. 222, with $F = X \setminus V$ that there exists $\varepsilon > 0$ such that

$$V' = \{y \in Y : d_Y(y, f(C)) < \varepsilon\} \subset V$$

and V' is open.

It follows from *Proposition 6.2.2* that C is contained in some K_j. Set $r = \varepsilon/2^j$ and consider $g \in \mathcal{C}(X,Y)$ such that $d(g,f) < r$. Then every term in the right hand side of (6.11) must be less than r. In particular,

$$\frac{1}{2^j} \sup\{d_Y(g(x), f(x)) : x \in K_j\} < r = \frac{\varepsilon}{2^j}$$

and

$$\sup\{d_Y(g(x), f(x)) : x \in K_j\} < \varepsilon.$$

If x is in C, then $d_Y(g(x), f(x)) < \varepsilon$ because $C \subset K_j$ from which it follows that $d_Y(g(x), f(C)) < \varepsilon$ and $g(x)$ is in $V' \subset V$. Therefore, g is in $N(C, V)$ and $B_r(f) \subset N(C, V)$, proving that $N(C, V)$ is an open subset of $\mathcal{C}(X, Y)$. □

The converse of *Proposition 6.2.3* is not true, but the intersections of finite collections of sets of the form $N(C, V)$ do provide a characterization of the open sets that is independent of the metric defined by (6.11).

Theorem 6.2.4 *Let X and Y be metric spaces with X locally compact and second countable. A subset U of $\mathcal{C}(X, Y)$ is an open subset for the metric defined by (6.11) if and only if for each f in U there exist compact sets $C_1, \ldots, C_p$ in X and open sets $V_1, \ldots, V_p$ in Y such that*

$$f \in \bigcap_{j=1}^{p} N(C_j, V_j) \subset U.$$

Proof. Let U be an open subset of the metric space $\mathcal{C}(X, Y)$ and let f be in U. There exists $r > 0$ such that $B_r(f) \subset U$ and it suffices to show the existence of compact sets $C_1, \ldots, C_p$ in X and open sets $V_1, \ldots, V_p$ in Y such that

$$f \in \bigcap_{j=1}^{p} N(C_j, V_j) \subset B_r(f).$$

Choose m such that

$$\sum_{j=m+1}^{\infty} \frac{1}{2^j} < r/2.$$

The function f is uniformly continuous on K_{m+1} by *Exercise* 3, p. 232. So there exists $\delta > 0$ such that $d_Y(f(x), f(x')) < r/4$ whenever $d_X(x, x') < \delta$ and both x and x' are in K_{m+1}. Using *Exercise* 3, p. 222, again, we can assume that δ is small enough so that

$$\{x \in X : d_X(x, K_m) < \delta\} \subset W_m \subset K_{m+1}.$$

By the compactness of K_m, there exist $x_1, \ldots, x_p$ in K_m such that

$$K_m \subset \bigcup_{i=1}^{p} B_{\delta/2}(x_i) \subset \{x \in X : d_X(x, K_m) < \delta\} \subset W_{m+1} \subset K_{m+1}.$$

It follows that $C_i = B_{\delta/2}(x_i)^{-}$ is compact. By uniform continuity

$$f(C_i) = f(B_{\delta/2}(x_i)^{-}) \subset f(B_\delta(x_i)) \subset B_{r/4}(f(x_i)).$$

Let $V_i = B_{r/4}(f(x_i))$. Clearly f is in $\bigcap_{i=1}^{p} N(C_i, V_i)$.

Now, let g be an arbitrary function in $\bigcap_{i=1}^{p} N(C_i, V_i)$. It must be shown that g is in $B_r(f)$. Consider x be in K_m. For some i, the point x is in $C_i = B_{\delta/2}(x_i)^{-}$. Note that $g(C_i) \subset V_i = B_{r/4}(f(x_i))$ because g in $N(C_j, V_j)$ for all j. Then

$$d_Y(g(x), f(x)) \leq d_Y(g(x), f(x_i)) + d_Y(f(x_i), f(x)) \leq \frac{r}{4} + \frac{r}{4} = \frac{r}{2},$$

and $\sup\{d_Y(f(x), g(x)) : x \in K_m\} \le r/2$. By equation (6.12)

$$d(f, g) \le \sup\{d_Y(f(x), g(x)) : x \in K_m\} + \frac{1}{2^m} < \frac{r}{2} + \frac{r}{2} = r,$$

and g is in $B_r(f)$. Thus

$$f \in \bigcap_{i=1}^{p} N(C_i, V_i) \subset B_r(f) \subset U.$$

This completes the proof that an open set satisfies the required condition. The converse is an immediate consequence of *Proposition 6.2.3.* □

Defining a subset U of $\mathcal{C}(X, Y)$ to be open if and only if for each f in W there exist compact sets $C_1, \dots, C_p$ in X and open sets $V_1, \dots, V_p$ in Y such that

$$f \in \bigcap_{j=1}^{p} N(C_j, V_j) \subset U.$$

always produces a topology on $\mathcal{C}(X, Y)$ called the *compact-open topology.* It is not generally a metric space, but *Theorem 6.2.4* shows that $\mathcal{C}(X, Y)$ with the compact-open topology is a metrizable space when X and Y are metric spaces with X locally compact and second countable. Our focus will be on locally compact second countable metric spaces. *Furthermore, $\mathcal{C}(X, Y)$ will always mean the set of continuous functions from X to Y with the compact-open topology.*

Let X and Y be metric spaces and let E be a subset of X. A sequence $f_k : X \to Y$ of continuous functions *converges uniformly on E* to a function $f : X \to Y$ provided that for every $\varepsilon > 0$ there exists $N > 0$ such that $d_Y(f_k(x), f(x)) < \varepsilon$ for all $x \in E$ and all $k > N$. (This is a slightly more general definition of uniform convergence than the one found on page 172.)

When X is compact, the metric for the compact-open topology on $\mathcal{C}(X, Y)$ is just $d(f, g) = \sup\{d_Y(f(x), g(x)) : x \in X\}$. Consequently, when X is compact, a sequence f_k in $\mathcal{C}(X, Y)$ converges in the compact-open topology to f in $\mathcal{C}(X, Y)$ if and only if f_k converges to f uniformly on X. The next theorem generalizes the compact case to locally compact second countable metric spaces using uniform convergence on every compact subset of X.

Theorem 6.2.5 *Let X be a locally compact second countable metric space and let Y be a metric space with a metric d_Y such that $d_Y(y, y') < 1$ for all y and y' in Y. A sequence f_k converges in $\mathcal{C}(X, Y)$ to f in $\mathcal{C}(X, Y)$ if and only if f_k converges uniformly to f on every compact subset of X.*

Proof. Let $d(f, g)$ be the metric for the compact-open topology on $\mathcal{C}(X, Y)$ given by equation (6.11) using d_Y. Thus f_k converges to f in the compact-open topology if and only if $d(f_k, f)$ converges to 0 as k goes to infinity. Now the proof is reduced to showing that $d(f_k, f)$ converges to 0 if and only if f_k converges uniformly to f with respect to d_Y on every compact subset of X.

First, suppose that $d(f_k, f)$ converges to 0. It suffices to prove that the convergence is uniform on each K_j because every compact subset of X is contained in some K_j.

Given $\varepsilon > 0$, there exist N such that $d(f_k, f) < \varepsilon/2^j$ for $k > N$ and so

$$\frac{1}{2^j}\sup\{d_Y(f_k(x), f(x)) : x \in K_j\} \le d(f_k, f) < \frac{\varepsilon}{2^j}.$$

It follows that $\sup\{d_Y(f_k(x), f(x)) : x \in K_j\} < \varepsilon$ for $k > N$, and the convergence is uniform on K_j.

Now, suppose the convergence is uniform on compact sets. Given $\varepsilon > 0$ choose m such that $1/2^m < \varepsilon/2$. There exists N such that $\sup\{d_Y(f_k(x), f(x)) : x \in K_m\} < \varepsilon/2$ when $k > N$. Then the estimate (6.12) implies that $d(f_k, f) < \varepsilon$ for $k > N$. □

Notice that $d(f_k, f) \to 0$ as $k \to \infty$ implies that $f_k(x) \to f(x)$ as $k \to \infty$ for every $x \in X$. The converse is not true in general (see *Exercise* 7, p. 242).

Theorem 6.2.5 is independent of the metric on Y. Applying (6.11) to a metric equivalent to the given metric d_Y that also satisfies the required boundedness condition, produces an equivalent metric on $C(X.Y)$ by *Theorem 6.2.4*. Furthermore, *Exercise* 1, p. 242 can then be used to extend the result to any metric equivalent to d_Y.

There is a natural function $\eta : C(X,Y) \times X \to Y$ called the *evaluation function* and defined by $\eta(f,x) = f(x)$.

Proposition 6.2.6 *Let X and Y be metric spaces with X locally compact second countable. If $C(X,Y) \times X$ has the product topology of the compact-open topology on $C(X,Y)$ and the metric topology on X, then the evaluation function $\eta : C(X,Y) \times X \to Y$ is continuous.*

Proof. Given (f,x) in $C(X,Y) \times X$, let V be an open subset of Y such that $\eta(f,x) = f(x) \in V$. There exists $r > 0$ such that $f(B_r(x)) \subset V$ and $B_r(x)^-$ is compact. So f is in $N(B_{r/2}(x)^-, V)$. If (g,y) is in $N(B_{r/2}(x)^-, V) \times B_{r/2}(x)$, then clearly $\eta(g,y)$ is in V and η is continuous at (f,x). □

The final topic in this section is to characterize the compact subsets of $C(X,Y)$ using properties of the functions in them. The key property is *equicontinuity.* A subset F of $C(X,Y)$ is *equicontinuous on a subset* K of X provided that for every $\varepsilon > 0$ there exists $\delta > 0$ such that $d_Y(f(x), f(x')) < \varepsilon$ for all f in F and all x and x' in K such that $d_X(x,x') < \delta$.

Theorem 6.2.7 (Ascoli) *Let X and Y be metric spaces with X locally compact and second countable. A subset F of $C(X,Y)$ is compact if and only if the following conditions are satisfied:*

(a) *For every x in X the set $\eta(F \times \{x\}) = \{f(x) : f \in F\}$ has a compact closure in Y.*

(b) *F is equicontinuous on every compact subset of X.*

(c) *F is a closed subset of $C(X,Y)$.*

Proof. Suppose F is a compact subset of $C(X,Y)$. Then F is closed because compact subsets of metric spaces are always closed. Furthermore, $\eta(F \times \{x\})$ is compact because η is continuous (*Proposition 6.2.6*) and $F \times \{x\}$ is a compact subset of $C(X,Y) \times X$. So $\eta(F \times \{x\})$ is compact and closed. Thus *(a)* and (c) hold when F is compact.

Suppose F is not equicontinuous on a compact subset K. Then there exists $\varepsilon > 0$ such that for all $k \in \mathbb{Z}^+$ there exist u_k and v_k in K and f_k in F such that $d_X(u_k, v_k) < 1/k$ and $d_Y(f_k(u_k), f_k(v_k)) \geq \varepsilon$. Since K and F are compact, we can assume without loss of generality that u_k and v_k converge to x in K and f_k converges to f in F. So both (f_k, u_k) and (f_k, v_k) converge to (f, x) in $C(X,Y) \times X$. Now by the continuity of η both $f_k(u_k)$ and $f_k(v_k)$ converge to $f(x)$. It follows that $d_Y(f_k(u_k), f_k(v_k))$ converges to 0 contradicting $d_Y(f_k(u_k), f_k(v_k)) \geq \varepsilon > 0$ for all k. Therefore, F is equicontinuous on every compact subset of X.

For the second half of the proof, assume the three conditions are satisfied. To prove that F is compact, it suffices by *Theorem 5.3.5* to prove that every sequence f_k in F has a convergent subsequence in the compact-open topology with its limit in F.

By *Proposition 6.2.1*, X is separable and there exists a countable dense subset $D = \{x_1, \ldots, x_n, \ldots\}$ of X. Denote the compact set $\eta(F \times \{x_n\})^-$ by F_n. Then $\prod_{j=1}^{\infty} F_j$ is a compact metric space by *Theorem 5.3.6*, and $\boldsymbol{\xi}_k = (f_k(x_1), \ldots, f_k(x_n), \ldots)$ is an element of $\prod_{n=1}^{\infty} F_n$. By compactness, $\boldsymbol{\xi}_k$ has a convergent subsequence. So by passing to a subsequence we can assume without loss of generality that $\boldsymbol{\xi}_k$ converges, and hence that $f_k(x_n)$ converges as $k \to \infty$ for every x_n in D.

Consider one of the compact sets K_j given by *Proposition 6.2.2.* Then the set of functions F is equicontinuous on the compact set K_j. Given $\varepsilon > 0$ there exists $\delta > 0$ such that $d_Y(f(u), f(v)) < \varepsilon/3$ for all f in F and all u and v in K_j such that $d_X(u, v) < \delta$. The set $D \cap W_j$ is dense in K_j. The open sets $B_\delta(x_n)$ such that $x_n \in D \cap W_j$ cover K_j. Hence, there exists a finite set $\{x_{n_1}, \ldots, x_{n_p}\}$ of $D \cap W_j$ such that the open balls $B_\delta(x_{n_i})$, $1 \leq i \leq p$ cover K_j.

Since the sequences $f_k(x_{n_i})$ converges for $i = 1, \ldots, p$, they are Cauchy sequences and because there are only a finite number of them, there exists $N > 0$ such that $d_Y(f_k(x_{n_i}), f_m(x_{n_i})) < \varepsilon/3$ when $k, m > N$ and $1 \leq i \leq p$. Now for x in K_j there exists x_{n_i} such that $d_X(x, x_{n_i}) < \delta$, and by the triangle inequality

$$\begin{aligned} d_Y(f_k(x), f_m(x)) &\leq \quad (6.13) \\ d_Y(f_k(x), f_k(x_{n_i})) + d_Y(f_k(x_{n_i}), f_m(x_{n_i})) + d_Y(f_m(x_{n_i}), f_m(x)) &< \\ \frac{\varepsilon}{3} + \frac{\varepsilon}{3} + \frac{\varepsilon}{3} &= \varepsilon. \end{aligned}$$

when $k, m > N$. So $f_k(x)$ is a Cauchy sequence in the compact set $\eta(F \times \{x\})^-$ for every x in K_j. Because compact spaces are complete metric spaces, the sequence $f_k(x)$ converges for $x \in K_j$. Because this argument applies to all K_j, the sequence $f_k(x)$ converges for every $x \in X$, and there exists a function $f : X \to Y$ defined by $f(x) = \lim_{k\to\infty} f_k(x)$.

Letting m go to infinity in the estimate (6.13) shows that $d_Y(f_k(x), f(x)) \leq \varepsilon$ when $k > N$. It follows that for each j the sequence f_k converges uniformly to f on the open set $W_j \subset K_j$. Applying *Proposition 4.4.3* to the metric spaces W_j with the relative topology and Y, proves that f is continuous on W_j. Thus f is continuous on X by *Exercise* 2, p. 179, and in $C(X, Y)$.

Since every compact subset K of X is contained in some W_j, the sequence f_k converges uniformly to f on K. Therefore, f_k converges to f in the compact-open topology by *Theorem 6.2.5*. It follows that f is in F because it is closed by hypothesis. In summary, we have now shown F is countably compact, which is equivalent to being compact in a metric space by *Theorem 5.3.5*. □

Corollary 6.2.8 *Let X and Y be metric spaces with X locally compact and second countable. Let F be a subset of $C(X,Y)$. The closure of F is a compact subset of $C(X,Y)$ if and only if the following conditions are satisfied:*

(a) For every x in X the set $\eta(F \times \{x\}) = \{f(x) : f \in F\}$ has a compact closure in Y.

(b) F is equicontinuous on every compact subset of X.

Proof. If F^- is compact, then the required conditions are immediate consequences of Ascoli's Theorem. Now assume the two conditions in the corollary are satisfied. The continuity of η implies that $\eta(F^- \times \{x\}) \subset (\eta(F \times \{x\})^-$. It follows from *(a)* that $\eta(F^- \times \{x\})$ also has a compact closure in Y. Since the F^- is an equicontinuous set of functions by *Exercise* 10, p. 243, Ascoli's Theorem implies that F^- is compact. □

Although uniform convergence and equicontinuity appear to be strictly metric concepts, they have natural generalizations to topological spaces determined by uniformities. A *uniformity* for a set X is a nonempty collection of sets containing the diagonal $\{(x, x) : x \in X\} \subset X \times X$ that satisfy a set of axioms. Like metrics, uniformities determine a topology on the set. If d is a metric on a set X, then the collection of subsets of $X \times X$ that contain a set of the form $\{(x, y) : d(x, y) < r\}$ for some $r > 0$ is a uniformity for X. This uniformity determines the same topology on X as the metric used to define it. So a uniformity is a generalization of a metric, and it is easy to extend the notions of uniform continuity, uniform convergence, and equicontinuity to uniform spaces, that is, a topological space determined by a uniformity. For a topological group, the collection of sets of the form $\{(x, y) : xy^{-1} \in U\}$, where U is neighborhood of the identity, is a uniformity, whose topology is the same as the topology of the topological group. An extensive treatment of uniform spaces and the compact-open topology can be found in Chapters 6 and 7 of [7] and Chapters 9 and 10 of

[16]. The general results for locally compact regular Hausdorff spaces are very similar to the results presented in this section for metric spaces.

The next three sections are devoted to groups of isometries, homeomorphisms, and homomorphisms that are metric groups with the compact-open topology. Sections 6.3 and 6.4, however, play no further role in subsequent chapters, while Section 6.5 contains the seminal theorem for Chapters 7 and 8.

EXERCISES

1. Let X and Y be metric spaces and let $f_k : X \to Y$ be a sequence of functions. Given a metric d for Y and a subset E of X, show that f_k converges uniformly on E to $f : X \to Y$ with respect to the metric d if and only if f_k converges uniformly on E to f with respect to the equivalent metric $d'(u, v) = d(u, v)/(1 + d(u, v))$. (One approach is to use the properties of the function $t \mapsto t/(1 + t)$ from the proof of *Theorem 1.2.7*.)

2. Let X be a compact metric space. Prove that the norm linear topology of $\| \cdot \|_s$ on $\mathcal{C}(X)$ equals the compact-open topology on $\mathcal{C}(X)$.

3. Let X be a compact metric space and let d and d' be equivalent metrics on X. Given $\varepsilon > 0$ show that there exists $\delta > 0$ such that $d'(x, y) < \varepsilon$ whenever $d(x, y) < \delta$.

4. Let $f_k : X \to Y$ be a sequence of functions from a metric space X to a compact metric space Y. Show that if f_k converges uniformly on X to $f : X \to Y$ for a metric d on Y, then f_k converges uniformly on X to f for every metric d' equivalent to d.

5. Let X, X', Y, and Y' be metric spaces with X and X' locally compact and separable. Suppose that X and X' are homeomorphic and that Y and Y' are homeomorphic. Show that $\mathcal{C}(X, Y)$ and $\mathcal{C}(X', Y')$ with their compact-open topologies are also homeomorphic.

6. Show that if X is a compact metric space with metric d, then there exists $M > 0$ such that $d(x, y) \leq M$ for all x and y in X.

7. Construct a sequence of continuous functions on $[0, 1]$ such that $f_k(x)$ converges to 0 for every x in $[0, 1]$, but does not converge uniformly on $[0, 1]$.

8. Prove that if X is locally compact and second countable metric space and Y is a complete metric space, then $\mathcal{C}(X, Y)$ is complete with the compact-open topology

9. Let X and Y be metric spaces with X locally compact and second countable. Show by constructing three counter examples that each of the three hypothesis in *Theorem 6.2.7* is needed to prove that F is compact.

10. Let X and Y be metric spaces with X locally compact and second countable. Let F be a subset of $\mathcal{C}(X,Y)$. Prove that if F is equicontinuous on every compact subset of X, then F^{-} is also equicontinuous on every compact subset of X.

6.3 Metric Groups of Isometries

Let X be a locally compact second countable metric space and consider $\mathcal{C}(X,X)$ with the compact-open topology as always. The binary operation of composition of functions gives $\mathcal{C}(X,X)$ a natural semigroup structure. Proving that this binary operation is continuous, is the first step toward finding metric groups of isometries and homeomorphisms in $\mathcal{C}(X,X)$.

Proposition 6.3.1 *Let X be a locally compact second countable metric space. The function $(f,g) \to f \circ g$ is a continuous function from $\mathcal{C}(X,X) \times \mathcal{C}(X,X)$ with the product topology to $\mathcal{C}(X,X)$.*

Proof. Let (f_k, g_k) be a sequence converging to (f,g) in $\mathcal{C}(X,X) \times \mathcal{C}(X,X)$. It must be shown that $f_k \circ g_k$ converges to $f \circ g$ in the compact-open topology. If this is not the case, by *Theorem 6.2.5* there exists a compact subset C of X on which $f_k \circ g_k$ does not converge uniformly to $f \circ g$. Thus for some $\varepsilon > 0$, for each positive integer m there exists $x_m \in C$ and $k_m > m$ such that

$$d_X\left(f_{k_m} \circ g_{k_m}(x_m), f \circ g(x_m)\right) \geq \varepsilon.$$

Because C is compact, we can assume without loss of generality that x_m converges to x in C. It follows from the continuity of f and g that $f \circ g(x_m)$ converges to $f \circ g(x)$ and $f_{k_m} \circ g_{k_m}(x_m)$ does not converge to $f \circ g(x)$.

Of course, f_{k_m} and g_{k_m} converge to f and g in $\mathcal{C}(X,X)$. Now *Proposition 6.2.6* implies first that $\eta(g_{k_m}, x_m) = g_{k_m}(x_m)$ converges to $g(x)$ and then

$$\eta(f_{k_m}, g_{k_m}(x_m)) = f_{k_m}(g_{k_m}(x_m)) = f_{k_m} \circ g_{k_m}(x_m)$$

converges to $f \circ g(x)$, contradiction. □

Given a metric space X with metric ρ, an *isometry of X* is a function f mapping X onto itself such that $\rho(f(x), f(y)) = \rho(x,y)$ for all x and y in X. This is a special case of the notion of an isometry between two metric spaces defined on page 29. The isometries of X are homeomorphisms of X onto itself and form a group under composition (*Exercise* 19, p. 29). Let $I_{(X,\rho)}$ denote the group of isometries of the metric space (X,ρ).

Theorem 6.3.2 *If X is a locally compact second countable metric space with metric ρ, then the group $I_{(X,\rho)}$ with the compact-open topology is a metric group. If X is compact, then $I_{(X,\rho)}$ is compact.*

Proof. Since the binary operation $f \circ g$ is continuous by *Proposition 6.3.1*, it remains to prove that $f \to f^{-1}$ is continuous on $I_{(X,\rho)}$. By *Exercise* 4, p. 39, it suffices to prove that $f \to f^{-1}$ is continuous at the identity ι.

A metric for $I_{(X,\rho)}$ is just the restriction of the metric

$$d(f_1, f_2) = \sum_{j=1}^{\infty} \frac{1}{2^j} \sup\{\rho(f_1(x), f_2(x)) : x \in K_j\}$$

for $\mathcal{C}(X, X)$. Let f_1 and f_2 be elements of $\mathcal{C}(X, X)$, and let g be an element of $I_{(X,\rho)}$. Then $\rho(g \circ f_1(x), g \circ f_2(x)) = \rho(f_1(x), f_2(x))$ for all x in X, and it follows that $d(g \circ f_1, g \circ f_2) = d(f_1, f_2)$. Thus the restriction of d to $I_{(X,\rho)}$ is a left-invariant metric for $I_{(X,\rho)}$. Consequently, $d(g^{-1}, \iota) = d(g \circ g^{-1}, g \circ \iota) = d(\iota, g)$. It is now obvious that $g \to g^{-1}$ is continuous at ι.

Ascoli's Theorem (*Theorem 6.2.7*) will be used to prove that $I_{(X,\rho)}$ is compact when X is compact. Obviously, any family of isometries of any metric space is equicontinuous. Also, the closure of $\eta(I_{(X,\rho)} \times \{x\})$ in X is compact because X is compact. So it remains to prove that $I_{(X,\rho)}$ is closed in the compact-open topology.

Let f_k be a sequence in $I_{(X,\rho)}$ converging to f in $\mathcal{C}(X, X)$. It suffices by *Proposition 1.2.8* to show that f is in $I_{(X,\rho)}$. Since η is continuous, $f_k(x)$ converges to $f(x)$ for every x in X. Consequently,

$$\rho(x, y) = \rho(f_k(x), f_k(y)) \to \rho(f(x), f(y))$$

as k goes to infinity because metrics are continuous (*Exercise* 1, p. 247). Hence $\rho(x, y) = \rho(f(x), f(y))$, and f preserves distance.

The last step is to prove that f is onto. Given y in X, for each k there exists x_k such that $f_k(x_k) = y$. Because X is compact, we can assume by passing to a subsequence that x_k converges to x. It follows from the continuity of η that $y = f_k(x_k) \to f(x)$. Hence, $f(x) = y$ and f is onto. Therefore, f is an isometry, $I_{(X,\rho)}$ is closed, and it is compact by *Theorem 6.2.7*. □

A *rigid motion* of $\mathbb{R}^n$ is an isometry of $\mathbb{R}^n$ for the Euclidean norm, that is, an onto function $f : \mathbb{R}^n \to \mathbb{R}^n$ such that $\|f(\mathbf{x}) - f(\mathbf{y})\| = \|\mathbf{x} - \mathbf{y}\|$ for all $\mathbf{x}$ and $\mathbf{y}$ in $\mathbb{R}^n$. Let $\mathrm{RM}(n)$ denote the metric group of rigid motions of $\mathbb{R}^n$ with the compact-open topology instead of the more cumbersome $I_{(\mathbb{R}^n, \|\cdot\|)}$.

When A is an orthogonal matrix, then $A^{\mathfrak{t}}$ is also orthogonal and the linear function $T_A(\mathbf{x}) = \mathbf{x}A^{\mathfrak{t}}$ is a rigid motion by *Proposition 4.2.1.* Notice that $T_A \circ T_B = T_{AB}$, which is the reason for using $A^{\mathfrak{t}}$ in the definition of T_A. The *translations* of $\mathbb{R}^n$ defined by $\theta_{\mathbf{a}}(\mathbf{x}) = \mathbf{x} + \mathbf{a}$ are also rigid motions as are the compositions of translations and orthogonal linear functions. Clearly, $\theta_{\mathbf{b}} \circ \theta_{\mathbf{a}} = \theta_{\mathbf{a}+\mathbf{b}}$. Although the rigid motions T_A and $\theta_{\mathbf{a}}$ do not commute, they do satisfy the following basic relationship

$$T_A \circ \theta_{\mathbf{a}} = \theta_{T_A(\mathbf{a})} \circ T_A, \tag{6.14}$$

which suggests that $\mathbb{R}^n \times \mathrm{O}(n)$ has a role to play in the study of rigid motions.

Define a binary operation on $\mathbb{R}^n \times \mathrm{O}(n)$ by setting

$$(\mathbf{a}, A) * (\mathbf{b}, B)) = (\mathbf{a} + \mathbf{b}A^{\mathrm{t}}, AB). \tag{6.15}$$

A routine calculation shows that this is an associative binary operation, the element $(\mathbf{0}, I)$ is an identity, and $(\mathbf{a}, A)^{-1} = (-\mathbf{a}(A^{-1})^{\mathrm{t}}, A^{-1})$. Thus it is a group.

This group structure on $\mathbb{R}^n \times \mathrm{O}(n)$ is an example of a *semidirect product.* To avoid confusion with the usual direct product of the groups $\mathbb{R}^n$ and $\mathrm{O}(n)$, the semidirect product will be denoted by $\mathbb{R}^n$Ⓢ $\mathrm{O}(n)$. The key to this group construction is the use of an algebraic homomorphism from the second group to a group of automorphisms of the first group. In this case, $A \to T_A$ is the required homomorphism from $\mathrm{O}(n)$ to the group of automorphisms of $\mathbb{R}^n$; the more general construction of a semidirect product appears in *Exercise* 3, p. 247. To obtain a metric group in this way, a continuity condition is required (see *Exercise* 4, p. 247).

Theorem 6.3.3 *The function* $\varphi : \mathbb{R}^n$Ⓢ $\mathrm{O}(n) \to \mathrm{RM}(n)$ *defined by* $\varphi(\mathbf{a}, A) = \theta_{\mathbf{a}} \circ T_A$ *is an algebraic isomorphism of the semidirect product* $\mathbb{R}^n$Ⓢ $\mathrm{O}(n)$ *defined by equation (6.15) onto the group* $\mathrm{RM}(n)$ *of rigid motions of* $\mathbb{R}^n$.

Proof. Let $(\mathbf{a}, A)$ and $(\mathbf{b}, B)$ be elements of $\mathbb{R}^n$Ⓢ$\mathrm{O}(n)$. Using equation (6.14), it follows that

$$\begin{aligned}
\varphi((\mathbf{a}, A) * (\mathbf{b}, B)) &= \varphi(\mathbf{a} + \mathbf{b}A^{\mathrm{t}}, AB) \\
&= \theta_{\mathbf{a}+\mathbf{b}A^{\mathrm{t}}} \circ T_{AB} \\
&= \theta_{\mathbf{a}} \circ \theta_{T_A(\mathbf{b})} \circ T_A \circ T_B \\
&= \theta_{\mathbf{a}} \circ T_A \circ \theta_{\mathbf{b}} \circ T_B \\
&= \varphi(\mathbf{a}, A) \circ \varphi(\mathbf{b}, B),
\end{aligned}$$

and φ is an algebraic homomorphism.

Suppose $\varphi(\mathbf{a}, A) = \theta_{\mathbf{a}} \circ T_A = \iota$, the identity function on $\mathbb{R}^n$. Then $\theta_{\mathbf{a}} \circ T_A(\mathbf{0}) = \mathbf{a}$ implies that $\mathbf{a} = \mathbf{0}$. Since $T_A = \iota$ if and only if $A = I$, we have $(\mathbf{a}, A) = (\mathbf{0}, I)$ and φ is one-to-one.

It remains to show that φ is onto, which is the most interesting part of the result. The proof proceeds by several reductions to simpler cases. Let f be an element of $\mathrm{RM}(n)$, and set $\mathbf{a} = f(\mathbf{0})$. Then $f_1 = \theta_{-\mathbf{a}} \circ f$ is a rigid motion satisfying $f_1(\mathbf{0}) = \mathbf{0}$.

The triangle with vertices $\mathbf{0}$, $\mathbf{e}_j$, and $\mathbf{e}_k$ such that $j \neq k$ is a right triangle with $\|\mathbf{e}_j - \mathbf{0}\| = 1$, $\|\mathbf{e}_k - \mathbf{0}\| = 1$, and $\|\mathbf{e}_j - \mathbf{e}_k\| = \sqrt{2}$. It follows that $\|f_1(\mathbf{e}_j) - \mathbf{0}\| = 1$, $\|f_1(\mathbf{e}_k) - \mathbf{0}\| = 1$, and $\|f_1(\mathbf{e}_j) - f(\mathbf{e}_k)\| = \sqrt{2}$. Now the law of cosines, $a^2 + b^2 = c^2 - 2ab\cos(\gamma)$, where γ is the angle opposite the side of length c and the lengths of the other two sides are a and b, implies that the triangle with vertices $f_1(\mathbf{0}) = \mathbf{0}$, $f_1(\mathbf{e}_j)$, and $f_1(\mathbf{e}_k)$ is also a right triangle. Therefore, $f_1(\mathbf{e}_1), \ldots, f_1(\mathbf{e}_n)$ is an orthonormal basis for $\mathbb{R}^n$.

Let A be the matrix whose columns are $f_1(\mathbf{e}_1), \ldots, f_1(\mathbf{e}_n)$ and observe that $T_A(\mathbf{e}_j) = f_1(\mathbf{e}_j)$ for $1 \leq j \leq n$. *Proposition 4.2.3* implies that A is in $\mathrm{O}(n)$. Set

$f_2 = T_{A^{-1}} \circ f_1$. Thus f_2 is a rigid motion such that $f_2(\mathbf{0}) = \mathbf{0}$ and $f_2(\mathbf{e}_j) = \mathbf{e}_j$ for $j = 1, \ldots, n$. With a little algebra it will be shown that $f_2 = \iota$.

Let $\mathbf{x}$ be in $\mathbb{R}^n$ and set $\mathbf{y} = f_2(\mathbf{x})$. Because $f_2(\mathbf{0}) = \mathbf{0}$ and $f_2(\mathbf{e}_j) = \mathbf{e}_j$ for $j = 1, \ldots, n$, it follows that

$$\begin{aligned} \|\mathbf{y}\| &= \|\mathbf{x}\| \\ \|\mathbf{y} - \mathbf{e}_j\| &= \|\mathbf{x} - \mathbf{e}_j\| \end{aligned}$$

for $j = 1, \ldots, n$. This leads to the following system of $n + 1$ equations:

$$\begin{aligned} \sum_{k=1}^{n} y_k^2 &= \sum_{k=1}^{n} x_k^2 \\ (y_j - 1)^2 + \sum_{k \neq j} y_k^2 &= (x_j - 1)^2 + \sum_{k \neq j} x_k^2 \end{aligned}$$

Using the first equation to simplify each of the remaining equation, implies that $y_j = x_j$ for $j = 1, \ldots, n$. Thus $\mathbf{y} = \mathbf{x}$ and $\iota = f_2 = T_{A^{-1}} \circ \theta_{-\mathbf{a}} \circ f$ or $f = \theta_{\mathbf{a}} \circ T_A = \varphi(\mathbf{a}, A)$. □

We now have a very satisfactory algebraic description of the metric group RM(n), but acquired an extra topology, namely, the product topology on $\mathbb{R}^n \times$ O(n). Using *Exercise* 4, p. 247, and *Exercise* 5, p. 99, it is not difficult to show that $\mathbb{R}^n$Ⓢ O(n) is a metric group with the product topology.

Proposition 6.3.4 *The function* $\varphi : \mathbb{R}^n$Ⓢ$\mathrm{O}(n) \to \mathrm{RM}(n)$ *defined by* $\varphi(\mathbf{a}, A) = \theta_{\mathbf{a}} \circ T_A$ *is a homeomorphism of* $\mathbb{R}^n \times \mathrm{O}(n)$ *with the product topology onto* RM(n)

Proof. Since every compact set of $\mathbb{R}^n$ is contained in $C_r = \{\mathbf{x} \in \mathbb{R}^n : \|x\| \le r\}$ for some $r > 0$, it suffices to prove that $\mathbf{a}_k$ converges to $\mathbf{a}$ in $\mathbb{R}^n$ and A_k converges to A in O(n) if and only if $\varphi(\mathbf{a}_k, A_k)$ converges uniformly to $\varphi(\mathbf{a}, A)$ on C_r when $r > 0$.

Suppose $\mathbf{a}_k$ converges to $\mathbf{a}$ in $\mathbb{R}^n$ and A_k converges to A in O(n). Then for $\mathbf{x}$ in C_r

$$\begin{aligned} \|\varphi(\mathbf{a}_k, A_k)(\mathbf{x}) - \varphi(\mathbf{a}, A)(\mathbf{x})\| &= \|\theta_{\mathbf{a}_k} \circ T_{A_k}(\mathbf{x}) - \theta_{\mathbf{a}} \circ T_A(\mathbf{x})\| \\ &= \|\mathbf{x}A_k^{\mathrm{t}} + \mathbf{a}_k - \mathbf{x}A^{\mathrm{t}} - \mathbf{a}\| \\ &= \|\mathbf{x}A_k^{\mathrm{t}} - \mathbf{x}A^{\mathrm{t}}\| + \|\mathbf{a}_k - \mathbf{a}\| \\ &\le \|A_k^{\mathrm{t}} - A^{\mathrm{t}}\|_B\, r + \|\mathbf{a}_k - \mathbf{a}\| \\ &\le \varepsilon \end{aligned}$$

when k is large, proving that the convergence is uniform on C_r.

Now suppose $\varphi(\mathbf{a}_k, A_k)$ converges uniformly to $\varphi(\mathbf{a}, A)$ on C_r when $r > 0$. In particular, $\varphi(\mathbf{a}_k, A_k)(\mathbf{0}) = \mathbf{a}_k$ converges to $\varphi(\mathbf{a}, A)(\mathbf{0}) = \mathbf{a}$. Then,

$$\begin{aligned} \|\mathbf{x}A_k^{\mathrm{t}} - \mathbf{x}A^{\mathrm{t}}\| &\le \|\mathbf{x}A_k^{\mathrm{t}} + \mathbf{a}_k - \mathbf{x}A^{\mathrm{t}} - \mathbf{a}\| + \|\mathbf{a}_k - \mathbf{a}\| \\ &= \|\varphi(\mathbf{a}_k, A_k)(\mathbf{x}) - \varphi(\mathbf{a}, A)(\mathbf{x})\| + \|\mathbf{a}_k - \mathbf{a}\| \end{aligned}$$

Hence, given $\varepsilon > 0$ there exists N such that $\|\mathbf{x}A_k^{\mathrm{t}} - \mathbf{x}A^{\mathrm{t}}\| < \varepsilon$ when $k > N$ and $\|\mathbf{x}\| \leq r$. Using $r = 1$,

$$\|A_k^{\mathrm{t}} - A^{\mathrm{t}}\|_{\mathcal{B}} = \sup\{\|\mathbf{x}A_k^{\mathrm{t}} - \mathbf{x}A^{\mathrm{t}}\| : \|\mathbf{x}\| \leq 1\} \leq \varepsilon,$$

and A_k^{t} converges A^{t} in $\mathrm{O}(n)$. It follows that A_k converges to A in $\mathrm{O}(n)$. □

Corollary 6.3.5 *The metric group* $\mathrm{RM}(n)$ *is a manifold of dimension* $n(n+1)/2$ *and has two components.*

Proof. The matrix group $\mathrm{O}(n)$ is a manifold of dimension $n(n-1)/2$ (*Theorem 4.2.10*) and has two components (*Corollary 5.2.5*). *Exercise* 6, p. 149, *Corollary 5.1.7*, and *Theorem 5.1.20* can now be applied to finish the proof. □

EXERCISES

1. Let X be a metric space with metric d. With the product topology on $X \times X$, show that $d : X \times X \to X$ is a continuous function.

2. Show that the compact metric group of isomorphisms for S^n with the metric $d(\mathbf{x}, \mathbf{y}) = \|\mathbf{x} - \mathbf{y}\|$ is isomorphic to $\mathrm{O}(n+1)$. (One of the approaches makes use of some of the ideas in the proof of *Theorem 6.3.3.*)

3. Let G_1 and G_2 be groups and let $y \to \psi_y$ be an algebraic homomorphism of G_2 to the group of algebraic automorphisms of G_1. Show that $G_1 \times G_2$ is a group, called the *semidirect product* and denoted by $G_1 Ⓢ\, G_2$, under the binary operation $(x, y) * (u, v) = (x\psi_y(u), yv)$. (The binary operation on $\mathrm{Auto}(G_2)$ is the composition of functions, and so $\psi_{yv} = \psi_y \circ \psi_v$.)

4. Let G_1 and G_2 be metric groups. Suppose $y \to \psi_y$ is an algebraic homomorphism of G_2 to the group of automorphisms of G_1. Prove That if the function $(x, y) \to \psi_y(x)$ is a continuous function from the product metric space $G_1 \times G_2$ to G_1, then $G_1 Ⓢ\, G_2$ is a metric group with the group multiplication $(x, y) * (u, v) = (x\psi_y(u), yv)$ from *Exercise 3* above.

5. Define $\theta : \mathbb{Z}_2 = \{0, 1\} \to \mathrm{DL}(n, \mathbb{R})$ by $\theta(\gamma) = \mathrm{Diag}\left[(-1)^\gamma, 1, \ldots, 1\right]$. Show that $GL^+(n, \mathbb{R})Ⓢ\, \mathbb{Z}_2$ is a metric group isomorphic to $GL(n, \mathbb{R})$ with $\psi_\gamma(A) = \theta(\gamma)A\theta(\gamma)^{-1}$.

6. Construct $z \to \psi_z$ such that $SL(n, \mathbb{C})Ⓢ\, (\mathbb{C} \setminus \{0\})$ is a metric group isomorphic to $GL(n, \mathbb{C})$.

7. For C in $\mathrm{DL}(n, \mathbb{S})$, let $\psi_C(I + A) = I + CAC^{-1}$ and show that $C \to \psi_C$ is a algebraic homomorphism of $\mathrm{DL}(n, \mathbb{S})$ into the group of automorphisms of $NL(n, \mathbb{S})$. Then prove that the product space $NL(n, \mathbb{S})Ⓢ\, \mathrm{DL}(n, \mathbb{S})$ with the binary operation

$$(I + A, C) * (I + B, D) = \big((I + A)\psi_C(I + B), CD\big)$$

is a metric group isomorphic to $\mathrm{TL}(n, \mathbb{S})$.

8. Let $\mathbf{u}$ and $\mathbf{v}$ be linearly independent vectors in $\mathbb{R}^2$. Consider the triangle with vertices $\mathbf{u}, \mathbf{v}, \mathbf{0}$ in the two-dimensional subspace Span $[\mathbf{u}, \mathbf{v}]$ and let γ denote the angle of the triangle at the vertex $\mathbf{0}$. Use the Law of Cosines to show that
$$\cos(\gamma) = \frac{\mathbf{u} \cdot \mathbf{v}}{\|\mathbf{u}\| \, \|\mathbf{v}\|}$$

9. Show that the metric group $\mathrm{RM}(n)$ is isomorphic to the closed subgroup
$$H = \left\{ \begin{bmatrix} A & \mathbf{a}^t \\ \mathbf{0} & 1 \end{bmatrix} : A \in O(n), \ \mathbf{a} \in \mathbb{R}^n, \text{ and } \mathbf{0} \in \mathbb{R}^n \right\}$$
of $GL(n+1, \mathbb{R})$.

10. Prove that two metric topologies on a set X are equal if and only if they have the same convergent sequences.

11. Prove that if V and V' are finite-dimensional normed linear spaces, then a sequence T_k in the normed linear space $\mathcal{L}(V, V')$ converges to T if and only if it converges uniformly on compact subsets.

12. Let V and V' be finite-dimensional normed linear spaces. Prove that the relative compact-open topology on $\mathcal{L}(V, V')$ as a subset of $\mathcal{C}(V, V')$ is the same as its normed linear topology.

6.4 Metric Groups of Homeomorphisms

When X is a locally compact second countable metric space, we know from *Proposition 6.3.1* that $\mathcal{C}(X, X)$ with the compact-open topology is a semigroup with a continuous binary operation, namely, composition of functions and with an identity element ι. Its group of units is the group H_X of homeomorphisms of X onto itself. Hence, H_X with the compact-open topology is a group with continuous multiplication. So H_X with the compact-open-topology is a metric group, if the inverse function is a continuous function of H_X onto itself.

Proposition 6.4.1 *If X is a compact metric space, then the group H_X of homeomorphisms of X onto itself with the compact-open topology is a metric group.*

Proof. Since composition of functions is continuous by *Proposition 6.3.1* it suffices to prove that $h \to h^{-1}$ is continuous on H_X at ι.

Let h_k be sequence in H_X converging to ι in H_X. If h_k^{-1} does not converge to ι in $\mathcal{C}(X, X)$, then h_k^{-1} does not converge uniformly to ι on X. Thus for some $\varepsilon > 0$, for each positive integer m there exists $x_m \in X$ and $k_m > m$ such that
$$d_X(h_{k_m}^{-1}(x_m), x_m) \geq \varepsilon.$$

Because X is compact, we can assume without loss of generality that x_m converges to x and $h_{k_m}^{-1}(x_m)$ converges to y. Thus $d_X(y,x) \geq \varepsilon$ and $y \neq x$.

Applying *Proposition 6.2.6* shows that

$$x = \lim_{m\to\infty} x_m = \lim_{m\to\infty} \eta(h_{k_m}, h_{k_m}^{-1}(x_m)) = \eta(\iota, y) = y,$$

a contradiction. □

For X locally compact and second countable, the inverse function is not always continuous (see *Exercise* 2, p. 254), but there are two interesting ways around this difficulty. The first places a connectivity hypothesis on the space X, and the second enlarges the compact-open topology to make H_X a metric group. Both will make use of the following definition: In a locally compact metric space X, a *sequence* x_k *goes to infinity* provided that for every compact subset C of X there exists a positive integer N such that x_k is not in C for $k > N$.

A topological space X is *locally connected* at x provided that every open neighborhood of x contains an open connected neighborhood of x. The topological space X is *locally connected* provided that it is locally connected at every point. For example, $\mathbb{R}^n$ is locally connected because $B_r(\mathbf{x})$ is connected for every $r > 0$ and $\mathbf{x} \in \mathbb{R}^n$.

Proposition 6.4.2 *A topological space is locally connected if and only if the components of open sets are open sets.*

Proof. Suppose X is locally connected. Let x be a point in C, a component of an open set U. Then U is an open neighborhood of x and there exists an open connected set V such that $x \in V \subset U$. It follows that $V \cup C$ is a connected subset of U by *Proposition 5.1.6*. By the definition of a component $x \in V \subset C$ and C is an open set in X.

Conversely, suppose the components of open sets are open. If U is an open neighborhood of x in X, then it has a component C containing x and C is an open neighborhood of x contained in U. Thus X is locally connected. □

Theorem 6.4.3 *Let X be a locally compact second countable metric space. If X is locally connected, then the group H_X of homeomorphisms of X onto itself with the compact-open topology is a metric group.*

Proof. As in the proof of *Proposition 6.4.1*, it suffices to show that if h_k is a sequence in H_X converging in the compact-open topology to ι in H_X, then h_k^{-1} converges to ι in the compact-open topology. If not, then for some compact subset C of X and $\varepsilon > 0$, there exists a sequence $y_m \in C$ and subsequence h_{k_m} of h_k such that

$$d_X(h_{k_m}^{-1}(y_m), y_m) \geq \varepsilon.$$

By passing to subsequences for convenience of notation, we can assume by the compactness of C that h_k is a sequence converging to ι in H_X and y_k is a sequence in C converging to y such that $d_X(h_k^{-1}(y_k), y_k) \geq \varepsilon$.

The first step is to show that $h_k^{-1}(y_k)$ goes to infinity. If not, there is another compact set C' and a subsequence y_{k_m} of y_k such that $h_{k_m}^{-1}(y_{k_m})$ is in C'. Without loss of generality, we can assume that $h_{k_m}^{-1}(y_{k_m})$ converges to z in C'. Clearly, $d_X(y, z) \geq \varepsilon$ and $y \neq z$. The same contradiction obtained in the proof of *Proposition 6.4.1* now occurs here, namely,

$$y = \lim_{m\to\infty} y_{k_m} = \lim_{m\to\infty} \eta(h_{k_m}, h_{k_m}^{-1}(y_{k_m})) = \eta(\iota, z) = z.$$

Thus y_k goes to infinity, which will be used later in the proof.

Let W_j and K_j be given by *Proposition 6.2.2.* There exists i such that $C \subset W_i$, and a component U of W_i containing y. Then, U is an open connected neighborhood of y, and U^- is a closed compact connected set such that $U^- \subset K_i \subset W_{i+1}$. The component V of W_{i+1} containing y is open and contains U^- because U^- is connected and contains y. Thus

$$y \in U \subset U^- \subset V \subset V^- \subset K_{i+1} \subset W_{i+2}.$$

Without loss of generality we can assume y_k is in U for all k.

Exercise 3, p. 222, applies to the compact set U^- and the closed set $X \setminus V$, and there exists $r > 0$ such that $\{x : d(x, U^-) < r\}$ and $\{x : d(x, X \setminus V) < r\}$ are disjoint open sets. In particular,

$$U \subset \{x : d(x, U^-) < r\} \subset V.$$

Since the sequence of homeomorphism h_k converges to ι in the compact-open topology, there exists N_1 such that $d(h_k(x), x) < r$ for all x in U^-, when $k \geq N_1$. In particular, $h_k(y_k) \in V$, when $k \geq N_1$.

Consider the sequence of connected open sets $h_k^{-1}(V)$. On the one hand, $h_k^{-1}(V)$ contains $h_k^{-1}(y_k)$ for all k, and there exists $N \geq N_1$ such that $h_k^{-1}(y_k)$ is in the open set $X \setminus K_{i+2}$ when $k \geq N$. Thus

$$h_k^{-1}(V) \cap (X \setminus K_{i+2}) \neq \phi$$

when $k > N$ because $h_k^{-1}(y_k)$ goes to infinity. On the other hand, $h_k(y_k)$ is in V when $k \geq N \geq N_1$, and hence $y_k = h_k^{-1}(h_k(y_k))$ is in $h_k^{-1}(V)$. Therefore,

$$h_k^{-1}(V) \cap W_{i+2} \supset h_k^{-1}(V) \cap U \neq \phi$$

when $k \geq N$. Because $h_k^{-1}(V)$ is connected and both W_{i+2} and $X \setminus K_{i+2}$ are open, it follows from the above that

$$h_k^{-1}(V) \cap \Big(X \setminus (W_{i+2} \cup (X \setminus K_{i+2}))\Big) \neq \phi.$$

Using de Morgan's formulas,

$$X \setminus (W_{i+2} \cup (X \setminus K_{i+2})) = (X \setminus W_{i+2}) \cap K_{i+2} = E.$$

Clearly, E is compact.

The last step is to show that the same contradiction occurs once more. For $k \geq N$, let w_k be an element of V such that $h_k^{-1}(w_k)$ is in E. By passing again to a subsequence, we can assume that w_k converges to w in the compact set V^- and that $h_k^{-1}(w_k)$ converges to z in E. Noting that $w \neq z$ because $V^- \subset W_{i+2}$, we again have the contradiction:

$$w = \lim_{k\to\infty} w_k = \lim_{k\to\infty} \eta(h_k, h_k^{-1}(w_k)) = \eta(\iota, z) = z.$$

Therefore, h_k^{-1} converges to ι in H_X, and H_X with the compact-open topology is a metric group. □

Corollary 6.4.4 *Let X be a second countable metric space. If X is a manifold, then H_X with the compact-open topology is a metric group.*

Proof. A manifold is locally compact and locally connected because open sets of $\mathbb{R}^n$ are locally compact and locally connected. □

Corollary 6.4.5 *Let G be a locally compact second countable metric group, and let A_G be its group of automorphisms. If G is locally connected, then A_G with the compact-open topology is a metric group.*

Proof. A_G is a subgroup of H_G.

Let Y be a compact metric space and let q be a point in Y. Then $\{q\}$ is a closed subset of Y and $X = Y \setminus \{q\}$ is an open subset of Y. A subset C of X is a compact subset of X if and only if it is a compact subset of Y. If x is in X, then there exists $r > 0$ such that $B_r(x) \cap B_r(q) = \phi$. Consequently, $B_r(x)^-$ is compact and contained in X. It follows that X with the relative topology is a locally compact metric space. Since Y is second countable by *Corollary 5.3.3*, X is second countable by *Exercise* 2, p. 213. Thus removing a point from a compact metric space is a simple way to construct a locally compact second countable metric space.

Conversely, if X is a locally compact second countable metric space, can we add a point q to X and make it a compact metric space? To prove this converse, we need a definition and some preliminary results.

Proposition 6.4.6 *Let Y be a compact metric space, let q be a point in Y, and let $X = Y \setminus \{q\}$. A sequence x_k of points in X converges to q in Y if and only if the sequence x_k goes to infinity in the locally compact space X.*

Proof. Suppose x_k converges to q, and let C be a compact subset of X. Then C is a compact and closed subset of Y. Hence, $Y \setminus C$ is an open neighborhood of q and there exists N such that x_k is in $Y \setminus C$ for $k > N$. Therefore, x_k goes to infinity in X.

Suppose x_k goes to infinity in X, and let U be an open neighborhood of q in Y. Then $C = Y \setminus U$ is a closed subset of Y, and thus a compact subset of Y and of X. So there exists N such that x_k is not in C for $k > N$. Thus x_k is in U for $k > N$ and x_k converges to q in Y. □

Corollary 6.4.7 *Let Y and Y' be compact metric spaces, and let q and q' be points in Y and Y' respectively. If f is a homeomorphism from $Y \setminus \{q\}$ onto $Y' \setminus \{q'\}$, then $\tilde{f} : Y \to Y'$ defined by*

$$\tilde{f}(y) = \begin{cases} f(y) & \text{if } y \in X \\ q' & \text{if } y = q \end{cases}$$

is a homeomorphism of Y onto Y'.

Proof. The function $\tilde{f}$ is obviously continuous at y when $y \neq q$. Observe that a sequence x_k in $Y \setminus \{q\}$ goes to infinity if and only if $f(x_k)$ goes to infinity in $Y' \setminus \{q'\}$. It follows from the proposition that $\tilde{f}$ is continuous at q and a homeomorphism by *Exercise* 2, p. 61. □

Theorem 6.4.8 *If X is a locally compact second countable metric space that is not compact, then there exists a compact metric space Y and a point q in Y such that X is homeomorphic to $Y \setminus \{q\}$. Furthermore, if Y' is a compact metric space with q' in Y' such that X is homeomorphic to $Y' \setminus \{q'\}$, then there exists a homeomorphism of Y onto Y' mapping q to q'.*

Proof. Since X is locally compact and second countable, it is separable, and there exists a countable dense subset $D = \{x_1, \ldots, x_k, \ldots\}$ of X. Recall the function

$$\nu(x) = \min\{n \in \mathbb{Z}^+ : B_{1/\nu(x)}(x)^- \text{ is compact}\}$$

used in the proof of *Proposition 6.2.1*. Set

$$\Lambda = \{(i, j) \in \mathbb{Z}^2 : i \geq 1; \text{and } j \geq \nu(x_i)\}.$$

Then Λ is countable and $[0, 1]^\Lambda$ is a compact metric space by *Theorem 5.3.6*.

For each (i, j) in Λ, let $f_{(i,j)} : X \to \mathbb{R}$ be a continuous function such that

(a) $0 \leq f_{(i,j)}(x) \leq 1$ for all $x \in X$

(b) $f_{(i,j)}(x) = 0$ if and only if $d(x, x_i) \geq 1/j$.

Such functions exist by *Exercise* 4, p. 27. Let $\mathbf{f} : X \to [0, 1]^\Lambda$ be the function whose (i, j) coordinate function is $f_{(i,j)}$. The function $\mathbf{f}$ is continuous by *Proposition 5.1.14*.

To show that $\mathbf{f}$ is one-to-one, consider distinct points x and x' in X. There exists $r > 0$ such that $B_r(x) \cap B_r(x') = \phi$ and $B_r(x)^-$ is compact. Choose $\beta \in \mathbb{Z}^+$ such that $1/\beta < r/2$. There exists $x_\alpha \in D \cap B_{1/\beta}(x)$. It follows that

$$x \in B_{1/\beta}(x_\alpha) \subset B_r(x),$$

and so $B_{1/\beta}(x_\alpha)^-$ is compact. Hence, $\beta \geq \nu(x_\alpha)$ and (α, β) is in Λ. Consequently, $f_{(\alpha,\beta)}(x) > 0 = f_{(\alpha,\beta)}(x')$ and $\mathbf{f}(x) \neq \mathbf{f}(x')$. Therefore, $\mathbf{f}$ is one-to-one.

To prove that $\mathbf{f}$ is a homeomorphism of X onto $\mathbf{f}(X)$ it suffices to show that $\mathbf{f}$ is an open function of X onto $\mathbf{f}(X)$. Consider an open set U of X and x in

U. There exists $r > 0$ such that $B_r(x)^- \subset U$ and $B_r(x)^-$ is compact. As in the previous paragraph, construct $(\alpha, \beta) \in \Lambda$ such that $x \in B_{1/\beta}(x_\alpha) \subset B_r(x)$. Set

$$V = \{\mathbf{w} \in [0,1]^\Lambda : w_{(\alpha,\beta)} > 0\}.$$

Then,

$$\mathbf{f}(x) \in V \cap \mathbf{f}(X) = \mathbf{f}\left(B_{1/\beta}(x_\alpha)\right) \subset \mathbf{f}(B_r(x)) \subset \mathbf{f}(U).$$

Note that V is an open subset of $[0,1]^\Lambda$ because $(0,1]$ is an open subset of $[0,1]$, and so $V \cap f(X)$ is an open subset of $\mathbf{f}(X)$. Thus $\mathbf{f}(U)$ is a neighborhood of $\mathbf{f}(x)$. Since x was an arbitrary point in an arbitrary open subset U of X, the set $\mathbf{f}(U)$ is an open subset of $\mathbf{f}(X)$ and $\mathbf{f}$ is a homeomorphism of X on $\mathbf{f}(X)$.

Set $Y = \mathbf{f}(X)^-$, which is compact. Because $\mathbf{f}(X)$ is homeomorphic to X, which is not compact, $\mathbf{f}(X) \neq Y$, and there exists at least one point in Y that is not in $\mathbf{f}(X)$. It will be shown that there exists $\mathbf{q}$ in Y such that $\mathbf{f}(X) = Y \setminus \{\mathbf{q}\}$.

Let $\mathbf{y}$ be a point in Y that is not in $\mathbf{f}(X)$. Then there exists a sequence x_k in X such that $\mathbf{f}(x_k)$ converges $\mathbf{y}$. If x_k does not go to infinity in X, there exists a compact subset C of X containing a subsequence of x_k. By passing to a subsequence, we can assume that x_k converges to x in C. It follows that $\mathbf{f}(x_k)$ converges to $\mathbf{f}(x)$ by continuity. Hence, $\mathbf{y} = \mathbf{f}(x)$, an element of $\mathbf{f}(X)$, contradicting our choice of $\mathbf{y}$. Therefore, x_k goes to infinity in X.

Since x_k goes to infinity in X, for any (i,j) in Λ there exists N such that x_k is not in $B_{1/j}(x_i)^-$ and $f_{(i,j)}(x_k) = 0$, when $k > N$. It follows that $y_{(i,j)} = 0$ for all (i,j) in Λ. Therefore, the only point in $Y \setminus \mathbf{f}(X)$ is the point $\mathbf{q}$ such that $q_{(i,j)} = 0$ for all $(i,j) \in \Lambda$. This completes the proof of the first statement of the theorem.

The second statement is an immediate consequence of *Corollary 6.4.7* □

Given a locally compact second countable metric space X, a compact metric space Y and a point q in Y such that X is homeomorphic to $Y \setminus \{q\}$ is called a *one-point compactification* of X and the point q is called the *point at infinity. Theorem 6.4.8* tells us not only does a locally compact second countable metric space always have a metric one-point compactification, but it is also topological invariant of X. One-point compactifications can be constructed far more generally, but they are not usually metric spaces.

One-point compactifications can sometimes be constructed more simply than in the proof of *Theorem 6.4.8.* For example, it is easily shown that the one point compactification of $\mathbb{R}^n$ is S^n. Define $\mathbb{R}^n \to S^n \subset \mathbb{R}^{n+1}$ by

$$y_i = \frac{2x_i}{\|\mathbf{x}\|^2 + 1}$$

and

$$y_{n+1} = \frac{\|\mathbf{x}\|^2 - 1}{\|\mathbf{x}\|^2 + 1}.$$

Then, $f(\mathbb{R}^n) = S^n \setminus \{(0, \ldots, 0, 1)\}$ and the coordinate functions of f^{-1} are given by

$$x_i = \frac{y_i}{1 - y_{n+1}}.$$

By *Theorem 6.4.8*, we can assume that $X = Y \setminus \{q\}$, where Y is a compact metric space with metric d_Y. So d_Y restricted to X is a metric for X, but a special one because not all metrics for X extend to metrics for Y. By *Proposition 6.4.1*, the group H_Y with the compact-open topology is a metric group with metric $d(f_1, f_2) = \sup\{d_Y(f_1(y), f_2(y)) : y \in Y\}$. Set $G = \{f \in H_Y : f(q) = q\}$. Observe that G is a closed subgroup of H_Y.

It is easy to see that $d'(f_1, f_2)) = \sup\{d_Y(f_1(x), f_2(x)) : x \in X\}$ defines a metric on H_X, but not a metric for the compact-open topology on $\mathcal{C}(X, X)$. In fact, the function $f \to \tilde{f}$ defined by *Corollary 6.4.7* is an isometry and an algebraic isomorphism of H_X onto G. An immediate consequence is that H_X is a metric group with the d' metric. This proves the following

Proposition 6.4.9 *If X is a locally compact second countable metrizable topological space, then there exists a metric d_X for X such that*

$$d'(f_1, f_2)) = \sup\{d_X(f_1(x), f_2(x)) : x \in X\}$$

defines a metric on H_X and with this metric H_X is a metric group.

For a more general and extensive treatment of topologies on groups of homeomorphisms see [2], Chapter X, Section 3.5.

EXERCISES

1. Use the unit interval to show that H_X need not be a closed set in $\mathcal{C}(X, X)$, when X is a compact metric space.

2. Let $X = \{0\} \cup \{2^n : n \in \mathbb{Z}\}$ with the relative topology from $\mathbb{R}$. Define a sequence of functions $f_k : X \to X$ by $f_k(0) = 0$ and

$$f_k(2^n) = \begin{cases} 2^n & |n| < k \\ 2^{-n} & n = k \\ 2^{n-1} & n > k \\ 2^{n-1} & n \leq -k. \end{cases}$$

   Show the following:

   (a) X is a locally compact second countable space.

   (b) f_k is a homeomorphism of X onto itself for $k \in \mathbb{Z}^+$.

   (c) f_k converges to ι in the compact-open topology on H_X.

   (d) f_k^{-1} does not converge to ι in the compact-open topology on H_X.

3. Let f be an element of $H_{\mathbb{R}}$. Prove the following:

   (a) If $a < b < c$ and $f(a) < f(c)$, then $f(a) < f(b) < f(c)$. (One approach is to use *Exercise* 5, p. 192.)

(b) If $a < c$ and $f(a) < f(c)$, then f is increasing on the closed interval $[a, c]$.

(c) The function f is either increasing or decreasing on $\mathbb{R}$.

4. Let f and g be increasing functions in $H_{\mathbb{R}}$. For $0 \leq t \leq 1$, set $h_t(x) = tf(x) + (1-t)g(x)$. Prove that $t \mapsto h_t$ is a continuous function from $[0,1]$ to $H_{\mathbb{R}}$.

5. Prove that $H_{\mathbb{R}}$ has two components.

6. Show that for $\mathbb{R}$ the topology given by *Proposition 6.4.9* does not equal the compact open topology on $H_{\mathbb{R}}$. (One approach is to use *Exercise 10*, p. 248.)

7. Let X be a locally compact topological space with topology $\mathcal{T}$, and let q be a point not in X. Set $Y = X \cup \{q\}$. Let $\mathcal{T}'$ denote the collection of subsets U of Y satisfying the following conditions:

   (a) $U \cap X$ is $\mathcal{T}$.

   (b) If q is in U, then $X \setminus U$ is a compact subset of X.

   Prove that $\mathcal{T}'$ is a topology for Y and that Y is compact with this topology. Also show that $\mathcal{T}'$ is a Hausdorff topology if and only if $\mathcal{T}$ s a Hausdorff topology.

8. Show that $\mathrm{GL}(n, \mathbb{R})$ is isomorphic to the group of automorphisms of the group $\mathbb{R}^n$ with the compact-open topology.

6.5 Metric Groups of Homomorphisms

Suppose G and G' are Abelian groups written multiplicatively and φ and ψ are algebraic homomorphisms of G to G'. Define $\varphi\psi : G \to G'$ by

$$(\varphi\psi)(x) = \varphi(x)\psi(x). \tag{6.16}$$

Then

$$\begin{aligned}
(\varphi\psi)(xy) &= \varphi(xy)\psi(xy) \\
&= \varphi(x)\varphi(y)\psi(x)\psi(y) \\
&= \varphi(x)\psi(x)\varphi(y)\psi(y) \\
&= (\varphi\psi)(x)(\varphi\psi)(y)
\end{aligned}$$

and $\varphi\psi$ is also an algebraic homomorphism of G to G'. Thus there is a natural binary operation on the algebraic homomorphisms of one Abelian Group to another. It is easily seen to be associative because multiplication is associative in groups. In fact, it is an Abelian group. The identity is the trivial algebraic homomorphism that sends every element of G to the identity of G' and $(\varphi^{-1})(x) = \varphi(x)^{-1}$.

If G and G' are Abelian metric groups and φ and ψ are homomorphisms of G to G', then $\varphi\psi$ is continuous because multiplication is continuous on $G' \times G'$ and $\varphi\psi$ is a homomorphism of G to G'. Similarly, φ^{-1} is continuous because the inverse function is continuous on G'. Thus the set of homomorphisms of G to G' is a subgroup of the group of algebraic homomorphisms of G to G' and will denoted by $\text{Hom}(G, G')$. Clearly, $\text{Hom}(G, G') \subset \mathcal{C}(G, G')$.

Proposition 6.5.1 *If G and G' are Abelian metric groups with G locally compact and second countable, then* $\text{Hom}(G, G')$ *is a closed subset of* $\mathcal{C}(G, G')$ *with the compact-open topology.*

Proof. Let φ be in the closure of $\text{Hom}(G, G')$ in $\mathcal{C}(G, G')$. If φ is not in $\text{Hom}(G, G')$, then there exist x and y in G such that $\varphi(xy) \neq \varphi(x)\varphi(y)$. So there exist disjoint open sets U and V containing $\varphi(xy)$ and $\varphi(x)\varphi(y)$, respectively. Then there exist open neighborhoods V_1 and V_2 of $\varphi(x)$ and $\varphi(y)$, respectively, such that $V_1V_2 \subset V$. Since finite sets are compact sets,

$$N(\{xy\}, U) \cap N(\{x\}, V_1) \cap N(\{y\}, V_2\})$$

is an open neighborhood of φ in the compact-open topology. Because $\varphi \in \text{Hom}(G, G')^-$, there exists ψ such that

$$\psi \in \text{Hom}(G, G') \cap N(\{xy\}, U) \cap N(\{x\}, V_1) \cap N(\{y\}, V_2\}).$$

It follows that the point $\psi(xy) = \psi(x)\psi(y)$ is in both U and V. This is a contradiction because U and V are disjoint. Therefore, φ is in $\text{Hom}(G, G')$. □

Proposition 6.5.2 *If G and G' are Abelian metric groups with G locally compact and second countable, then the group* $\text{Hom}(G, G')$ *with the compact-open topology is a metric group.*

Proof. Obviously, $\text{Hom}(G, G')$ with the compact-open topology is a metric space because it is subset of the metric space $\mathcal{C}(G, G')$. Suppose φ_k and ψ_k are sequences in $\text{Hom}(G, G')$ converging to φ and ψ respectively in $\text{Hom}(G, G')$. It suffices to prove that $\varphi_k\psi_k$ converges to $\varphi\psi$ and φ_k^{-1} converges to φ^{-1} in $\text{Hom}(G, G')$, using uniform convergence on compact subsets of G (*Theorem 6.2.5*).

By *Theorem 1.3.5*, there exists a bounded invariant metric d' for G'. Let C be a compact subset of G. Given $\varepsilon > 0$, there exists N such that $d'(\varphi_k(x), \varphi(x)) < \varepsilon$ and $d'(\psi_k(x), \psi(x)) < \varepsilon$ for all $x \in C$ when $k \geq N$. Then,

$$\begin{aligned} d'(\varphi_k(x)\psi_k(x), \varphi(x)\psi(x)) &\leq d'(\varphi_k(x)\psi_k(x), \varphi_k(x)\psi(x)) \\ &+ d'(\varphi_k(x)\psi(x), \varphi(x)\psi(x)) \\ &= d'(\psi_k(x), \psi(x)) + d'(\varphi_k(x), \varphi(x)) \\ &\leq \varepsilon + \varepsilon = 2\varepsilon, \end{aligned}$$

when x is in C and $k \geq N$, and $\varphi_k(x)\psi_k(x)$ converges uniformly to $\varphi(x)\psi(x)$ on C. Thus multiplication is continuous on $\text{Hom}(G, G')$.

We can assume that φ is the identity of $\text{Hom}(G, G')$ (*Exercise* 5, p. 39), that is, $\varphi(x) = e'$, the identity of G', for all x in G. There exists N such that $d'(\varphi_k(x), \varphi(x)) = d'(\varphi_k(x), e') < \varepsilon$ when x is in C and $k \geq N$. Then,

$$\begin{aligned} d'(\varphi_k^{-1}(x), \varphi^{-1}(x)) &= d'(\varphi_k(x)^{-1}, e') \\ &= d'(\varphi_k(x)\varphi_k(x)^{-1}, \varphi_k(x)e') \\ &= d'(e', \varphi_k(x)) < \varepsilon \end{aligned}$$

when x is in C and $k \geq N$. Thus the inverse function is also continuous on $\text{Hom}(G, G')$. □

Let V and V' be finite-dimensional normed linear spaces over $\mathbb{R}$. In particular, V is locally compact and second countable because it is homeomorphic to $\mathbb{R}^n$ for some n. It follows from *Propositions 2.4.1 and 2.4.7* that

$$\mathcal{L}(V, V') = \text{Hom}(V, V').$$

Furthermore, the norm linear topology and the compact-open topology are equal by *Exercise* 12, p. 248. Thus $\text{Hom}(G, G')$ with the compact-open topology can be viewed as a generalization of the normed linear space $\mathcal{L}(V, V')$ when V is a finite-dimensional linear space over $\mathbb{R}$.

Henceforth, $\text{Hom}(G, G')$ will denote the metric group obtained by endowing $\text{Hom}(G, G')$ with the compact-open topology. The identity of $\text{Hom}(G, G')$ will be denoted by ζ, where $\zeta(x) = e'$ for all $x \in G$. It is obviously continuous and will be referred to as the *trivial homomorphism.*

Theorem 6.5.3 *If G_1 and G_2 are Abelian locally compact second countable metric groups, and let H is an Abelian metric group, then the metric groups* $\text{Hom}(G_1 \times G_2, H)$ *and* $\text{Hom}(G_1, H) \times \text{Hom}(G_2, H)$ *are isomorphic.*

Proof. The metric group $G_1 \times G_2$ is locally compact and second countable. So $\text{Hom}(G_1 \times G_2, H)$ is a metric group by *Proposition 6.5.2*.

Let e_1 and e_2 denote the identity elements of G_1 and G_2. Given $\varphi \in \text{Hom}(G_1 \times G_2, H)$, define $\theta_1(\varphi)$ by $\theta_1(\varphi)(x) = \varphi(x, e_2)$. It is easy to see that $\theta_1(\varphi)$ is in $\text{Hom}(G_1, H)$ and that the function $\theta_1 : \text{Hom}(G_1 \times G_2, H) \to \text{Hom}(G_1, H)$ is an algebraic homomorphism.

To show that θ_1 is continuous, consider a convergent sequence φ_k in the group $\text{Hom}(G_1 \times G_2, H)$ with limit φ. Given C_1 a compact subset of G_1, it must be shown that $\theta_1(\varphi_k)$ converges uniformly to $\theta_1(\varphi)$ on C_1. Since $C = C_1 \times \{e_2\}$ is a compact subset of $G_1 \times G_2$, the sequence φ_k converges uniformly on C to φ. Given $\varepsilon > 0$, there exists N such that for $k \geq N$ and $x \in C_1$

$$d_H(\theta_1(\varphi_k)(x), \theta_1(\varphi)(x_1)) = d_H(\varphi_k(x, e_2), \varphi(x, e_2)) < \varepsilon$$

for $x \in C_1$ and $k \geq N$. Therefore, $\theta_1(\varphi_k)$ converges to $\theta_1(\varphi)$ in $\text{Hom}(G_1, H)$ and θ_1 is continuous.

Similarly, $\theta_2(\varphi)(x) = \varphi(e_1, y)$ defines a homomorphism $\theta_2 : \mathrm{Hom}(G_1 \times G_2, H) \to \mathrm{Hom}(G_2, H)$. It follows that $\Theta(\varphi) = (\theta_1(\varphi), \theta_2(\varphi))$ is homomorphism of $\mathrm{Hom}(G_1 \times G_2, H)$ to $\mathrm{Hom}(G_1, H) \times \mathrm{Hom}(G_2, H)$.

Let $\boldsymbol{\varphi} = (\varphi_1, \varphi_2)$ be an element of $\mathrm{Hom}(G_1, H) \times \mathrm{Hom}(G_2, H)$. An easy calculation shows that $\Theta^{-1}(\boldsymbol{\varphi})$ is the algebraic homomorphism defined by

$$\Theta^{-1}(\boldsymbol{\varphi})(x, y) = \varphi_1(x)\varphi_2(y).$$

So Θ is an algebraic isomorphism of $\mathrm{Hom}(G_1 \times G_2, H)$ onto $\mathrm{Hom}(G_1, H) \times \mathrm{Hom}(G_2, H)$.

To complete the proof it suffices to show that Θ^{-1} is continuous. Without loss of generality, it can be assumed that d_H is a bounded invariant metric on H. Let (φ_k, ψ_k) be a convergent sequence in $\mathrm{Hom}(G_1, H) \times \mathrm{Hom}(G_2, H)$ with limit (φ, ψ). It follows that φ_k and ψ_k converge to φ and ψ in $\mathrm{Hom}(G_1, H)$ and $\mathrm{Hom}(G_2, H)$, respectively. Hence φ_k converges uniformly to φ on every compact subset of G_1 and likewise for ψ_k.

Every compact subset of $G_1 \times G_2$ is contained in a compact set of the form $C_1 \times C_2$ with C_1 and C_2 compact subsets of G_1 and G_2, respectively. Consequently, it suffices to show that $\Theta^{-1}(\varphi_k, \psi_k)$ converges to $\Theta^{-1}(\varphi, \psi)$ uniformly on every compact set of the form $C_1 \times C_2$. Because

$$\begin{aligned} d_H\left(\Theta^{-1}(\varphi_k, \psi_k)(x, y), \Theta^{-1}(\varphi, \psi)(x, y)\right) &= \\ d_H\left(\varphi_k(x)\psi_k(y), \varphi(x)\psi(y)\right) &\leq \\ d_H\left(\varphi_k(x)\psi_k(y), \varphi(x)\psi_k(y)\right) + d_H\left(\varphi(x)\psi_k(y), \varphi(x)\psi(y)\right) &= \\ d_H\left(\varphi_k(x), \varphi(x)\right) + d_H\left(\psi_k(y), \psi(y)\right), & \end{aligned}$$

the uniform convergence of φ_k and ψ_k to φ and ψ on the compact sets C_1 and C_2, respectively, implies that $\Theta^{-1}(\varphi_k, \psi_k)$ converges uniformly to $\Theta^{-1}(\varphi, \psi)$ on the compact set $C_1 \times C_2$. Therefore, $\Theta^{-1}(\varphi_k, \psi_k)$ converges to $\Theta^{-1}(\varphi, \psi)$ in $\mathrm{Hom}(G_1 \times G_2, H)$ and Θ^{-1} is continuous. □

Corollary 6.5.4 *If $G_1, \ldots, G_m$ are Abelian locally compact second countable metric groups, and H is an Abelian metric group, then* $\mathrm{Hom}(G_1 \times \ldots \times G_m, H)$ *and* $\mathrm{Hom}(G_1, H) \times \ldots \times \mathrm{Hom}(G_m, H)$ *are isomorphic the metric groups.*

Proof. Use induction. □

A subset F of $\mathcal{C}(X, Y)$ is *equicontinuous at a point z* of X provided that for every $\varepsilon > 0$ there exists $\delta > 0$ such that $d_Y(f(x), f(z)) < \varepsilon$ for all f in F and all x in X such that $d_X(x, y) < \delta$.

Proposition 6.5.5 *Let G and G' be metric groups with left-invariant metrics d and d'. A set F of homomorphisms of G to G' is equicontinuous on G if and only if it is equicontinuous at e, the identity of G.*

Proof. Suppose F is equicontinuous at e. So given $\varepsilon > 0$, there exists $\delta > 0$ such that $d(\varphi(x), e) < \varepsilon$ for all φ in F and all x in G such that $d_X(x, e) < \delta$.

Observe that $d(x, y) < \delta$ if and only if $d(y^{-1}x, e) < \delta$, and $d'(\varphi(x), \varphi(y)) < \varepsilon$ if and only if $d'(\varphi(y^{-1}x), e') < \varepsilon$. Thus $d(x, y) < \delta$ implies $d'(\varphi(x), \varphi(y)) < \varepsilon$ for all φ in F. The other half is trivial. □

Proposition 6.5.6 *Let G and G' be Abelian metric groups with G locally compact and second countable, let d and d' be invariant metrics for G and G', and let F be a subset of* $\text{Hom}(G, G')$. *The closure of F is a compact subset of* $\text{Hom}(G, G')$ *if and only if the following conditions are satisfied:*

(a) For every x in X the set $\eta(F \times \{x\}) = \{\varphi(x) : \varphi \in F\}$ has a compact closure in G'.

(b) F is equicontinuous at the identity of G.

Proof. Use *Corollary 6.2.8*, *Proposition 6.5.5*, and *Proposition 6.5.1*. □

The next theorem is the most important theorem in this chapter. It is the foundation for the study of character groups and duality of Abelian local compact second countable metric groups in Chapters 7 and 8. More generally, the same result holds with the words " Abelian locally compact second countable metric group" replace by " Abelian locally compact Hausdorff topological group." One place a proof of this result, which plays the same foundational role in this more general context, can be found is in Chapter 3 of [11].

Theorem 6.5.7 *If G is an Abelian locally compact second countable metric group, then the group* $\text{Hom}(G, \mathbb{K})$ *is also an Abelian locally compact second countable metric group.*

Proof. Since $x \to e^{2\pi i x}$ is a homeomorphism of the open interval $(-1/2, 1/2)$ onto $\mathbb{K} \setminus \{-1\}$, we can make use of the interval structure on the real line. For $r \leq 1/2$, set

$$U_r = \{e^{2\pi i x} : |x| < r\}.$$

So $U_{1/2} = \mathbb{K} \setminus \{-1\}$ and $U_{1/4}$ is the open arc of $\mathbb{K}$ from $-i$ to i that includes 1. (This family of open sets in $\mathbb{K}$ will be used in several other proofs.)

If C is a compact neighborhood of e in G, then $V = N(C, U_\alpha) \cap \text{Hom}(G, \mathbb{K})$ is an open neighborhood of ζ in $\text{Hom}(G, \mathbb{K})$, because $\zeta(x) = 1$ for all $x \in G$. Note that $V^- \subset \text{Hom}(G, \mathbb{K})$ because $\text{Hom}(G, \mathbb{K})$ is closed in $\mathcal{C}(G, \mathbb{K})$ by *Proposition 6.5.1*. The heart of the proof is using *Proposition 6.5.6* to show that V^- is a compact subset of $\text{Hom}(G, \mathbb{K})$, when $0 < \alpha < 1/4$.

Because $\mathbb{K}$ is compact, condition *(a)* is obviously satisfied. So it remains to prove that V is an equicontinuous family of functions at e using invariant metrics for G and $\mathbb{K}$. Let d be an invariant metric for G and let $d'(z, w) = |z - w|$, which is an invariant metric for $\mathbb{K}$.

Given $\varepsilon > 0$, there exists β such that $0 < \beta < 1/4$ and $U_\beta \subset B_\varepsilon(1)$. To prove that V is an equicontinuous family of functions at e, it suffices to find $\delta > 0$ such that $\varphi(x) \in U_\beta$ for all φ in V and all x in G such that $d(x, e) < \delta$.

Observe that there exists a positive integer m such that $\alpha < m\beta < 1/2$ because $1/4 < 1/2 - \alpha$ and $0 < \beta < 1/4$. There exists $\delta > 0$ such that

$$\underbrace{B_\delta(e)\ldots B_\delta(e)}_{m \text{ terms}} \subset C$$

by *Exercise* 9, p. 39. Obviously, $B_\delta(e) \subset C$ and for $1 < k < m$

$$\underbrace{B_\delta(e)\ldots B_\delta(e)}_{k \text{ terms}} \subset C.$$

To complete the proof that V is an equicontinuous family of functions at e, it will be shown that $d(x,e) < \delta$ implies $\varphi(x) \in U_\beta$ for all $\varphi \in V$. If this not the case, there exists $\varphi \in V$ and $x \in B_\delta(e)$ such that $\varphi(x)$ is not in U_β. Note that $\varphi(x)$ is in U_α because $B_\delta(e) \subset C$ and φ is in V. So $\varphi(x) = e^{2\pi\gamma}$ with $\beta < \gamma < \alpha$.

As in the previous paragraph, there exists a positive integer k such that $\alpha < k\gamma < 1/2$. If $k > m$, then $\alpha < m\gamma < 1/2$ because $\beta < \gamma$ implies that $\alpha < m\beta < m\gamma < k\gamma < 1/2$. Consequently, $\alpha < k\gamma < 1/2$ for some positive integer k such that $k \leq m$. It follows that x^k is in C and

$$\varphi(x^k) = \varphi(x)^k = (e^{2\pi\gamma})^k = e^{2\pi k\gamma} \in U_\alpha.$$

But this is impossible because $\alpha < k\gamma < 1/2$. Therefore, given a compact neighborhood C of the identity in G and $0 < \alpha < 1/4$, the family of functions $V = N(C, U_\alpha) \cap \operatorname{Hom}(G, \mathbb{K})$ is equicontinuous at e, and V^- is a compact neighborhood of ζ in $\operatorname{Hom}(G, \mathbb{K})$ by *Proposition 6.5.6.* It follows that $\operatorname{Hom}(G, \mathbb{K})$ is a locally compact Abelian metric group.

To prove that $\operatorname{Hom}(G, \mathbb{K})$ is second countable, let $\nu = \nu(e) = \min\{n \in \mathbb{Z}^+ : B_{1/n}(e)^- \text{ is compact}\}$ and let $C_m = B_{1/m}(e)^-$ for $m \geq \nu$. Consider $V_m = N(C_m, U_\alpha) \cap \operatorname{Hom}(G, \mathbb{K})$ for $0 < \alpha < 1/4$. It follows from the first part of the proof that V_m^-, $m \geq \nu$ is a sequence of compact subsets of $\operatorname{Hom}(G, \mathbb{K})$.

Fix α satisfying $0 < \alpha < 1/4$. Let φ be an element of $\operatorname{Hom}(G, \mathbb{K})$. By continuity, there exists $r > 0$ such that $\varphi(B_r(e)) \subset U_\alpha$. There exists $m \geq \nu$ such that $1/m < r$. Hence, $\varphi(C_m) \subset U_\alpha$ and φ is in V_m. Therefore, $G = \bigcup_{m=\nu}^{\infty} V_m^-$ and G is σ-compact. Since $\operatorname{Hom}(G, \mathbb{K})$ is locally compact, it follows that $\operatorname{Hom}(G, \mathbb{K})$ is second countable by *Proposition 6.2.1.* □

The group $\operatorname{Hom}(G, \mathbb{K})$ is particularly interesting because starting with G, an Abelian locally compact second countable metric group, $\operatorname{Hom}(G, \mathbb{K})$ is also an Abelian locally compact second countable metric group. So the process can be iterated and $\operatorname{Hom}(\operatorname{Hom}(G, \mathbb{K}), \mathbb{K})$ is also an Abelian locally compact second countable metric group. When G is an Abelian locally compact second countable metric group, $\operatorname{Hom}(G, \mathbb{K})$ is by definition the *character group* or *dual group* of G. The character group is frequently denoted by $\widehat{G}$ and the elements of $\widehat{G}$ are called *characters. Corollary 6.5.4* provides us with one important basic fact about character groups, and we close the section with the character group version of this result.

Theorem 6.5.8 *If $G_1, \ldots, G_m$ are Abelian locally compact second countable metric groups, then $G = G_1 \times \ldots \times G_m$ is an Abelian locally compact second countable metric group and $\widehat{G}$ is isomorphic to $\widehat{G_1} \times \ldots \times \widehat{G_m}$.*

EXERCISES

1. Let G be an Abelian locally compact separable metric group and let H_1 and H_2 be Abelian metric groups. Prove that $\mathrm{Hom}(G, H_1 \times H_2)$ is isomorphic to $\mathrm{Hom}(G, H_1) \times \mathrm{Hom}(G, H_2)$ with the product topology.

2. Let G be an Abelian locally compact separable metric group. Show that the groups $\mathrm{Hom}(G, \mathbb{K}^n)$, for $n \in \mathbb{Z}^+$, are Abelian locally compact separable metric groups.

3. Prove that $\mathrm{Hom}(\mathbb{R}^n, \mathbb{K}^m)$ is isomorphic to $\mathbb{R}^{nm}$.

4. Prove that $\mathrm{Hom}(\mathbb{K}^n, \mathbb{K}^m)$ is isomorphic to $\mathbb{Z}^{nm}$.

5. Let X and Y be metric spaces with X compact, and let F be a subset of $\mathcal{C}(X, Y)$ Prove that F is equicontinuous on X if and only if F is equicontinuous at every point of X.

6. Let G and G' be Abelian locally compact second countable metric groups, and let H be an Abelian metric group. Show that if $\theta : G \to G'$ is a homomorphism, then setting $\widehat{\theta}(\varphi') = \varphi' \circ \theta$ for $\varphi' \in \mathrm{Hom}(G', H)$ defines a homomorphism of $\mathrm{Hom}(G', H)$ to $\mathrm{Hom}(G, H)$ called the *dual homomorphism.*

7. Let G_1, G_2, and G_3 be Abelian locally compact second countable metric groups, and let H be an Abelian metric group. Using the definition of $\widehat{\theta}$ from *Exercise 6* above, show that if $\theta : G_1 \to G_2$ and $\psi : G_2 \to G_3$ are homomorphisms, then $\widehat{\psi \circ \theta} = \widehat{\theta} \circ \widehat{\psi}$.

8. Let G and G' be Abelian locally compact second countable metric groups and let H be an Abelian metric group. Show that if $\theta : G \to G'$ is an isomorphism of G onto G', then $\widehat{\theta}$ is an isomorphism of $\mathrm{Hom}(G', H)$ onto $\mathrm{Hom}(G, H)$.

Chapter 7

Compact Groups

The primary goal in this chapter is to prove in Section 7.4 that the characters of a compact Abelian metric group separate the points of the group. This is a particularly deep result and will require a substantive study of integral equations (Section 7.2) and their eigenfunctions (Section 7.3). Moreover, this theorem will be absolutely critical for the study of duality in Chapter 8, the next and final chapter.

This chapter begins with the development of an invariant integration theory for continuous real-valued functions on a compact metric group G. Specifically, we prove the existence of an integral $\int f(x)\,dx$ with the same basic properties as the Riemann integral of a continuous real-valued function on a closed interval. This integral has the important additional property that $\int f(xa)\,dx = \int f(x)\,dx$ for all a in G. It is the primary tool used in the rest of the chapter to obtain fundamental results about compact metric groups.

The study of the integral equation $\lambda\theta(x) = \int \kappa(x,y)\theta(x)\,dy$, starts in Section 7.2. In this equation, κ is a given continuous function and the unknowns are the continuous function θ, called an *eigenfunction*, and the real number λ that is required to be nonzero. The main result of the section is that there is at least one solution when κ is not the zero function. The section concludes with a constructive description of all the solutions.

Section 7.3 addresses the question of which continuous real-valued functions can be approximated by eigenfunctions of a given integral equation. When the initial approximation results are applied to an integral equation such that $\kappa(xa,ya) = \kappa(x,y)$ for all x, y, a in a compact metric group G, a connection between the eigenfunctions and homomorphisms of G to the orthogonal group $\mathrm{O}(n)$ emerges, making it possible to approximate every continuous real-valued function using homomorphisms of G to $\mathrm{O}(n)$.

In Section 7.4, the final theorem of Section 7.3 is applied to compact Abelian groups to prove that their characters separate their points. After also showing that the character groups of a compact Abelian metric group is a countable discrete group, the character groups of familiar compact Abelian groups are determined.

The first three sections of this chapter can also be used as a foundation for the study of homomorphisms of G to the unitary group $\mathrm{U}(n)$, called *unitary representations.* The last section takes advantage of this opportunity to at least establish a number of basic properties of unitary representations of compact metric groups including the Peter-Weyl Theorem.

7.1 Invariant Means

Let G be a compact metric group. As before $\mathcal{C}(G)$ denotes the normed linear space of continuous real-valued functions on G with norm $\|f\|_s = \sup\{|f(x)| : x \in G\}$. For functions f and g in $\mathcal{C}(G)$, the inequality $f \leq g$ or equivalently $g \geq f$ means $f(x) \leq g(x)$ for all x in G. It is also convenient to abuse notation and denote constant functions by their value. Depending on the context, 0 or 1 could denote the number 0 or 1 or the constant function with value 0 or 1.

A *right-invariant integral* on a compact metric group G is a function $f \to \int f(x)\,dx$ from $\mathcal{C}(G)$ to $\mathbb{R}$ the following properties:

(a) $\int cf(x)\,dx = c\int f(x)\,dx$ for all $f \in \mathcal{C}(G)$ and $c \in \mathbb{R}$.

(b) $\int (f(x) + g(x))\,dx = \int f(x)\,dx + \int g(x)\,dx$ for all $f, g \in \mathcal{C}(G)$.

(c) $\int 1\,dx = 1$.

(d) $\int f(x)\,dx \geq 0$, when $f \geq 0$.

(e) $f(xa)\,dx = \int f(x)\,dx$ for all $f \in \mathcal{C}(G)$ and $a \in G$.

The first proposition contains several easy consequences of the definition of a right-invariant integral that will be useful later in the chapter. The proof is left to the reader.

Proposition 7.1.1 *If $\int f(x)\,dx$ is a right-invariant integral on a compact metric group G, then the following hold:*

(a) If c is a real number, then $\int c\,dx = c$.

(b) If $f \leq g$, then $\int f(x)\,dx \leq \int g(x)\,dx$.

(c) If $f > 0$, then $\int f(x)\,dx > 0$.

(d) For all f in $\mathcal{C}(G)$,

$$\left|\int f(x)\,dx\right| \leq \int |f(x)|\,dx.$$

Proof. Exercise.

The goal of this section is to prove the existence and uniqueness of a right-invariant integral on a compact metric group along with a few other important properties of such an integral. The results can be extended to compact Hausdorff topological groups, and with minor modification, the proofs can also be used in this more general setting. Specifically, the general Ascoli Theorem is needed, and the invariant metrics must be replaced with the natural invariant uniformities of the topological group.

An integral is typically defined by taking finite averages of a function and passing to a limit, usually by using infimums and supremums. For example, in the construction of the Riemann integral of a continuous function on a closed interval, it is shown that the infimum of the upper Riemann sums is equal to the supremum of the lower Riemann sums. If the interval of integration is partitioned into N intervals of equal length L, then the upper and lower sums are averages of the form

$$\frac{1}{N}\sum_{j=1}^{N} f(x_j)L$$

where x_j is an appropriately selected point in the jth interval.

For a compact metric group G, we need a very different kind of average. Given f in $\mathcal{C}(G)$ and a in G, the functions $x \to f(xa)$ and $x \to f(ax)$ are called *right and left translations of the function* f. Instead of taking an average of minimal or maximal values of a function on small intervals, we take averages of a finite sequence of either right or left translations of a function and then use the minimum oscillation of these right and left averages of a function to construct an invariant integral. This construction is due to Von Neumann.

Given $\alpha = \{a_1, \ldots, a_m\}$ a finite sequence of elements from G with repeats allowed, define a linear function $R_\alpha : \mathcal{C}(G) \to \mathcal{C}(G)$ by setting

$$R_\alpha(f)(x) = \frac{1}{m}\sum_{i=1}^{m} f(xa_i).$$

Similarly, let

$$L_\alpha(f)(x) = \frac{1}{m}\sum_{i=1}^{m} f(a_i x).$$

Obviously, R_α and L_α are linear functions from $\mathcal{C}(G)$ to itself such that $R_\alpha(1) = 1$ and $L_\alpha(1) = 1$. The interplay between these two types of averages of translations of a function will play an important role in the construction of an invariant integral for G.

To iterate the R_α and L_α constructions, it is necessary to define the product of two finite sequences. Although the order of the elements in a finite sequence does not effect R_α and L_α, it is convenient for calculations to specify a particular ordering for the product of two finite sequences. Given two finite sequences $\alpha = \{a_1, \ldots, a_m\}$ and $\beta = \{b_1, \ldots, b_n\}$, define $\alpha\beta$ using the lexicographic ordering by

$$\alpha\beta = \{a_1b_1, a_1b_2, \ldots, a_1b_n, \ldots, a_mb_1, a_mb_2 \ldots, a_mb_n\}.$$

Proposition 7.1.2 *If $\alpha = \{a_1, \ldots, a_m\}$ and $\beta = \{b_1, \ldots, b_n\}$ are two finite sequences of elements from a compact metric group G, then the following hold:*

(a) $R_\alpha \circ R_\beta = R_{\alpha\beta}$ *and* $L_\alpha \circ L_\beta = L_{\beta\alpha}$.

(b) $R_\alpha \circ L_\beta = L_\beta \circ R_\alpha$.

(c) $\|R_\alpha\|_{\mathcal{B}} = 1$ *and* $\|L_\alpha\|_{\mathcal{B}} = 1$.

Proof. For the first part,

$$\begin{aligned} R_\alpha \circ R_\beta(f)(x) &= \frac{1}{m}\sum_{i=1}^{m} R_\beta(f)(xa_i) \\ &= \frac{1}{m}\sum_{i=1}^{m}\frac{1}{n}\sum_{j=1}^{n} f(xa_ib_j) \\ &= R_{\alpha\beta}(x) \end{aligned}$$

Similarly, one shows that $L_\alpha \circ L_\beta = L_{\beta\alpha}$.

Next,

$$\begin{aligned} R_\alpha \circ L_\beta(f)(x) &= \frac{1}{m}\sum_{i=1}^{m} L_\beta(f)(xa_i) \\ &= \frac{1}{m}\sum_{i=1}^{m}\frac{1}{n}\sum_{j=1}^{n} f(b_jxa_i) \\ &= \frac{1}{n}\sum_{j=1}^{n}\frac{1}{m}\sum_{i=1}^{m} f(b_jxa_i) \\ &= L_\beta \circ R_\alpha(f)(x). \end{aligned}$$

Finally,

$$\begin{aligned} \|R_\alpha\|_{\mathcal{B}} &= \sup\left\{\|R_\alpha(f)\|_s : \|f\|_s \le 1\right\} \\ &= \sup\left\{|R_\alpha(f)(x)| : \|f\|_s \le 1 \text{ and } x \in G\right\} \\ &= \sup\left\{\left|\frac{1}{m}\sum_{i=1}^{m} f(xa_i)\right| : \|f\|_s \le 1 \text{ and } x \in G\right\} \\ &\le \sup\left\{\frac{1}{m}\sum_{i=1}^{m} |f(xa_i)| : \|f\|_s \le 1 \text{ and } x \in G\right\} \\ &\le 1. \end{aligned}$$

It follows that $\|R_\alpha\|_{\mathcal{B}} = 1$ because $R_\alpha(1) = 1$. □

Part *(c)* implies that R_α and L_α are continuous linear functions and

$$\|R_\alpha(f)\|_s \le \|f\|_s \text{ and } \|L_\alpha(f)\|_s \le \|f\|_s \tag{7.1}$$

Given f in $\mathcal{C}(G)$, let $\mathcal{R}_f$ denote the collection of all the functions of the form $R_\alpha(f)$, where α runs through all the finite sequences of elements from G. Note that $R_\alpha(f)$ and $R_\beta(f)$ can be equal when $\alpha \neq \beta$. In fact, $\mathcal{R}_f = \{f\}$ when f is a constant function. Define $\mathcal{L}_f$ similarly. Obviously, $\mathcal{R}_f$ and $\mathcal{L}_f$ are well-defined subsets of $\mathcal{C}(G)$. Their key property is the following:

Proposition 7.1.3 *The sets $\mathcal{R}_f^-$ and $\mathcal{L}_f^-$ are compact subsets of $\mathcal{C}(G)$.*

Proof. Consider $\mathcal{R}_f$ and let $d(x, y)$ be a right-invariant metric for G. *Corollary 6.2.8* to Ascoli's Theorem can be used here because the norm linear topology on $\mathcal{C}(X)$ coincides with the compact-open topology (*Exercise* 2, p. 242).

Since $\|R_\alpha(f)\|_s \leq \|f\|_s$, it follows that $-\|f\|_s \leq R_\alpha(f)(x) \leq \|f\|_s$ for all x in G. Hence,

$$\eta(\mathcal{R}_f \times \{x\}) \subset \{t \in \mathbb{R} : |t| \leq \|f\|_s\},$$

and condition *(a)* of *Corollary 6.2.8* is satisfied.

It remains to prove that $\mathcal{R}_f$ is equicontinuous. Because f is continuous and G is compact, f is uniformly continuous on G, so given $\varepsilon > 0$ there exists $\delta > 0$ such that $|f(x) - f(y)| < \varepsilon$, whenever $d(x, y) < \delta$. Since $d(xa, ya) = d(x, y)$,

$$|R_\alpha(f)(x) - R_\alpha(f)(y)| \leq \frac{1}{m}\sum_{i=1}^{m} |f(xa_i) - f(ya_i)| < \frac{1}{m} m\varepsilon = \varepsilon$$

when $d(x, y) < \delta$. Thus $\mathcal{R}_f$ is equicontinuous and $\mathcal{R}_f^-$ is compact.

The proof for $\mathcal{L}_f$ is similar and will be omitted. □

The *oscillation of a function* f in $\mathcal{C}(G)$ is defined by

$$\omega(f) = \max\{f(x) : x \in G\} - \min\{f(x) : x \in G\}.$$

Obviously, $\omega(f) \geq 0$. It is easy to see that

$$\max\{R_\alpha(f)(x) : x \in G\} \leq \max\{f(x) : x \in G\}$$

and

$$\min\{R_\alpha(f)(x) : x \in G\} \geq \min\{f(x) : x \in G\}.$$

Consequently,

$$\omega(R_\alpha(f)) \leq \omega(f).$$

The function $f \mapsto \omega(f)$ is a continuous real-valued function on $\mathcal{C}(G)$ by *Exercise* 7, p. 232. Also note that $\omega(f) = 0$ if and only if f is a constant function.

Proposition 7.1.4 *For every nonconstant function f in $\mathcal{C}(G)$, there exist finite sequences α and β in G such that*

$$\omega(R_\alpha(f)) < \omega(f)$$

and

$$\omega(L_\beta(f)) < \omega(f).$$

Proof. Because f is a nonconstant function, there exists y in G such that $f(y) < \max\{f(x) : x \in G\} = M$. Hence, there exists an open neighborhood U of y and a real number c such that $f(x) \leq c < M$ for all x in U. Since the collection of sets $\{Ua^{-1} : a \in G\}$ is an open cover of G, there exists a finite sequence $\alpha = \{a_1, \ldots, a_m\}$ such that $G = Ua_1^{-1} \cup \ldots \cup Ua_m^{-1}$.

For x in G, there exists a_j such that $x \in Ua_j^{-1}$ or $xa_j \in U$ and hence $f(xa_j) < M$. Consequently, for all x in G

$$
\begin{aligned}
R_\alpha(f)(x) &= \frac{1}{m}\sum_{i=1}^{m} f(xa_i) \\
&= \frac{f(xa_j)}{m} + \frac{1}{m}\sum_{i \neq j} f(xa_i) \\
&\leq \frac{c}{m} + \frac{M(m-1)}{m} \\
&= \frac{c + M(m-1)}{m}
\end{aligned}
$$

and

$$\max\{R_\alpha(f)(x) : x \in G\} < \max\{f(x) : x \in G\},$$

because

$$\frac{c + M(m-1)}{m} < M = \max\{f(x) : x \in G\}.$$

Using

$$\min\{R_\alpha(f)(x) : x \in G\} \geq \min\{f(x) : x \in G\},$$

it follows that

$$
\begin{aligned}
\omega(R_\alpha(f) &= \max\{R_\alpha(f)(x) : x \in G\} - \min\{R_\alpha(f)(x) : x \in G\} \\
&< \max\{f(x) : x \in G\} - \min\{f(x) : x \in G\} \\
&= \omega(f)
\end{aligned}
$$

The proof of the second inequality is now similar. □

The value of a constant function in $\mathcal{R}_f^-$ will be called a *right mean* for f. Similarly, a *left mean* of f is the value of a constant function in $\mathcal{L}_f^-$.

Theorem 7.1.5 *If G is a compact metric group, then for every function in $C(G)$ there exist unique right and left means and they are equal.*

Proof. Given f in $C(G)$, let

$$r = \inf\{\omega(R_\alpha(f)) : R_\alpha(f) \in \mathcal{R}_f\}$$

Note that $0 \leq r \leq \omega(f)$. There exists a sequence of finite sequences β_k such that $\omega(R_{\beta_k}(f))$ converges to r. By *Proposition 7.1.3*, it can be assumed without loss of generality that the functions $R_{\beta_k}(f)$ converge to a function g in $C(G)$.

If g is not a constant function, then there exists a finite sequence α such that $\omega(R_\alpha(g)) < \omega(g) = r$ by *Proposition 7.1.4*. Because R_α is continuous, the sequence $R_\alpha \circ R_{\beta_k}(f) = R_{\alpha\beta_k}(f)$ converges to $R_\alpha(g)$. Now the continuity of the oscillation function ω on $\mathcal{C}(G)$ implies that $\omega(R_{\alpha\beta_k}(f))$ converges to $\omega(R_\alpha(g))$. Thus $\omega(R_{\alpha\beta_k}(f)) < r$ for large k, contradicting the choice of r. Therefore, g is a constant function in $\mathcal{R}_f$, and every function in $\mathcal{C}(G)$ has at least one right mean. A similar proof establishes the existence of left means.

Let p and q be right and left means, respectively, for f, and think of them as constant functions. To complete the proof, it suffices to show that $p = q$. Given $\varepsilon > 0$, there exist finite sequences α and β such that $\|R_\alpha(f) - p\,\|_s < \varepsilon$ and $\|L_\beta(f) - q\,\|_s < \varepsilon$.

Because $R_\alpha(1) = 1 = L_\alpha(1)$ for any finite sequence α, we have $R_\alpha(c) = c = L_\alpha(c)$ for any constant function c. Now, it follows from (7.1) that

$$\|L_\beta \circ R_\alpha(f) - p\,\|_s = \|L_\beta\big(R_\alpha(f) - p\big)\|_s \le \|R_\alpha(f) - p\,\|_s < \varepsilon.$$

Similarly, $\|R_\alpha \circ L_\beta(f) - q\,\|_s < \varepsilon$. By part *(b)* of the *Proposition 7.1.2*

$$\begin{aligned} \|p - q\| &= \|p - L_\beta \circ R_\alpha(f) + R_\alpha \circ L_\beta(f) - q\|_s \\ &\le \|p - L_\beta \circ R_\alpha(f))\|_s + \|R_\alpha \circ L_\beta(f)) - q\,\|_s \\ &< 2\varepsilon \end{aligned}$$

Since ε was an arbitrary positive number, $p = q$. □

Let $\mu(f)$ denote the unique and equal right and left means for f in $\mathcal{C}(G)$.

Proposition 7.1.6 *If α is a finite sequence of elements from a compact group G, then*

$$\mu(R_\alpha(f)) = \mu(f) = \mu(L_\alpha(f)). \tag{7.2}$$

for all f in $\mathcal{C}(G)$.

Proof. Given $f \in \mathcal{C}(G)$, there exists a sequence β_k of finite sequences from G such that $L_{\beta_k}(f)$ converges to the constant function $\mu(f)$ in $\mathcal{C}(G)$. If α is any finite sequence from G, then the continuity of R_α and part *(b)* of *Proposition 7.1.2* imply that

$$L_{\beta_k} \circ R_\alpha(f)) = R_\alpha \circ L_{\beta_k}(f)$$

converges to the function $R_\alpha(\mu(f)) = \mu(f)$. Therefore, the constant function $\mu(f)$ is in $\left(\mathcal{L}_{R_\alpha(f)}\right)^-$. By the uniqueness of the left mean, $\mu(R_\alpha(f)) = \mu(f)$. Similarly, one shows that $\mu(L_\alpha(f)) = \mu(f)$. □

Theorem 7.1.7 *If G is a compact metric group, then $\int f(x)\,dx = \mu(f)$ is a unique right-invariant integral on G. Furthermore, it is a left-invariant integral and satisfies $\int f(x^{-1})\,dx = \int f(x)\,dx$ for all $f \in C(X)$.*

Proof. The first step is to verify the required five properties of a right-invariant integral.

(a) $\int cf(x)\,dx = c\int f(x)\,dx$ for all $f \in \mathcal{C}(G)$ and $c \in \mathbb{R}$: There exists a sequence of finite sequences α_k of elements G such that $R_{\alpha_k}(f)$ converges to $\mu(f)$ in $\mathcal{C}(G)$. It follows from the linearity of R_{α_k} that $R_{\alpha_k}(cf)$ converges to the constant function $c\mu(f)$ in $\mathcal{C}(G)$ and by the uniqueness of the right mean $\mu(cf) = c\mu(f)$.

(b) $\int (f(x) + g(x))\,dx = \int f(x)\,dx + \int g(x)\,dx$ for all $f, g \in \mathcal{C}(G)$: For any $\varepsilon > 0$, there exists a finite sequence β such that $\|R_\beta(g) - \mu(g)\|_s < \varepsilon$. It follows from equation (7.2) that there exists another finite sequence α such that $\|R_{\alpha\beta}(f)] - \mu(f)\|_s = \|R_\alpha[R_\beta(f)] - \mu(f)\|_s < \varepsilon$. Then $R_\alpha(c) = c$ and equation (7.1) imply that

$$\begin{aligned}\|R_{\alpha\beta}(g)] - \mu(g)\|_s &= \left\|R_\alpha\left[R_\beta(g)\right] - \mu(g)\right\|_s \\ &= \left\|R_\alpha\left[R_\beta(g) - \mu(g)\right]\right\|_s \\ &\leq \|R_\beta(g) - \mu(g)\|_s \\ &\leq \varepsilon.\end{aligned}$$

Thus

$$\begin{aligned}\|R_{\alpha\beta}(f+g) - \mu(f) - \mu(g)\|_s &= \\ \|R_{\alpha\beta}(f) + R_{\alpha\beta}(g) - \mu(f) - \mu(g)\|_s &\leq \\ \|R_{\alpha\beta}(f)] - \mu(f)\|_s + \|R_{\alpha\beta}(g)] - \mu(g)\|_s &\leq \\ 2\varepsilon,\end{aligned}$$

and the uniqueness of $\mu(f+g)$ implies that $\mu(f+g) = \mu(f) + \mu(g)$.

(c) $\int 1\,dx = 1$: Since $R_\alpha(1) = 1$ for all finite sequences α, it follows that $\mathcal{R}_1 = \{1\}$ and $\mu(1) = 1$.

(d) $\int f(x)\,dx \geq 0$ when all $f \geq 0$: If $f \geq 0$, then $R_\alpha(f) \geq 0$ for every finite sequence α and $g \geq 0$ for every g in $\overline{\mathcal{R}_f}$. In particular, $\mu(f) \geq 0$.

(e) $\int f(xa)\,dx = \int f(x)\,dx$ for all $f \in \mathcal{C}(G)$ and $a \in G$: Let α be a finite sequence of length 1 with $a_1 = a$. Then $R_\alpha(f)(x) = f(xa)$ and $\mu(R_\alpha(f)) = \mu(f)$ by *Proposition 7.1.6*. So $\int f(xa)\,dx = \int R_\alpha(f)(x)\,dx = \mu(R_\alpha(f)) = \mu(f) = \int f(x)\,dx$.

It has now been shown that $\int f(x)\,dx = \mu(f)$ defines a right-invariant integral on $\mathcal{C}(G)$. Suppose $\int f(x)\,d^*x$ is another right-invariant integral on $\mathcal{C}(G)$. It follows from the definition of a right-invariant integral that $\int R_\alpha(f)(x)\,d^*x = \int f(x)\,d^*x$ for all f in $\mathcal{C}(G)$ and all finite sequences α of elements from G. Let $\varepsilon > 0$. There exists a finite sequence α such that $\|R_\alpha(f) - \mu(f)\|_s < \varepsilon$, so $|R_\alpha(f)(x) - \mu(f)| < \varepsilon$ for all x in G. Using the definition of right-invariant integral and *Proposition 7.1.1*, it follows that

$$\left|\int f(x)\,d^*x - \mu(f)\right| = \left|\int R_\alpha(f)(x)\,d^*x - \mu(f)\right|$$

$$\begin{aligned} &= \left| \int [R_\alpha(f)(x) - \mu(f)]\, d^*x \right| \\ &\leq \int |R_\alpha(f)(x) - \mu(f)|\, d^*x \\ &\leq \varepsilon. \end{aligned}$$

Since ε was any positive number, $\int f(x)\, d^*x = \mu(f)$ for all f in $C(G)$ and $\int f(x)\, dx = \mu(f)$ is the unique right-invariant integral.

The left invariance of $\int f(x)\, dx = \mu(f)$ follows from *Proposition 7.1.6* as did the right invariance in part (e) of the first part of the proof.

To prove that $\int f(x^{-1})\, dx = \int f(x)\, dx$ for all $f \in C(X)$, set $\int f(x)\, d^*x = \int f(x^{-1})\, dx$. Clearly, $\int f(x^{-1})\, dx$ satisfies conditions *(a)* through *(d)* in the definition of right-invariant integral. Moreover,

$$\begin{aligned} \int f(xa)\, d^*x &= \int f(x^{-1}a)\, dx \\ &= \int f((a^{-1}x)^{-1})\, dx \\ &= \int f(x^{-1})\, dx \\ &= \int f(x)\, d^*x \end{aligned}$$

by the left invariance of $\int f(x)\, dx = \mu(f)$. Thus $\int f(x)\, d^*x = \int f(x^{-1})\, dx$ is a right-invariant integral, and $\int f(x^{-1})\, dx = \int f(x)\, dx$ follows from the uniqueness of the right-invariant integral. □

Proposition 7.1.8 *Let G and H be compact metric groups. If f is in $C(G\times H)$, then $g(y) = \int f(x, y)\, dx$ is in $C(H)$.*

Proof. Let d_G and d_H be metrics for G and H, respectively. Then,

$$d((x, y), (x', y')) = d_G(x, x') + d_H(y, y')$$

is a metric for $G \times H$. By compactness, $f(x, y)$ is uniformly continuous on $G \times H$. So given $\varepsilon > 0$ there exists $\delta > 0$ such that $|f(x, y) - f(x', y')| < \varepsilon$ whenever $d((x, y), (x', y')) < \delta$. In particular,

$$\begin{aligned} |g(y) - g(y')| &= \left| \int f(x, y) - f(x, y')\, dx \right| \\ &\leq \int |f(x, y) - f(x, y')|\, dx \\ &\leq \varepsilon \end{aligned}$$

when $d_H(y, y') = d((x, y), (x, y')) < \delta$, proving that $g(y)$ is continuous. □

Corollary 7.1.9 *If G and H are compact metric groups, then*

$$\int\int f(x,y)\,dx\,dy = \int f(x,y)\,d(x,y) = \int\int f(x,y)\,dy\,dx$$

for all f is in $C(G \times H)$

Proof. It follows from the proposition that $\int\int f(x,y)\,dx\,dy$ is defined. Moreover, it is easy to verify that $\int\int f(x,y)\,dx\,dy$ is a right-invariant integral on $G \times H$. The conclusion then follows by the uniqueness of right-invariant integrals. □

Set

$$\langle\langle f, g\rangle\rangle = \int f(x)g(y)\,dx.$$

It is a function from $C(G) \times C(G)$ to $\mathbb{R}$ with the following properties:

(a) $\langle\langle f, g\rangle\rangle = \langle\langle g, f\rangle\rangle$ (*symmetric*)

(b) $\langle\langle c_1 f_1 + c_2 f_2, g\rangle\rangle = c_1\langle\langle f_1, g\rangle\rangle + c_2\langle\langle f_2, g\rangle\rangle$ (*bilinear*)

(c) $\langle\langle f, f\rangle\rangle \geq 0$ (*positive*)

(d) $\langle\langle f, f\rangle\rangle = 0$ if and only if $f = 0$ (*strictly positive*)

The first three are obvious, and the last one follows from *Exercise* 2, p. 274. The function $\langle\langle f, g\rangle\rangle$ is an example of an *inner product* on a linear space over $\mathbb{R}$. A more familiar example of an inner product is the dot product $\mathbf{x} \cdot \mathbf{y}$ for $\mathbf{x}$ and $\mathbf{y}$ in $\mathbb{R}^n$. A linear space with a specified inner product is called an *inner-product space.* It will be clear from our discussion of $\langle\langle f, g\rangle\rangle$, which will play a crucial role in Section 7.2, how to extend the basic properties of the dot product to an arbitrary inner-product space.

A more common notation than $\langle\langle f, g\rangle\rangle$ for an inner product is (f, g), but that already has two possible meanings, a point in a Cartesian product of two sets or an open interval. And $\langle f, g\rangle$ already denotes the subgroup generated f and g. So we will stick with the notation $\langle\langle f, g\rangle\rangle$ for $\int f(x)g(x)\,dx$.

Several important results about the dot product carry over to the more general inner product $\langle\langle f, g\rangle\rangle$. First, the Cauchy-Schwarz Inequality holds in this context. Although the proof used in *Proposition 1.2.1* works here, we will instead prove Bessel's Inequality, which we need, and then show that the Cauchy-Schwarz Inequality follows from it.

Two functions f and g in $C(G)$ are said to be *orthogonal* when $\langle\langle f, g\rangle\rangle = 0$. Functions $u_1, \ldots, u_n$ are *orthonormal* if $\langle\langle u_i, u_i\rangle\rangle = 1$ and $\langle\langle u_i, u_j\rangle\rangle = 0$, when $i \neq j$ and $1 \leq i, j \leq n$. The Gram-Schmidt Process (p. 153) works in any finite-dimensional subspace of $C(G)$. In particular, given a finite-dimensional subspace V of $C(G)$, we can construct an orthonormal basis $u_1(x), \ldots, u_m(x)$ for V. If f is in V, then mimicking the calculations from *Exercises 5 and 6*, p.158, we obtain

$$f(x) = \sum_{j=1}^{m} \langle\langle f, u_j\rangle\rangle u_j(x)$$

for all $x \in G$ and

$$\langle\langle f, f \rangle\rangle = \sum_{j=1}^{m} \langle\langle f, u_j \rangle\rangle^2.$$

Now we can state and prove Bessel's Inequality. Although it holds in finite-dimensional inner-product spaces, the more interesting setting in an infinite-dimensional setting.

Proposition 7.1.10 (Bessel's Inequality) *If $u_1, \ldots$ is a finite or infinite sequence of orthonormal functions in $C(G)$, then*

$$\sum_i \langle\langle f, u_i \rangle\rangle^2 \leq \langle\langle f, f \rangle\rangle \tag{7.3}$$

for f in $C(G)$.

Proof. For the finite case, simply compute and simplify

$$0 \leq \left\langle\!\!\left\langle f - \sum_{i=1}^{m} \langle\langle f, u_i \rangle\rangle u_i, f - \sum_{j=1}^{m} \langle\langle f, u_j \rangle\rangle u_j \right\rangle\!\!\right\rangle$$

to get

$$\sum_{i=1}^{m} \langle\langle f, u_i \rangle\rangle^2 \leq \langle\langle f, f \rangle\rangle.$$

Since the right side is independent of m and $\langle\langle f, u_i \rangle\rangle^2 \geq 0$, it follows in the case of an infinite sequence of orthonormal functions that the series is convergent and

$$\sum_{i=1}^{\infty} \langle\langle f, u_i \rangle\rangle^2 \leq \langle\langle f, f \rangle\rangle$$

to complete the proof. $\square$

The Cauchy-Schwarz Inequality

$$\langle\langle f, g \rangle\rangle^2 \leq \langle\langle f, f \rangle\rangle \, \langle\langle g, g \rangle\rangle \tag{7.4}$$

is a consequence of Bessel's Inequality. To see this, first note that the Cauchy-Schwarz Inequality is trivial, if $g = 0$. So we can assume that $g \neq 0$. Now apply Bessel's Inequality with $m = 1$ and

$$u_1 = \frac{g}{\langle\langle g, g \rangle\rangle^{1/2}},$$

and simplify to get (7.4). Using the definition of $\langle\langle f, g \rangle\rangle$, the Cauchy-Schwarz Inequality can also be stated as

$$\left(\int f(x) g(x) \, dx \right)^2 \leq \left(\int f(x)^2 \, dx \right) \left(\int g(x)^2 \, dx \right). \tag{7.5}$$

Setting
$$\|f\|_2 = \sqrt{\langle\langle f, f\rangle\rangle}$$
defines a norm on $\mathcal{C}(G)$ because by the Cauchy-Schwarz Inequality

$$\begin{aligned}
\|f+g\|_2^2 &= \\
\langle\langle f, f\rangle\rangle + 2\langle\langle f, g\rangle\rangle + \langle\langle g, g\rangle\rangle &\leq \\
\langle\langle f, f\rangle\rangle + 2\langle\langle f, f\rangle\rangle^{1/2}\langle\langle g, g\rangle\rangle^{1/2} + \langle\langle g, g\rangle\rangle &= \\
\left(\langle\langle f, f\rangle\rangle^{1/2} + \langle\langle g, g\rangle\rangle^{1/2}\right)^2 &= \\
(\|f\|_2 + \|g\|_2)^2 &
\end{aligned}$$

or $\|f+g\| \leq \|f\|_2 + \|g\|_2$. The Cauchy-Schwarz can now also be stated as
$$|\langle\langle f, g\rangle\rangle| \leq \|f\|_2 \, \|g\|_2. \tag{7.6}$$

The subscript 2 in the notation indicates that we are integrating the square of a function and then taking the square root. Following the usual nomenclature, we will refer to $\|\cdot\|_2$ as the "L-two norm".

Note well that we now have two norm topologies on $\mathcal{C}(G)$ and they are not equivalent (see *Exercises 4 and 5*, p. 274). Of course, on any finite-dimensional subspace of $\mathcal{C}(G)$ these two norms are equivalent (*Proposition 2.3.6*), but they are not equivalent on $\mathcal{C}(G)$. Consequently, we must be careful henceforth to always clearly indicate which metric on $\mathcal{C}(G)$ is being used. To this end, we will let $\mathcal{C}(G)_s$ and $\mathcal{C}(G)_2$ denote $\mathcal{C}(G)$ with the metrics $d_s(f,g) = \|f-g\|_s$ and $d_2(f,g) = \|f-g\|_2$, respectively, when the topology is playing an active role.

Every locally compact Hausdorff topological group has a right- and left-invariant integral, which are not always the same. When the group is not compact, however, the constant functions cannot be integrated, and the integrals are only unique up to multiplication by a positive constant. The continuous functions that vanish off a compact subset of the group are the primary linear space of integrable functions. One method of constructing an invariant integral of such functions uses linear combinations of translates to approximate it, as is done in [5] and [12].

The other approach is to use measure theory on locally compact Hausdorff topological spaces to prove the existence of a right-invariant measure on a group called *Haar measure*. The integral with respect to this measure is the desired invariant integral, and it integrates a much larger class of functions than just the continuous ones that vanish off a compact set. This the approach is taken in [4] and [14]. These invariant integrals are the beginning of abstract harmonic analysis, which includes classical Fourier analysis. Lebesgue measure on $\mathbb{R}$ (Chapter 3 in [14]) is a left- and right-invariant measure and the Haar measure for $\mathbb{R}$, and the Lebesgue integral (Chapter 4 of [14]) is a left- and right-invariant integral for $\mathbb{R}$.

EXERCISES

1. Show that for $f \in \mathcal{C}(\mathbb{K})$, the usual Riemann integral $\int_0^1 f(e^{2\pi i x})\,dx$ defines an invariant integral on $\mathbb{K}$.

2. Show that if $f \geq 0$ and $f(y) > 0$ for some $y \in G$, then $\int f(x)\,dx > 0$. (One approach is modify the proof of *Proposition 7.1.4*.)

3. Let G be a compact metric group. Show that the invariant integral $\mu(f) = \int f(x)\,dx$ is a bounded linear function from $\mathcal{C}(G)_s$ to $\mathbb{R}$ and $\|\mu\|_{\mathcal{B}} = \sup\{|\mu(f)| : \|f\|_s \leq 1\} = 1$.

4. Let G be a compact metric group. Show that $\|f\|_2 \leq \|f\|_s$ and that every open set in $\mathcal{C}(G)_2$ is an open set in $\mathcal{C}(G)_s$.

5. Show that the topologies for $\mathcal{C}(\mathbb{K})_2$ and $\mathcal{C}(\mathbb{K})_s$ are not equal. (One approach is to show that the sequence of functions $f_k(z) = [(z+\overline{z})/2]^{2k}$ converges in one of the spaces but not in the other and use *Exercise* 10, p. 248.)

6. Let E be a subset of $\mathcal{C}(G)$. Show that $E^{\perp} = \{f \in \mathcal{C}(G) : \langle\langle f, g\rangle\rangle = 0 \text{ for all } g \in E\}$ is a subspace. Show that $E^{\perp}$ is closed in the $\|\cdot\|_2$ topology and hence in the $\|\cdot\|_s$ topology.

7. Let $u_1, \ldots, u_n$ be an orthonormal basis for a finite-dimensional subspace V of $\mathcal{C}(G)$. Given f in $\mathcal{C}(G)$, set $f_1 = \sum_{i=1}^{n} \langle\langle f, u_i\rangle\rangle u_i$ and show that $f_2 = f - f_1$ is in $V^{\perp}$. Show that if $f = g_1 + g_2$ with $g_1 \in V$ and $g_2 \in V^{\perp}$, then $g_1 = f_1$ and $g_2 = f_2$. Also show that $V \cap V^{\perp}$ is the zero function.

8. Prove that if V is a finite-dimensional subspace of $\mathcal{C}(G)$, then $V = V^{\perp\perp}$.

9. Prove that $\mathcal{C}(G)_2$ is not complete.

7.2 Integral Equations

Continuing the context of Section 7.1, G will remain a compact metric group and $\int f(x)\,dx$ will be the unique right-invariant integral, which is also left invariant, given by *Theorem 7.1.7*. Recall that there are now two norms on $\mathcal{C}(G)$ denoted by $\|\cdot\|_s$ and $\|\cdot\|_2$. Thus $\mathcal{B}(\mathcal{C}(G)_s, \mathcal{C}(G)_s)$ and $\mathcal{B}(\mathcal{C}(G)_2, \mathcal{C}(G)_2)$ are two different normed linear spaces of bounded linear functions from $\mathcal{C}(G)$ to itself. Continuing our notational pattern, their usual $\|T\|_{\mathcal{B}}$ norms will be denoted by $\|T\|_s$ and $\|T\|_2$. Both these normed linear spaces are subspaces of $\mathcal{L}(\mathcal{C}(G), \mathcal{C}(G))$, the linear space of all liner functions from $\mathcal{C}(G)$ to itself. We begin by describing a type of linear function from $\mathcal{C}(G)$ to itself that is in

$$\mathcal{B}(\mathcal{C}(G)_s, \mathcal{C}(G)_s) \cap \mathcal{B}(\mathcal{C}(G)_2, \mathcal{C}(G)_2)$$

and hence also has two different norms.

Proposition 7.2.1 *Let G be a compact metric group, and let $\kappa : G \times G \to \mathbb{R}$ be a continuous function. If $T_\kappa(f)(x) = \int \kappa(x,y)f(y)\,dy$ for f in $C(G)$, then the following hold:*

(a) T_κ is a bounded linear function from $C(G)_s$ to itself such that

$$\|T_\kappa\|_s = \sup\{\|T_\kappa(f)\|_s : \|f\|_s \leq 1\} \leq \|\kappa\|_s.$$

(b) T_κ is bounded linear function from $C(G)_2$ to itself such that

$$\|T_\kappa\|_2 = \sup\{\|T_\kappa(f)\|_2 : \|f\|_2 \leq 1\} \leq \|\kappa\|_2.$$

Proof. It follows from *Proposition 7.1.8* and the linearity of the integral that T_κ is a linear function from the linear space $C(G)$ to itself. What remains is to estimate the norm of T_κ using the two norms on $C(G)$.

The first one follows from the following estimates:

$$\begin{aligned}\|T_\kappa(f)\|_s &= \sup\left\{\left|\int \kappa(x,y)f(y)\,dy\right| : x \in G\right\}\\ &\leq \sup\left\{\int |\kappa(x,y)|\,|f(y)|\,dy : x \in G\right\}\\ &\leq \int \|f\|_s\,\|\kappa\|_s\,dy\\ &\leq \|f\|_s\,\|\kappa\|_s\end{aligned}$$

For the second one, the Cauchy-Schwarz Inequality (7.5) implies that for fixed x

$$\begin{aligned}(T_\kappa(f)(x))^2 &= \left(\int \kappa(x,y)f(y)\,dy\right)^2\\ &\leq \left(\int f(y)^2\,dy\right)\left(\int \kappa(x,y)^2\,dy\right),\end{aligned}$$

and

$$(T_\kappa(f)(x))^2 \leq \|f\|_2^2 \int \kappa(x,y)^2\,dy.$$

Consequently,

$$\begin{aligned}\|T_\kappa(f)\|_2 &= \sqrt{\int (T_\kappa(f)(x))^2\,dx}\\ &\leq \|f\|_2\sqrt{\int\int \kappa(x,y)^2\,dy\,dx}\\ &= \|f\|_2\,\|\kappa\|_2\end{aligned}$$

by *Corollary 7.1.9.* □

Corollary 7.2.2 *Let G be a compact metric group. Given κ in $\mathcal{C}(G \times G)$, let $T_\kappa(f)(x) = \int \kappa(x,y)f(y)\,dy$ for f in $\mathcal{C}(G)$. Then the following hold:*

(a) The function $\kappa \mapsto T_\kappa$ is a bounded linear function from $\mathcal{C}(G \times G)_s$ to $\mathcal{B}(\mathcal{C}(G)_s, \mathcal{C}(G)_s)$.

(b) The function $\kappa \mapsto T_\kappa$ is a bounded linear function from $\mathcal{C}(G \times G)_2$ to $\mathcal{B}(\mathcal{C}(G)_2, \mathcal{C}(G)_2)$.

Proof. The linearity $\kappa \mapsto T_\kappa$ follows from the linearity of the integral. Since $\|T_\kappa\|_s \le \|\kappa\|_s$,

$$\sup\{\|T_\kappa\|_s : \|\kappa\|_s \le 1\} \le 1$$

and $\kappa \mapsto T_\kappa$ is a bounded linear function from $\mathcal{C}(G \times G)_s$ to $\mathcal{B}(\mathcal{C}(G)_s, \mathcal{C}(G)_s)$. The same reasoning applies to the second part. □

This section is devoted to studying the *integral equation*

$$\lambda\theta = T_\kappa(\theta) \tag{7.7}$$

or

$$\lambda\theta(x) = \int \kappa(x,y)\theta(y)\,dy \tag{7.8}$$

where $\kappa : G \times G \to$ is a continuous function and $T_\kappa(\theta)(x) = \int \kappa(x,y)\theta(y)\,dy$. The continuous function κ in the integral equation (7.8) is called the *kernel of the integral equation.* A solution of equation (7.7) or equivalently (7.8) is a nonzero continuous function $\theta : G \to G$ called a *eigenfunction* and a nonzero real number λ called a *eigenvalue* that satisfy it.

Given a finite-dimensional linear space matrix V over $\mathbb{C}$ and a linear function $T : V \to V$, we can consider the equation $\lambda\mathbf{v} = T(\mathbf{v})$ as in the proof of *Theorem 2.5.11.* If λ and $\mathbf{v} \neq \mathbf{0}$ satisfy this equation, then λ is a characteristic value for a matrix of T with respect to a basis for V, that is, it is a root of the polynomial $p(z) = \operatorname{Det}[A - zI]$. Thus the eigenvalues and eigenfunctions that solve equation (7.7) can be thought of as generalizations of characteristic values and characteristic vectors. In our terminology, however, a characteristic value always refers to a root of the characteristic polynomial $f(z) = \operatorname{Det}[A - zI]$ of a matrix A, and there is no characteristic polynomial associated with (7.7).

Our study of equation $T_\kappa(\theta) = \lambda\theta$ will follow the approach used by Pontryagin [13] for symmetric kernels. We will, however, emphasize the general continuity properties of linear and bilinear functions that underlie Pontryagin's more ad hoc approach. Using T_κ instead of writing out everything in terms of integrals will provide cleaner arguments.

The kernel $\kappa(x,y)$ is *symmetric* provided $\kappa(x,y) = \kappa(y,x)$ for all x and y in G. If κ is symmetric, then

$$\begin{aligned}\langle\langle T_\kappa(f), g\rangle\rangle &= \int\int \kappa(x,y)f(y)g(x)\,dy\,dx \\ &= \int f(y)\int \kappa(y,x)g(x)\,dx\,dy \\ &= \langle\langle f, T_\kappa(g)\rangle\rangle.\end{aligned} \tag{7.9}$$

The proof of *Proposition 5.2.11* can now be reused to prove the following:

Proposition 7.2.3 *Let κ be a symmetric kernel for the compact metric group G. If θ_1 and θ_2 are eigenfunctions of T_κ for distinct eigenvalues λ_1 and λ_2 respectively, then θ_1 and θ_2 are orthogonal.*

Next, *Proposition 7.2.4* will reduce many considerations to finite-dimensional subspaces of $\mathcal{C}(G)$, where the norms $\|\cdot\|_s$ and $\|\cdot\|_2$ are equivalent by *Proposition 2.3.6*.

Proposition 7.2.4 *Let $\kappa(x, y)$ be a symmetric kernel for the compact metric group G. If θ_i, $1 \leq i \leq m$ is a finite sequence of orthonormal eigenfunctions with nonzero eigenvalues λ_i (allowing repeats) for $\lambda\theta = T_\kappa(\theta)$, then*

$$\sum_{i=1}^{m} \lambda_i^2 \theta_i(x)^2 \leq \int \kappa(x, y)^2 \, dy \tag{7.10}$$

and

$$\sum_{i=1}^{m} \lambda_i^2 \leq \int\int \kappa(x, y)^2 \, dy \, dx \tag{7.11}$$

Proof. Fixed x. Then the first equation is just Bessel Inequality (*Proposition 7.1.10*) for the function $f(y) = \kappa(x, y)$ because $\lambda_i\theta_i(x) = \int \kappa(x, y)\theta_i(y) \, dy$. And then integrating the first inequality with respect to x yields the second one because $\langle\langle \theta_i, \theta_i \rangle\rangle = 1$ $\square$

Corollary 7.2.5 *Let $\kappa(x, y)$ be a symmetric kernel for the compact metric group G. If λ is a nonzero eigenvalue for T_κ and*

$$V_\lambda = \{f \in \mathcal{C}(G) : T_\kappa(f) = \lambda f\},$$

then V_λ is a finite-dimensional subspace of $\mathcal{C}(G)$.

Proof. Suppose $f_1, \ldots, f_m$ are linearly independent functions in V_λ. Using the Gram Schmidt Process, there exist orthonormal functions $\theta_1, \ldots, \theta_m$ in V_λ. Then equation (7.11) implies that $m\lambda^2 \leq \int\int \kappa(x, y)^2 \, dy \, dx$, and it follows that

$$\text{Dim}\,[V_\lambda] \leq \frac{1}{\lambda^2} \int\int \kappa(x, y)^2 \, dy \, dx < \infty$$

to complete the proof. $\square$

Corollary 7.2.6 *Let $\kappa(x, y)$ be a symmetric kernel for the compact metric group G. If $a > 0$, then the number of distinct eigenvalues λ for $\lambda\theta = T_\kappa(\theta)$ such that $|\lambda| \geq a$ is finite. In particular, an infinite sequence of distinct eigenvalues λ_n converges to 0.*

Proof. Suppose $\lambda_1, \ldots, \lambda_m$ are distinct eigenvalues such that $\lambda_i \geq a$ for $i = 1, \ldots, m$. Then

$$ma^2 \leq \sum_{i=1}^{m} \lambda_i^2 \leq \int\int \kappa(x,y)^2\, dy\, dx$$

by equation (7.11), and there are at most

$$\frac{1}{a^2} \int \kappa(x,y)^2\, dy\, dx$$

distinct eigenvalues λ for (7.7) such that $|\lambda| \geq a$. $\square$

Let

$$\mathcal{A}_p = \left\{ \sum_{i=1}^{m} f_i(x) g_i(y) : f_i, g_i \in \mathcal{C}(G) \text{ for } 1 \leq i \leq m \text{ and } m \in \mathbb{Z}^+ \right\}$$

By *Exercise* 8, p. 232, $\mathcal{A}_p$ is a dense subalgebra of $\mathcal{C}(G \times G)_s$. So functions from $\mathcal{A}_p$ can be used to approximate any kernel $\kappa(x, y)$ in $\mathcal{C}(G \times G)_s$. Our strategy will be to first study symmetric kernels in $\mathcal{A}_p$.

Let $\kappa(x,y) = \sum_{i=1}^{m} f_i(x) g_i(y)$ be an element of $\mathcal{A}_p$, and let

$$V = \text{Span}\,[f_1, \ldots, f_m, g_1, \ldots, g_m].$$

Then V is a finite-dimensional subspace of $\mathcal{C}(G)$, and by the Gram-Schmidt process there exists an orthonormal basis $u_1, \ldots, u_n$ for V. And it is easy to see that there exists an $n \times n$ real matrix A such that

$$\kappa(x,y) = \sum_{1 \leq i,j \leq n} A(i,j) u_i(x) u_j(y) = \sum_{j=1}^{n} \left(\sum_{i=1}^{n} u_i(x) A(i,j) \right) u_j(y)$$

Following our earlier notational conventions, it will be convenient to let $\mathbf{u}(x) = (u_1(x), \ldots, u_n(x))$. So $\mathbf{u}(x)$ is an element of $\mathcal{C}(G, \mathbb{R}^n)$, and we can write

$$\kappa(x,y) = \mathbf{u}(x) A \cdot \mathbf{u}(y).$$

It follows from *Exercise* 2, p. 290, that $\kappa(x, y)$ is symmetric if and only if the matrix A is symmetric.

Proposition 7.2.7 *If $\kappa(x, y)$ is a symmetric kernel in $\mathcal{A}_p$, then there exist orthonormal functions $v_1(x), \ldots, v_n(x)$ in $\mathcal{C}(G)$ and real numbers $\lambda_1, \ldots, \lambda_n$ such that $\kappa(x,y) = \sum_{i=1}^{n} \lambda_i v_i(x) v_i(y)$.*

Proof. It follows from the previous discussion that there exist orthonormal functions $u_1(x), \ldots, u_n(x)$ and a symmetric matric A such that $\kappa(x, y) = \mathbf{u}(x) A \cdot \mathbf{u}(y)$. By *Theorem 5.2.12*, there exists an orthogonal matrix B and a diagonal matrix $D = \text{Diag}[\lambda_1, \ldots, \lambda_n]$ such that $BDB^t = A$. Define $\mathbf{v}(x) = (v_1(x), \ldots, v_n(x))$ by $\mathbf{v}(x) = \mathbf{u}(x)B$. Then

$$\begin{aligned}\kappa(x,y) &= \mathbf{u}(x)A \cdot \mathbf{u}(y)\\ &= \mathbf{u}(x)BDB^t \cdot \mathbf{u}(y)\\ &= \mathbf{u}(x)BD \cdot \mathbf{u}(y)B\\ &= \mathbf{v}(x)D \cdot \mathbf{v}(y)\\ &= \sum_{i=1}^{n} \lambda_i v_i(x) v_i(y)\end{aligned}$$

and $v_1(x), \ldots, v_n(x)$ are orthonormal by *Exercise* 3, p. 291. □

Proposition 7.2.8 *Let V be a a finite-dimensional subspace of $\mathcal{C}(G)$, and let $v_1, \ldots, v_n$ be an orthonormal basis for V. If $\kappa(x,y) = \sum_{i=1}^{n} \lambda_i v_i(x) v_i(y)$ with $\lambda_1, \ldots, \lambda_n$ nonzero real numbers, then*

$$\lambda_j v_j(x) = \int \kappa(x,y) v_j(y)\, dy$$

and

$$\int\int \kappa(x,y) v_j(y) v_j(x)\, dy\, dx = \lambda_j.$$

Furthermore, $\lambda_1, \ldots, \lambda_n$ are the only nonzero eigenvalues of equation $\lambda\theta = T_\kappa(\theta)$.

Proof. The two equations are straightforward integration calculations. Now suppose $\lambda\theta(x) = \int \kappa(x,y)\theta(y)\, dy$ with $\lambda \neq 0$. It follows from *Exercise* 4, p. 291, that $\lambda\theta$ is in V. Thus θ is in V because $\lambda \neq 0$. If $\lambda \neq \lambda_i$ for all i, then θ is in $V^\perp$. Hence θ is in $V \cap V\perp$ and must be the zero function by *Exercise* 7, p. 274. Since eigenfunctions are not the zero function by definition, it follows that λ must equal some λ_i. □

We have just seen the first indication that the function

$$\beta_\kappa(f,g) = \int\int \kappa(x,y) f(y) g(x)\, dy\, dx = \langle\!\langle T_\kappa(f), g \rangle\!\rangle \tag{7.12}$$

might be a useful tool for studying the basic integral equation (7.7).

To put $\beta_\kappa(f,g)$ in proper context, let V_1 and V_2 be normed linear spaces over $\mathbb{R}$ with norms $\|\cdot\|_a$ and $\|\cdot\|_b$. A function $\beta : V_1 \times V_1 \to V_2$ is *bilinear* provided

$$\beta(c\mathbf{x} + c'\mathbf{x}', \mathbf{y}) = c\beta(\mathbf{x}, \mathbf{y}) + c'\beta(\mathbf{x}', \mathbf{y})$$

and

$$\beta(\mathbf{x}, c\mathbf{y} + c'\mathbf{y}') = c\beta(\mathbf{x}, \mathbf{y}) + c'\beta(\mathbf{x}, \mathbf{y}')$$

for all $c, c' \in \mathbb{R}$ and $\mathbf{x}, \mathbf{x}', \mathbf{y}, \mathbf{y}' \in V_1$. Of course, a bilinear function β is said to be *symmetric* when $\beta(\mathbf{x}, \mathbf{y}) = \beta(\mathbf{y}, \mathbf{x})$ for all $\mathbf{x}$ and $\mathbf{y}$ in V_1. The bilinear functions $\beta : V_1 \times V_1 \to V_2$ clearly form a linear space over $\mathbb{R}$ and the symmetric ones are a subspace.

A bilinear function $\beta : V_1 \times V_1 \to V_2$ is said to be *bounded*, provided that the set

$$\{\|\beta(\mathbf{x},\mathbf{y})\|_b : \|\mathbf{x}\|_a \leq 1 \text{ and } \|\mathbf{y}\|_a \leq 1\}$$

is a bounded subset of $\mathbb{R}$. Like linear functions, bilinear functions are continuous if and only if they are bounded (see *Exercise* 5, p. 291, which is the analogue for bilinear functions of *Proposition 2.4.3*). Obviously, the set of continuous bilinear functions $\beta : V_1 \times V_1 \to V_2$, which will be denoted by $\mathcal{B}^2(V_1, V_2)$, is a subspace of the linear space of all bilinear functions from $V_1 \times V_1$ to V_2. It is normed linear space by *Exercise* 6, p. 291, with norm

$$\|\beta\|_{\mathcal{B}^2} = \sup\{\|\beta(\mathbf{x},\mathbf{y})\|_b : \|\mathbf{x}\|_a \leq 1 \text{ and } \|\mathbf{y}\|_a \leq 1\}..$$

Clearly, β_κ is a bilinear function from $\mathcal{C}(G) \times \mathcal{C}(G)$ to $\mathbb{R}$. If the kernel of the integral equation κ is symmetric, then using equation (7.9)

$$\begin{aligned}
\beta_\kappa(f,g) &= \langle\langle T_\kappa(f), g\rangle\rangle \\
&= \langle\langle f, T_\kappa(g)\rangle\rangle \\
&= \langle\langle T_\kappa(g), f\rangle\rangle \\
&= \beta_\kappa(g,f),
\end{aligned}$$

and β_κ is a symmetric bilinear function. We can now prove an analogue of *Proposition 7.2.1* and its corollary. Again we us the subscripts s and 2 to distinguish the two relevant norms for bounded bilinear functions.

Proposition 7.2.9 *Let G be a compact metric group, and let $\kappa : G \times G \to$ be a continuous function. If β_κ is the bilinear function defined by equation (7.12), then the following hold:*

(a) β_κ is a bounded bilinear linear function from $\mathcal{C}(G)_s \times \mathcal{C}(G)_s$ to $\mathbb{R}$ such that

$$\|\beta_\kappa\|_s = \sup\{|\beta_\kappa(f,g)| : \|f\|_s \leq 1 \text{ and } \|g\|_s \leq 1\} \leq \|\kappa\|_s.$$

(b) β_κ is bounded bilinear linear function from $\mathcal{C}(G)_2 \times \mathcal{C}(G)_2$ to $\mathbb{R}$ such that

$$\|\beta_\kappa\|_2 = \sup\{|\beta_\kappa(f,g)| : \|f\|_2 \leq 1 \text{ and } \|g\|_2 \leq 1\} \leq \|\kappa\|_2.$$

Proof. Using the Cauchy-Schwarz Inequality first, then part *(b)* of *Proposition 7.2.1*, and finally *Exercise* 4, p. 274, yields

$$\begin{aligned}
|\beta_\kappa(f,g)| &= |\langle\langle T_\kappa(f), g\rangle\rangle| \\
&\leq \|T_\kappa(f)\|_2 \, \|g\|_2 \\
&\leq \|\kappa\|_2 \, \|f\|_2 \, \|g\|_2 \\
&\leq \|\kappa\|_s \, \|f\|_s \, \|g\|_s.
\end{aligned}$$

Parts *(a)* and *(b)* follow immediately from the last and second last lines, respectively. □

Corollary 7.2.10 *Let G be a compact metric group and let $\kappa : G \times G \to$ be a continuous function. If β_κ is the bilinear function defined by equation (7.12), then the following hold:*

(a) The function $\kappa \mapsto \beta_\kappa$ is a bounded linear function from $\mathcal{C}(G \times G)_s$ to $\mathcal{B}^2(\mathcal{C}(G)_s, \mathbb{R})$.

(b) The function $\kappa \mapsto \beta_\kappa$ is a bounded linear function from $\mathcal{C}(G \times G)_2$ to $\mathcal{B}^2(\mathcal{C}(G)_2, \mathbb{R})$.

(c) The function $\kappa \mapsto \beta_\kappa$ is a bounded linear function from $\mathcal{C}(G \times G)_s$ to $\mathcal{B}^2(\mathcal{C}(G)_2, \mathbb{R})$.

Proof. The linearity $\kappa \mapsto \beta_\kappa$ follows from the linearity of the integral. Since $\|\beta_\kappa\|_s \leq \|\kappa\|_s$,

$$\sup\{\|\beta_\kappa\|_s : \|\kappa\|_s \leq 1\} \leq 1$$

and $\kappa \mapsto T_\kappa$ is a bounded linear function from $\mathcal{C}(G \times G)_s$ to $\mathcal{B}^2(\mathcal{C}(G)_s, \mathbb{R})$. The same reasoning applies to part *(b)* using $\|\beta_\kappa\|_2 \leq \|\kappa\|_2$. Since $\|\kappa\|_2 \leq \|\kappa\|_s$, we also have $\|\beta_\kappa\|_2 \leq \|\kappa\|_s$. Hence, $\sup\{\|\beta_\kappa\|_2 : \|\kappa\|_s \leq 1\} \leq 1$ proving part *(c)* □

Returning to the special case when κ is a symmetric kernel in $\mathcal{A}_p$, the maximal eigenvalue can be calculated directly from β_κ.

Proposition 7.2.11 *Let V be a a finite-dimensional subspace of $\mathcal{C}(G)$, and let $v_1, \ldots, v_n$ be an orthonormal basis for V. If $\kappa(x, y) = \sum_{i=1}^n \lambda_i v_i(x) v_i(y)$ with $\lambda_1 \geq \lambda_2 \geq \ldots \geq \lambda_n$ and $\lambda_1 > 0$ is a symmetric kernel in $\mathcal{A}_p$, then*

$$\lambda_1 = \beta_\kappa(v_1, v_1) = \sup\left\{\beta_\kappa(f, f) : \|f\|_2 \leq 1\right\}$$

Proof. *Proposition 7.2.8* implies that $\beta_\kappa(v_1, v_1) = \lambda_1 > 0$, and consequently $\sup\{\beta_\kappa(f, f) : \|f\|_2 \leq 1\} \geq \lambda_1 > 0$.

To prove that $\sup\{\beta_\kappa(f, f) : \|f\|_2 \leq 1\} \leq \lambda_1$, let f be an element of $\mathcal{C}(G)$ with $\|f\|_2 = 1$. Using *Exercise* 7, p. 274, write $f = f_1 + f_2$ with $f_1 = \sum_{i=1}^n \langle\langle f, v_i\rangle\rangle v_i$ and f_2 in $V^\perp$. Then, $\beta_\kappa(f, f) = \beta_\kappa(f_1, f_1) + 2\beta_\kappa(f_1, f_2) + \beta_\kappa(f_2, f_2)$. It follows readily from *Exercise* 4, p. 291, that $\beta_\kappa(f_1, f_2) = 0 = \beta_\kappa(f_2, f_2)$. Thus $\beta_\kappa(f, f) = \beta_\kappa(f_1, f_1)$.

Now by *Proposition 7.2.8*,

$$\begin{aligned} T_\kappa(f_1) &= T_\kappa\left(\sum_{i=1}^n \langle\langle f, v_i\rangle\rangle v_i\right) \\ &= \sum_{i=1}^n \langle\langle f, v_i\rangle\rangle T_\kappa(v_i) \\ &= \sum_{i=1}^n \langle\langle f, v_i\rangle\rangle \lambda_i v_i. \end{aligned}$$

Consequently,

$$\begin{aligned}\beta_\kappa(f_1, f_1) &= \langle\langle T_\kappa(f_1), f_1\rangle\rangle \\ &= \left\langle\left\langle \sum_{i=1}^{n} \langle\langle f, v_i\rangle\rangle \lambda_i v_i, \sum_{j=1}^{n} \langle\langle f, v_j\rangle\rangle v_j \right\rangle\right\rangle \\ &= \sum_{i=1}^{n} \lambda_i \langle\langle f, v_i\rangle\rangle^2.\end{aligned}$$

Therefore,

$$\beta_\kappa(f, f) = \sum_{i=1}^{n} \lambda_i \langle\langle f, v_i\rangle\rangle^2. \tag{7.13}$$

Now using Bessel's Inequality [equation (7.3)], it follows that

$$\begin{aligned}\beta_\kappa(f, f) &= \sum_{i=1}^{n} \lambda_i \langle\langle f, v_i\rangle\rangle^2 \\ &\leq \lambda_1 \sum_{i=1}^{n} \langle\langle f, v_i\rangle\rangle^2 \\ &\leq \lambda_1 \|f\|_2^2.\end{aligned}$$

Hence $\sup\{\beta_\kappa(f, f) : \|f\|_2 \leq 1\} \leq \lambda_1$. □

Define $\rho : \mathcal{B}^2(\mathcal{C}(G)_2, \mathbb{R}) \to \mathbb{R}$ by setting

$$\rho(\beta) = \sup\{\beta(f, f) : \|f\|_2 \leq 1\}.$$

Note that $\rho(\beta) \geq 0$ because $\beta(0, 0) = 0$. It follows from *Exercise* 7, p. 291, that ρ is a continuous non-negative function. Since $\kappa \mapsto \beta_\kappa$ is a continuous function from $\mathcal{C}(G \times G)_s$ to $\mathcal{B}^2(\mathcal{C}(G)_2, \mathbb{R})$ by *Proposition 7.2.10*, the function $\kappa \mapsto \rho(\beta_\kappa)$ is a continuous real-valued function on $\mathcal{C}(G \times G)_s$.

The continuous function $\rho(\beta_\kappa)$ on $\mathcal{C}(G \times G)_s$ will play a major role in the remainder of this section. Having shown that it is the composition of continuous functions and thus continuous, it will be convenient to write $\rho(\kappa)$ instead of $\rho(\beta_\kappa)$. In summary,

$$\begin{aligned}\rho(\kappa) &= \sup\{\beta_\kappa(f, f) : \|f\|_2 \leq 1\} \\ &= \sup\{\langle\langle T_\kappa(f), f\rangle\rangle : \|f\|_2 \leq 1\} \\ &= \sup\left\{\int\int \kappa(x, y) f(x) f(y)\, dx\, dy : \|f\|_2 \leq 1\right\}\end{aligned}$$

is a continuous non-negative real-valued function on $\mathcal{C}(G \times G)_s$.

In preparation for *Theorem 7.2.12*, we want to recast the statement of *Proposition 7.2.11*. If κ is a symmetric kernel in $\mathcal{A}_p$, then by *Proposition 7.2.7* there exist orthonormal functions $v_1(x), \ldots, v_n(x)$ in $\mathcal{C}(G)$ and real numbers $\lambda_1, \ldots, \lambda_n$ such that $\kappa(x,y) = \sum_{i=1}^{n} \lambda_i v_i(x) v_i(y)$. Using equation (7.13), it is easy to see that $\lambda_i > 0$ for some i if and only if $\beta_\kappa(f,f) > 0$ for some f if and only if $\rho(\kappa) > 0$. Now *Proposition 7.2.11* can be restated as follows: *If κ is a symmetric kernel in $\mathcal{A}_p$ and $\rho(\kappa) > 0$, then $\rho(\kappa)$ is the largest positive eigenvalue of the integral equation (7.7).* The next step and main theorem in this section is to extend this result to all symmetric kernels.

Theorem 7.2.12 *Let $\kappa(x,y)$ be a symmetric kernel for the compact metric group G. If $\rho(\kappa) > 0$, then $\rho(\kappa)$ is the largest positive eigenvalue of the integral equation $\lambda\theta = T_\kappa(\theta)$.*

Proof. Suppose $\lambda\theta = T_\kappa(\theta)$ holds for a specific nonzero θ and $\lambda > 0$. Without loss of generality it can be assumed that $\|\theta\|_2 = 1$. Then,

$$\begin{aligned} \beta_\kappa(\theta,\theta) &= \langle\!\langle T_\kappa(\theta), \theta \rangle\!\rangle \\ &= \langle\!\langle \lambda\theta, \theta \rangle\!\rangle \\ &= \lambda. \end{aligned}$$

Therefore, $\lambda \leq \rho(\kappa)$. It remains to show that $\rho(\kappa)$ is a eigenvalue, which is the heart of the proof.

Because $\mathcal{A}_p$ is dense in $\mathcal{C}(G \times G)_s$, there exists a sequence of kernels κ_m in $\mathcal{A}$ converging to κ in $\mathcal{C}(G \times G)_s$. If κ_m is not symmetric, then replace it with the symmetric kernel $(\kappa_m(x,y) + \kappa_m(y,x))/2$ and observe that

$$\begin{aligned} \left| \kappa(x,y) - \frac{\kappa_m(x,y) + \kappa_m(y,x)}{2} \right| &= \\ \left| \frac{\kappa(x,y) + \kappa(y,x)}{2} - \frac{\kappa_m(x,y) + \kappa_m(y,x)}{2} \right| &\leq \\ \frac{1}{2}\left|\kappa(x,y) - \kappa_m(x,y)\right| + \frac{1}{2}\left|\kappa(y,x) - \kappa_m(y,x)\right| \end{aligned}$$

from which it follows that

$$\left\| \kappa(x,y) - \frac{\kappa_m(x,y) + \kappa_m(y,x)}{2} \right\|_s \leq \|\kappa - \kappa_m\|_s.$$

Thus there exists a sequence of symmetric kernels in $\mathcal{A}_p$ converging to κ in $\mathcal{C}(G \times G)_s$. Moreover, $\rho(\kappa_m)$ converges to $\rho(\kappa)$ because ρ is continuous on $\mathcal{C}(G \times G)_s$. *Propositions 7.2.7, 7.2.8, and 7.2.11* imply that for each m there exists a characteristic function θ_m with eigenvalue $\rho(\kappa_m)$, that is

$$\rho(\kappa_m)\theta_m(x) = \int \kappa_m(x,y)\theta_m(y)\, dy.$$

Without loss of generality it can be assumed that $\|\theta_m\|_2 = 1$.

Let d be a metric for the compact metric group G. To prove that a subsequence of the sequence of characteristic functions θ_m converges in $\mathcal{C}(G)_s$, we will make use of both the "if" and "only if" part of *Corollary 6.2.8* to Ascoli's Theorem. First, $\{\kappa_m : m \geq 1\} \cup \{\kappa\}$ is a compact subset of $\mathcal{C}(G \times G)_s$, and hence an equicontinuous family of functions on the compact space $G \times G$. Second, we will use the equicontinuity of the sequence κ_m to apply *Corollary 6.2.8* to the sequence of functions θ_m.

Without loss of generality it can be assumed that $\rho(\kappa_m) \geq \rho(\kappa)/2$ or

$$\frac{1}{\rho(\kappa_m)} \leq \frac{2}{\rho(k)}.$$

Because $\{\kappa_m : m \geq 1\} \cup \{\kappa\}$ is an equicontinuous family of functions on $G \times G$, given $\varepsilon > 0$ there exists $\delta > 0$ such that $|\kappa_m(x, y) - \kappa_m(x', y')| < \varepsilon$ for all m when $d(x, x') + d(y, y') < \delta$. Consequently, when $d(x, x') < \delta$

$$\begin{aligned}
|\theta_m(x) - \theta_m(x')| &= \frac{1}{\rho(\kappa_m)} \left| \int (\kappa_m(x, y) - \kappa_m(x', y))\theta_m(y)\, dy \right| \\
&\leq \frac{2}{\rho(\kappa)} \left(\int (\kappa_m(x, y) - \kappa_m(x', y))^2\, dy \right)^{\frac{1}{2}} \left(\int \theta_m(y)^2\, dy \right)^{\frac{1}{2}} \\
&\leq \frac{2}{\rho(\kappa)} \left(\int \varepsilon^2\, dy \right)^{\frac{1}{2}} \\
&\leq \frac{2\varepsilon}{\rho(\kappa)}
\end{aligned}$$

for all m. Therefore, $\{\theta_m : m \geq 1\}$ is an equicontinuous family of functions in $\mathcal{C}(G)_s$.

Because κ_m is a convergent sequence in $\mathcal{C}(G)_s$, there exists $K > 0$ such that $\|\kappa_m\|_s \leq K$ for all m. Hence

$$\begin{aligned}
|\theta_m(x)| &= \frac{1}{\rho(\kappa_m)} \left| \int \kappa_m(x, y)\theta_m(y)\, dy \right| \\
&\leq \frac{2}{\rho(\kappa)} \left(\int \kappa_m(x, y)^2\, dy \right)^{\frac{1}{2}} \left(\int \theta_m(y)^2\, dy \right)^{\frac{1}{2}} \\
&\leq \frac{2}{\rho(\kappa)} \left(\int K^2\, dy \right)^{\frac{1}{2}} \\
&\leq \frac{2K}{\rho(\kappa)}
\end{aligned}$$

for all m. So for each x in G, the set of real numbers $\{\theta_m(x) : m \geq 1\}$ is contained in the closed interval $[-2K/\rho(\kappa), -2K/\rho(\kappa)]$ and has a compact closure. Therefore, *Corollary 6.2.8* applies to the set of functions $\{\theta_m; m \geq 1\}$, and its closure in $\mathcal{C}(G)_s$ is compact. We can assume without loss of generality by passing to a subsequence that θ_m converges in $\mathcal{C}(G)_s$ to a function θ.

By *Proposition 7.2.2*, the sequence of bounded linear functions T_{κ_m} converges to T_κ in $\mathcal{B}(\mathcal{C}(G)_s, \mathcal{C}(G)_s)$. It then follows from *Exercise* 5, p. 99, that $T_{\kappa_m}(\theta_m)$

converges to $T_\kappa(\theta)$. Since $\rho(\kappa_m)\theta_m$ equals $T_{\kappa_m}(\theta_m)$ and converges to $\rho(\kappa)\theta$,

$$\rho(\kappa)\theta = T_\kappa(\theta) = \int \kappa(x,y)\theta(y)\,dy$$

Thus $\rho(\kappa)$ is a eigenvalue for the integral equation. □

Corollary 7.2.13 *Let $\kappa(x,y)$ be a symmetric kernel for the compact metric group G. If $\rho(-\kappa) > 0$, then $-\rho(-\kappa)$ is the smallest negative eigenvalue of the integral equation $\lambda\theta = T_\kappa(\theta)$.*

Proof. λ is a eigenvalue of $T_{-\kappa}$ if and only if $-\lambda$ is a eigenvalue of T_κ. □

Corollary 7.2.14 *If κ is a symmetric kernel, then every eigenvalue of the integral equation $\lambda\theta = T_\kappa(\theta)$ is in the closed interval $[-\rho(-\kappa), \rho(\kappa)]$.*

Proof. Obviously, $-\rho(-\kappa) \le 0 \le \rho(\kappa)$. If λ is a nonzero eigenvalue for T_κ, then there exists a eigenfunction θ for λ such that $\|\theta\|_2 = 1$. It follows that $\beta_\kappa(\theta,\theta) = \lambda$ and either $0 < \lambda \le \rho(\kappa)$ or $-\rho(-\kappa) \le \lambda < 0$. □

Corollary 7.2.15 *If κ is a nonzero symmetric kernel, then the integral equation $\lambda\theta = T_\kappa(\theta)$ has a nonzero eigenvalue.*

Proof. Use part (a) of *Exercise* 8, p. 291, to show that $\beta_\kappa(f,f) \ne 0$ for some f if and only if β_κ is nonzero. Then apply *Exercise* 9, p. 292, to show either the theorem or *Corollary 7.2.13* applies to every nonzero kernel. □

The next two propositions provides the foundation for an effective constructive process. In some sense, this process allows us to optimally decompose a symmetric kernel by symmetric kernels in $\mathcal{A}_p$, obtaining the eigenvalues and orthonormal eigenfunctions along the way.

Proposition 7.2.16 *Let κ be a symmetric kernel for the compact metric group G, and let $v_1(x), \ldots, v_p(x)$ be orthonormal eigenfunctions with eigenvalues $\lambda_j \ne 0$ for $1 \le j \le p$. Set $V = \operatorname{Span}[v_1, \ldots, v_p]$ and*

$$\gamma(x,y) = \kappa(x,y) - \sum_{j=1}^{p} \lambda_j v_j(x) v_j(y).$$

Then the following hold:

(a) $T_\gamma(f) = 0$ for all $f \in V$.

(b) $T_\kappa(f) \in V$ for all $f \in V$.

(c) $T_\gamma(f) \in V^\perp$ for all $f \in V^\perp$.

(d) $T_\kappa(f) \in V^\perp$ for all $f \in V^\perp$,

(e) $T_\gamma(f) = T_\kappa(f)$ for all $f \in V^\perp$.

Proof. Starting with part *(a)*,

$$\begin{aligned} T_\gamma(v_i)(x) = \int \gamma(x,y)v_i(y)\,dy &= \\ \int \Big(\kappa(x,y) - \sum_{j=1}^{p} \lambda_j v_j(x)v_j(y)\Big)v_i(y)\,dy &= \\ \int \kappa(x,y)v_i(y)\,dy - \sum_{j=1}^{p} \lambda_j v_j(x) \int v_j(y)v_i(y)\,dy &= \\ T_\kappa(v_i) - \lambda_i v_i(x) &= 0. \end{aligned}$$

It follows that $T_\gamma(f) = 0$ for all f in V, proving *(a)*.

Part *(b)* is obvious because V has a basis consisting of eigenfunctions functions for T_κ. Parts *(c)* and *(d)* follow from parts *(a)* and *(b)*, respectively, using $\langle\langle T_\gamma(f), g\rangle\rangle = \langle\langle f, T_\gamma(g)\rangle\rangle$ and $\langle\langle T_\kappa(f), g\rangle\rangle = \langle\langle f, T_\kappa(g)\rangle\rangle$.

Part *(e)* is the following integration calculation:

$$\begin{aligned} T_\gamma(f) &= \int \Big(\kappa(x,y) - \sum_{j=1}^{p} \lambda_j v_j(x)v_j(y)\Big)f(y)\,dy \\ &= \int \kappa(x,y)f(y)\,dy - \sum_{j=1}^{p} \lambda_j v_j(x) \int v_j(y)f(y)\,dy \\ &= \int \kappa(x,y)f(y)\,dy \\ &= T_\kappa(f). \end{aligned}$$

This completes the proof. □

Proposition 7.2.17 *Let κ be a symmetric kernel for the compact metric group G, and let $v_1(x), \ldots, v_p(x)$ be orthonormal eigenfunctions with eigenvalues $\lambda_j \neq 0$ for $1 \leq j \leq p$. Set $V = \text{Span}\,[v_1, \ldots, v_p]$ and*

$$\gamma(x,y) = \kappa(x,y) - \sum_{j=1}^{p} \lambda_j v_j(x)v_j(y).$$

Then the following are equivalent for any nonzero function f in $C(G)$ and nonzero real number λ:

(a) $T_\gamma(f) = \lambda f$.

(b) $f \in V^\perp$ *and* $T_\kappa(f) = \lambda f$.

Furthermore, $\rho(\gamma) \leq \rho(\kappa)$ *and* $\rho(-\gamma) \leq \rho(-\kappa)$.

Proof. Suppose f is a nonzero function in $\mathcal{C}(G)$ and λ is a nonzero real number such that $T_\gamma(f) = \lambda f$. Write $f = f_1 + f_2$ with $f_1 \in V$ and $f_2 \in V^\perp$ (*Exercise* 7, p. 274). Then by the previous proposition $\lambda(f_1 + f_2) = T_\gamma(f_1 + f_2) = T_\gamma(f_2) \in V^\perp$. Hence $\lambda f_1 = 0$ and $f_1 = 0$ because $\lambda \neq 0$, proving that $f = f_2 \in V^\perp$. It now follows from part (e) of the *Proposition 7.2.16* that $T_\kappa(f) = \lambda f$. The converse is an immediate consequence of part (e) of the same proposition.

Consequently, if $\lambda \neq 0$ is a eigenvalue for T_γ, then λ is a eigenvalue for T_κ and $-\rho(-\kappa) \leq \lambda \leq \rho(\kappa)$. It follows that $\rho(\gamma) \leq \rho(\kappa)$ and $\rho(-\gamma) \leq \rho(-\kappa)$ because $\rho(\gamma)$ and $-\rho(-\gamma)$are eigenvalues for T_γ. □

In preparation for proving that there is a constructive process that completely describes the solutions of $\lambda\theta = T_\kappa(\theta)$, we need to put in place some notation that will be used as part of the constructive process. Set

$$\begin{aligned} \gamma_1 &= \kappa \\ \lambda_1 &= \rho(\gamma_1) \\ \lambda_{-1} &= -\rho(-\gamma_1) \\ \Upsilon_1 &= \{j : |j| = 1 \text{ and } \lambda_j \neq 0\} \\ \Lambda_1 &= \Upsilon_1 \end{aligned}$$

For $k > 1$, the sets Υ_k and Λ_k will not be equal and Λ_k will provide an indexing set for the eigenvalues occurring in the first k steps. The next proposition summarizes our key results in the above notation. The proof is left to the reader.

Proposition 7.2.18 *If κ is a nonzero symmetric kernel for the compact metric group G, then the following statements hold:*

(a) λ_1 and λ_{-1} are not both zero and $\Upsilon_1 \neq \phi$.

(b) If j is in Λ_1, then λ_j is a nonzero eigenvalue of T_κ.

(c) For each $j \in \Upsilon_1$ there exists a eigenfunction θ_j for λ_j such that $\|\theta_j\|_2 = 1$.

(d) If $\Lambda_1 = \{-1, 1\}$, then θ_{-1} and θ_1 are orthogonal.

(e) $\{\theta_j : j \in \Lambda_1\}$ is an orthonormal set of eigenfunctions.

(f) The open intervals $(-\infty, \lambda_{-1})$ and (λ_1, ∞) do not contain any eigenvalues of T_κ.

The stage is now set for an inductive process that provides a complete structural description of the solutions of the integral equation (7.7). *The careful use of Proposition 7.2.17 in both directions is the key to understanding the construction, which requires lots of notation and details.*

Theorem 7.2.19 *If κ is a nonzero symmetric kernel for the compact metric group G, then for all $m \geq 2$ there exists a symmetric kernel $\gamma_m(x,y)$, a subset Υ_m of $\{-m,m\}$, a subset Λ_m of nonzero integers, and for each j in Λ_m an eigenvalue λ_j and eigenfunction θ_j such that the following hold:*

(a)

$$\begin{aligned}\gamma_m(x,y) &= \kappa(x,y) - \sum_{j\in\Lambda_{m-1}} \lambda_j\theta_j(x)\theta_j(y) \\ &= \gamma_{m-1}(x,y) - \sum_{j\in\Upsilon_{m-1}} \lambda_j\theta_j(x)\theta_j(y).\end{aligned}$$

(b)

$$0 \leq \lambda_m = \rho(\gamma_m) \leq \rho(\gamma_{m-1})$$

and

$$0 \geq \lambda_{-m} = -\rho(-\gamma_{-m}) \geq -\rho(-\gamma_{m-1}).$$

(c)

$$\Upsilon_m = \{j : |j| = m \text{ and } \lambda_j \neq 0\},$$

and for each j in Υ_m, the number λ_j is a nonzero eigenvalue of T_κ such that there are no eigenvalues for T_κ in the open intervals $(\lambda_{-(m-1)}, \lambda_{-m})$ and $(\lambda_m, \lambda_{m-1})$.

(d)

$$\Lambda_m = \bigcup_{j=1}^{m} \Upsilon_j,$$

and for each $j \in \Lambda_m$ there exists a eigenfunction θ_j with eigenvalue λ_j for T_κ such that

$$\{\theta_j : j \in \Lambda_m\}$$

is an orthonormal set of eigenfunctions for T_κ.

Furthermore, if λ is a nonzero eigenvalue for T_κ, then $\lambda = \lambda_j$ for some

$$j \in \Lambda = \bigcup_{m=1}^{\infty} \Upsilon_m$$

and

$$\{\theta_j : \lambda_j = \lambda \text{ with } j \in \Lambda\}$$

is an orthonormal basis for $V_\lambda = \{f \in \mathcal{C}(G) : T_\kappa(f) = \lambda f\}$.

Proof. The proof proceeds by induction starting with $m = 2$ and building on *Proposition 7.2.18.* Clearly,

$$\gamma_2(x,y) = \kappa(x,y) - \sum_{j\in\Lambda_1} \lambda_j\theta_j(x)\theta_j(y)$$

is a symmetric kernel for G, and both

$$0 \leq \lambda_2 = \rho(\gamma_2) \leq \rho(\gamma_1)$$

and

$$0 \geq \lambda_{-2} = -\rho(-\gamma_{-2}) \geq -\rho(-\gamma_{-1}).$$

follow from the last part of *Proposition 7.2.17.* Thus *(a)* and *(b)* hold for $m = 2$.

Continuing to part *(c)*, consider j in Υ_2. Since λ_j is an eigenvalue for T_{γ_2} by *Theorem 7.2.12*, it is also an eigenvalue for T_κ by *Proposition 7.2.17.*

Suppose λ is a positive eigenvalue of T_κ not equal to λ_1. Then there exists an eigenfunction θ for λ, and θ is orthogonal to θ_j for $j \in \Upsilon_1$ by *Proposition 7.2.3.* Hence, λ is a eigenvalue for T_{γ_2} by *Proposition 7.2.17*, and $0 < \lambda \leq \rho(\gamma_2) = \lambda_2$. So λ is not in (λ_2, λ_1). Similarly, there are no negative eigenvalues of T_κ in $(\lambda_{-(m-1)}, \lambda_{-m})$. This proves that *(c)* holds for $m = 2$.

For $j \in \Upsilon_2$, there exists an eigenfunction θ_j with eigenvalue λ_j for T_{γ_2} such that $\|\theta_j\|_2^2 = 1$. Then by *Proposition 7.2.17* again θ_j is an eigenfunction for T_κ and orthogonal to θ_i when $i \in \Upsilon_1$. When $\Upsilon_2 = \{-2, 2\}$, obviously $\lambda_{-2} \neq \lambda_2$ and θ_{-2} and θ_2 are orthogonal by *Proposition 7.2.3.* Thus part *(d)* holds and completes the $m = 2$ case. (Note that this argument works perfectly even when $\lambda_2 = \lambda_1$ or $\lambda_{-2} = \lambda_{-1}$. Thus the construction allows the dimensions of $\{f \in \mathcal{C}(G) : T_\kappa(f) = \lambda_1 f\}$ and $\{f \in \mathcal{C}(G) : T_\kappa(f) = \lambda_{-1} f\}$ to grow, but in a controlled way.)

Suppose *(a)* through *(d)* hold for $2 \leq k \leq m$. Showing that they also hold for $m+1$ closely parallels the $m = 2$ case with repeated uses of *Proposition 7.2.17.* The key to the induction step is that

$$\gamma_{m+1}(x,y) = \kappa(x,y) - \sum_{j \in \Lambda_m} \lambda_j \theta_j(x)\theta_j(y) \tag{7.14}$$

$$= \gamma_m(x,y) - \sum_{j \in \Upsilon_m} \lambda_j \theta_j(x)\theta_j(y). \tag{7.15}$$

Using (7.15), parts *(a)* and *(b)* follow exactly as in the $m = 2$ case.

When j is in Υ_{m+1}, it follows, as in the $m = 2$ case, from (7.14) that λ_j is a nonzero eigenvalue for T_κ and there exists an eigenfunction θ_j such that $\|\theta_j\|_2^2 = 1$ and θ_j is orthogonal to θ_i for i in Λ_m.

If λ is a positive eigenvalue for T_κ not equal to λ_i such that $1 \leq i \leq m$, then it follows that λ is a eigenvalue for $T_{\gamma_{m+1}}$ and $\lambda \leq \rho(\gamma_{m+1})$, as in the $m = 2$ case. The negative case is similar and *(c)* holds for $m + 1$.

Part *(d)* follows as in the $m = 2$ case. This completes the induction and proves that *(a)* through *(d)* hold for all $m \geq 2$.

Observe that if m is not in Υ_m, then $\lambda_p = 0$ for all $p \geq m$. Similarly, $\lambda_{-p} = 0$ for all $p \geq m$ when $-m$ is not in Υ_m. If $\Upsilon_m = \phi$, then $\rho(\gamma_m) = 0 = \rho(-\gamma_m)$ and κ is in $\mathcal{A}_p$ by *Exercise* 10, p. 292.

If j is in Υ_m, then $\text{Dim}\,[\{f : T_\kappa(f) = \lambda_j f\}] < \infty$ by *Corollary 7.2.5.* In particular, given λ_k, the set $\{j : \lambda_j = \lambda_k\}$ is a finite set of consecutive integers because the sequence λ_i is non increasing and the λ_{-i} is non-decreasing.

Therefore, either the sequence λ_m $\{\lambda_{-m}\}$ is eventually 0 or contains infinitely many distinct positive {negative} terms. Now, *Corollary 7.2.6* implies that

$$\lim_{m\to\infty} \lambda_m = 0$$

and

$$\lim_{m\to\infty} \lambda_{-m} = 0.$$

Turning to the final statement of the theorem, let λ be a nonzero eigenvalue for T_κ. So λ satisfies $\lambda_{-1} \le \lambda \le \lambda_1$ by *Proposition 7.2.18.* Let $E = \{j : \lambda = \lambda_j \text{ with } j \in \Lambda\}$. It follows that E is either empty or a finite sequence of consecutive integers.

If E is empty, then λ must be in one of the open intervals $(\lambda_{-(m-1)}, \lambda_{-m})$ and $(\lambda_m, \lambda_{m-1})$ for some m, contradicting property *(c)*. Therefore, E is nonempty and a finite sequence of consecutive integers in Λ.

Assuming that $\lambda > 0$, there exist positive integers p and q such that $E = \{k : p \le k \le q\}$. It suffices to show that

$$\text{Span}\,[\{\theta_k : k \in E\}] = \{f : T_\kappa(f) = \lambda f\} = V_\lambda.$$

Obviously, the left side is contained in the right side. If they are not equal, then by the Gram-Schmidt Process there exists θ in V_λ such that $\|\theta\|_2 = 1$ and θ is orthogonal to every θ_j such that $p \le j \le q$. Because $\lambda_j > \lambda$ for $j < p$, the eigenfunction θ is also orthogonal to θ_j for $j < p$. Hence θ is in

$$(\text{Span}\,[\{\theta_j : |j| \le q\}])^\perp.$$

Now, *Proposition 7.2.17* implies that θ is a eigenfunction for $T_{\gamma_{q+1}}$. Consequently,

$$\lambda \le \rho(\gamma_{q+1}) = \lambda_{q+1} \le \lambda_q = \lambda$$

and $q + 1$ is in E, a contradiction. Thus $\text{Span}\,[\{\theta_k : k \in E\}] = V_\lambda$. □

EXERCISES

1. Show that when $\kappa(x, y)$ is a symmetric kernel for a compact group, then the set of eigenvalues for T_κ is either finite or countable.

2. Let $u_1, \ldots, u_n$ be linear independent functions in $C(G)$, and let A be an $n \times n$ real matrix. Define a function $\kappa : G \times G \to \mathbb{R}$ by

$$\kappa(x, y) = \sum_{1 \le i,j \le n} A(i, j) u_i(x) u_j(y).$$

Prove the following about the function κ:

(a) $\kappa(x, y) = 0$ for all x and y in X if and only if A is the zero matrix.

(b) $\kappa(x, y) = \kappa(y, x)$ for all x and y in X if and only if the matrix A is symmetric.

3. Let $u_1, \ldots, u_n$ be orthonormal functions in $\mathcal{C}(G)_2$, and let B be an $n \times n$ real matrix. Set $v_p(x) = \sum_{i=1}^{n} u_i(x)B(i, p)$. Prove that the functions $v_1(x), \ldots, v_n(x)$ are orthonormal if and only if B is an orthogonal matrix.

4. Let $v_1, \ldots, v_n$ be an orthonormal basis for a finite-dimensional subspace V of $\mathcal{C}(G)_2$. If $\kappa(x, y) = \sum_{i=1}^{n} \lambda_i v_i(x) v_i(y)$ with $\lambda_1, \ldots, \lambda_n$ nonzero real numbers, then

$$T_\kappa\left(\mathcal{C}(G)_2\right) \subset V.$$

5. Let V_1 and V_2 be normed linear spaces over $\mathbb{R}$. Prove that the following are equivalent for a bilinear function $\beta : V_1 \times V_1 \to V_2$:

   (a) β is continuous.

   (b) β is continuous at $(\mathbf{0}, \mathbf{0})$.

   (c) β is bounded.

6. Let V_1 and V_2 be normed linear spaces over $\mathbb{R}$ with norms $\|\cdot\|_a$ and $\|\cdot\|_b$. Building on the previous exercise, let $\mathcal{B}^2(V_1, V_2)$ be the set of all bounded bilinear functions $\beta : V_1 \times V_1 \to V_2$. Show that $\mathcal{B}^2(V_1, V_2)$ is real normed linear space with norm

$$\|\beta\|_{\mathcal{B}^2} = \sup\{\|\beta(\mathbf{x}, \mathbf{y})\|_b : \|\mathbf{x}\|_a \leq 1 \text{ and } \|\mathbf{y}\|_a \leq 1\}.$$

7. Let V be a real normed linear space with norm $\|\cdot\|_a$. Given $\beta \in \mathcal{B}^2(V, \mathbb{R})$, set $\rho(\beta) = \sup\{\beta(\mathbf{x}, \mathbf{x}) : \|\mathbf{x}\|_a \leq 1\}$. Prove the following:

   (a) Given $\varepsilon > 0$, there exists $\mathbf{x}$ in V such that

$$\rho(\beta) - \rho(\beta') < \beta(\mathbf{x}, \mathbf{x}) - \beta'(\mathbf{x}, \mathbf{x}) + \varepsilon.$$

   (b) $|\rho(\beta) - \rho(\beta')| \leq \|\beta - \beta'\|_{\mathcal{B}^2}$ for all β and β' in $\mathcal{B}^2(V, \mathbb{R})$.

   (c) ρ is a uniformly continuous non-negative function on the normed linear space $\mathcal{B}^2(V, \mathbb{R})$.

   (d) $-\rho(-\beta)\|\mathbf{x}\|_a^2 \leq \beta(\mathbf{x}, \mathbf{x}) \leq \rho(\beta)\|\mathbf{x}\|_a^2$.

8. Let V be a real normed linear space with norm $\|\cdot\|_a$ and let β be a symmetric bilinear function in $\mathcal{B}^2(V, \mathbb{R})$. Prove the following:

   (a) $2\beta(\mathbf{x}, \mathbf{y}) = \beta(\mathbf{x} + \mathbf{y}, \mathbf{x} + \mathbf{y}) - \beta(\mathbf{x}, \mathbf{x}) - \beta(\mathbf{y}, \mathbf{y})$.

   (b) If $\|\mathbf{x}\| \leq 1$ and $\|\mathbf{y}\| \leq 1$, then

$$-\rho(\beta) - \rho(-\beta)) \leq \beta(\mathbf{x}, \mathbf{y}) \leq \rho(\beta) + \rho(-\beta).$$

   (c) $\|\beta\|_{\mathcal{B}^2} \leq \rho(\beta) + \rho(-\beta)$.

9. Let κ be a kernel function for a compact metric group G. Prove that $\beta_\kappa = \mathbf{0}$ if and only if $\kappa = \mathbf{0}$. (One approach is to use *Exercise* 4, p. 27 and *Exercise* 2, p. 274.)

10. Let κ be a symmetric kernel for the compact metric group G. Prove that $\kappa = 0$ if and only if $\rho(\kappa) = \rho(\beta_\kappa) = 0 = \rho(-\beta_\kappa) = \rho(-\kappa)$.

11. Let V_1 and V_2 be a real normed linear spaces. Show that the bounded symmetric bilinear functions are a closed subspace of $\mathcal{B}^2(V_1, V_2)$.

7.3 Eigenfunctions

The results in this section build on the work of Section 7.2. The goal is to approximate continuous functions on a compact metric group G using eigenfunctions. There are two distinct pieces. First, it will be shown that for a symmetric nonzero kernel κ every function in the range of T_κ can be written as a finite or infinite sum of eigenfunctions. After introducing right-invariant kernels, it will be shown that by varying the kernel, any function in $\mathcal{C}(G)$ can be approximated using eigenfunctions. The symmetric right-invariant kernels are particularly interesting because their eigenfunctions can be expressed in terms of homomorphisms of G to $\mathrm{O}(n)$.

Throughout this section the notation and conclusions of *Proposition 7.2.18* and *Theorem 7.2.19* will be used. When κ is a nonzero symmetric kernel for G, there is a nonempty collection of eigenfunctions θ_j with eigenvalues λ_j indexed by

$$\Lambda = \bigcup_{k=1}^{\infty} \Upsilon_k$$

where Υ_j is a subset of $\{-j, j\}$. Let $\Lambda^+ = \Lambda \cap \mathbb{Z}^+$ and $\Lambda^- = \Lambda \cap (-\mathbb{Z}^+)$. Zero is never in Λ, so $\Lambda = \Lambda^+ \cup \Lambda^-$. If $\Lambda^+ \neq \phi$, then Λ^+ is either a finite sequence of consecutive positive integers starting with 1 or all of $\mathbb{Z}^+$. Similarly, $\Lambda^- \neq \phi$ implies it is either a finite sequence of consecutive negative integers starting at -1 or all of $-\mathbb{Z}^+$.

Also recall that

$$\begin{aligned}\gamma_m(x, y) &= \kappa(x, y) - \sum_{j \in \Lambda_{m-1}} \lambda_j \theta_j(x) \theta_j(y) \\ &= \gamma_{m-1}(x, y) - \sum_{j \in \Upsilon_{m-1}} \lambda_j \theta_j(x) \theta_j(y),\end{aligned}$$

defines a symmetric kernel such that

$$0 \leq \lambda_m = \rho(\gamma_m) \leq \rho(\gamma_{m-1})$$

and

$$0 \geq \lambda_{-m} = -\rho(-\gamma_{-m}) \geq -\rho(-\gamma_{m-1}).$$

Proposition 7.3.1 *If κ is a symmetric nonzero kernel, then $\|\beta_{\gamma_m}\|_2$ is eventually equals zero or converges to zero in $\mathcal{B}^2(\mathcal{C}(G)_2, \mathbb{R})$ as m goes to infinity.*

Proof. By *Proposition 7.2.4*

$$\sum_{m\geq 1} \rho(\gamma_m)^2 = \sum_{m\geq 1} \lambda_m^2 \leq \int\int \kappa(x,y)^2\, dy\, dx = \|\kappa\|_2^2$$

Thus either $\rho(\gamma_m)$ is eventually zero or converges to zero as m goes to infinity. The same holds for $\rho(-\gamma_m)$. The conclusion now follows from part *(c)* of *Exercise* 8, p. 291. □

Let κ be a nonzero symmetric kernel, and let θ be an eigenfunction of T_κ with eigenvalue λ. The symmetry of κ implies that $\int \kappa(x,y)\theta(x)\, dx = \lambda\theta(y)$. If $f = T_\kappa(g)$, then using the invariant integral for $G \times G$ (*Corollary 7.1.9*)

$$\begin{aligned}
\langle\!\langle f, \theta\rangle\!\rangle &= \int f(x)\theta(x)\, dx \\
&= \int\int \kappa(x,y)g(y)\theta(x)\, dy\, dx \\
&= \int g(y) \int \kappa(x,y)\theta(x)\, dx\, dy \\
&= \int g(y)\lambda\theta(y)\, dy \\
&= \lambda\langle\!\langle g, \theta\rangle\!\rangle
\end{aligned}$$

and

$$\langle\!\langle T_\kappa(g), \theta\rangle\!\rangle = \lambda\langle\!\langle g, \theta\rangle\!\rangle. \tag{7.16}$$

Proposition 7.3.2 *Let κ be a nonzero symmetric kernel, and let g be in $\mathcal{C}(G)$. If $\Lambda^+ = \mathbb{Z}^+$, then the infinite series*

$$\sum_{i=1}^{\infty} \lambda_i \langle\!\langle g, \theta_i\rangle\!\rangle \theta_i(x)$$

converges in $\mathcal{C}(G)_s$.

Proof. The first task is to show that the series converges for every x in G. Given positive integers p and q with $p < q$,

$$\begin{aligned}
\left|\sum_{i=1}^{q} \lambda_i \langle\!\langle g, \theta_i\rangle\!\rangle \theta_i(x) - \sum_{i=1}^{p} \lambda_i \langle\!\langle g, \theta_i\rangle\!\rangle \theta_i(x)\right| &= \\
\left|\sum_{i=p+1}^{q} \lambda_i \langle\!\langle g, \theta_i\rangle\!\rangle \theta_i(x)\right| &\leq
\end{aligned}$$

$$\left(\sum_{i=p+1}^{q} \langle\!\langle g, \theta_i \rangle\!\rangle^2\right)^{\frac{1}{2}} \left(\sum_{i=p+1}^{q} \lambda_i^2 \theta_i(x)^2\right)^{\frac{1}{2}} \leq$$

$$\left(\sum_{i=p+1}^{q} \langle\!\langle g, \theta_i \rangle\!\rangle^2\right)^{\frac{1}{2}} \left(\int \kappa(x,y)^2\, dy\right)^{\frac{1}{2}} \leq$$

$$\left(\sum_{i=p+1}^{q} \langle\!\langle g, \theta_i \rangle\!\rangle^2\right)^{\frac{1}{2}} \|\kappa\|_s$$

by using the Cauchy-Schwarz Inequality for $\mathbb{R}^{q-p}$, equation (7.10), p. 277, and $\int \kappa(x,y)^2\, dy \leq \|\kappa\|_s^2$ for all $x \in G$. The infinite series $\sum_{i=1}^{\infty} \langle\!\langle g, \theta_i \rangle\!\rangle^2$ converges by Bessel's Inequality. So given $\varepsilon > 0$, there exists $P > 0$ such that

$$\sum_{i=p+1}^{q} \langle\!\langle g, \theta_i \rangle\!\rangle^2 < \left(\frac{\varepsilon}{\|\kappa\|_s}\right)^2$$

when $q > p \geq P$. It follows that for all x in G,

$$\left|\sum_{i=1}^{q} \lambda_i \langle\!\langle g, \theta_i \rangle\!\rangle \theta_i(x) - \sum_{i=1}^{p} \lambda_i \langle\!\langle g, \theta_i \rangle\!\rangle \theta_i(x)\right| < \varepsilon$$

when $q > p \geq P$. Therefore, the series is Cauchy for every x in G and converges to a function g' because $\mathbb{R}$ is complete. Letting q go to infinity in the preceding inequality yields

$$\left|g'(x) - \sum_{i=1}^{p} \lambda_i \langle\!\langle g, \theta_i \rangle\!\rangle \theta_i(x)\right| \leq \varepsilon$$

when $p \geq P$. Thus the convergence to g' is uniform on G. Hence, g' is continuous on G, and $\sum_{i=1}^{\infty} \lambda_i \langle\!\langle g, \theta_i \rangle\!\rangle \theta_i(x)$ converges to g' in $\mathcal{C}(G)_s$. □

Since $-\lambda_{-i}$ is a positive eigenvalue for $T_{-\kappa}$, the series

$$\sum_{i=1}^{\infty} \lambda_{-i} \langle\!\langle g, \theta_{-i} \rangle\!\rangle \theta_{-i}(x)$$

converges in $\mathcal{C}(G)_s$ when $\Lambda^- = -\mathbb{Z}^+$. Therefore, in all cases

$$\sum_{i \in \Lambda} \lambda_i \langle\!\langle g, \theta_i \rangle\!\rangle \theta_i(x)$$

is a finite sum in $\mathcal{C}(G)_s$ or converges in $\mathcal{C}(G)_s$ as $|i| \to \infty$.

Theorem 7.3.3 *If κ is a nonzero symmetric kernel and g is in $\mathcal{C}(G)$, then*

$$\sum_{i \in \Lambda} \lambda_i \langle\!\langle g, \theta_i \rangle\!\rangle \theta_i \tag{7.17}$$

equals $T_\kappa(g)$ when Λ is finite and converges to $T_\kappa(g)$ in $\mathcal{C}(G)_s$ when Λ is infinite.

Proof. The finite case is part of *Exercise* 1, p. 301. We will assume that $\Lambda^+ = \mathbb{Z}^+ = -\Lambda^-$ and leave the cases when Λ^+ is infinite and Λ^- is finite and visa versa to the reader.

Let $f = T_\kappa(g)$ and consider

$$f(x) - \sum_{i=1}^{m} \lambda_i \langle\langle g, \theta_i \rangle\rangle \theta_i(x) - \sum_{i=-1}^{-m} \lambda_i \langle\langle g, \theta_i \rangle\rangle \theta_i(x) =$$

$$f(x) - \sum_{i=1}^{m} \lambda_i \left(\int g(y)\theta_i(y)\,dy \right) \theta_i(x) - \sum_{i=-1}^{-m} \lambda_i \left(\int g(y)\theta_i(y)\,dy \right) \theta_i(x) =$$

$$\int \kappa(x,y)g(y)\,dy - \int \left(\sum_{i=1}^{m} \lambda_i\theta_i(y)\theta_i(x) + \sum_{i=-1}^{-m} \lambda_i\theta_i(y)\theta_i(x) \right) g(y)\,dy =$$

$$\int \left(\kappa(x,y) - \sum_{i\in\Lambda_m} \lambda_i\theta_i(y)\theta_i(x) \right) g(y)\,dy =$$

$$\int \gamma_{m+1}(x,y)g(y)\,dy.$$

Let h be an arbitrary element of $\mathcal{C}(G)$. With the help of the previous calculation,

$$\left| \left\langle\!\!\left\langle f - \sum_{i=1}^{m} \lambda_i \langle\langle g, \theta_i \rangle\rangle \theta_i - \sum_{i=-1}^{-m} \lambda_i \langle\langle g, \theta_i \rangle\rangle \theta_i, h \right\rangle\!\!\right\rangle \right| =$$

$$\left| \int \left(f(x) - \sum_{i=1}^{m} \lambda_i \langle\langle g, \theta_i \rangle\rangle \theta_i(x) - \sum_{i=-1}^{-m} \lambda_i \langle\langle g, \theta_i \rangle\rangle \theta_i(x) \right) h(x)\,dx \right| =$$

$$\left| \int\int \gamma_{m+1}(x,y)g(y)h(x)\,dy\,dx \right| =$$

$$\left| \beta_{\gamma_{m+1}}(g,h) \right| \leq$$

$$\|\beta_{\gamma_{m+1}}\|_2 \, \|g\|_2 \, \|h\|_2.$$

Since $\|\beta_{\gamma_{m+1}}\|_2$ goes to zero as m goes to infinity by *Proposition 7.3.1*,

$$\lim_{m\to\infty} \left\langle\!\!\left\langle f - \sum_{i=1}^{m} \lambda_i \langle\langle g, \theta_i \rangle\rangle \theta_i - \sum_{i=-1}^{-m} \lambda_i \langle\langle g, \theta_i \rangle\rangle \theta_i, h \right\rangle\!\!\right\rangle = 0.$$

The previous calculation was done using $\|\cdot\|_2$. The next step is to calculate the same limit in a different way using $\|\cdot\|_s$. By *Proposition 7.3.2*

$$\sum_{i=1}^{m} \lambda_i \langle\langle g, \theta_i \rangle\rangle \theta_i(x) + \sum_{i=-1}^{-m} \lambda_i \langle\langle g, \theta_i \rangle\rangle \theta_i(x)$$

converges in $\mathcal{C}(G)_s$ to a continuous function g' and

$$g' = \sum_{m=1}^{\infty} \lambda_i \langle\langle g, \theta_i \rangle\rangle \theta_i + \sum_{i=-1}^{-\infty} \lambda_i \langle\langle g, \theta_i \rangle\rangle \theta_i.$$

Since the invariant integral is a continuous linear function on $\mathcal{C}(G)_s$ (*Exercise* 3, p. 274),

$$\begin{aligned}
\lim_{m\to\infty}\left\langle\!\!\left\langle f-\sum_{i=1}^{m}\lambda_i\langle\!\langle g,\theta_i\rangle\!\rangle\theta_i-\sum_{i=-1}^{-m}\lambda_i\langle\!\langle g,\theta_i\rangle\!\rangle\theta_i, h\right\rangle\!\!\right\rangle &= \\
\lim_{m\to\infty}\int\left(f(x)-\sum_{i=1}^{m}\lambda_i\langle\!\langle g,\theta_i\rangle\!\rangle\theta_i(x)-\sum_{i=-1}^{-m}\lambda_i\langle\!\langle g,\theta_i\rangle\!\rangle\theta_i(x)\right)h(x)\,dx &= \\
\int\lim_{m\to\infty}\left(f(x)-\sum_{i=1}^{m}\lambda_i\langle\!\langle g,\theta_i\rangle\!\rangle\theta_i(x)-\sum_{i=-1}^{-m}\lambda_i\langle\!\langle g,\theta_i\rangle\!\rangle\theta_i(x)\right)h(x)\,dx &= \\
\int (f(x)-g'(x))h(x)\,dx &= \\
\langle\!\langle f-g',h\rangle\!\rangle.
\end{aligned}$$

Therefore, $\langle\!\langle f-g',h\rangle\!\rangle=0$ because the two calculations of the limit must yield the same answer. Since h was an arbitrary continuous function on G,

$$\|f-g'\|_2=\langle\!\langle f-g',f-g'\rangle\!\rangle^{1/2}=0$$

and

$$T_\kappa(g)=f=g'=\sum_{m=1}^{\infty}\lambda_i\langle\!\langle g,\theta_i\rangle\!\rangle\theta_i+\sum_{i=-1}^{-\infty}\lambda_i\langle\!\langle g,\theta_i\rangle\!\rangle\theta_i \tag{7.18}$$

to complete the proof □

Corollary 7.3.4 *If κ is a nonzero symmetric kernel and g is in $\mathcal{C}(G)$, then given $\varepsilon>0$, there exists N such that*

$$\left\|T_\kappa(g)-\sum_{i\in\Lambda_k}\lambda_i\langle\!\langle g,\theta_i\rangle\!\rangle\theta_i\right\|_s<\varepsilon$$

for all $k\geq N$.

A kernel κ in $\mathcal{C}(G\times G)$ is *right-invariant* provided $\kappa(xa,ya)=\kappa(x,y)$ for all a in G. The next simple proposition has far reaching consequences.

Proposition 7.3.5 *If κ is a symmetric right-invariant kernel and θ is a eigenfunction for T_κ with eigenvalue $\lambda\neq 0$, then for a in G the function $x\mapsto\theta(xa)$ is also a eigenfunction for T_κ with eigenvalue λ.*

Proof. Clearly, $x\to\theta(xa)$ is continuous. Then,

$$\begin{aligned}
\int\kappa(x,y)\theta(ya)\,dy &= \int\kappa(xa,ya)\theta(ya)\,dy \\
&= \int\kappa(xa,y)\theta(y)\,dx \\
&= \lambda\theta(xa)
\end{aligned}$$

by the right invariance of both κ and the integral. $\square$

Let κ be a symmetric right-invariant kernel, and let λ be a nonzero eigenvalue for T_κ. Recall that

$$V_\lambda = \{\theta \in \mathcal{C}(G) : T_\kappa(\theta) = \lambda\theta\}$$

is a finite-dimensional subspace of $\mathcal{C}(G)$ (*Corollary 7.2.5*) and that on V_λ the norms $\|\cdot\|_s$ and $\|\cdot\|_2$ are equivalent. If $\theta_1(x)$ and $\theta_2(x)$ are in V_λ, then

$$\int \theta_1(xa)\theta_2(xa)\,dx = \int \theta_1(x)\theta_2(x)\,dx$$

In particular, if $\theta_1(x), \ldots, \theta_n(x)$ is an orthonormal basis for V_λ, then so is $\theta_1(xa), \ldots, \theta_n(xa)$ for each $a \in G$.

Let $n = \mathrm{Dim}\,[V_\lambda]$ and let $\theta_1(x), \ldots, \theta_n(x)$ be an orthonormal basis for V_λ. It follows that for each $a \in G$, there exists a $n \times n$ real matrix $\Phi(a)$ such that

$$\theta_j(xa) = \sum_{i=1}^{n} \Phi(a)(i,j)\theta_i(x) \tag{7.19}$$

for all $x \in G$. Note well that the matrix $\Phi(a)$ is unique because $\theta_1(x), \ldots, \theta_n(x)$ is a basis for V_λ. The function $\Phi : G \to \mathcal{M}_n(\mathbb{R})$ is worth further study.

A homomorphism of a topological group into $\mathrm{GL}(n, \mathbb{S})$ is called *a matrix representation*. The integer n is the *dimension of the representation.* If the values of a matrix representation are in $\mathrm{O}(n)$ or $\mathrm{U}(n)$, it is called an *orthogonal representation* or *unitary representation*, respectively. The idea behind the study of representations of a topological group is that they contain useful information about the topological group.

Proposition 7.3.6 *Let κ be a symmetric right-invariant kernel for a compact metric group G, and let λ be a nonzero eigenvalue for T_κ. If $\theta_1(x), \ldots, \theta_n(x)$ is an orthonormal basis for V_λ, the function Φ defined by (7.19) is an orthogonal representation of G.*

Proof. Equation (7.19) can be used to calculate $\langle\langle \theta_k(xa), \theta_m(xa)\rangle\rangle$ as follows:

$$\begin{aligned}
\langle\langle \theta_k(xa), \theta_m(xa)\rangle\rangle &= \\
\int \theta_k(xa)\theta_m(xa)\,dx &= \\
\int \left(\sum_{i=1}^{n} \Phi(a)(i,k)\theta_i(x)\right)\left(\sum_{j=1}^{n} \Phi(a)(j,m)\theta_j(x)\right) dx &= \\
\sum_{i=1}^{n} \Phi(a)(i,k)\Phi(a)(i,m) &= \\
[\Phi(a)^t\Phi(a)](k,m) &
\end{aligned}$$

Since $\theta_1(xa), \ldots, \theta_n(xa)$ is an orthonormal basis for V_λ for each $a \in G$, it follows that $[\Phi(a)^t\Phi(a)](k,m)$ equals 0 or 1 according as $k \neq m$ or $k = m$, in other words, $\Phi(a)^t\Phi(a) = I$. Therefore, $\Phi(a)$ is in $\mathrm{O}(n)$ for all $a \in G$.

To prove that Φ is an algebraic homomorphism, consider a and b in G. On the one hand,

$$\theta_j(xab) = \sum_{i=1}^{n} \Phi(ab)(i,j)\theta_i(x).$$

On the other hand,

$$\begin{aligned}
\theta_j(xab) &= \sum_{k=1}^{n} \Phi(b)(k,j)\theta_k(xa) \\
&= \sum_{k=1}^{n}\sum_{i=1}^{n} \Phi(b)(k,j)\Phi(a)(i,k)\theta_i(x) \\
&= \sum_{i=1}^{n}\sum_{k=1}^{n} \Phi(a)(i,k)\Phi(b)(k,j)\theta_i(x) \\
&= \sum_{i=1}^{n} [\Phi(a)\Phi(b)](i,j)\theta_i(x).
\end{aligned}$$

Since the matrix $\Phi(ab)$ is unique, $\Phi(a)\Phi(b) = \Phi(ab)$ and Φ is an algebraic homomorphism of G into $\mathrm{O}(n)$.

To prove that Φ is continuous, it suffices to prove that it is continuous at e the identity of G (*Exercise* 5, p. 39). If Φ is not continuous at e, there exists a sequence a_k converging to e in G such that $\Phi(a_k)$ does not converge to I. Because $\mathrm{O}(n)$ is compact (*Theorem 4.2.4*), we can assume without loss of generality that $\Phi(a_k)$ converges to A in $\mathrm{O}(n)$ such that $A \neq I$. Then,

$$\begin{aligned}
\theta_j(x) &= \lim_{k\to\infty} \theta_j(xa_k) \\
&= \lim_{k\to\infty} \sum_{i=1}^{n} \Phi(a_k)(i,j)\theta_i(x) \\
&= \sum_{i=1}^{n} A(i,j)\theta_i(x).
\end{aligned}$$

Now, the uniqueness of $\Phi(a)$ implies that $A = \Phi(e) = I$, a contradiction. Therefore, Φ is continuous at e and a homomorphism of G to $\mathrm{O}(n)$. □

Corollary 7.3.7 *Let κ be a symmetric right-invariant kernel for a compact metric group G, and let λ be a nonzero eigenvalue for T_κ. There exists an orthogonal representation Φ of G such that every eigenfunction for λ is a linear combination of the functions $\Phi(x)(i,j)$.*

Proof. Let $\theta_1, \ldots, \theta_n$ be an orthonormal basis of V_λ, and let $\Phi : G \to \mathrm{O}(n)$ be the orthogonal representation from the proposition. Since both x and a can vary in (7.19), setting $x = e$ it can be written as yields

$$\theta_j(a) = \sum_{i=1}^{n} \Phi(a)(i,j)\theta_i(e) = \sum_{i=1}^{n} \theta_i(e)\Phi(a)(i,j)$$

or

$$\theta_j = \sum_{i=1}^{n} \theta_i(e)\Phi(i,j),$$

showing that θ_j is a linear combination of the entries of Φ. Because every eigenfunction is a liner combination of the orthogonal basis $\theta_1, \ldots, \theta_n$ for V_λ, every eigenfunction is a linear combination of the entries of $\Phi(x)$. $\square$

Theorem 7.3.8 *Let f be in $\mathcal{C}(G)$ with G a compact metric group. Given $\varepsilon > 0$, there exists a symmetric right-invariant kernel κ such that $\|f - T_\kappa(f)\|_s < \varepsilon$.*

Proof. Let d be a right-invariant metric for G. Because G is compact, f is uniformly continuous on G. So there exists $\delta > 0$ such that $|f(x) - f(y)| < \varepsilon$ when $d(x,y) < \delta$ or equivalently $d(e, yx^{-1}) < \delta$. Using *Exercise* 4, p. 27, with $y = e$, there exists $\gamma \in \mathcal{C}(G)$ with the following properties:

(a) $0 \le \gamma(x) \le 1$ for all $x \in G$

(b) $\gamma(x) = 1$ if and only if $d(x,e) \le \delta/2$

(c) $\gamma(x) = 0$ if and only if $d(x,y) \ge \delta$

(d) $\gamma(x^{-1}) = \gamma(x)$.

[The last property follows from the construction because $d(x^{-1}, e) = d(x,e)$.]

Since $\int \gamma(x)\,dx > 0$ (*Exercise* 2, p. 274), there exists $\alpha > 0$ such that $\int \alpha\gamma(x)\,dx = 1$. Set $\kappa(x,y) = \alpha\gamma(yx^{-1})$. Then,

$$\kappa(y,x) = \alpha\gamma(xy^{-1}) = \alpha\gamma\big((yx^{-1})^{-1}\big) = \alpha\gamma(yx^{-1}) = \kappa(x,y)$$

and

$$\kappa(xa, ya) = \alpha\gamma(ya(xa)^{-1}) = \alpha\gamma(yaa^{-1}x^{-1}) = \alpha\gamma(yx^{-1}) = \kappa(x,y).$$

So κ is a non-negative symmetric right-invariant kernel. Furthermore,

$$\int \kappa(x,y)\,dy = \alpha \int \gamma(yx^{-1})\,dy = \alpha \int \gamma(y)\,dy = 1,$$

and

$$\int \kappa(x,y) f(x)\,dy = f(x) \int \kappa(x,y)\,dy = f(x).$$

Notice that

$$\gamma(yx^{-1})|f(x)-f(y)| \begin{cases} =0 & \text{when } d(e,yx^{-1}) \geq \delta \\ \leq \varepsilon\gamma(yx^{-1}) & \text{when } d(e,yx^{-1}) < \delta \end{cases},$$

and hence

$$\kappa(x,y)|f(x)-f(y)| = \alpha\gamma(yx^{-1})|f(x)-f(y)| < \alpha\varepsilon\gamma(yx^{-1})$$

for all x and y in G. It follows that

$$\int \kappa(x,y)|f(x)-f(y)|\,dy < \int \alpha\varepsilon\gamma(yx^{-1})\,dy = \varepsilon\int \alpha\gamma(y)\,dy = \varepsilon.$$

Now for all x in G,

$$\begin{aligned} |f(x)-T_\kappa(f)(x)| &= \left|f(x) - \int \kappa(x,y)f(y)\,dy\right| \\ &= \left|\int \kappa(x,y)\,[f(x)-f(y)]\,dy\right| \\ &\leq \int \kappa(x,y)|f(x)-f(y)|\,dy. \\ &< \varepsilon \end{aligned}$$

and $\|f-T_\kappa(f)\|_s < \varepsilon$. □

The results of this section can now be put together in a single key theorem.

Theorem 7.3.9 *Let G be a compact metric group, and let f be in $C(G)_s$. Given $\varepsilon > 0$, there exists a finite sequence $\Phi_1, \ldots, \Phi_p$ of orthogonal representations of G with dimensions $n_1, \ldots, n_p$ and a function g in the finite-dimensional subspace*

$$\text{Span}\left[\{\Phi_k(i,j) : 1 \leq k \leq p, 1 \leq i \leq n_k, \text{ and } 1 \leq j \leq n_k\}\right]$$

of $C(G)_s$ such that $\|f-g\|_s < \varepsilon$.

Proof. It follows from *Theorem 7.3.8* that there exists a symmetric right-invariant kernel κ such that

$$\|f-T_\kappa(f)\|_s < \frac{\varepsilon}{2}.$$

Then, applying *Corollary 7.3.4*, there exists k such that

$$\left\|T_\kappa(f) - \sum_{i\in\Lambda_k} \lambda_i \langle\langle f,\theta_i\rangle\rangle\theta_i\right\|_s < \frac{\varepsilon}{2}.$$

Therefore,

$$\left\|f - \sum_{i\in\Lambda_k} \lambda_i \langle\langle f,\theta_i\rangle\rangle\theta_i\right\|_s < \varepsilon,$$

and f can be approximated by a linear combination of eigenfunctions. *Corollary 7.3.7* now completes the proof. □

EXERCISES

1. Let κ be a symmetric kernel for the compact metric group G, and let Λ be defined as in *Theorem 7.2.19.*

   (a) Prove that the set Λ is finite if and only if κ is in $\mathcal{A}_p$.

   (b) Use *Exercise* 4, p. 291, to prove *Theorem 7.3.3* when Λ is finite.

2. Show that the symmetric right-invariant kernels for a compact metric group G are a closed subspace of $\mathcal{C}(G \times G)_s$

3. Show that if κ is a symmetric right-invariant kernel for a compact metric group G, then $\beta_\kappa(f_a, g_a) = \beta_\kappa(f, g)$, where $f_a(x) = f(xa)$ and so on.

4. Let κ be a symmetric right-invariant kernel for a compact metric group G, and let $\lambda \neq 0$ be eigenvalue for T_κ. Let $\theta_1(x), \ldots, \theta_n(x)$ be an orthonormal basis for V_λ, and let Φ be the representation of G given by equation (7.19). Show that

$$\Phi(a)(k, j) = \int \theta_j(xa)\theta_k(x)\, dx$$

   and use it to give alternate proof of the continuity of Φ.

5. Let x and y be distinct points in a compact metric group G. Prove that there exists an orthogonal representation Φ of G such that $\Phi(x) \neq \Phi(y)$.

6. Let G be a compact metric group with metric d. Prove that for $r > 0$ there exist a finite number of orthogonal representations $\Phi_1, \ldots, \Phi_m$ such that for x in $G \setminus B_r(e)$, there exists j, $1 \leq j \leq m$ such that $\Phi_j(x) \neq \Phi_j(e) = I$. Then prove that there exists a finite or infinite sequence of representations Φ_k of G such that for $x \neq e$ there exists j such that $\Phi_j(x) \neq I$.

7. Let G be a compact metric group. By using the previous exercise, prove that there exists a finite or infinite sequence of positive integers n_k such that G is isomorphic to a closed subgroup of

$$\prod_k \mathrm{O}(n_k).$$

7.4 Compact Abelian Groups

Let G be a topological group, and let $\Phi : G \to \mathrm{GL}(n, \mathbb{S})$ be a matrix representation of G, that is, a homomorphism of G to $\mathrm{GL}(n, \mathbb{S})$. A subspace V of $\mathbb{S}^n$ is an *invariant subspace* of the matrix representation Φ provided $\mathbf{v}\Phi(x)$ is in V for all $\mathbf{v}$ in V and all x in G. Obviously, $\{\mathbf{0}\}$ and $\mathbb{S}^n$ are invariant subspaces of every matrix representation. A matrix representation is *irreducible* provided that $\{\mathbf{0}\}$ and $\mathbb{S}^n$ are its only invariant subspaces. Every one-dimensional representation is a irreducible. We will be be interested primarily in unitary representations and their invariant subspaces, which are subspaces of $\mathbb{C}^n$. In this context we will use the Hermitian product, so $z \cdot w = \overline{w \cdot z}$.

Proposition 7.4.1 *Let $\Phi : G \to U(n)$ be a unitary representation of a topological group G. If V is an invariant subspace for Φ, then $V^\perp$ is an invariant subspace.*

Proof. Suppose $\mathbf{u}$ is in $V^\perp$. Then,

$$\mathbf{u}\,\Phi(x) \cdot \mathbf{v} = \mathbf{u} \cdot \mathbf{v}\,\Phi(x)^* = \mathbf{u} \cdot \mathbf{v}\,\Phi(x)^{-1} = \mathbf{u} \cdot \mathbf{v}\,\Phi(x^{-1}) = 0$$

because $\mathbf{v}\,\Phi(x^{-1}) \in V$ for all $\mathbf{v} \in V$ and $x \in G$. Thus $\mathbf{u}\Phi(x)$ is in $V^\perp$ for all $\mathbf{u} \in V^\perp$ and $x \in G$. □

Theorem 7.4.2 *If $\Phi : G \to U(n)$ is an irreducible unitary representation of an Abelian topological group G, then $n = 1$ and Φ is a homomorphism of G to $\mathbb{K}$.*

Proof. Consider the unitary matrix $\Phi(y)$ for a given element of G. By the Fundamental Theorem of Algebra, there exists a characteristic value λ of $\Phi(y)$, and then by *Corollary 2.5.10* there exists $\mathbf{u} \neq \mathbf{0}$ in $\mathbb{C}^n$ such that $\mathbf{u}\,\Phi(y) = \lambda\mathbf{u}$. Let $V_\lambda = \{\mathbf{v} \in \mathbb{C}^n : \mathbf{v}\,\Phi(y) = \lambda\mathbf{v}\}$, and consider arbitrary x in G and $\mathbf{v}$ in V_λ. Then,

$$\begin{aligned}
\lambda\mathbf{v}\,\Phi(x) &= \mathbf{v}\,\Phi(y)\,\Phi(x) \\
&= \mathbf{v}\,\Phi(yx) \\
&= \mathbf{v}\,\Phi(xy) \\
&= \mathbf{v}\,\Phi(x)\,\Phi(y).
\end{aligned}$$

Thus $\mathbf{v}\,\Phi(x)$ is in V_λ for all $\mathbf{v} \in V_\lambda$ and $x \in G$. Therefore, V_λ is an invariant subspace for the irreducible representation Φ. It follows that $V_\lambda = \mathbb{C}^n$ because $V_\lambda \neq \{\mathbf{0}\}$. Consequently, $\Phi(y) = \lambda I$.

Since there were no restrictions on y, for every element y of G there exists λ_y such that $\Phi(y) = \lambda_y I$. It follows that every subspace of $\mathbb{C}^n$ is an invariant subspace for Φ. If $n > 1$, then the irreducible representation Φ would have an invariant subspace not equal to either $\{\mathbf{0}\}$ or $\mathbb{C}^n$. Therefore, $n = 1$. Finally, $\mathrm{U}(1) = \mathbb{K}$ and Φ is homomorphism of G to $\mathbb{K}$. □

If Φ is an algebraic homomorphism of a group G to $\mathrm{U}(n)$, then Φ is a unitary representation of G with the discrete topology and *Proposition 7.4.1* and *Theorem 7.4.2* apply to it.

A homomorphism of a topological group G to $\mathbb{K}$ is in the linear space $\mathcal{C}(G, \mathbb{C})$, and we can form linear combinations of homomorphisms of G to $\mathbb{K}$. In fact, all unitary representations are built from linear combinations of these homomorphisms.

Theorem 7.4.3 *If $\Phi : G \to \mathrm{U}(n)$ is a unitary representation of an Abelian topological group G, then the coordinate functions $\Phi(i, j)$, $1 \leq i \leq n$ and $1 \leq j \leq n$ are linear combinations of homomorphisms of G to $\mathbb{K}$.*

Proof. The proof proceeds by induction on the dimension of the representation, starting with $n = 1$, which is obvious because $\mathrm{U}(1) = \mathbb{K}$.

Suppose the conclusion holds for $1, \ldots, n$ and consider a unitary representation $\Phi : G \to \mathrm{U}(n+1)$. It is reducible by *Theorem 7.4.2* because $n + 1 > 1$, and so there exists an invariant subspace V such that $1 \leq \mathrm{Dim}\,[V] \leq n$. *Proposition 7.4.1* implies that $V^{\perp}$ is also an invariant subspace such that $1 \leq \mathrm{Dim}\,[V^{\perp}] \leq n$. Let $\mathrm{Dim}\,[V] = m$.

By applying the Gram-Schmidt Process (*Proposition 4.2.5*), there exists an orthonormal basis $\mathbf{u}_1, \ldots, \mathbf{u}_{n+1}$ of $\mathbb{C}^{n+1}$ such that $\mathbf{u}_1, \ldots, \mathbf{u}_m$ is an orthonormal basis for V and $\mathbf{u}_{m+1}, \ldots, \mathbf{u}_{n+1}$ is an orthonormal basis for $V^{\perp}$. Let A be the $n + 1 \times n + 1$ matrix whose rows are the coordinates of $\mathbf{u}_1, \ldots, \mathbf{u}_{n+1}$. Then A is a unitary matrix (*Proposition 4.2.3*) and

$$A\Phi(x)A^{-1} = \begin{bmatrix} \Phi_1(x) & \mathrm{O} \\ \mathrm{O} & \Phi_2(x) \end{bmatrix} \tag{7.20}$$

where $\Phi_1(x)$ and $\Phi_2(x)$ are the $m \times m$ and $(n + 1 - m) \times (n + 1 - m)$ matrices of $\Phi(x)$ restricted to the invariant subspaces V and $V^{\perp}$ with respect to the orthonormal basis $\mathbf{u}_1, \ldots, \mathbf{u}_m$ of V and the orthonormal basis $\mathbf{u}_{m+1}, \ldots, \mathbf{u}_{m+1}$ of $V^{\perp}$. Clearly, $A\Phi(x)A^{-1}$ is a unitary representation of G because A is unitary. It follows (using *Proposition 4.2.3* again) that $\Phi_1(x)$ and $\Phi_2(x)$ are unitary representations of G. By the induction assumption, the coordinate functions $\Phi_1(i, j)$ and $\Phi_2(i, j)$ are linear combinations of homomorphisms of G to $\mathbb{K}$. Since

$$\Phi(x) = A^{-1} \begin{bmatrix} \Phi_1(x) & \mathrm{O} \\ \mathrm{O} & \Phi_2(x) \end{bmatrix} A,$$

the functions $\Phi(i, j)$ are linear combinations of homomorphisms of G to $\mathbb{K}$. □

Recall that for an Abelian locally compact second countable metric group G, the homomorphisms of G to $\mathbb{K}$ are called *characters*, and the characters form a group $\widehat{G}$ that is also a locally compact second countable Abelian metric group (*Theorem 6.5.7*). We can now bring together the basic facts about unitary representations in the beginning of this section and the approximation of continuous functions using orthogonal representations from the end Section7.3 to prove the following:

Theorem 7.4.4 *If G is a compact Abelian metric group, then the characters of G separate the points of G.*

Proof. Let a and b be distinct points in G and let $f(x) = d(x, b)$, where d is a metric for G. So f is a continuous function such that $f(a) = d(a, b) > 0$ and $f(b) = 0$. Let $\alpha = d(a, b) = f(a) - f(b)$. By *Theorem 7.3.9*, there exists a finite sequence $\Phi_1, \ldots, \Phi_p$ of orthogonal representations of dimension $n_1, \ldots, n_p$ and a function g in the finite-dimensional subspace

$$W = \text{Span}\left[\{\Phi_k(i,j) : 1 \leq k \leq p,\ 1 \leq i \leq n_k, \text{ and } 1 \leq j \leq n_k\}\right]$$

of $\mathcal{C}(G)_s$ such that $\|f - g\|_s < \alpha/3$. It follows that $g(a) - g(b) > \alpha/3 > 0$.

Since $\mathrm{O}(n) \subset U(n)$, orthogonal representations are unitary representations. Consequently the functions in

$$\{\Phi_k(i,j) : 1 \leq k \leq p, 1 \leq i \leq n_k, \text{ and } 1 \leq j \leq n_k\}$$

are linear combinations of characters by *Theorem 7.4.3*. Therefore, $W \subset \text{Span}\left[\widehat{G}\right]$ and $g = c_1\chi_1 + \ldots + c_k\chi_k$ with $c_1, \ldots, c_k$ in $\mathbb{C}$ and $\chi_1, \ldots, \chi_k$ in $\widehat{G}$. If $\chi_j(a) = \chi_j(b)$ for $j = 1, \ldots, k$, then $g(a) = g(b)$. So $\chi_j(a) \neq \chi_j(b)$ for some j. □

Corollary 7.4.5 *If G is a compact Abelian group, then $\mathcal{A}_c = \text{Span}\left[\widehat{G}\right]$ is a subalgebra of $\mathcal{C}(G)_s$ such that $\mathcal{A}_c^- = \mathcal{C}(G, \mathbb{C})_s$.*

Proof. It is obvious that $\mathcal{A}_c$ is a subalgebra of $\mathcal{C}(G, \mathbb{C})_s$ because $\widehat{G}$ is closed under the multiplication of characters. Since the identity of $\widehat{G}$ is $\zeta(x) = 1$ for all $x \in G$, the constant functions are in $\mathcal{A}_c$. Because the complex conjugate $\overline{\chi}$ of a character χ is a character, the algebra $\mathcal{A}_c$ is closed under conjugation of functions. Lastly, characters separate points by the theorem. So the Complex Stone-Weierstrass Theorem (*Theorem 6.1.6*) applies to complete the proof. □

The remainder of this section will be devoted to proving additional results about the character groups of compact Abelian groups. These results will be useful in the more general study of the character groups of Abelian locally compact second countable metric groups in Chapter 8.

Proposition 7.4.6 *If G is compact metric group, then $\widehat{G}$ is a countable discrete group.*

Proof. This proof and several others make use of the open neighborhoods $U_r = \left\{e^{2\pi i x} : |x| < r\right\}$ of 1 in $\mathbb{K}$. In particular, when C is a compact subset of G, the set $V = N(C, U_r) \cap \widehat{G}$ is an open neighborhood of ζ, the identity of $\widehat{G}$. Since G is compact by hypothesis, $V = N(G, U_r) \cap \widehat{G}$ is an open neighborhood of ζ.

If γ is a character in V, then $\gamma(G)$ is a compact, and hence closed subgroup of $\mathbb{K}$ contained in U_r. When $r < 1/4$, the only closed subgroup of $\mathbb{K}$ contained in U_r is $\{1\}$. Thus $V = \{\zeta\}$, when $r < 1/4$, and $\widehat{G}$ is discrete.

It follows from *Theorem 6.5.7* that $\widehat{G}$ is second countable and hence separable. Since the only dense subset of a discrete space is the space itself, $\widehat{G}$ must be countable. □

Theorem 7.4.7 *Every compact Abelian metric group is isomorphic to a closed subgroup of* $\mathbb{K}^\infty$.

Proof. Let G be a compact Abelian group. Since the character group of G is countable by *Proposition 7.4.6*, the characters can be indexed by $\mathbb{Z}^+$ as a sequence $\chi_1, \chi_2, \ldots$. Define $\varphi : G \to \mathbb{K}^\infty$ by $\varphi(x) = (\chi_1(x), \chi_2(x), \ldots)$. Clearly φ is a homomorphism because its coordinate functions are characters. It follows that $\varphi(G)$ is a compact subgroup of $\mathbb{K}^\infty$. The homomorphism φ is one-to-one because characters separate points by *Theorem 7.4.4*. Therefore, φ is a homeomorphism (*Exercise* 2, p. 61) and an isomorphism of G onto closed subgroup $\varphi(G)$ of $\mathbb{K}^\infty$. □

Next, we want to determine the character groups of familiar compact Abelian groups. Because the character group of a compact Abelian group is discrete by *Proposition 7.4.6*, it is only necessary to determine the algebraic structure of $\widehat{G}$.

Proposition 7.4.8 *The character group* $\widehat{\mathbb{K}^n}$ *of* $\mathbb{K}^n$ *is isomorphic to* $\mathbb{Z}^n$.

Proof. Let χ be a character of $\mathbb{K}^n$. By *Exercise* 7, p. 131, there exists $\mathbf{k} = (k_1, \ldots, k_n))$ in $\mathbb{Z}^n$ such that

$$\chi(\mathbf{z}) = \chi(z_1, \ldots z_n) = \prod_{j=1}^{n} z_j^{k_j}.$$

Conversely, every function of this form is a character for $\mathbb{K}^n$. It follows from the properties of exponents that setting

$$\varphi(\mathbf{k})(\mathbf{z}) = \prod_{j=1}^{n} z_j^{k_j} \tag{7.21}$$

defines an onto algebraic homomorphism $\varphi : \mathbb{Z}^n \to \widehat{\mathbb{K}^n}$. If $\mathbf{k} \neq \mathbf{0}$, then it is easy to construct $\mathbf{z}$ such that $\varphi(\mathbf{k})(\mathbf{z}) \neq 1$. Therefore, the kernel of φ is $\{\mathbf{0}\}$, and it is an isomorphism of $\mathbb{Z}^n$ onto $\widehat{\mathbb{K}^n}$, because both groups are discrete. □

To determine the character group of $\mathbb{K}^\infty$ requires a more careful analysis and best approached by considering the more general problem of determining the character group of $G = \prod_{k=1}^\infty G_k$ where G_k is a sequence of compact Abelian metric groups. The identity elements of G_k and $\widehat{G_k}$ will be denoted by e_k and ζ_k, respectively. We know from *Theorem 5.1.11*, *Proposition 5.1.15*, and *Theorem 5.3.6* that G is a compact Abelian metric group.

Theorem 7.4.9 *Let* G_k *be a sequence of compact Abelian metric groups, and let* $G = \prod_{k=1}^\infty G_k$. *Then,* $\widehat{G}$ *is isomorphic to the subgroup* H *of* $\prod_{j=1}^\infty \widehat{G_j}$ *defined by*

$$H = \bigcup_{m=1}^{\infty} \{\chi = (\chi_1, \chi_2 \ldots) : \chi_j = \zeta_j \text{ for } j \geq m\}.$$

with the discrete topology.

Proof. It is easy to verify that H is a subgroup of $\prod_{j=1}^{\infty} \widehat{G_j}$. ($H$ is sometimes called the *weak or restricted direct product.*) Because G is compact, $\widehat{G}$ is discrete and it suffices to find an algebraic isomorphism of $\widehat{G}$ onto H.

As in the proof of *Theorem 6.5.3* for each k there exists an algebraic homomorphism $\theta_k : \widehat{G} \to \widehat{G_k}$ defined by

$$\theta_k(\chi)(x_k) = \chi(e_1, \ldots, e_{k-1}, x_k, e_{k+1}, \ldots)$$

for χ in $\widehat{G}$. It follows that $\Theta(\chi) = (\theta_1(\chi), \theta_2(\chi), \ldots)$ is an algebraic homomorphism of $\widehat{G}$ to $\prod_{j=1}^{\infty} \widehat{G_j}$.

Given χ in $\widehat{G}$, consider the open neighborhood $\chi^{-1}(U_r)$ of $\mathbf{e} = (e_1, \ldots, e_j, \ldots)$ the identity of G. There exists an open neighborhood V of $\mathbf{e}$ of the form

$$V = \prod_{j=1}^{\infty} V_j \subset \chi^{-1}(U_r)$$

such that V_j is an open neighborhood of e_j for all j and $V_j = G_j$ for all $j \geq m$.

Let $r < 1/4$. Observe that

$$G' = \{e_1\} \times \ldots \times \{e_{m-1}\} \times \prod_{j=m}^{\infty} G_m$$

is a compact closed subgroup of G such that $G' \subset V \subset \chi^{-1}(U_r)$. It follows that $\chi(G')$ is a compact closed subgroup of $\mathbb{K}$ contained in U_r. Therefore, $\chi(G') = \{1\}$, the only closed subgroup of $\mathbb{K}$ contained in U_r, when $r < 1/4$. Thus G' is contained in the kernel of χ. Consequently, $\theta_j(\chi) = \zeta_j$ for $j \geq m$ and $\Theta(\chi)$ is in H.

Furthermore, it is now easy to see that

$$\chi(x_1, \ldots, x_j, \ldots) = \prod_{j=1}^{m-1} \theta_j(\chi)(x_j).$$

If $\Theta(\chi) = (\zeta_1, \ldots, \zeta_k, \ldots)$, then $\chi(\mathbf{x}) = 1$ for all $\mathbf{x}$ in G and $\chi = \zeta$. Thus the kernel of Θ is $\{\zeta\}$, and Θ is an algebraic isomorphism of $\widehat{G}$ to H.

To see that Θ is onto H consider $(\chi_1, \ldots, \chi_k, \ldots)$ such that $\chi_j = \zeta_j$ for $j \geq m$. Define χ in $\widehat{G}$ by

$$\chi(x_1, \ldots, x_j, \ldots) = \prod_{j=1}^{m-1} \chi_j(x_j).$$

It follows that $\theta_j(\chi) = \chi_j$ for $1 \leq j \leq m-1$ and $\Theta(\chi) = (\chi_1, \ldots, \chi_k, \ldots)$. Therefore, Θ is the required algebraic isomorphism of $\widehat{G}$ onto H. □

Corollary 7.4.10 *The group* $\widehat{\mathbb{K}^\infty}$ *is isomorphic to the subgroup* H *of* $\mathbb{Z}^\infty$ *defined by*

$$H = \bigcup_{m=1}^{\infty} \{(a_1, a_2, \ldots) \in \mathbb{Z}^\infty : a_j = 0 \text{ for } j \geq m\}.$$

Proposition 7.4.11 *Let n be a positive integer. The character group of the finite cyclic group $\mathbb{Z}_n$ is isomorphic to $\mathbb{Z}_n$. Furthermore, if w is an element of $\mathbb{K}$ such that $w^n = 1$, then there exists a unique character χ in $\widehat{\mathbb{Z}_n}$ such that $\chi(1) = w$.*

Proof. Recall that $\mathbb{Z}_n = \{0, 1, \ldots, n-1\}$ and $i + j = r$ if and only if r is the remainder of the usual sum $i + j$ divided by n. Also the nth roots of 1 in $\mathbb{K}$ are the elements of the cyclic subgroup

$$H = \left\{e^{2\pi ik/n} : 1 \leq k \leq n\right\} = \langle e^{2\pi i/n} \rangle$$

of $\mathbb{K}$, and $\gamma(j) = e^{2\pi ij/n}$ is an isomorphism of $\mathbb{Z}_n$ onto H.

Because $H \subset \mathbb{K}$ and $\mathbb{Z}_n$ is discrete, γ is a character of $\mathbb{Z}_n$. Hence, γ^k is also a character of $\mathbb{Z}_n$ for $1 \leq k \leq n$ and $\gamma^n = \zeta$, the identity character. It follows that $\varphi(k) = \gamma^k$ is a homomorphism of $\mathbb{Z}_n$ to $\widehat{\mathbb{Z}_n}$. Furthermore, φ is one-to-one because $\gamma^k(1) = e^{2\pi ik/n} \neq e^{2\pi im/n} = \gamma^m(1)$ when $1 \leq k, m \leq n$ and $k \neq m$.

Let χ be an arbitrary character of $\mathbb{Z}_n$. Keeping in mind that 1 is the identity of $\mathbb{K}$ and $\mathbb{Z}_n = \langle 1 \rangle$,

$$\chi(1)^n = \chi(n1) = \chi(0) = 1$$

and $\chi(1) = e^{2\pi ik/n} = \gamma^k(1)$ for some k such that $1 \leq k \leq n$. Thus $\chi = \gamma^k = \varphi(k)$ and that γ^k is the unique character in $\widehat{\mathbb{Z}^n}$ such that $\gamma^k(1) = e^{2\pi ik/n}$ for $1 \leq k \leq n$. Therefore, φ is an isomorphism of $\mathbb{Z}_n$ onto $\widehat{\mathbb{Z}_n}$ □

Let $\mathbb{K}_d$ denote $\mathbb{K}$ with the discrete metric. It is an Abelian locally compact metric group, but it is not separable (and hence not second countable by *Proposition 6.2.1*) because it is uncountable. Its countable subgroups, however, will be useful.

Proposition 7.4.12 *If G is a compact monothetic metric group with $G = \langle g \rangle^-$, then $\chi \mapsto \chi(g)$ defines an isomorphism of $\widehat{G}$ onto a countable subgroup of $\mathbb{K}_d$.*

Proof. Since $(\chi_1\chi_2)(g) = \chi_1(g)\chi_2(g)$ and $\widehat{G}$ is discrete, the function $\chi \mapsto \chi(g)$ is a homomorphism. This function is one-to-one by *Exercise* 3, p. 309. The image is countable because $\widehat{G}$ is countable. □

Proposition 7.4.13 *Let G be a compact Abelian totally disconnected metric group. If χ is in $\widehat{G}$, then $\chi(G)$ is a finite cyclic subgroup of $\mathbb{K}$ and $\chi^n = \zeta$ for some positive integer n.*

Proof. By *Theorem 5.4.7*, the group G has a neighborhood base at the identity e consisting of subgroups that are both open and closed. Choose r such that $0 < r < 1/4$. Then $\chi^{-1}(U_r)$ contains an open-and-closed subgroup H of G. Because G is compact, H is compact, and $\chi(H) = \{1\}$ because it is a compact closed subgroup of $\chi^{-1}(U_r)$ with $0 < r < 1/4$. Thus the kernel K of χ is open and closed in G (*Exercise* 2, p. 50).

Since the cosets of K are open and G is compact, a finite number of them cover G. Consequently, G/K is a finite group and by *Corollary 1.5.19* it is isomorphic to the compact and closed subgroup $\chi(G)$ of $\mathbb{K}$. *Theorem 1.4.20* now implies that $\chi(G) = \{e^{2\pi ik/n} : 1 \le k \le n\}$ for some n in $\mathbb{Z}^+$. Thus $\chi(x)^n = 1$ for all $x \in G$ and $\chi^n = \zeta$. □

Corollary 7.4.14 *If $G = \langle g \rangle^-$ is a totally disconnected compact monothetic group, then $\chi \to \chi(g)$ is an isomorphism of $\widehat{G}$ to a subgroup of*

$$L = \{z \in \mathbb{K} : z^k = 1 \text{ for some } k \in \mathbb{Z}^+\}$$

with the discrete topology.

Proof. Clearly, L is a subgroup of $\mathbb{K}$. It follows from the proposition that $\chi(g)$ is in L for every $\chi \in \widehat{G}$. Hence, $\widehat{G}$ is isomorphic to a subgroup of L by *Proposition 7.4.12.* □

To conclude this section, we want to determine the character group of the **a**-adic groups $\Gamma_{\mathbf{a}}$ constructed in Section 5.4. Recall that $\mathbf{a} = (a_0, a_1, \ldots, a_k, \ldots)$ is an infinite sequence of positive integers,

$$\Gamma_{\mathbf{a}} = \prod_{k=0}^{\infty} \{0, \ldots, a_k - 1\},$$

and the group operation on $\Gamma_{\mathbf{a}}$ is defined by equations (5.8) and (5.9), p. 219. *Theorem 5.4.10* showed that the **a**-adic groups were totally disconnected compact monothetic groups and $\langle \mathbf{g} \rangle^- = \Gamma_{\mathbf{a}}$ for $\mathbf{g} = (1, 0, \ldots, 0, \ldots)$. Also recall that $\Lambda_m = \{\mathbf{x} \in \Gamma_{\mathbf{a}} : x_j = 0 \text{ for } j = 0, \ldots, m\}$ is a closed and open subgroup of $\Gamma_{\mathbf{a}}$ and $\Gamma_{\mathbf{a}}/\Lambda_m$ is isomorphic to $\mathbb{Z}_{p_m}$ where $p_m = a_0 a_1 \cdots a_m$ (*Proposition 5.4.11*).

Proposition 7.4.15 *If $\Gamma_{\mathbf{a}}$ is an **a**-adic group, then $\widehat{\Gamma_{\mathbf{a}}}$ is isomorphic to the subgroup*

$$D_{\mathbf{a}} = \{z \in \mathbb{K}_d : z^{p_m} = 1 \text{ for some } m \in \mathbb{Z}^+\}$$

of $\mathbb{K}_d$.

Proof. Note that *Corollary 7.4.14* implies that $\varphi(\chi) = \chi(\mathbf{g})$ is an isomorphism of $\widehat{G}$ to a subgroup of L. It remains to show that the range of φ is $D_{\mathbf{a}}$.

Let m be a positive integer, and let π be the canonical homomorphism of $\Gamma_{\mathbf{a}}$ onto $\Gamma_{\mathbf{a}}/\Lambda_m$. There exists an isomorphism ψ of $\Gamma_{\mathbf{a}}/\Lambda_m$ onto $\mathbb{Z}_{p_m}$ such that $\psi \circ \pi(\mathbf{g}) = 1$ because $\Gamma_{\mathbf{a}}/\Lambda_m = \langle \Lambda_m + \mathbf{g} \rangle$. If χ is a character of $\mathbb{Z}_{p_m}$, then $\chi \circ \psi \circ \pi$ is a character of $\Gamma_{\mathbf{a}}$. It follows from *Proposition 7.4.11* that if w is an element of $\mathbb{K}$ such that $w^{p_m} = 1$, then there exists a unique character χ in $\widehat{\mathbb{Z}_{p_m}}$ such that $\chi(1) = w$. So, $\chi \circ \psi \circ \pi(\mathbf{g}) = w$ and $D_{\mathbf{a}}$ is contained in the range of φ.

Consider χ in $\widehat{\Gamma_{\mathbf{a}}}$. Because $\chi(G)$ is a finite cyclic group (*Proposition 7.4.13*), the kernel K of χ is an open-and-closed subgroup of $\Gamma_{\mathbf{a}}$ and there exists a positive integer m such that $\Lambda_m \subset K$. Hence, there exists a homomorphism $\tilde{\chi}$ of $\Gamma_{\mathbf{a}}/\Lambda_m$ to $\mathbb{K}$ such that $\chi = \tilde{\chi} \circ \pi$ (*Exercise* 15, p. 51). Since $\tilde{\chi}$ is a character of the cyclic group $\Gamma_{\mathbf{a}}/\Lambda_m$, which is isomorphic to $\mathbb{Z}_{p_m}$, the range of $\tilde{\chi}$ is contained in $D_{\mathbf{a}}$. Thus $\varphi(\chi) = \chi(\mathbf{g}) = \tilde{\chi}(\Lambda_m + \mathbf{g}) = w \in D_{\mathbf{a}}$. □

If $G_1, \ldots, G_m$ are compact Abelian metric groups and $G = G_1 \times \ldots \times G_m$, then $\widehat{G}$ is isomorphic to $\widehat{G_1} \times \ldots \times \widehat{G_m}$ (*Theorem 6.5.8*). Consequently, we can now easily compute the character groups for a wide variety of compact Abelian metric groups. As Chapter 8 unfolds, our ability to calculate character groups of locally compact second countable Abelian groups will grow further.

EXERCISES

1. If A is a unitary matrix, then every characteristic value of A is in $\mathbb{K}$.

2. Prove that if $\Phi : G \to U(n)$ is a unitary representation of an Abelian metric group G, then there exists a unitary matrix A and characters $\chi_1(x), \ldots, \chi_n$ such that $A\Phi(x)A^{-1} = \text{Diag}\,[\chi_1(x), \ldots, \chi_n]$.

3. Consider a monothetic metric group $G = \langle g \rangle^{-}$. Let χ_1 and χ_2 be two characters of G. Prove that $\chi_1 = \chi_2$ if and only if $\chi_1(g) = \chi_2(g)$.

4. Let G be a compact metric group. If f is in $\mathcal{C}(G, \mathbb{C})$, then $f(x) = u(x) + iv(x)$, where $u(x)$ and $v(x)$ are in $\mathcal{C}(G)$ and a right- and left-invariant integral of f can be defined by setting

$$\int u(x) + iv(x)\, dx = \int u(x)\, dx + i \int v(x)\, dx.$$

   Show that $f \mapsto \int f(x)\, dx$ is a complex linear function.

5. Using the integral defined in *Exercise 4* above, prove the following for G a compact Abelian metric group and χ an element of $\widehat{G}$:

   (a) If $\chi(a) \neq 1$ for some a in G, then $(1 - \chi(a)) \int \chi(x)\, dx = 0$.

   (b) If $\chi \neq \zeta$, then $\int \chi(x)\, dx = 0$.

6. Let G be a compact metric group. Using the integral defined in *Exercise 4* above, extend $\langle\langle f, g\rangle\rangle$ to $\mathcal{C}(G, \mathbb{C})$ by setting $\langle\langle f, g\rangle\rangle = \int f(x)\overline{g(x)}\, dx$.

   (a) Show that in this context $\langle\langle f, g\rangle\rangle$ has the same properties as the Hermitian product $\mathbf{z} \cdot \mathbf{w}$ listed in *Exercise* 8, p. 82.

   (b) Prove Bessel's Inequality in this context.

   (c) Prove the Cauchy-Schwarz Inequality in this context.

   (d) Show that $\|f\|_2 = \langle\langle f, \overline{f}\rangle\rangle^{1/2}$ defines a norm on $\mathcal{C}(G, \mathbb{C})$.

7.5 Matrix Representations

The construction of the invariant integral, the determination of the eigenvalues for the integral equation, and appearance of matrix representations in the analysis of the eigenfunctions in Sections 7.1, 7.2, and 7.3 provide ready access to a number of fundamental results about unitary representations of compact metric groups. Keeping in mind that representation theory is a large subject in its own right, this section provides a glimpse of one corner of this topic.

Let G be a topological group. Since $\mathrm{GL}(n,\mathbb{R})$ is naturally a subgroup of $\mathrm{GL}(n,\mathbb{C})$, we can without loss of generality assume every matrix representation of G has values in $\mathrm{GL}(n,\mathbb{C})$. Two matrix representations $\Phi, \Psi : G \to \mathrm{GL}(n,\mathbb{C})$ are *equivalent* provided there exists A in $\mathrm{GL}(n,\mathbb{C})$ such that $\Phi(x) = A\Psi(x)A^{-1}$ for all x in G. The equivalence of two matrix representations Φ and Ψ will be denoted by $\Phi \sim \Psi$.

In this section, we use the extension of the integral of continuous real-valued functions to complex-valued functions on a compact metric group G. Every f in $\mathcal{C}(G,\mathbb{C})$ can be written as $f(x) = u(x) + iv(x)$, where $u(x)$ and $v(x)$ are continuous real-valued functions. By definition $\int f(x)\,dx = \int u(x)\,dx + i\int v(x)\,dx$, which is a complex linear function (*Exercise* 4, p. 309). In addition, when $\Theta : G \to \mathcal{M}_n(\mathbb{C})$ is a continuous function, $\int \Theta(x)\,dx$ denotes the matrix whose entries are given by

$$\int \Theta(x)\,dx(i,j) = \int \Theta(x)(i,j)\,dx.$$

Proposition 7.5.1 *Every matrix representation of a compact metric group G is equivalent to a unitary representation of G.*

Proof. Given a matrix representation $\Phi : G \to \mathrm{GL}(n,\mathbb{C})$, using the Hermitian product, the function $\mathbf{z}\Phi(x)\cdot\mathbf{w}\Phi(x)$ is clearly a continuous complex-valued function and can be integrated. Set

$$\alpha(\mathbf{z},\mathbf{w}) = \int \mathbf{z}\Phi(x)\cdot\mathbf{w}\Phi(x)\,dx = \int \mathbf{z}\Phi(x)\Phi(x)^*\cdot\mathbf{w}\,dx.$$

It is a complex-valued bilinear function on $\mathbb{C}^n$ with the following invariance property:

$$\begin{aligned}
\alpha(\mathbf{z}\Phi(y),\mathbf{w}\Phi(y)) &= \int \mathbf{z}\Phi(y)\Phi(x)\cdot\mathbf{w}\Phi(y)\Phi(x)\,dx \\
&= \int \mathbf{z}\Phi(yx)\cdot\mathbf{w}\Phi(yx)\,dx \\
&= \int \mathbf{z}\Phi(x)\cdot\mathbf{w}\Phi(x)\,dx \\
&= \alpha(\mathbf{z},\mathbf{w}).
\end{aligned}$$

When $\mathbf{z} \neq \mathbf{0}$, the function $\mathbf{z}\Phi(x)\cdot\mathbf{z}\Phi(x)$ is real-valued and positive and $\alpha(\mathbf{z},\mathbf{z}) > 0$. An alternate formula for α is the following:

$$\alpha(\mathbf{z},\mathbf{w}) = \int \mathbf{z}\Phi(x)\Phi(x)^*\cdot\mathbf{w}\,dx$$

$$= \mathbf{z}\left(\int \Phi(x)\Phi(x)^* \, dx\right) \cdot \mathbf{w}.$$

Setting $B = \int \Phi(x)\Phi(x)^* \, dx$, we have $\alpha(\mathbf{z}, \mathbf{w}) = \mathbf{z}B \cdot \mathbf{w}$ with B positive definite because $\alpha(\mathbf{z}, \mathbf{z}) > 0$, when $\mathbf{z} \neq \mathbf{0}$. By *Theorem 5.2.20*, there exists a unique positive definite matrix A such that $A^2 = B$. In particular, A is Hermitian and

$$\alpha(\mathbf{z}, \mathbf{w}) = \mathbf{z}A^2 \cdot \mathbf{w} = \mathbf{z}A \cdot \mathbf{w}A.$$

Recalling that positive definite matrices are invertible, set $\Psi(x) = A^{-1}\Phi(x)A$. Then,

$$\begin{aligned} \mathbf{z}\Psi(x) \cdot \mathbf{w}\Psi(x) &= \mathbf{z}A^{-1}\Phi(x)A \cdot \mathbf{w}A^{-1}\Phi(x)A \\ &= \alpha(\mathbf{z}A^{-1}\Phi(x), \mathbf{w}A^{-1}\Phi(x)) \\ &= \alpha(\mathbf{z}A^{-1}, \mathbf{w}A^{-1}) \\ &= \mathbf{z}A^{-1}A \cdot \mathbf{w}A^{-1}A \\ &= \mathbf{z} \cdot \mathbf{w} \end{aligned}$$

for all $\mathbf{z}$ and $\mathbf{w}$ in $\mathbb{C}^n$. It follows from *Exercise* 1, p. 158, that $\Psi(x)$ is a unitary matrix for all x in G. □

The matrix representations of dimension n of a topological group G are a subset $\mathcal{R}_n$ of $\mathcal{C}(G, \mathrm{GL}(n, \mathbb{C}))$. Since equivalent matrix representations have the same dimension, the set of matrix representations that are equivalent to a given n-dimensional matrix representation is a subset of $\mathcal{R}_n$ called an *equivalence class.* If $\Phi_1 \sim \Phi_2$ and $\Phi_2 \sim \Phi_3$, then obviously $\Phi_1 \sim \Phi_3$. It follows that two equivalence classes of matrix representations are either equal or disjoint. Thus they decompose $\mathcal{R}_n$ into a collection of disjoint sets.

The equivalence classes of representations can be added. Given an equivalence class of matrix representations in $\mathcal{R}_m$ and another in $\mathcal{R}_n$, let $\Phi : G \to \mathrm{GL}(m, \mathbb{C})$ and $\Psi : G \to \mathrm{GL}(n, \mathbb{C})$ be one matrix representation from each of these two equivalence class. Define a matrix representation of dimension $m + n$ by

$$(\Phi + \Psi)(x) = \begin{bmatrix} \Phi(x) & \mathrm{O} \\ \mathrm{O} & \Psi(x) \end{bmatrix}.$$

The sum of the two equivalence classes is then defined to be the equivalence class of $\Phi + \Psi$. It follows from *Exercise* 2, p. 317, that this definition is unambiguous.

Equation (7.20) in the proof of *Theorem 7.4.3* shows that every reducible unitary matrix representation is equivalent to the sum of two unitary matrix representations of dimension at least one. Since every one-dimensional representation is irreducible, the proof of the next proposition follows immediately by induction.

Proposition 7.5.2 *Every unitary representation of a compact metric group is irreducible or equivalent to a sum of irreducible unitary representations.*

If Φ and Ψ are equivalent representations, then

$$\begin{aligned}\text{Span}\,[\{\Phi(i,j) : 1 \le i \le n, \text{ and } 1 \le j \le n\}] &= \\ \text{Span}\,[\{\Psi(i,j) : 1 \le i \le n, \text{ and } 1 \le j \le n\}] & \qquad (7.22)\end{aligned}$$

because every $\Phi(p,q)$ is a linear combination of the functions in $\{\Psi(i,j) : 1 \le i \le n, \text{ and } 1 \le j \le n\}$, and visa versa. Consequently, the equivalence class of a matrix representation has an associated a subspace of $C(G,\mathbb{C})$.

Proposition 7.5.3 *If Φ is a unitary representation of a compact group G, then there exist a finite collection of irreducible unitary representations $\Phi_1, \ldots, \Phi_m$ of dimension $n_1, \ldots, n_m$ such that*

$$\begin{aligned}&\text{Span}\,[\{\Phi(i,j) : 1 \le i \le n, \text{ and } 1 \le j \le n\}] = \\ &\text{Span}\,[\{\Phi_k(i,j) : 1 \le k \le m, 1 \le i \le n_k, \text{ and } 1 \le j \le n_k\}]. \qquad (7.23)\end{aligned}$$

Proof. Apply *Proposition 7.5.2*, and equation (7.22) □

A compact Hausdorff topological group has enough irreducible unitary representations so that collectively these subspaces can be used to approximate any continuous complex-valued function on the group. This result is known as the Peter-Weyl Theorem or the Peter-Weyl-van Kampen Theorem and implies *Theorem 7.4.4*. It in turn became a corollary of the more general Gelfand-Raikov Theorem for locally compact Hausdorff groups (Section 22 of [4]). For compact metric groups, the Peter-Weyl-van Kampen Theorem (*Theorem 7.5.4*) follows easily from the results in the first three sections of this chapter. One way to prove it for compact Hausdorff topological groups is simply to extend the results in the first three sections of this chapter to this larger class of groups. Doing so only requires using more sophisticated general topology to extend the proofs as presented. Another way is to approximate the compact Hausdorff topological group with compact metric groups and then apply the compact metric group version of the Peter-Weyl-van Kampen Theorem. A more detailed outline of this method appears in [11] on page 63.

Theorem 7.5.4 *Let G be a compact metric group, and let f be in $C(G,\mathbb{C})_s$. Given $\varepsilon > 0$, there exists a finite sequence $\Phi_1, \ldots, \Phi_m$ of irreducible unitary representations of dimension $n_1, \ldots, n_m$ and a function g in the finite-dimensional subspace*

$$\text{Span}\,[\{\Phi_k(i,j) : 1 \le k \le m, 1 \le i \le n_k, \text{ and } 1 \le j \le n_k\}]$$

of $C(G,\mathbb{C})_s$ such that $\|f - g\|_s < \varepsilon$.

Proof. *Theorem 7.3.9* can be applied to the real and complex parts of the function f using entries from a finite number of orthogonal representations. Since $\mathrm{O}(n) \subset \mathrm{U}(n)$, *Proposition 7.5.3* implies that every entry of an orthogonal representation used in this approximation can be replaced by a linear combination of entries from irreducible unitary representations. □

Corollary 7.5.5 *If G is a compact metric group, then for each a in G such that $a \neq e$ there exists an irreducible unitary representation Φ of G such that $\Phi(a) \neq I$.*

Proof. Let d be a metric for G. If $\Phi(a) = I$ for every irreducible unitary representation of G, then linear combinations of the entries of the irreducible representations of G could not be used to approximate the continuous function $f(x) = d(x, e)$ because every approximation would have the same value at both e and a. □

We turn now to studying the complex-valued functions $\Phi(x)(i, j)$ that make up an irreducible representation Φ of a compact metric group G. We begin with a simple version of Schur's Lemma, which appears in a variety of more general forms in the literature, such as [3] and [13].

Lemma 7.5.6 (Schur) *Let Φ and Ψ be irreducible unitary matrix representations of a metric group G having dimensions m and n, respectively. If A is a $m \times n$ matrix such that $\Phi(x)A = A\Psi(x)$ for all x in G, then either A is the zero matrix or $m = n$ and A is invertible.*

Proof. Denoting the rows of A again by $A_1, \ldots, A_m$, let $V = \text{Span}\,[A_1, \ldots, A_m]$, a subspace of $\mathbb{C}^n$. On the one hand, the rows of $\Phi(x)A$ are linear combinations of the rows of A, specifically,

$$(\Phi(x)A)_i = \sum_{j=1}^{m} \Phi(x)(i, j)A_j.$$

Thus the rows of $\Phi(x)A$ are in V for all x in G. On the other hand, the rows of $A\Psi(x)$ are $A_1\Psi(x), \ldots, A_m\Psi(x)$. Now $\Phi(x)A = A\Psi(x)$ implies that $A_j\Psi(x)$ is in V for all x in G, making V is an invariant subspace for Ψ. Because Ψ is irreducible, V can only be $\{\mathbf{0}\}$ or $\mathbb{C}^n$. Therefore, either A is the zero matrix or $m \geq n$ and the rows of A span $\mathbb{C}^n$.

Taking the conjugate transpose of $\Phi(x)A = A\Psi(x)$, yields $A^*\Phi(x)^* = \Psi(x)^*A^*$ or simply $\Psi(x^{-1})A^* = A^*\Phi(x^{-1})$ for all $x \in G$ because Φ and Ψ are unitary representations. Thus $\Psi(x)A^* = A^*\Phi(x)$ for all $x \in G$, and the initial argument can be repeated to conclude that A^* is the zero matrix or $n \geq m$ and the rows of A^* span $\mathbb{C}^m$.

Putting the conclusions together either A is the zero matrix or $m = n$ and the rows of A span $\mathbb{C}^n$. If A is not the zero matrix, then the rows of A are linearly independent by *Exercise* 8, p. 69, and A is invertible by *Exercise* 10, p. 114. □

Proposition 7.5.7 *Let Φ and Ψ be irreducible unitary representations of a compact metric group G. If Φ and Ψ are not equivalent, then the functions $\Phi(x)(i, j)$ and $\Psi(x)(p, q)$ are orthogonal functions in $\mathcal{C}(G, \mathbb{C})_2$.*

Proof. Let m and n be the dimensions of the representations Φ and Ψ, respectively, and let B be a matrix in $\mathcal{M}_{mn}(\mathbb{C})$. The entries of $\Phi(x)B\Psi(x^{-1})$

are continuous complex-valued functions on G, and $A = \int \Phi(x)B\Psi(x^{-1})\,dx$ is another $m \times n$ matrix. Then, by the left invariance of the integral

$$\begin{aligned}\Phi(y)A\Psi(y^{-1}) &= \int \Phi(y)\Phi(x)B\Psi(x^{-1})\Psi(y^{-1})\,dx \\ &= \int \Phi(yx)B\Psi((yx)^{-1})\,dx \\ &= \int \Phi(x)B\Psi(x^{-1})\,dx \\ &= A.\end{aligned}$$

It follows that $\Phi(y)A\Psi(y^{-1}) = A$ or $\Phi(y)A = A\Psi(y)$ for all y in G. If A is not the zero matrix, then by Schur's Lemma $m = n$ and A is invertible, implying that $\Phi(x) = A\Psi(x)A^{-1}$ for all $x \in G$. Therefore, A is the zero matrix because Φ and Ψ are not equivalent by hypothesis. Of course, A depends on the choice of B. The strategy is to choose B so that the desired orthogonality follows from $A = \mathrm{O}$. There are m^2n^2 orthogonal relationships to be established.

If B be the $m \times n$ matrix of all zeros except a 1 in the (μ, ν) position, then the (i, q) entry of the integrand is

$$\begin{aligned}\Phi(x)B\Psi(x^{-1})(i,q) &= \Phi(x)B\Psi(x)^*(i,q) \\ &= \sum_{p=1}^{n}\sum_{j=1}^{m} \Phi(x)(i,j)B(j,p)\overline{\Psi}(x)(q,p) \\ &= \Phi(x)(i,\mu)\overline{\Psi}(x)(q,\nu).\end{aligned}$$

Since $A = \mathrm{O}$ for all B, it follows that $\int \Phi(x)(i,\mu)\overline{\Psi}(x)(q,\nu)\,dx = 0$ Thus each choice of (μ, ν) produces mn orthogonality conditions. Varying (μ, ν) produces all of them. □

Proposition 7.5.8 *If Φ is a n-dimensional irreducible unitary representation of a compact metric group G, then $\Phi(x)(i,j)$ and $\Phi(x)(p,q)$ are orthogonal when $(i,j) \neq (p,q)$ and $\int \Phi(x)(i,j)\overline{\Phi}(x)(i,j)\,dx = 1/n$.*

Proof. Let B be $n \times n$ matrix, and set $A = \int \Phi(x)B\Phi(x^{-1})\,dx$. Then as in the proof of *Proposition 7.5.7*, $\Phi(x)A = A\Phi(x)$ for all $x \in G$, but A is now a square matrix and has at least one characteristic value. Let a be a characteristic value of A and observe that $\Phi(x)(A - aI) = (A - aI)\Phi(x)$ for all $x \in G$. Because $A - aI$ is not invertible by the choice of a, Schur's Lemma implies that $A - aI = \mathrm{O}$ and $A = aI$.

The trace of a matrix is defined as the sum of the diagonal elements, and $\mathrm{Tr}\,[CBC^{-1}] = \mathrm{Tr}\,[B]$ when C is in $\mathrm{GL}(n, \mathbb{C})$ (*Exercise* 11, p. 180). It follows that

$$\begin{aligned}\mathrm{Tr}\,[A] &= \mathrm{Tr}\left[\int \Phi(x)B\Phi(x)^{-1}\,dx\right] \\ &= \int \mathrm{Tr}\,[\Phi(x)B\Phi(x)^{-1}]\,dx\end{aligned}$$

$$= \int \mathrm{Tr}\,[B]\,dx$$
$$= \mathrm{Tr}\,[B].$$

Now $A = aI$ implies that $na = \mathrm{Tr}\,[B]$ and $a = \mathrm{Tr}\,[B]/n$.

If B is the $n \times n$ matrix of all zeros except a 1 in the (μ, ν) position, then the (i, q) entry of the integrand is

$$\begin{aligned} \Phi(x)B\Phi(x^{-1})(i,q) &= \Phi(x)B\Phi(x)^*(i,q) \\ &= \sum_{p=1}^{n}\sum_{j=1}^{n} \Phi(x)(i,j)B(j,p)\overline{\Phi}(x)(q,p) \\ &= \Phi(x)(i,\mu)\overline{\Phi}(x)(q,\nu). \end{aligned}$$

If $\mu \neq \nu$, then $\mathrm{Tr}\,[B] = 0$ and $A = O$. It follows that

$$\int \Phi(x)(i,\mu)\overline{\Phi}(x)(q,\nu)\,dx = 0.$$

If $\mu = \nu$ and $i \neq q$, then $A(i,q) = 0$ and

$$\int \Phi(x)(i,\mu)\overline{\Phi}(x)(q,\mu)\,dx = 0.$$

If $\mu = \nu$ and $i = q$, then $A(i,q) = A(i,i) = 1/n$ and

$$\int \Phi(x)(i,\mu)\overline{\Phi}(x)(i,\mu)\,dx = 1/n$$

to complete the proof. □

The *character of a matrix reprresentation* Φ of a topological group G is the function

$$\chi(\Phi(x)) = \mathrm{Tr}\,[\Phi(x)] = \sum_{j=1}^{n} \Phi(x)(j,j)$$

in $\mathcal{C}(G, \mathbb{C})$. It depends only on the equivalence class of the matrix representation because $\mathrm{Tr}\left[A\Phi(x)A^{-1}\right] = \mathrm{Tr}\left[\Phi(x)\right]$ and captures important information about the matrix representation.

Theorem 7.5.9 *A unitary representation Φ of a compact metric group G with character χ is irreducible if and only if $\int \chi(x)\overline{\chi}(x)\,dx = 1$.*

Proof. Assume first that Φ is irreducible. Then, using *Proposition 7.5.8*,

$$\begin{aligned} \int \chi(x)\overline{\chi}(x)\,dx &= \int \sum_{i=1}^{n}\sum_{j=1}^{n} \Phi(i,i)\overline{\Phi}(j,j)\,dx \\ &= \sum_{i=1}^{n}\sum_{j=1}^{n} \int \Phi(i,i)\overline{\Phi}(j,j)\,dx \end{aligned}$$

$$\begin{aligned} &= \sum_{i=1}^{n} \int \Phi(i,i)\overline{\Phi}(i,i)\,dx \\ &= n(1/n) \\ &= 1. \end{aligned}$$

Next, assume that $\int \chi(x)\overline{\chi}(x)\,dx = 1$. By *Proposition 7.5.2*, $\Phi \sim \Phi_1 + \ldots + \Phi_m$, where each Φ_k is an irreducible unitary representation with character χ_k. Notice that $\chi = \chi_1 + \ldots + \chi_m$. Using the first part of the proof

$$\begin{aligned} \int \chi(x)\overline{\chi}(x)\,dx &= \int \sum_{i=1}^{m}\sum_{j=1}^{m} \chi_i(x)\overline{\chi}_j(x)\,dx \\ &= \sum_{i=1}^{m} \int \chi_i(x)\overline{\chi}_i(x)\,dx + \sum_{i\neq j} \int \chi_i(x)\overline{\chi}_j(x)\,dx \\ &= m + \sum_{i\neq j} \int \chi_i(x)\overline{\chi}_j(x)\,dx \end{aligned}$$

By *Exercise* 4, p. 318, $\int \chi_i(x)\overline{\chi}_j(x)\,dx = 0$, if Φ_i and Φ_j are not equivalent. Therefore, $\int \chi(x)\overline{\chi}(x)\,dx = m$, the number of irreducible representations that sum to Φ, and Φ is irreducible because $\int \chi(x)\overline{\chi}(x)\,dx = 1$ by hypothesis. □

Theorem 7.5.10 *If G is a compact metric group, then there are a finite or countable number of equivalences classes of irreducible unitary representations.*

Proof. Let E denote the characters of the irreducible unitary representations of the compact metric group G with metric d. Let Φ and Ψ be irreducible unitary representations of G. If χ is the character of Φ, then $\chi(e) = \mathrm{Tr}\,[\Phi(e)] = Tr[I] \in \mathbb{Z}^+$ and χ is not identically zero. It has already been pointed out that equivalent irreducible unitary representations have the same character because $\mathrm{Tr}\,[A\Phi(x)A^{-1}] = \mathrm{Tr}\,[\Phi(x)]$. If Φ and Ψ are not equivalent, then their characters are orthogonal (*Exercise* 4, p. 318). Since their characters are also nonzero, they can not be equal. Thus the function that assigns to each equivalence class of irreducible unitary representations its character is a well-defined one-to-one function onto E. Therefore, it suffices to show that E is finite or countable.

Given χ in E, write $\chi(x) = u(x)+iv(x)$, where $u(x)$ and $v(x)$ are continuous real valued functions. Since $u(e) > 0$, there exists $k \in \mathbb{Z}^+$ such that $u(x) > 0$ for all x in $B_{1/k}(e)$. Set

$$\nu(\chi) = \min\left\{k \in \mathbb{Z}^+ : u(x) > 0 \text{ for all } x \in B_{1/k}(e)\right\},$$

and for $k \in \mathbb{Z}^+$ set

$$E_k = \left\{\chi \in E : \nu(\chi) = k\right\}.$$

Clearly, $E_p \cap E_q = \phi$ when $p \neq q$ and $E = \bigcup_{k=1}^{\infty} E_k$. This reduces the proof to showing that each E_k is finite or countable.

For each $k \in \mathbb{Z}^+$, there exists a continuous real-valued function f_k on G such that

(a) $0 \leq f_k(x) \leq 1$ for all $x \in G$

(b) $f_k(x) = 1$ if and only if $d(x, e) \leq 1/2k$

(c) $f_k(x) = 0$ if and only if $d(x, e) \geq 1/k$.

For all χ in E_k, it follows that $\int f_k(x)u(x)\,dx > 0$ and

$$\langle\langle f_k, \chi\rangle\rangle = \int f_k(x)\overline{\chi}(x)\,dx = \int f_k(x)u(x)\,dx - i\int f_k(x)v(x)\,dx \neq 0.$$

In particular, $|\langle\langle f_k, \chi\rangle\rangle| > 0$ for all $\chi \in E_k$.

Now we can use the same technique to partition E_k into smaller pieces by setting

$$\mu(\chi) = \min\{m \in \mathbb{Z}^+ : |\langle\langle f_k, \chi\rangle\rangle| \geq 1/m\},$$

and

$$E_{(k,m)} = \{\chi \in E_k : \mu(\chi) = m\}.$$

Suppose $\chi_1, \ldots, \chi_p$ are distinct elements of $E_{(k,m)}$. Then $\chi_1, \ldots, \chi_p$ are orthonormal (*Theorem 7.5.9* and *Exercise* 4, p. 318, again). By Bessel's Inequality for $C(G, \mathbb{C})_2$ from part (b) of *Exercise* 6, p. 309,

$$\frac{p}{m^2} \leq \sum_{j=1}^{p} \langle\langle f_k, \chi_j\rangle\rangle^2 \leq \langle\langle f, f\rangle\rangle$$

and

$$p \leq m^2\langle\langle f, f\rangle\rangle.$$

Therefore, $E_{(k,m)}$ is a finite set of functions. It follows that E_k is finite or countable, and then E is finite or countable. □

Corollary 7.5.11 *If G is a compact metric group, then there exists a finite or infinite sequence of irreducible unitary representations $\Phi_1, \ldots$ such that for every irreducible unitary representation Φ of G there exists exactly one Φ_k satisfying $\Phi \sim \Phi_k$.*

EXERCISES

1. Let Φ and Ψ be equivalent matrix representations of a topological group. Show that Φ is irreducible if and only if Ψ is irreducible.

2. Let Φ, Φ', Ψ, and Ψ' are matrix representations of a topological group G. Show that if $\Phi \sim \Phi'$ and $\Psi \sim \Psi'$, then $\Phi + \Psi \sim \Phi' + \Psi'$.

3. Show that if Φ are Ψ are matrix representations of a topological group G, then $\Phi + \Psi \sim \Psi + \Phi$.

4. Let Φ and Ψ be irreducible matrix representations of a compact metric group G. Show that if Φ and Ψ are not equivalent, then their characters are orthogonal.

5. Let Ψ be a reducible unitary representation of a compact metric group G with character γ, and let Φ be an irreducible unitary representation of G with character χ. Show that if $\int \chi(x)\overline{\gamma}(x)\,dx \neq 0$, then $\int \chi(x)\overline{\gamma}(x)\,dx = p$ is a positive integer and there exists a unitary representation Ψ' of G such that
$$\Psi \sim \underbrace{\Phi + \ldots + \Phi}_{p} + \Psi'.$$

6. Let G be a compact metric group, and let n_k be the dimension of Φ_k in the sequence of irreducible unitary representation from *Corollary 7.5.11.* Prove that G is isomorphic to a closed subgroup of $\prod_k \mathrm{U}(n_k)$.

7. Show that if A and B are $n \times n$ matrices, then $\mathrm{Tr}\,[AB] = \mathrm{Tr}\,[BA]$.

Chapter 8

Character Groups

As part of our study of compact metric groups, we proved that the character group of a compact Abelian metric group was a countable discrete Abelian group (*Proposition 7.4.6*). This reduced the calculation of the character groups of compact Abelian metric groups to determining their algebraic structure and enabled the calculation of various character groups of specific compact Abelian metric groups. Building on this knowledge of the character groups of compact Abelian metric groups, this chapter studies the character groups of all Abelian locally compact second countable metric groups.

For convenience, Abelian locally compact second countable metric groups will be referred to simply as ALS-*groups.* Besides avoiding the more cumbersome "ALS-metric group", these groups are usually metrizable. Since second countable implies first countable (*Exercise* 3, p. 213), a second countable topological group is metrizable by *Theorem 1.3.8*, if its points are closed sets. Rarely does one consider topological groups without the property that points are not closed sets. The ALS-groups as defined here are the same as the Abelian locally compact second countable topological groups whose points are closed sets.

Along with the compact Abelian metric groups, the ALS-groups include the Euclidean groups $\mathbb{R}^n$ and the discrete groups $\mathbb{Z}^n$, but not $\mathbb{R}^\infty$ and $\mathbb{Z}^\infty$ because they are not locally compact. The discrete metric group $\mathbb{R}_d$ is not an ALS-group because it not second countable. Although it is a locally compact Abelian metric group, its character group is a topological group that is not metrizable. So $\mathbb{R}_d$ cannot fit into a duality theorem in the context of only metrizable groups. Thus restricting our attention to the subclass of ALS-groups is essential.

The most important property of ALS-groups for this chapter is that the character group of an ALS-group is an ALS-group by *Theorem 6.5.7.* Thus the ALS-groups provide a closed system for studying character groups. Of course, information about an ALS-group is encoded in its character group. The big questions are how much information and how do you retrieve it? The amazing answer is that an ALS-group can be recovered from its character group. Simply put, an ALS-group is isomorphic to the character group of its character group. This duality theorem will be the center piece of this finial chapter, although along the way to it there will be a variety of interesting results about ALS-groups.

Section 8.1 compliments Section 7.4 with a detailed study of countable discrete Abelian groups and their character groups. The duality homomorphism from an ALS-group to the character group of its character group is introduced in Section 8.2 and its basic properties developed. By definition, duality means that the duality homomorphism is an onto isomorphism. Using the results of Sections 7.4 and 8.1, it is shown that compact Abelian metric groups and countable discrete Abelian groups satisfy duality. The proof in Section 8.4 that all ALS-groups satisfy duality depends heavily on the deep results about the structure of compactly generated ALS-groups obtained in Section 8.3. Finally, Section 8.4 concludes with a collection of important consequences of duality that demonstrate its fundamental importance in the study of locally compact Abelian topological groups.

8.1 Countable Discrete Abelian Groups

Every discrete space is locally compact and metric, but not necessarily second countable. In fact, a discrete metric space is clearly second countable if and only if it is countable. Consequently, a discrete Abelian group is an ALS-group if and only if it is countable. *This section is devoted to discrete ALS-groups or equivalently countable Abelian groups with the discrete topology.*

Since the character group of a compact Abelian metric group is a countable discrete Abelian group, it is natural to begin with the opposite question.

Proposition 8.1.1 *If G is a discrete ALS-group, then $\widehat{G}$ is a compact metric group.*

Proof. Because G is discrete, $\{e\}$ is a compact neighborhood of e in G. Using the open neighborhoods $U_r = \{e^{2\pi i x} : |x| < r\}$ of 1 in $\mathbb{K}$ again, recall from the proof of *Theorem 6.5.7* that the closure of $V = N(\{e\}, U_r) \cap \widehat{G}$ is a compact subset of $\widehat{G}$ when $0 < r < 1/4$. Obviously, $V = N(\{e\}, U_r) \cap \widehat{G} = \widehat{G}$. Because $\widehat{G}$ is closed in $C(G, \mathbb{K})$, it follows that $\widehat{G} = V^-$ is compact. □

Notice that we have not yet proved that an ALS-group is compact if and only if its character group is countable discrete. Or that an ALS-group is countable discrete if and only if its character group is compact. At this point, we do not know whether or not there are noncompact ALS-groups with countable discrete character groups or non-discrete ALS-groups with compact character groups. These issues will eventually be resolved.

Proposition 8.1.2 *The function $\varphi(\chi) = \chi(1)$ defines an isomorphism of $\widehat{\mathbb{Z}}$ onto $\mathbb{K}$.*

Proof. Because $\mathbb{Z}$ is discrete, an algebraic homomorphism from $\mathbb{Z}$ to $\mathbb{K}$ is automatically continuous. Hence, the characters of $\mathbb{Z}$ are just the algebraic homomorphisms from $\mathbb{Z}$ to $\mathbb{K}$. For $z \in \mathbb{K}$, define $\gamma_z : \mathbb{Z} \to \mathbb{K}$ by $\gamma_z(n) = z^n$. Clearly, γ_z is an algebraic homomorphism of $\mathbb{Z}$ to $\mathbb{K}$ and every character has this form because $\mathbb{Z} = \langle 1 \rangle$. Thus $\{\gamma_z : z \in \mathbb{K}\} = \widehat{\mathbb{Z}} \subset C(\mathbb{Z}, \mathbb{K})$.

It is now easy to check that $\varphi(\chi) = \chi(1)$ is an algebraic isomorphism of $\widehat{\mathbb{Z}}$ onto $\mathbb{K}$. To prove that φ is a homeomorphism, it now suffices to show that φ is continuous because $\widehat{\mathbb{Z}}$ is compact by *Proposition 8.1.1* (*Exercise* 2, p. 61).

Let χ_k be a sequence in $\widehat{\mathbb{Z}}$ converging to $\chi \in \widehat{\mathbb{Z}}$. It must be shown that $\varphi(\chi_k) = \chi_k(1)$ converges to $\varphi(\chi) = \chi(1)$. Because convergence in the compact-open topology is equivalent to uniform convergence on compact sets and $\{1\}$ is a compact set, $\chi_k(1)$ converges to $\chi(1)$ and φ is continuous. Therefore, φ is a homeomorphism, and hence an isomorphism of $\mathbb{K}$ onto $\widehat{\mathbb{Z}}$. □

Corollary 8.1.3 *The character group of $\mathbb{Z}^n$ is isomorphic to $\mathbb{K}^n$.*

Proof. Apply *Theorem 6.5.8.* □

Because groups like $\mathbb{Z}_n$ and $\mathbb{Z}^n$ are the most prominent examples of finite and countable Abelian groups, it is convenient to use the additive notation with 0 as the identity for countable Abelian groups. The rest of this section will be devoted to an algebraic discussion of these groups. With the discrete metric they are, of course, ALS-groups, and their algebraic isomorphisms and algebraic homomorphisms will automatically be continuous.

Countable groups, however, can also be metric groups with topologies other than the discrete topology. Such topologies are not locally compact by *Exercise* 1, p. 328. For example, the rational numbers $\mathbb{Q}$ are a countable dense subgroup of $\mathbb{R}$ and hence a metric group with the metric $d(x, y) = |x - y|$, but $\mathbb{Q}$ is not locally compact with this metric. So the adjective "algebraic" is important in the remainder of this section.

A group G is *divisible* provided that for all k in $\mathbb{Z}^+$ in additive notation $G = \{kx : x \in G\}$ or in multiplicative notation $G = \{x^k : x \in G\}$. Examples of divisible groups are $\mathbb{K}$ and $\mathbb{R}$. It follows that all groups of the form $\mathbb{R}^m \times \mathbb{K}^n$ are divisible (see *Exercise* 5, p. 328). The groups $\mathbb{Z}^n$ are not divisible for $n > 0$.

Proposition 8.1.4 *Let H be a subgroup of an Abelian group G such that G/H is finite or countable. If $\varphi : H \to G'$ is an algebraic homomorphism and G' is a divisible group, then there exists an algebraic homomorphism $\widetilde{\varphi} : G \to G'$ that extends φ to G, that is, $\widetilde{\varphi}(x) = \varphi(x)$ for all x in H.*

Proof. First, suppose $G = \langle a, H \rangle$, the group generated by H and some $a \notin H$. Note that $ka \in H$ if and only if $-ka \in H$. If ka is not in H for all $k > 1$, then every element of G can be uniquely written in the form $ka + x$ with $k \in \mathbb{Z}$ and $x \in H$. It follows that $\widetilde{\varphi}(ka+x) = \varphi(x)$ is the required algebraic homomorphism.

If ka is in H for some $k > 1$, then clearly $K = \{k \in \mathbb{Z} : ka \in H\}$ is a subgroup of $\mathbb{Z}$ not equal to $\{0\}$ and $H = \{km : k \in \mathbb{Z}\}$ for some positive integer m (*Corollary 1.4.19*). Because G' is divisible, there exists $w \in G'$ such that $w^m = \varphi(ma)$. (The group operation in G' is being written multiplicatively because G' will be $\mathbb{K}$ in the applications of the proposition.) Define $\widetilde{\varphi}$ by $\widetilde{\varphi}(ka+x) = w^k\varphi(x)$. If $ka + x = k'a + x'$, then $k - k' = dm$ and $x = -dma + x'$ for some $d \in \mathbb{Z}$. It follows that

$$\begin{aligned} \widetilde{\varphi}(ka + x) &= w^k\varphi(x) \\ &= w^{k'+dm}\varphi(x) \\ &= w^{k'}\varphi(dma)\varphi(-dma + x') \\ &= w^{k'}\varphi(x'). \end{aligned}$$

Thus the definition of $\widetilde{\varphi}$ is not ambiguous, and $\widetilde{\varphi}$ is an algebraic homomorphism. For the general case, we will repeatedly use the special case that $G = \langle a, H\rangle$.

Since G/H is finite or countable, there exists a finite or infinite sequence a_k in G with $a_1 = 0$ such that given x in G there exists a unique a_k such that x is in the coset $a_k + H$. We will inductively define a function $\lambda : \mathbb{Z}^+ \to \mathbb{Z}^+$, a sequence of subgroups H_k containing H for $k \geq 1$, and algebraic homomorphisms $\varphi_k : H_k \to G'$ extending φ such that

(a) $\lambda(k) \leq \lambda(k+1)$ for all $k \geq 1$

(b) $H_k \subset H_{k+1}$ for all $k \geq 1$

(c) $\varphi_{k+1}(x) = \varphi_k(x)$ for all $x \in H_k$

(d) $a_j + H \subset H_k$ for $1 \leq j \leq \lambda(k)$.

Let $\lambda(1) = 1$, $H_1 = a_1 + H = H$, and $\varphi_1 = \varphi$. To start the induction process, set $\lambda(2) = 2$, $H_2 = \langle a_2, H_1\rangle$, and φ_2 is an extension of φ to H_2 obtained by applying the special case of the proof to H_2. Note that a_k is in H_2 if and only if $a_k + H \subset H_2$.

Assume that $\lambda(j)$, H_j, and φ_j have been constructed as required for $1 \leq j \leq k$. If $H_k = G$, set $H_{k+1} = G$, $\lambda(k+1) = \lambda(k)$, and $\varphi_{k+1} = \varphi_k$. Otherwise, $\{j : a_j \notin H_k\}$ is nonempty. In this case, set $\lambda(k+1) = \min\{j : a_j \notin H_k\}$ and $H_{k+1} = \langle a_{\lambda(k+1)}, H_k\rangle$. Then use the special case of the proof again to construct an algebraic homomorphism extending φ_k to H_{k+1} and check that the required properties are satisfied. This completes the induction argument.

It is easy to verify that $G = \bigcup_{k=1}^{\infty} H_k$ and that setting $\widetilde{\varphi}(x) = \varphi_k(x)$ when x is in H_k unambiguously defines the required algebraic homomorphism. □

The assumption that G/H is finite or countable can be removed by using the Axiom of Choice in its equivalent form of Zorn's Lemma. The special case $G = \langle a, H\rangle$ in the beginning of the proof shows that unless $H = G$, the homomorphism φ can always be extended to a larger group, making the proof of the general result a very standard Zorn's Lemma argument.

Corollary 8.1.5 *Let H be a subgroup of a finite or countable Abelian group G. If $\varphi : H \to G'$ is an algebraic homomorphism and G' is a divisible group, then there exists an algebraic homomorphism $\widetilde{\varphi} : G \to G'$ that extends φ to G.*

Proof. Since G is finite or countable, G/H is also finite or countable. □

If G is a group and a is an element of G, then $\langle a \rangle$ is a cyclic group. When $ka = 0$ (or in multiplicative notation, when $a^k = e$) for some $k \in \mathbb{Z}^+$, the element a is said to be of *finite order*. In this case, the function $\varphi(k) = ka$ is an algebraic homomorphism of $\mathbb{Z}$ to G and its kernel $K = \{k \in \mathbb{Z} : kx = 0\}$ is a subgroup of $\mathbb{Z}$ that is not equal to $\{0\}$. Hence, there exists a positive number $o(a)$, called the *order* of a, such that $K = \{n\,o(a) : n \in \mathbb{Z}\}$. *Thus $o(a)$ divides every integer k such that $ka = 0$ and is the smallest positive integer such that $ka = 0$.* Note that $o(a) = 1$ if and only if a is the identity. Moreover, $\langle a \rangle$ is algebraically isomorphic to $\mathbb{Z}_{o(a)}$, when a is of finite order.

For an Abelian group G, if $kx = 0$ and $my = 0$, then $km(x + y) = 0$ and $k(-x) = -kx = 0$. It follows that

$$T_G = \{x \in G : kx = 0 \text{ for some } k \in \mathbb{Z}^+\}$$

is a subgroup of G called the *torsion subgroup*. A group G is said to be *torsion-free* when $T_G = \{0\}$.

Recall that a group G is *finitely generated* (p. 118), provided that there exists a finite subset $\{a_1, \ldots, a_n\}$ of G such that $G = \langle a_1, \ldots, a_n \rangle$. When $G = \langle a_1, \ldots, a_n \rangle$, it will be understood that $a_1, \ldots, a_n$ are distinct elements of G. If $G = \langle a_1, \ldots, a_n \rangle$ is Abelian, then in additive notation

$$G = \left\{ \sum_{j=1}^{n} c_j a_j : c_j \in \mathbb{Z} \right\}.$$

by *Exercise* 4, p. 69. In preparation for Section 8.2, we will determine the algebraic structure of all finitely generated Abelian groups.

Proposition 8.1.6 *If G is a finitely generated torsion-free Abelian group, then G is algebraically isomorphic to $\mathbb{Z}^q$ for some $q \geq 0$.*

Proof. The proof proceeds by induction on the number of generators. Suppose $G = \langle g \rangle$ is torsion-free. If $g = 0$, then $G = \{0\}$ and G is isomorphic to $\mathbb{Z}^0$. If $g \neq 0$, then clearly $\varphi(n) = ng$ defines an algebraic isomorphism of $\mathbb{Z}$ onto G.

Suppose that the result holds for torsion-free Abelian groups generated by n elements with $n \geq 1$. Suppose $G = \langle a_1, \ldots, a_{n+1} \rangle$ is Abelian and torsion-free. Set $H_1 = \langle a_1, \ldots, a_n \rangle$. Then H_1 is algebraically isomorphic to $\mathbb{Z}^q$ for some $q > 0$ by the induction assumption. Set $H_2 = \langle a_{n+1} \rangle$. We can assume that $a_{n+1} \neq 0$, otherwise there is nothing to prove. Thus H_2 is algebraically isomorphic to $\mathbb{Z}$. If ka_{n+1} is not in H_1 for all integers $k \neq 0$, then G is algebraically isomorphic to $\mathbb{Z}^q \times \mathbb{Z} = \mathbb{Z}^{q+1}$ by *Exercise* 6, p. 329, with $m = 2$. (*Exercise 6* will be used several more times in this section.)

Suppose ka_{n+1} is in H_1 for some nonzero integer k. Then, as in the proof of *Proposition 8.1.4*, there exists a smallest positive integer m such that ma_{n+1} is in H_1. Because G is torsion-free, $\varphi(x) = mx$ defines an algebraic isomorphism of G into H_1. Thus G is algebraically isomorphic to a subgroup of $\mathbb{Z}^q$, which is a discrete subgroup of $\mathbb{R}^q$. So G is algebraically isomorphic to a discrete subgroup of $\mathbb{R}^q$. Therefore, G is algebraically isomorphic to $\mathbb{Z}^r$ for some r by *Corollary 3.1.7.* □

Proposition 8.1.7 *If G is a finitely generated Abelian group, then T_G is a finite group and G is algebraically isomorphic to $\mathbb{Z}^q \times T_G$ for some $q \geq 0$.*

Proof. If $G = T_G$, then $q = 0$ and T_G is finite by *Exercise* 2, p. 328.

Assume $G \neq T_G$ and consider G/T_G. Observe that $x + T_G$ is in the torsion group of G/T_G if and only if kx is in T_G for some k in $\mathbb{Z}^+$ if and only if $mkx = 0$ for some k and m in $\mathbb{Z}^+$ if and only if x is in T_G. Therefore, G/T_G is torsion-free and algebraically isomorphic to $\mathbb{Z}^q$ for some $q > 0$ by *Proposition 8.1.6.*

Let φ be an algebraic isomorphism of $\mathbb{Z}^q$ onto G/T_G. For each of the standard generators $\mathbf{e}_1, \ldots, \mathbf{e}_q$ of $\mathbb{Z}^q$, let $a_1, \ldots, a_q$ be elements of G such that $\pi(a_j) = \varphi(\mathbf{e}_j)$, where π is the canonical algebraic homomorphism. Set $H_1 = \langle a_1, \ldots, a_q \rangle$. Then,

$$\pi(k_1 a_1 + \ldots + k_q a_q) = \varphi(k_1 \mathbf{e}_1 + \ldots + k_q \mathbf{e}_q).$$

It follows that $\pi|H_1$ is an algebraic isomorphism of H_1 onto G/T_G and $H_1 \cap T_G = \{0\}$. Given x in G, there exists y in H_1 such that $\pi(x) = \pi(y)$. Consequently, $x - y = z$ is in T_G and $x = y + z$ is in $H_1 + T_G$. Again *Exercise* 6, p. 329, can be applied to show that G is algebraically isomorphic to $H_1 \times T_G$. Since H_1 and $\mathbb{Z}^q$ are algebraically isomorphic to G/T_G, the group H_1 is algebraically isomorphic to $\mathbb{Z}^q$ and G is algebraically isomorphic to $\mathbb{Z}^q \times T_G$.

To complete the proof, it remains to show that T_G is finite. Let $a_1, \ldots, a_m$ be a finite set of generators for G. Since each generator can be written as $a_i = b_i + \tau_i$ with b_i is in H_1 and τ_i is in T_G, it is easy to verify that $T_G = \langle \tau_1, \ldots, \tau_m \rangle$ and T_G is a finite. □

Given an Abelian group G and a prime number p, set

$$G_p = \left\{ x \in G : o(x) = p^j \text{ for some } j \geq 0 \right\}.$$

(Prime numbers are integers greater than 1 with no divisors except 1 and the prime itself. We will assume that the reader is familiar with the elementary properties of prime numbers.) If x and y are in G_p, then $p^i x = 0 = p^j y$ for positive integers i and j and $p^{i+j}(x+y) = 0$. So $o(x+y)$ divides p^{i+j} and equals p^k for some $k \geq 0$. Hence, $x+y$ is in G_p, and $-x$ is in G_p because $o(x) = o(-x)$. Thus G_p is a subgroup of G.

If x is in $G_{p_1} \cap G_{p_2}$ for two different primes p_1 and p_2, then $o(x) = p_1^i = p_2^j$ and $i = j = 0$. Consequently, $G_{p_1} \cap G_{p_2} = \{0\}$ and the sets $G_p \setminus \{0\}$ are disjoint or empty subsets of $G \setminus \{0\}$. When G is finite, it follows that $F = \{p : G_p \neq \{0\}\}$ is a finite set.

Proposition 8.1.8 *If G is a finite Abelian group, then G is algebraically isomorphic to $\prod_{p \in F} G_p$ where $F = \{p : G_p \neq \{0\}\}$.*

Proof. It suffices to show that the finite collection of subgroups G_p of G with p in F satisfy the two required conditions in *Exercise* 6, p. 329.

Given $x \neq 0$ in G, there exist a positive integer β, primes $p_1, \dots, p_\beta$, and positive integers $n_1, \dots, n_\beta$ such that

$$o(x) = \prod_{i=1}^{\beta} p_i^{n_i}.$$

If $\beta = 1$, then x is in G_{p_1}. Assume $\beta > 1$. For $1 \le j \le \beta$, set

$$q_j = \prod_{i \neq j} p_i^{n_i}$$

and consider the subgroup $H = \langle q_1, \dots, q_\beta \rangle$ of $\mathbb{Z}$. Clearly $H \neq \{0\}$, so there exists a positive integer m such that $H = \{km : k \in \mathbb{Z}\}$. Since q_j is in H, m divides q_j. But a moments reflection shows that no positive integer other than 1 can divide every q_j for $1 \le j \le \beta$. Therefore, $m = 1$ and there exists integers $c_1, \dots, c_\beta$ such that

$$1 = c_1 q_1 + \dots + c_\beta q_\beta.$$

Setting $x_j = c_j q_j x$,

$$x = 1x = (c_1 q_1 + \dots + c_\beta q_\beta)x = x_1 + \dots + x_\beta.$$

Since $p_j^{n_j} c_j \, q_j = c_j \, o(x)$, it follows that

$$p_j^{n_j} x_j = p_j^{n_j} c_j \, q_j x = c_j \, o(x) x = 0.$$

Therefore, $o(x_j)$ divides $p_j^{n_j}$ and x_j is in G_{p_j}. So p_j is in F when $x_j \neq 0$. Thus $G = \sum_{p \in F} G_p$, and the first condition is satisfied.

Given p a prime in F, we can assume by reordering F that $p = p_1$. Suppose x is in $G_p \cap (G_{p_2} + \dots + G_{p_\beta})$. Then $x = y_2 + \dots + y_\beta$ with y_j in G_{p_j}, and $o(y_j) = p_j^{k_j}$ for some $k_j \ge 0$. Let

$$\sigma = \prod_{j=2}^{\beta} p_j^{k_j}.$$

Clearly $\sigma x = 0$ and the $o(x)$ divides σ. Because x is also in G_p, $o(x) = p^k$ for some $k \ge 0$. But p^k does not divide σ for $k > 0$. So $k = 0$ and $o(x) = 1$. Thus $x = 0$ and $G_p \cap (G_{p_2} + \dots + G_{p_\beta}) = \{0\}$, which is the second condition. ☐

Lemma 8.1.9 *Given a countable Abelian group G, a subgroup H of G, and a prime number p, the following hold:*

(a) If a is in G and $o(a) = p$, then $\langle a \rangle = \langle ka \rangle$ for $k = 1, 2, \dots, p-1$.

(b) If a is in G and $o(a) = p$, then either a is in H or $\langle a \rangle \cap H = \{0\}$.

(c) If a is in $G \setminus H$ and $o(a) = p$, then $\langle a, H\rangle$ is algebraically isomorphic to $\mathbb{Z}_p \times H$.

Proof. Since $\langle a\rangle$ is algebraically isomorphic to $\mathbb{Z}_p$, its only subgroups are $\{0\}$ and $\langle a\rangle$ (*Exercise* 18, p. 12), and the nontrivial subgroups $\langle ka\rangle$ for $k = 1, 2, \ldots, p-1$ must equal $\langle a\rangle$. This proves part *(a)*, and part *(b)* is an immediate consequence of it. Then part *(c)* follows from *(b)*. □

The next proposition is really the first step in the induction proof of the fundamental algebraic result that all finite Abelian groups are isomorphic to a product of cyclic groups.

Proposition 8.1.10 *Let p be a prime number. If G is a finite Abelian group such that $px = 0$ for all x in G, then G is algebraically isomorphic to $\mathbb{Z}_p^n$ for some $n \geq 0$.*

Proof. If G is the trivial group, then $n = 0$. Consider $a_1 \neq 0$ in G. Then $o(a) = p$ and $\langle a_1\rangle$ is algebraically isomorphic to $\mathbb{Z}_p$. Either $\langle a_1\rangle = G$ and the proof is complete or there exists a_2 in $G \setminus \langle a_1\rangle$. Then it follows from the lemma that $\langle a_1, a_2\rangle$ is isomorphic to $\mathbb{Z}_p^2$. Either $\langle a_1, a_2\rangle = G$ or the process can be repeated, but must eventually terminate because G is finite. □

Theorem 8.1.11 (Fundamental Theorem for Finite Abelian Groups) *If G is a finite Abelian group, then G is algebraically isomorphic to a group of the form*

$$\prod_{j=1}^{N} \mathbb{Z}_{m_j}$$

where each m_j is a positive integral power of a prime number.

Proof. By *Proposition 8.1.8*, it suffices to prove the following: If G is a finite Abelian group such that $G = G_p$ for some prime p, then G is algebraically isomorphic to a group of the form

$$\prod_{j=1}^{N} \mathbb{Z}_{m_j}$$

where each m_j is a positive integral power of the prime number p.

Because G is finite, the set

$$\{k \in \mathbb{Z}^+ : o(x) = p^k \text{ for some } x \in G\}$$

is finite. The proof proceeds by induction on

$$n = \max\{k \in \mathbb{Z}^+ : o(x) = p^k \text{ for some } x \in G\}.$$

The $n = 1$ case follows from *Proposition 8.1.10*.

Assume the conclusion holds for n. If $n+1 = \max\{k \in \mathbb{Z}^+ : o(x) = p^k$ for some $x \in G\}$, then the function $x \to px$ is an algebraic homomorphism of G onto a subgroup of G denoted by pG such that

$$n = \max\{k \in \mathbb{Z}^+ : o(x) = p^k \text{ for some } x \in pG\}.$$

Thus by the induction assumption, pG is algebraically isomorphic to a finite product of groups of the form $\mathbb{Z}_{p^k}$. So there exist $a_1, \ldots, a_m$ in pG with none of them equal to the identity and such that pG is algebraically isomorphic to $\langle a_1 \rangle \times \ldots \times \langle a_m \rangle$.

There exist b_i in G such that $p\, b_i = a_i$ for $i = 1, \ldots, m$. Note that $o(b_i) = p\, o(a_i) = p^k$ for some k. The next step is to show that $\langle b_1, \ldots, b_m \rangle$ is algebraically isomorphic to $\langle b_1 \rangle \times \ldots \times \langle b_m \rangle$. If not, the second condition in *Exercise* 6, p. 329, is violated because the first obviously holds. By reordering the a_i and b_i, we can assume that $\langle b_1 \rangle \cap \langle b_2, \ldots, b_m \rangle$ contains a nonzero element $q_1 b_1$ of G. It follows that there exist integers $q_2, \ldots q_m$ such that

$$q_1 b_1 = q_2 b_2 + \ldots + q_m b_m, \tag{8.1}$$

$1 \le q_1 < o(b_1)$ and $0 \le q_i < o(b_i)$ for $i = 2, \ldots, m$.

Applying the homomorphism $x \to px$ to the equation (8.1) yields

$$q_1 a_1 = q_2 a_2 + \ldots + q_m a_m.$$

Because the second condition holds for pG, every term in this equation must equal 0 and $o(a_i)$ divides q_i for each i. In particular, p divides q_i for $i = 1, \ldots, m$. Now (8.1) can be rewritten as

$$\frac{q_1}{p} p\, b_1 = \frac{q_2}{p} p\, b_2 + \ldots + \frac{q_m}{p} p\, b_m$$

or equivalently

$$\frac{q_1}{p}\, a_1 = \frac{q_2}{p}\, a_2 + \ldots + \frac{q_m}{p}\, a_m.$$

Again because the second condition holds for pG, every term in this equation must also equal 0. Since $p\, o(a_1) = o(b_1)$ and $1 \le q_1 < o(b_1)$, it follows that $1 \le q_1/p < o(a_1)$ and $(q_1/p)a_1 \ne 0$. This contradiction completes the proof that $H = \langle b_1, \ldots, b_m \rangle$ is algebraically isomorphic to $\langle b_1 \rangle \times \ldots \times \langle b_m \rangle$ and algebraically isomorphic to a finite product of groups of the form $\mathbb{Z}_{p^k}$.

If $H = G$, we are done. If not, let y be an element of $G \setminus H$. Since $x \to px$ maps H onto pG, there exists y' in H such that $py = py'$ or $p(y - y') = 0$. Set $b_{m+1} = y - y'$ Observe that b_{m+1} is not in H because y is not in H, and that $o(b_{m+1}) = p$ because $b_{m+1} \ne 0$. It follows from *Lemma 8.1.9* that $H_1 = \langle b_{m+1}, H \rangle$ is algebraically isomorphic to $\mathbb{Z}_p \times H$. Either $H_1 = G$ or the process can be repeated, but because G is finite, must eventually terminate with G algebraically isomorphic to a finite product of groups of the form $\mathbb{Z}_{p^k}$. □

Corollary 8.1.12 *If G is a finite Abelian group with the discrete topology, then then $\widehat{G}$ is isomorphic to G.*

Proof. *Proposition 7.4.11* and *Theorem 6.5.8* can now be applied to a finite Abelian group with the discrete topology. □

Together *Proposition 8.1.7* and *Theorem 8.1.11* determine the general structure of all finitely generated Abelian groups. The proof of this structure theorem really began with *Proposition 8.1.6* and was completed with the proof of *Theorem 8.1.11*, both of which it now includes. Because of its importance we state the result as a theorem, although there is nothing left to prove.

Theorem 8.1.13 *If G is a finitely generated Abelian group, then G is algebraically isomorphic to*

$$\mathbb{Z}^q \times \prod_{j=1}^{N} \mathbb{Z}_{m_j}$$

where $q \geq 0$ and each m_j is a positive integral power of a prime number.

As a corollary we also know the structure of the character group of all finitely generated Abelian groups.

Corollary 8.1.14 *If G is finitely generated Abelian group with the discrete topology, then $\widehat{G}$ is isomorphic to*

$$\mathbb{K}^q \times \prod_{j=1}^{N} \mathbb{Z}_{m_j}.$$

where $q \geq 0$ and each m_j is a positive integral power of a prime number.

Proof. Finitely generated groups are countable (*Exercise* 3, p. 328). So G is an ALS-group. *Theorem 6.5.8*, *Corollary 8.1.3*, and *Proposition 7.4.11* apply to complete the proof. □

EXERCISES

1. Let G be a countable metric group. Prove that G is locally compact if and only if it is discrete. (One approach is to use the Baire Category Theorem.)

2. Prove that if G is a finitely generated group such that $T_G = G$, then G is finite.

3. Show that finitely generated Abelian groups are countable or finite.

4. Give an example of a countable Abelian group that is not finitely generated.

5. Show that $G_1 \times G_2$ is divisible, when both G_1 and G_2 are divisible groups.

6. Let $H_1, \ldots, H_m$ be subgroups of an Abelian group G. Prove that the algebraic homomorphism $\varphi : H_1 \times \ldots \times H_m \to G$ defined by $\varphi(x_1, \ldots, x_m) = x_1 + \ldots + x_m$ is an algebraic isomorphism of $H_1 \times \ldots \times H_m$ onto G if and only if

   (a)
$$G = H_1 + \ldots + H_m$$

   (b)
$$H_i \cap (H_1 + \ldots + H_{i-1} + H_{i+1} + \cdots + H_m) = \{0\}$$
   for $1 \leq i \leq m$.

7. Let F be a finite Abelian group. Show that F is isomorphic to a closed subgroup of $\mathbb{K}^n$ for suitable $n > 0$.

8. Prove that if H is a closed subgroup of an ALS-group G, then H is an ALS-group.

9. Prove that if H is a closed subgroup of an ALS-group G, then G/H is an ALS-group.

10. Prove that the product of a sequence of ALS-groups is an ALS-group if and only if all but a finite number of the groups in the sequence are compact.

8.2 The Duality Homomorphism

The beginning of this section depends heavily on basic properties of the compact-open topology in Section 6.2. In particular, the evaluation function $\eta(f, x) = f(x)$ is continuous on $\mathcal{C}(X, Y) \times X$ when X is a locally compact second countable metric space and the topology on $\mathcal{C}(X, Y)$ is the compact-open topology (*Corollary 6.2.6*). Thus the function $\eta(\chi, x) = \chi(x)$ is a continuous function from $\widehat{G} \times G$ to $\mathbb{K}$ when G is an ALS-group. Consequently, for fixed $x \in G$,the function $\chi \to \chi(x)$ is a continuous function from $\widehat{G}$ to $\mathbb{K}$. So

$$\eta^*(x)(\chi) = \eta(\chi, x) = \chi(x) \tag{8.2}$$

defines a function η^* from G to $\mathcal{C}(\widehat{G}, \mathbb{K})$. In fact, $\eta^*(x)$ is in $G^* = \widehat{\widehat{G}}$ because

$$\eta^*(x)(\chi_1\chi_2) = (\chi_1\chi_2)(x) = \chi_1(x)\chi_2(x) = \eta^*(x)(\chi_1)\eta^*(x)(\chi_2).$$

Similarly, η^* is an algebraic homomorphism because

$$\eta^*(xy)(\chi) = \chi(xy) = \chi(x)\chi(y) = \eta^*(x)(\chi)\eta^*(y)(\chi)$$

and $\eta^*(xy) = \eta^*(x)\eta^*(y)$.

Before we can prove that η^* is continuous, we need to describe a neighborhood base at e^*, the identity character in G^*. Notice that we are working at three levels here: group, functions on a group, and functions of functions on a group. Using *Theorem 6.2.4*, every neighborhood of e^* contains an open subset of the form

$$\bigcap_{j=1}^{p} N(C_j, U_j) \cap G^*,$$

where $C_1, \ldots, C_p$ are compact sets in $\widehat{G}$ and $U_1, \ldots, U_p$ are open subsets of $\mathbb{K}$ such that $e^*(C_j) \subset U_j$. Since $e^*(C_j) = 1$, we can assume without loss of generality that $U_j = \{e^{2\pi i x} : |x| < \varepsilon\}$ for $1 \leq j \leq p$ and some ε such that $0 < \varepsilon < 1/4$ because making U_j smaller makes $N(C_j, U_j)$ smaller.

Similarly, making C_j larger, makes $N(C_j, U_\varepsilon)$ smaller. Without loss of generality we can assume $C_1 = \ldots = C_p = C$. Thus every neighborhood of e^* in G^* contains an open subset of the form $N(C, U_\varepsilon) \cap G^*$, where C is a compact subset of $\widehat{G}$ and $0 < \varepsilon < 1/4$.

Proposition 8.2.1 *If G is an ALS-group, then the function η^* defined by equation (8.2) is a homomorphism of G to G^*.*

Proof. Because it has already been shown that η^* is an algebraic homomorphism, it suffices to show that η^* is continuous at e. Let W be a neighborhood of e^* in G^*. Using the definition of continuity at a point, it must be shown that there exists a neighborhood V of e in G such that $\eta^*(V) \subset W$. It follows from the discussion preceding the statement of the proposition that without loss of generality it can be assumed that $W = N(C, U_\varepsilon) \cap G^*$ with $0 < \varepsilon < 1/4$.

Let V' be a compact neighborhood of e in G. Then, $\{\chi N(V', U_{\varepsilon/2}) : \chi \in \widehat{G}\}$ is an open cover of C in $\widehat{G}$, and there exist $\gamma_1, \ldots, \gamma_m$ in $\widehat{G}$ such that

$$C \subset \bigcup_{i=1}^{m} \gamma_i N(V', U_{\varepsilon/2}).$$

There exists a neighborhood V of e such that $V \subset V'$ and $\gamma_i(V) \subset U_{\varepsilon/2}$ for $i = 1, \ldots, m$. We will show that $\eta^*(V) \subset W = N(C, U_\varepsilon) \cap G^*$.

Clearly, $\eta^*(V) \subset G^*$. Now let x be in V and χ be in C. There exists γ_i such that $\chi \in \gamma_i N(V', U_{\varepsilon/2})$ or $\gamma_i^{-1}\chi \in N(V', U_{\varepsilon/2})$. Hence, $\gamma_i(x)^{-1}\chi(x) \in U_{\varepsilon/2}$ and $\gamma_i(x) \in U_{\varepsilon/2}$, implying that $\chi(x) \in U_\varepsilon$. Thus $\eta^*(x)(\chi) = \chi(x) \in U_\varepsilon$ for all $\chi \in C$ and $x \in V$ or $\eta^*(V) \subset N(C, U_\varepsilon)$. So η^* is continuous at the identity. □

The function η^* will be called the *duality homomorphism*. By definition, G *satisfies duality* when η^* is an isomorphism of G onto G^*. Satisfying duality is an invariant property of an ALS-group (*Exercise* 4, p. 336).

Once the main theorems about the compact-open topology in Section 6.2 have been extended to at least locally compact Hausdorff topological spaces, it can be shown that $\widehat{G} = \text{Hom}(G, \mathbb{K})$ with the compact-open topology is an LCA-group (locally compact Abelian Hausdorff group), when G is a LCA-group. In this context, η^* can be defined as above and shown to be a homomorphism using essentially the same arguments we just used for an ALS-group. Then using the same definition, a LCA-group satisfies duality when η^* is an isomorphism of G onto G^*. Thus everything we prove about the duality of ALS-groups is part of the theory of duality for LCA-groups. Furthermore, most of the proofs we present are adaptable to the more general setting of LCA-groups. In a few cases, some extra work is required to extend a proof, but the main ideas are all there.

As an example we will show that $\mathbb{Z}$ satisfies duality. Recall from *Proposition 8.1.2* that the function $\varphi(\chi) = \chi(1)$ is an isomorphism of $\widehat{\mathbb{Z}}$ onto $\mathbb{K}$. The group $\widehat{\mathbb{K}}$ is isomorphic to $\mathbb{Z}$, and every character of $\mathbb{K}$ has the form $z \to z^k$ by *Proposition 7.4.8.* Thus every character of $\widehat{\mathbb{Z}}$ has the form $\varphi(\chi)^k$. It follows that φ is in $\mathbb{Z}^*$ and $\langle\varphi\rangle = \mathbb{Z}^*$. It suffices to show that $\eta^*(1) = \varphi$. By definition, $\eta^*(1)(\chi) = \chi(1) = \varphi(\chi)$ and $\eta^*(1) = \varphi$, proving that $\mathbb{Z}$ satisfies duality.

Proposition 8.2.2 *If* $\mathbf{a}$ *is in* $\mathbb{R}^n$*, then* $\gamma_{\mathbf{a}}(\mathbf{x}) = \exp(2\pi i \mathbf{a} \cdot \mathbf{x})$ *is a character of* $\mathbb{R}^n$ *and* $\varphi(\mathbf{a}) = \gamma_{\mathbf{a}}$ *defines an isomorphism of* $\mathbb{R}^n$ *onto* $\widehat{\mathbb{R}^n}$.

Proof. Since for every linear function $T : \mathbb{R}^n \to \mathbb{R}$ there exists a unique $\mathbf{a}$ in $\mathbb{R}^n$ such that $T(\mathbf{x}) = \mathbf{a} \cdot \mathbf{x}$, it follows from *Theorem 3.2.6* that not only is $\gamma_{\mathbf{a}}$ a character, but every character equals $\gamma_{\mathbf{a}}$ for some $\mathbf{a}$ in $\mathbb{R}^n$. So φ is onto. Clearly, $\gamma_{\mathbf{a}}\gamma_{\mathbf{b}} = \gamma_{\mathbf{a}+\mathbf{b}}$ and φ is an algebraic homomorphism. It is equally clear that $\gamma_{\mathbf{a}}(\mathbf{x}) = 1$ for all $\mathbf{x}$ in $\mathbb{R}^n$ if and only if $\mathbf{a} = \mathbf{0}$. Thus the kernel of φ is $\{\mathbf{0}\}$ and φ is an algebraic isomorphism of $\mathbb{R}^n$ onto $\widehat{\mathbb{R}^n}$.

To prove that φ is an isomorphism, it suffices to show that φ and φ^{-1} are continuous at the identities of $\mathbb{R}^n$ and $\widehat{\mathbb{R}^n}$, respectively. First, consider a sequence $\mathbf{a}_k$ converging to $\mathbf{0}$ in $\mathbb{R}^n$. It is sufficient to show that $\gamma_{\mathbf{a}_k}(\mathbf{x})$ converges uniformly on compact subsets of $\mathbb{R}^n$ to ζ. Using *Exercise* 1, p. 336, it is easy to verify that $e^{2\pi i s} = \cos(2\pi s) + i\sin(2\pi s)$ is a uniformly continuous function from $\mathbb{R}$ to $\mathbb{K}$. Hence, it suffices to show that the sequence of linear functions $T_k(\mathbf{x}) = \mathbf{a}_k \cdot \mathbf{x}$ converges uniformly on compact subsets of $\mathbb{R}^n$ to the zero linear function $T(\mathbf{x}) = \mathbf{0} \cdot \mathbf{x} = 0$. By the Cauchy-Schwarz Inequality

$$|T_k(\mathbf{x}) - T(\mathbf{x})| = |\mathbf{x} \cdot \mathbf{a}_k| \leq \|\mathbf{x}\| \, \|\mathbf{a}_k\|,$$

and the convergence is uniform on compact sets of $\mathbb{R}^n$ because they are bounded sets in $\mathbb{R}^n$. Thus $\varphi(\mathbf{a}_k) = \gamma_{\mathbf{a}_k}$ converges to $\varphi(\mathbf{0}) = \zeta$ in $\widehat{\mathbb{R}^n}$ and φ is continuous.

Next, suppose $\gamma_{\mathbf{a}_k}$ is a sequence of characters converging to ζ in $\widehat{\mathbb{R}^n}$. It must be shown that $\varphi^{-1}(\gamma_{\mathbf{a}_k}) = \mathbf{a}_k$ converges to $\varphi^{-1}(\zeta) = \mathbf{0}$. Set $C = \{\mathbf{x} \in \mathbb{R}^n : \|\mathbf{x}\| \leq 1\}$, which is compact. Given ε such that $0 < \varepsilon < 1/4$, the set $N(C, U_\varepsilon)$ is an open neighborhood of ζ, and there exists M such that $\gamma_{\mathbf{a}_k}$ is in $N(C, U_\varepsilon)$ for $k \geq M$.

It suffices to show that $\|\mathbf{a}_k\| < \varepsilon$ when $k \geq M$. Suppose $\|\mathbf{a}_k\| \geq \varepsilon$ for some $k \geq M$. Set $t = \varepsilon/\|\mathbf{a}_k\|$ and $\mathbf{u}_k = \mathbf{a}_k/\|\mathbf{a}_k\|$. Then $0 < t \leq 1$ and $\|\mathbf{u}_k\| = 1$, so that $t\mathbf{u}_k$ is in C and $\gamma_{\mathbf{a}_k}(t\mathbf{u}_k)$ is in U_ε. An easy calculation shows that $\mathbf{a}_k \cdot (t\mathbf{u}_k) = \varepsilon$ and implies that $\gamma_{\mathbf{a}_k}(t\mathbf{u}_k) = e^{2\pi i \varepsilon}$, which is not in U_ε. Therefore, $\|\mathbf{a}_k\| < \varepsilon$ for $k > M$ to complete the proof. □

Theorem 8.2.3 *The metric group $\mathbb{R}^n$ satisfies duality.*

Proof. Given $\mathbf{x}$ in $\mathbb{R}^n$ and $\gamma_{\mathbf{a}}$ in $\widehat{\mathbb{R}^n}$,

$$\eta^*(x)(\gamma_{\mathbf{a}}) = \gamma_{\mathbf{a}}(x) = e^{2\pi i \mathbf{a}\cdot\mathbf{x}}.$$

If $\mathbf{x} \neq \mathbf{0}$, then there exists $\mathbf{a}$ such that $e^{2\pi i \mathbf{a}\cdot\mathbf{x}} \neq 0$ and $\eta^*(\mathbf{x})(\gamma_{\mathbf{a}}) \neq 0$. Thus, the kernel of η^* equals $\{\mathbf{0}\}$ and η^* is one-to-one.

To show that η^* is onto, consider ξ is in $(\mathbb{R}^n)^*$ and let φ be the isomorphism $\varphi(\mathbf{a}) = \gamma_{\mathbf{a}}$ of $\mathbb{R}^n$ onto $\widehat{\mathbb{R}^n}$ from *Proposition 8.2.2*. Then, $\xi \circ \varphi$ is in $\widehat{\mathbb{R}^n}$ and $\xi \circ \varphi = \gamma_{\mathbf{x}}$ for some $\mathbf{x}$ in $\mathbb{R}^n$. If χ is an arbitrary element of $\widehat{\mathbb{R}^n}$, then there exists $\mathbf{a}$ in $\mathbb{R}^n$ such that $\chi = \gamma_{\mathbf{a}} = \varphi(\mathbf{a})$ and

$$\xi(\chi) = \xi(\varphi(\mathbf{a})) = (\xi \circ \varphi)(\mathbf{a}) = \gamma_{\mathbf{x}}(\mathbf{a}).$$

Since $\gamma_{\mathbf{x}}(\mathbf{a}) = e^{2\pi i \mathbf{a}\cdot\mathbf{x}} = \gamma_{\mathbf{a}}(\mathbf{x})$,

$$\xi(\chi) = \gamma_{\mathbf{x}}(\mathbf{a}) = \gamma_{\mathbf{a}}(\mathbf{x}) = \chi(\mathbf{x}) = \eta^*(\mathbf{x})(\chi)$$

for all χ in $\widehat{\mathbb{R}^n}$. Therefore, $\eta^*(\mathbf{x}) = \xi$ and η^* is onto. Finally, η^* is open and an isomorphism by the Open Homomorphism Criterion. □

Returning to the general question of when an ALS-group satisfies duality, *Proposition 8.2.4* will help us break the question down into smaller pieces.

Proposition 8.2.4 *If the ALS-groups G_1 and G_2 satisfy duality, then $G_1 \times G_2$ satisfies duality.*

Proof. It follows from *Theorem 6.5.3* that $\widehat{G_1} \times \widehat{G_2}$ and $\widehat{G_1 \times G_2}$ are isomorphic. Setting $\beta = \Theta^{-1}$ from the proof of the just quoted theorem,

$$\beta(\chi_1, \chi_2)(x_1, x_2) = \chi_1(x_1)\chi_2(x_2)$$

is an isomorphism of $\widehat{G_1} \times \widehat{G_2}$ onto $\widehat{G_1 \times G_2}$. Analogously, an isomorphism of $G_1^* \times G_2^*$ onto

$$\Gamma = \widehat{\widehat{G_1} \times \widehat{G_2}}$$

is defined by setting $\beta'(\xi_1, \xi_2)(\chi_1, \chi_2) = \xi_1(\chi_1)\xi_2(\chi_2)$.

By hypothesis, $\eta_1^* \times \eta_2^*$ is an isomorphism of $G_1 \times G_2$ onto $G_1^* \times G_2^*$ and so $\beta' \circ (\eta_1^* \times \eta_2^*)$ is an isomorphism of $G_1 \times G_2$ onto Γ. Moreover, $\widehat{\beta}$ is an isomorphism of $(G_1 \times G_2)^*$ onto Γ because β is an isomorphism of $\widehat{G_1} \times \widehat{G_2}$ onto $\widehat{G_1 \times G_2}$ (*Exercise* 8, p. 261). It follows that $\widehat{\beta} \circ \eta^*$ is a homomorphism of $G_1 \times G_2$ to Γ. If it can be shown that $\widehat{\beta} \circ \eta^* = \beta' \circ (\eta_1^* \times \eta_2^*)$, then $\eta^* = \widehat{\beta}^{-1} \circ \beta' \circ (\eta_1^* \times \eta_2^*)$ is an isomorphism of $G_1 \times G_2$ onto $(G_1 \times G_2)^*$ to complete the proof.

Let (x_1, x_2) be in $G_1 \times G_2$ and (χ_1, χ_2) in $\widehat{G_1} \times \widehat{G_2}$. Then,

$$\begin{aligned}
\widehat{\beta} \circ \eta^*(x_1, x_2)(\chi_1, \chi_2) &= \eta^*(x_1, x_2) \circ \beta(\chi_1, \chi_2) \\
&= \beta(\chi_1, \chi_2)(x_1, x_2) \\
&= \chi_1(x_1)\chi_2(x_2) \\
&= \eta_1^*(x_1)(\chi_1)\eta_2^*(x_2)(\chi_2) \\
&= \beta' \circ (\eta_1^* \times \eta_2^*)(x_1, x_2)(\chi_1, \chi_2).
\end{aligned}$$

Thus $\widehat{\beta} \circ \eta^* = \beta' \circ (\eta_1^* \times \eta_2^*)$ and the proof is complete. □

Corollary 8.2.5 *If the ALS-groups $G_1, \ldots, G_m$ satisfy duality, then $G_1 \times \ldots \times G_m$ satisfies duality.*

Proof. Use induction.

Corollary 8.2.6 *If G is a finitely generated Abelian group with the discrete topology, than G satisfies duality.*

Proof. A finitely generated group is algebraically isomorphic to a product of cyclic groups by *Theorem 8.1.13*. Thus a finitely generated Abelian group with the discrete topology is isomorphic to a product of cyclic groups each with the discrete topology. It was shown at the beginning of the section that $\mathbb{Z}$ satisfied duality, and finite cyclic groups satisfy duality by *Exercise* 3, p. 336. So *Corollary 8.2.5* applies. □

Proposition 8.2.7 *Let G be an ALS-group. The characters of G separate points of G if and only if the duality homomorphism η^* is one-to-one.*

Proof. Let x and y be distinct points in G. Then $\eta^*(x) = \eta^*(y)$ if and only if $\chi(x) = \chi(y)$ for all χ in $\widehat{G}$ if and only if no character separates x and y. □

The next step is to prove that compact Abelian metric groups and countable discrete Abelian groups satisfy duality. The proof presented here is based on a proof given by Morris [11]. It makes use of several preliminary density results for subgroups of $\widehat{G}$ that separate points of G.

Proposition 8.2.8 *Let G be an ALS-group that satisfies duality, and let A be a subgroup of $\widehat{G}$ that separates points of G. If $\widehat{G}$ has the property that whenever H is a proper closed subgroup of $\widehat{G}$, the character group of $\widehat{G}/H$ is nontrivial, then A is dense in $\widehat{G}$.*

Proof. Let $H = A^-$. It must be shown that $H = \widehat{G}$. If $H \neq \widehat{G}$, then by hypothesis there exists a nontrivial character $\xi : \widehat{G}/H \to \mathbb{K}$. Letting $\pi : \widehat{G} \to \widehat{G}/H$ be the canonical homomorphism, $\xi' = \xi \circ \pi$ is a nontrivial character of $\widehat{G}$ such that H, and hence also A is contained in the kernel of ξ'. Because G satisfies duality, there exists x in $G \setminus \{e\}$ such that $\eta^*(x) = \xi'$ and $\xi'(\chi) = \eta^*(x)(\chi) = \chi(x)$ for all $\chi \in \widehat{G}$. It follows that $\chi(x) = 1$ for all $\chi \in A$, and A does not separate x from e in G. This contradiction completes the proof. □

Proposition 8.2.9 *If G is a finitely generated Abelian group with the discrete topology, and A is a subgroup of $\widehat{G}$ that separates the points of G, then A is dense in $\widehat{G}$.*

Proof. Since G is finitely generated, it satisfies duality by *Corollary 8.2.6*, and the dual group $\widehat{G}$ of G is isomorphic to $\mathbb{K}^p \times \prod_{j=1}^N \mathbb{Z}_{m_j}$ with $p \geq 0$, $N \geq 0$, and $m_j \geq 2$ for all j. *Proposition 8.2.8* applies to complete the proof, if it can be shown that for every closed subgroup H of $\mathbb{K}^p \times \prod_{j=1}^N \mathbb{Z}_{m_j}$, the metric group $(\mathbb{K}^p \times \prod_{j=1}^N \mathbb{Z}_{m_j})/H$ has a nontrivial character.

Let H be a proper closed subgroup of $\mathbb{K}^p \times \prod_{j=1}^N \mathbb{Z}_{m_j}$. Using *Exercise 7*, p. 329, $\widehat{G}$ can be viewed as a closed subgroup of $\mathbb{K}^n$ for suitable n, making H a closed subgroup of $\mathbb{K}^n$. *Corollary 3.2.2* implies that $\mathbb{K}^n/H$ is isomorphic to $\mathbb{K}^q$. There exists $\gamma \in \widehat{G} \setminus H$ because $H \neq \widehat{G}$. Since characters of $\mathbb{K}^q$ separate points of $\mathbb{K}^q$ (*Exercise* 6, p. 336), there exists a character ξ of $\mathbb{K}^n/H$ such that $\xi(\gamma H) \neq 1$. Then ξ restricted to the subgroup $\widehat{G}/H$ of $\mathbb{K}^n/H$ (*Corollary 1.4.9*) is the required character. □

If H is a proper closed subgroup of G, an ALS-group, then H is also an ALS-group (*Exercise* 8, p. 329) and there is a function $\iota : H \to G$ defined by $\iota(x) = x$. (This is a slight extension of our earlier use of ι for the identity map. The formula is the same, but the domain of the function is allowed to be a subset of the range. The image still equals the domain.) Clearly, ι is a homomorphism of H into G, and $\widehat{\iota}(\chi) = \chi \circ \iota = \chi|H$, the restriction of χ to the subgroup H, is a homomorphism of $\widehat{G}$ to $\widehat{H}$.

Proposition 8.2.10 *If G is a discrete ALS-group, and A is a subgroup of $\widehat{G}$ that separates the points of G, then A is dense in $\widehat{G}$.*

Proof. Let $C_1, \ldots, C_p$ be compact subsets of G, and let $V_1, \ldots, V_p$ be open sets in $\mathbb{K}$. Set

$$U = \widehat{G} \cap \left(\bigcap_{j=1}^{p} N(C_j, V_j) \right),$$

and assume U is not empty. Since the topology on $\widehat{G}$ is the compact-open topology, it suffices by *Theorem 6.2.4* to show that $A \cap U$ is not the empty set.

Because G is discrete, the compact sets $C_1, \ldots, C_p$ are finite. Let $C = \bigcup_{j=1}^p C_j$ and $H = \langle C \rangle$. So H is a finitely generated Abelian subgroup of G. Because $\widehat{\iota}(\chi) = \chi|H$, it follows from *Corollary 8.1.5* that $\widehat{\iota}$ is onto. The Open Homomorphism Criterion implies that $\widehat{\iota}$ is an open function because $\widehat{G}$ and $\widehat{H}$ are compact metric groups by *Proposition 8.1.1.*

Clearly, the restrictions of the characters in A to H separate points in H, and *Proposition 8.2.9* applies to H and $\widehat{\iota}(A) \subset \widehat{H}$. Therefore, $\widehat{\iota}(A)$ is dense in $\widehat{H}$ and $\widehat{\iota}(U) \cap \widehat{\iota}(A) \neq \phi$. Thus there exists $\gamma \in U$ and $\gamma' \in A$ such that $\gamma|H = \gamma'|H$. Because $C \subset H$, it follows that $\gamma' \in U$ and $U \cap A \neq \phi$. □

Proposition 8.2.11 *If G is an ALS-group, then $\eta^*(G)$ is a group of characters of $\widehat{G}$ that separate points of $\widehat{G}$.*

Proof. Let χ and χ' be two elements of $\widehat{G}$. Observe that $\eta^*(x)(\chi) = \eta^*(x)(\chi')$ for all $x \in G$ if and only if $\chi(x) = \chi'(x)$ for all $x \in G$ if and only if $\chi = \chi'$. Therefore, if $\chi \neq \chi'$, there exists $x \in G$ such that $\eta^*(x)(\chi) \neq \eta^*(x)(\chi')$. □

Theorem 8.2.12 *Every compact Abelian metric group satisfies duality.*

Proof. Let G be a compact Abelian metric group. Then $\widehat{G}$ is a discrete ALS-group by *Proposition 7.4.6*, and G^* is a compact metric group by *Proposition 8.1.1.* Since characters of a compact Abelian metric group separate points by *Theorem 7.4.4*, the duality homomorphism η^* is one-to one by *Proposition 8.2.7* .

Since $\eta^*(G)$ is a group of characters of $\widehat{G}$ that separate the points of $\widehat{G}$ by *Proposition 8.2.11*, $\eta^*(G)$ is dense in G^* by *Proposition 8.2.10.* Because G is compact, $\eta^*(G)$ is also compact and a closed subgroup of G^*. Therefore, $\eta^*(G) = G^*$ and η^* is onto.

Finally, the Open Homomorphism Criterion implies that η^* is an open function and thus and isomorphism of G onto G^*. □

Theorem 8.2.13 *Every discrete ALS-group satisfies duality.*

Proof. Let G be a discrete ALS-group. Then $\widehat{G}$ is a compact metric group, and G^* is a discrete ALS-group. So it suffices to show that η^* is one-to-one and onto. It is an immediate consequence of *Exercise* 7, p. 336, and *Proposition 8.2.7* that η^* is one-to-one.

Now $\widehat{G}$ satisfies duality (*Theorem 8.2.12*), and $\eta^*(G)$ is a group of characters of $\widehat{G}$ that separate the points of $\widehat{G}$ by *Proposition 8.2.11.* If H is a closed subgroup of G^*, then G^*/H is also a nontrivial discrete ALS-group and, like G, has nontrivial characters. Thus *Proposition 8.2.8* can be applied to $\widehat{G}$ to conclude that $\eta^*(G)$ is dense in G^*. Since G^* is discrete, $\eta^*(G) = G^*$. □

The following theorem summarizes what we know about duality thus far.

Theorem 8.2.14 *If K is a compact ALS-group, D a discrete ALS-group, and n a non-negative integer, then $K \times D \times \mathbb{R}^n$ satisfies duality.*

Proof. *Theorems 8.2.12, 8.2.13, and 8.2.3* imply that K, D, and $\mathbb{R}^n$, respectively satisfy duality. *Corollary 8.2.5* applies to complete the proof. □

Proposition 8.2.15 *If G is a monothetic locally compact metric group, then either G is compact or isomorphic to $\mathbb{Z}$.*

Proof. Let a be an element of G such that $G = \langle a \rangle^-$. If $a^k = e$ for some $k \neq 0$, then the group $\langle a \rangle$ is finite. It follows that $G = \langle a \rangle^- = \langle a \rangle$ is compact. So we can assume that $a^k \neq e$ for all $k \neq 0$

If G is discrete, then $G = \{a^k : k \in \mathbb{Z}\}$ because the only dense subset of a discrete space is the space itself. It follows $k \to a^k$ is an isomorphism of $\mathbb{Z}$ onto G. Thus we can also assume that G is not discrete.

Because G is locally compact by hypothesis, there exists a compact symmetric neighborhood U of e. Let U' be an open neighborhood of e such that $U' \subset U$. Given x in G, there exists $k \in \mathbb{Z}$ such that xa^k is in U' because $\{xa^k : k \in \mathbb{Z}\}$ is dense in G. It follows that there exists a symmetric neighborhood V of e such that $xa^kV \subset U' \subset U$. Since G is not discrete, the set $E = \{m \in \mathbb{Z} : a^m \in V\}$ is infinite. Moreover, m is in E if and only if $-m$ is in E. Thus there exist j in

E such that $j < -k$. Consequently, $xa^k a^j = xa^{j+k}$ is in U or x is in Ua^{-j-k} with $0 < -j-k$. Therefore,

$$G = \bigcup_{i=1}^{\infty} Ua^i$$

and there exists $N > 0$ such that

$$U \subset \bigcup_{i=1}^{N} Ua^i.$$

because U is compact.

To show that G is compact, it suffices to show that

$$G = \bigcup_{i=1}^{N} Ua^i.$$

Given y in G, there exists a smallest positive integer m such that y is in Ua^m or ya^{-m} is in U. It suffices to show that $1 \leq m \leq N$. Suppose $m > N$. Because U is symmetric, $y^{-1}a^m$ is also in U and there exists i satisfying $1 \leq i \leq N$ such that $y^{-1}a^m$ is in Ua^i. Thus $y^{-1}a^{m-i} \in U$ and $ya^{-m+i} \in U$ by symmetry again. Therefore, y is in Ua^{m-i} and $0 < m-i < m$ contradicting the choice of m. So $1 \leq m \leq N$. □

Corollary 8.2.16 *Every locally compact monothetic metric group is an ALS-group that satisfies duality.*

Proof. Recall that every monothetic group is Abelian. Both $\mathbb{Z}$ and compact Abelian metric groups are ALS-groups that satisfy duality. □

Corollary 8.2.17 *If G is an ALS-group and a is an element of G, then $\langle a \rangle^-$ is either compact or isomorphic to $\mathbb{Z}$.*

EXERCISES

1. Show that a continuous periodic function $f : \mathbb{R} \to \mathbb{R}$ is uniformly continuous on $\mathbb{R}$.
2. Prove directly that $\mathbb{K}$ satisfies duality.
3. Show that the finite cyclic group $\mathbb{Z}_m$ satisfies duality.
4. Let G_1 and G_2 be isomorphic ALS-groups. Show that G_1 satisfies duality if and only if G_2 satisfies duality.
5. Let G be an ALS-group. Show that the characters of G separate points of G if and only if for every x in G there exists a χ in $\widehat{G}$ such that $\chi(x) \neq 1$.
6. Show directly that the characters of $\mathbb{Z}^n$, $\mathbb{R}^n$, and $\mathbb{K}^n$ separate points.
7. Use *Proposition 8.1.5* to show that the characters of a nontrivial discrete ALS-group separate points.

8.3 Compactly Generated Abelian Groups

A topological group G is *compactly generated*, if there exists a compact set C in G such that $G = \langle C \rangle$. Of course, all compact topological groups are compactly generated. Note that if C is a compact subset of G such that $G = \langle C \rangle$ and C' is another compact subset of G such that $C \subset C'$, then $G = \langle C' \rangle$. Thus we can assume without loss of generality that C is a symmetric subset of G by replacing C with $C \cup C^{-1}$. When $G = \langle C \rangle$ and C is symmetric,

$$G = \bigcup_{k=1}^{\infty} C^k$$

by *Exercise* 9, p. 192.

Similarly, if G is a compactly generated locally compact topological group, then there exists a compact symmetric neighborhood U of e such that $G = \langle U \rangle$. It follows that a locally compact group G is compactly generated if and only if there exists a compact symmetric neighborhood U of e such that

$$G = \bigcup_{k=1}^{\infty} U^k. \tag{8.3}$$

A discrete group is compactly generated if and only if it is finitely generated. Thus a simple example of an ALS-group that is not compactly generated is

$$\widehat{\mathbb{K}^{\infty}} = \bigcup_{k=1}^{\infty} \{(a_1, a_2, \ldots) \in \mathbb{Z}^{\infty} : a_j = 0 \text{ for } j \geq k\},$$

with the discrete topology.

Proposition 8.3.1 *The following hold for a metric group G:*

(a) If G is compactly generated, then G is σ-compact.

(b) If G is compactly generated and locally compact, then G is also second countable and separable.

(c) If G is a compactly generated locally compact Abelian, then G is an ALS-group.

(d) If G is a compactly generated ALS-group and H is a closed subgroup of G, then G/H is a compactly generated ALS-group.

(e) If G is a compactly generated ALS-group and H is an open subgroup of G, then G/H is a finitely generated Abelian group.

Proof. For the first part, there exists a compact symmetric set C such that $G = \bigcup_{k=1}^{\infty} C^k$. Obviously, C^k is compact for all k, proving that G is σ-compact.

The second part now follows from the first part and *Proposition 6.2.1.*

The third part is an immediate consequence of the second part.

Turning to the fourth part, it now suffices by part *(c)* to show that G/H is compactly generated and locally compact. Let U be a compact symmetric neighborhood of the identity such that (8.3) holds. The canonical homomorphism $\pi : G \to G/H$ is an open function, so $\pi(U)$ is a compact neighborhood of the identity in G/H such that $\pi(U)^{-1} = \pi(U^{-1}) = \pi(U)$. It follows that G/H is locally compact and that $G/H = \bigcup_{k=1}^{\infty} \pi(U)^k$, making G/H compactly generated.

Since G/H is discrete when H is open, the fifth part follows from the fourth because a discrete group is compactly generated if and only if it is finitely generated. □

Proving that a closed subgroup of a compactly generated group is compactly generated is more difficult (see [11], page 93).

Subgroups isomorphic to $\mathbb{Z}^n$ will play a key role in this section. A compact metric group cannot have a subgroup isomorphic to $\mathbb{Z}^n$ with $n > 0$ (*Exercise* 4, p. 343). For compactly generated ALS-groups that are not compact, not only do there exists closed subgroups isomorphic to $\mathbb{Z}^n$ for positive n, but there exist closed subgroups isomorphic to $\mathbb{Z}^n$ with compact quotients. Thus all compactly generated ALS-groups have a basic structural similarity with $\mathbb{R}^n$.

Theorem 8.3.2 *If G is a compactly generated ALS-group, then there exists a discrete subgroup L isomorphic to $\mathbb{Z}^r$ for some $r \geq 0$ such that G/L is compact. Moreover, if U is a specified compact neighborhood of e, then L can be constructed so that $L \cap U = \{e\}$.*

Proof. If G is compact, just set $L = \{e\}$ and $r = 0$. Assume G is not compact for the rest of the proof. There exists a symmetric compact neighborhood U of e such that $G = \bigcup_{k=1}^{\infty} U^k$. Since U^2 is compact, there exist $a_1, \ldots, a_m$ in G such that

$$U^2 \subset \bigcup_{j=1}^{m} Ua_j.$$

Consider the finitely generated Abelian group, $H = \langle a_1, \ldots, a_m \rangle$. Clearly $U \subset UH$ and $U^2 \subset UH$. If $U^k \subset UH$, then $U^{k+1} \subset UUH = U^2H \subset UH$. Thus by induction $U^k \subset UH$ for all $k \geq 1$ and $G = UH$.

By *Proposition 8.1.7*, H is algebraically isomorphic to $F \times \mathbb{Z}^n$ for some finite Abelian group F and non-negative integer n. Note that if $n = 0$, then H is finite and G is compact because $G = UH$ and U is compact. Therefore, $n > 0$ and there exist subgroups F' and K of H algebraically isomorphic to F and $\mathbb{Z}^n$, respectively, such that $F' \cap K = \{e\}$ and $H = F'K$. Clearly, $V = UF'$ is a compact symmetric neighborhood of e such that $G = VK$. Let φ be an algebraic isomorphism of $\mathbb{Z}^n$ onto K and think of $\mathbb{Z}^n$ as a discrete subgroup of $\mathbb{R}^n$ with $n > 0$.

There exists a subgroup L' of $\mathbb{R}^n$ of maximal rank $r \geq 0$ such that $L' \subset \mathbb{Z}^n$ and $L = \varphi(L')$ is a discrete subgroup of G. Obviously, L' is a discrete subgroup of $\mathbb{R}^n$, and L' is isomorphic to $\mathbb{Z}^r$ for some $r \geq 0$ (*Corollary 3.1.7*). Because L is a discrete subgroup of G, *Exercise* 5, p. 343, implies that $\varphi|L'$ is an isomorphism of L' onto L. Thus L is isomorphic $\mathbb{Z}^r$. The next task is to show that $r \geq 1$.

Let $\mathbf{e}_1, \ldots, \mathbf{e}_n$ be the usual set of linear independent generators for $\mathbb{Z}^n$. Set $K_j = \langle \varphi(\mathbf{e}_j) \rangle^-$. If K_j is compact for $j = 1, \ldots, n$, then $K_1 K_2 \ldots K_n$ is a compact set containing K. Hence, K^- is compact, and G is compact because V is compact and $G = VK \subset VK^- \subset G$, contradicting the assumption that G is not compact. Therefore, at least one K_j is not compact. It follows from *Corollary 8.2.17* that at least one K_j is isomorphic to $\mathbb{Z}$, and $K_j = \langle \varphi(\mathbf{e}_j) \rangle$ is a discrete subgroup of G, proving that $r \geq 1$.

Because L is discrete, it is a closed subgroup (*Proposition 1.4.1*), and $G' = G/L$ is a compactly generated ALS-group by part *(d)* of *Proposition 8.3.1*. Let $\pi : G \to G'$ be the canonical homomorphism. Then $W = \pi(V)$ is a compact symmetric neighborhood of e', the identity of G', such that $G' = \bigcup_{k=1}^{\infty} W^k$. Also $K' = \pi(K) = \pi \circ \varphi(\mathbb{Z}^n)$ is a finitely generated subgroup of G' such that $G' = WK'$. The next step is to show that G' is compact.

If $r = n$, then $\mathbb{R}^n / L'$ is isomorphic to $\mathbb{K}^n$ and compact (*Theorem 3.2.1*). It follows that $\mathbb{Z}^n / L'$ is a discrete subgroup of a compact group and finite by *Exercise* 4, p. 343. Then K/L is also finite. Therefore, $G' = \pi(G) = \pi(VK) = \pi(V)\pi(K)$ is compact because $\pi(V)$ is compact and $\pi(K) = K/L$ is finite.

If $r < n$, set $K_j' = \langle \pi \circ \varphi(\mathbf{e}_j) \rangle^-$. By using the earlier argument, either G' is compact or there exists j such that K_j' is isomorphic to $\mathbb{Z}$. We will show that the second possibility contradicts the choice of L'.

Suppose K_j' is isomorphic to $\mathbb{Z}$, or in other words, $\langle \pi \circ \varphi(\mathbf{e}_j) \rangle$ is a discrete subgroup of G' isomorphic to $\mathbb{Z}$. It follows that $k\mathbf{e}_j$ is not in L' for all $k \neq 0$ because $\pi \circ \varphi(\mathbf{e}_j)$ is not in the torsion group of G'. Consequently, $L' \cap \langle \mathbf{e}_j \rangle = \{\mathbf{0}\}$ and $\langle L', \mathbf{e}_j \rangle$ is isomorphic to $L' \times \langle \mathbf{e}_j \rangle$, which is isomorphic to $\mathbb{Z}^{r+1}$. Since $\langle L', \mathbf{e}_j \rangle$ is a subgroup of $\mathbb{Z}^n$, it is discrete and hence closed subgroup of $\mathbb{R}^n$. It follows from *Exercise* 9, p. 124, that the rank of $\langle L', \mathbf{e}_j \rangle$ must be $r + 1$.

Suppose there exist sequences m_k in $\mathbb{Z}$ and $\mathbf{v}_k$ in L' such that $\varphi(m_k \mathbf{e}_j + \mathbf{v}_k)$ converges to e. Then, $\pi \circ \varphi(m_k \mathbf{e}_j + \mathbf{v}_k) = \pi \circ \varphi(m_k \mathbf{e}_j)$ converges to e' in G'. Because $\langle \pi \circ \varphi(\mathbf{e}_j) \rangle$ is a discrete subgroup of G', this is only possible if $m_k = 0$ for large k. Hence, $\varphi(\mathbf{v}_k)$ converges to e in G and then $\mathbf{v}_k = \mathbf{0}$ for large j because $L = \varphi(L')$ is a discrete subgroup of G. Consequently, $\varphi(\langle L', \mathbf{e}_j \rangle)$ must be a discrete subgroup of G, but the rank of $\varphi(\langle L', \mathbf{e}_j \rangle)$ is $r + 1$, contradicting the choice of L'. Therefore, G' is compact.

If U is a specified compact neighborhood of the identity, then $L \cap U$ must be finite because L is discrete. So there exists a positive integer p such $pL' \cap \varphi^{-1}(L \cap U) = \phi$. Since the ranks of pL' and L' are both r, the above argument shows that $G/(pL)$ is also compact. □

Proposition 8.3.3 *If G is an ALS-group, then the characters of G separate points in G and the duality homomorphism η^* is one-to-one.*

Proof. If G is compact, then *Theorem 7.4.4* applies and the characters of G separate points. Suppose G is a noncompact ALS-group with identity e. Let a

be an element of G not equal to e and U a compact symmetric neighborhood of e containing a. Then, $H = \bigcup_{k=1}^{\infty} U^k$ is an open-and-closed compactly generated subgroup of G. Applying *Theorem 8.3.2* to H, which is also an ALS-group, there exists a discrete subgroup L of H isomorphic to $\mathbb{Z}^r$ for some $r > 0$ such that H/L is compact and $L \cap U = \{e\}$. Let π' be the canonical homomorphism of H onto the compact group H/L. Then, $\pi'(a) \neq \pi'(e)$ because $L \cap U = \{e\}$ and a is in U. *Theorem 7.4.4* implies that there exists a character χ for H/L such that $\chi(\pi'(a)) \neq 1$. Obviously, $\chi \circ \pi'$ is a character of H such that $\chi(a) \neq 1$.

Note that G/H is a discrete ALS-group (*Exercise* 9, p. 329) and thus finite or countable. It follows from *Proposition 8.1.4* that there exists an algebraic homomorphism $\varphi : G \to \mathbb{K}$ such that $\varphi(x) = \chi \circ \pi'(x)$ for all x in H. Since $\varphi = \chi \circ \pi'$ is continuous on H, which is an open subgroup, φ is continuous at the identity and a homomorphism. Thus φ is a character of G such that $\varphi(a) \neq 1$. Since a was any element of G not equal to e, the characters of G separate points of G by *Exercise* 5, p. 336, and η^* is one-to-one by *Proposition 8.2.7*. □

Proposition 8.3.4 *If H is an open divisible subgroup of an Abelian metric group G such that G/H is countable, then G is isomorphic to $H \times G/H$.*

Proof. *Proposition 8.1.4* applies to the identity isomorphism $\iota : H \to H$, so there exist an algebraic homomorphism $\varphi : G \to H$ such that $\varphi(x) = x$ for all x in H. As usual, let π be the canonical homomorphism of G onto G/H. Then $\theta(x) = (\varphi(x), \pi(x))$ defines an algebraic homomorphism of G to $H \times G/H$. It remains to show that θ is an isomorphism.

If $\theta(x) = (e, H)$, then x is in H and $e = \varphi(x) = x$. Thus θ is one-to-one. When x is in H, observe that $\varphi(a\varphi(a)^{-1}x) = x$ because $\varphi(a) \in H$ and $\varphi(\varphi(a)) = \varphi(a)$. Hence, $\theta(a\varphi(a)^{-1}x) = (x, aH)$ because $\varphi(a)^{-1}x$ is in H. Thus θ is onto and an algebraic isomorphism.

To prove that θ is an isomorphism, it suffices to show that θ is continuous and open at the identity of G. Since ι is continuous on H and H is open, φ is continuous at the identity. It follows that θ is continuous at e. Since H is an open subgroup, ι and hence φ are open at the identity. And π is always an open function. If U is a neighborhood of the identity of G, then $\theta(U) = \varphi(U) \times \pi(U)$ is a neighborhood of the identity of $H \times G/H$, and θ is open at the identity. □

Corollary 8.3.5 *If H is a closed subgroup of $\mathbb{K}^n$, then H is isomorphic to $\mathbb{K}^m \times F$, where $0 \leq m \leq n$ and F is a finite Abelian group.*

Proof. *Theorem 3.2.3* shows that the proposition can be applied to a closed subgroup of the torus. □

Proposition 8.3.6 *Let G be a compact Abelian metric group. If U is a neighborhood of the identity of G, then there exists a closed subgroup H contained in U such that G/H is isomorphic to $\mathbb{K}^m \times F$, where $m \geq 0$ and F is a finite Abelian group.*

Proof. It suffices by *Theorem 7.4.7* to prove the result for a closed subgroup G of $\mathbb{K}^\infty$. If U is a neighborhood of the identity in G, then there exists $\varepsilon > 0$ and $m > 0$ such that

$$\{\mathbf{z} \in \mathbb{K}^\infty : |z_j| < \varepsilon \text{ for } 1 \leq j \leq m\} \cap G \subset U.$$

Consider the closed subgroup of $\mathbb{K}^\infty$ given by

$$H = \{\mathbf{z} \in \mathbb{K}^\infty : z_j = 1 \text{ for } 1 \leq j \leq m.\}$$

Clearly, H is contained in U, and $\mathbb{K}^\infty/H$ is isomorphic to $\mathbb{K}^m$. Thus G/H is isomorphic to a closed subgroup of $\mathbb{K}^m$, and the conclusion follows from *Corollary 8.3.5.* □

Corollary 8.3.7 *Let G be a compact Abelian group. If there exists a neighborhood of the identity in G that contains no subgroup other than the identity itself (no small subgroups), then G is isomorphic to $\mathbb{K}^m \times F$, where $m \geq 0$ and F is a finite Abelian group.*

Proposition 8.3.8 *If G is a compactly generated ALS-group, then there exists a compact subgroup K and a discrete subgroup L isomorphic to $\mathbb{Z}^r$ for some $r \geq 0$ with the following properties:*

(a) G/L is a compact metric group.

(b) KL is a closed subgroup of G isomorphic to $K \times L$.

(c) KL/K is a discrete subgroup of G/K isomorphic to $\mathbb{Z}^r$.

(d) $(G/K)/(KL/K)$ is isomorphic to $\mathbb{K}^m \times F$, where m is a non-negative integer and F is a finite Abelian group.

Proof. Let L be given by *Theorem 8.3.2.* Because L is discrete, it is closed and there exists a compact symmetric neighborhood U of e in G such that $U^3 \cap L = \{0\}$. Then $\pi(U)$ is a neighborhood of the identity in the compact group G/L. By *Proposition 8.3.6*, there exists a closed subgroup H of G/L contained in $\pi(U)$ such that $(G/L)/H$ is isomorphic to $\mathbb{K}^m \times F$, where $m \geq 0$ and F is a finite Abelian group. Set $H' = \pi^{-1}(H)$, a closed subgroup of G such that $H = H'/L$. Set $K = U \cap H'$. Note that K is compact and $\pi(K) = H$.

To show that K is a subgroup of G, consider xy^{-1} for x and y in K. Then $\pi(xy^{-1})$ is in H and there exists z in K such that $\pi(z) = \pi(xy^{-1})$. It follows that $xy^{-1}z^{-1}$ is in $U^3 \cap L = \{e\}$. Therefore, $xy^{-1} = z$ is in K, and K is a compact subgroup of G such that $K \cap L = \{e\}$.

Clearly, $\pi^{-1}(H) = H' = KL$ is a closed subgroup of G and an ALS-group. In addition, the function $y \mapsto Ky$ is an algebraic isomorphism of L onto KL/K. It follows from *Exercise* 6, p. 343, that $(x, y) \mapsto xy$ is an isomorphism of $K \times L$ onto $KL = H'$. Since $K \times \{e\}$ is an open subset of $K \times L$ because L is discrete, K is an open subgroup of KL and KL/K is a discrete group isomorphic to L and $\mathbb{Z}^r$ for some $r \geq 0$. Because the quotient topology on KL/L equals the relative topology from G/K (*Corollary 1.4.9*), KL/L is a discrete subgroup of G/K.

By *Corollary 1.4.16* to the Second Isomorphism Theorem, the metric groups $(G/K)/(KL/K)$ and $(G/L)/H = (G/L)/(H'/L)$ are isomorphic to $G/KL = G/H'$. Therefore, $(G/K)/(KL/K)$ is isomorphic to $\mathbb{K}^m \times F$, where $m \geq 0$ and F is a finite Abelian group. □

Theorem 8.3.9 *If G is a compactly generated ALS-group, then there exist a compact subgroup K of G such that G/K is isomorphic to a group of the form $\mathbb{R}^p \times \mathbb{K}^q \times \mathbb{Z}^n \times F$, where $p \geq 0$, $q \geq 0$, $n \geq 0$, and F is a finite Abelian group.*

Proof. The first half of the proof is devoted to setting the stage for applying *Theorem 3.2.5*. To start, let $\pi_m : \mathbb{R}^m \to \mathbb{K}^m$ by $\pi_m(\mathbf{x}) = \left(e^{2\pi i x_1}, \ldots, e^{2\pi i x_m}\right)$. Recall that π_m is an open homomorphism of $\mathbb{R}^m$ onto $\mathbb{K}^m$ with kernel $\mathbb{Z}^m$ that is one-to-one on the set $\{\mathbf{x} : \|x\| < 1/2\}$. Let $W = \pi_m(\{\mathbf{x} : \|x\| < 1/2\})$, which is an open subset of $\mathbb{K}^m$.

Let K and L be given by *Proposition 8.3.8*, and set $G' = G/K$ and $D = KL/K$. So D is a discrete subgroup of G' isomorphic to $\mathbb{Z}^r$ for some $r \geq 0$, and $G'/D = (G/K)/(KL/K)$ is isomorphic to $\mathbb{K}^m \times F'$, where m is a non-negative integer and F' is a finite Abelian group. It follows that there exists an open homomorphism φ from G' onto $\mathbb{K}^m \times F'$ with its kernel equal to D.

Because D is a discrete subgroup of G' and $\mathbb{K}^m$ is an open subgroup of $\mathbb{K}^m \times F'$, there exists an open symmetric neighborhood U of e' in G' such $U^3 \cap D = \{e'\}$ and $\varphi(U) \subset W \subset \mathbb{K}^m$. Clearly, $\varphi|U$ is a homeomorphism of U onto $\varphi(U) = V$, an open subset of $\mathbb{K}^m$.

Let $\theta = (\varphi|U)^{-1} : V \to U$ and suppose $\mathbf{z}$, $\mathbf{w}$, and $\mathbf{zw}$ are in V. Then $\theta(\mathbf{z})$, $\theta(\mathbf{w})$, and $\theta(\mathbf{zw})$ are in U. Clearly, $\theta(\mathbf{z})\theta(\mathbf{w})\theta(\mathbf{zw})^{-1} \in U^3 \cap D = \{e'\}$ and $\theta(\mathbf{z})\theta(\mathbf{w}) = \theta(\mathbf{zw})$, when $\mathbf{z}$, $\mathbf{w}$, and $\mathbf{zw}$ are in V.

Set $V' = \pi_m^{-1}(V) \cap \{\mathbf{x} : \|x\| < 1/2\}$. Then V' is an open subset of $\mathbb{R}^m$ and $\pi_m|V'$ is a homeomorphism of V' onto V. Consider the function $f = \theta \circ (\pi_m|V')$. It is a homeomorphism of V' onto U. If $\mathbf{x}$, $\mathbf{y}$, and $\mathbf{x}+\mathbf{y}$ are in V', then $\pi_m(\mathbf{x}+\mathbf{y}) = \pi_m(\mathbf{x})\pi_m(\mathbf{y})$ and the points $\pi_m(\mathbf{x})$, $\pi_m(\mathbf{y})$, $\pi_m(\mathbf{x}+\mathbf{y})$ are all V. It follows from the previous paragraph that $f(\mathbf{x}+\mathbf{y}) = f(\mathbf{x})f(\mathbf{y})$. Now *Theorem 3.2.5* applies and there exist a unique homomorphism $\psi : \mathbb{R}^m \to G'$ such that $\psi(\mathbf{x}) = f(\mathbf{x})$ whenever $\mathbf{x} \in V'$.

Since $U = f(V') \subset \psi(\mathbb{R}^m)$, the group $\psi(\mathbb{R}^m)$ is an open-and-closed subgroup of G'. Because f is a homeomorphism of V', a neighborhood of the identity of $\mathbb{R}^m$, onto U, the homomorphism ψ is an open homomorphism. Letting E denote the kernel of ψ and applying *Theorem 1.4.14*, shows that $\psi(\mathbb{R}^m)$ is isomorphic to $\mathbb{R}^m/E$. If $E \cap V' \neq \{\mathbf{0}\}$, then f would not be a homeomorphism of V' onto U. So E is a discrete subgroup of $\mathbb{R}^m$, and $\mathbb{R}^m/E$ is isomorphic to $\mathbb{R}^p \times \mathbb{K}^q$ with $p + q = m$ by *Theorem 3.2.1*. Therefore, the open subgroup $\psi(\mathbb{R}^m)$ of G' is isomorphic to $\mathbb{R}^p \times \mathbb{K}^q$ and is a divisible subgroup of G'.

Now part *(d)* of *Proposition 8.3.1* implies that $G' = G/K$ and $G'/\psi(\mathbb{R}^m)$ are compactly generated ALS-groups. Because $\psi(\mathbb{R}^m)$ is an open subgroup of G', part *(e)* also applies to $\psi(\mathbb{R}^m)$. So $G'/\psi(\mathbb{R}^m)$ is a finitely generated

Abelian group and isomorphic to a group of the form $\mathbb{Z}^n \times F$, where $n \geq 0$ and F is a finite Abelian group by *Theorem 8.1.13*. In particular, $\mathbb{Z}^n \times F$ is countable and *Proposition 8.3.4* applies. Therefore, $G' = G/K$ is isomorphic to $\psi(\mathbb{R}^m) \times (G'/\psi(\mathbb{R}^m))$ and hence to $\mathbb{R}^p \times \mathbb{K}^q \times \mathbb{Z}^n \times F$. □

Metric groups of the form $\mathbb{R}^p \times \mathbb{K}^q \times \mathbb{Z}^r \times F$, where $p \geq 0$, $q \geq 0$, $r \geq 0$, and F is a finite Abelian group are often called *elementary groups.* The elementary groups are obviously compactly generated ALS-groups and satisfy duality by *Theorem 8.2.14*. Moreover, given an ALS-group G, it is an elementary group if and only if $\widehat{G}$ is an elementary group.

EXERCISES

1. Let G_1 and G_2 be compactly generated topological groups. Show that $G_1 \times G_2$ is compactly generated.

2. Show that a connected locally compact topological group is compactly generated.

3. Let G be a compactly generated topological group. Show that if $\varphi : G \to G'$ is a homomorphism of G onto a topological group G', then G' is compactly generated.

4. Let H be a subgroup of compact Hausdorff topological group. Show that H is a discrete subgroup of G if and only if H is finite.

5. Let G be a Hausdorff topological group and let D be a discrete group. Show that an algebraic isomorphism of D into G is an isomorphism of D onto the subgroup $\varphi(D)$ of G if and only if $\varphi(D)$ is a discrete subgroup of G.

6. Let H_1 and H_2 be closed subgroups of an ALS-group G. Prove that the homomorphism $\varphi : H_1 \times H_2 \to G$ defined by $\varphi(x, y) = xy$ is an isomorphism of $H_1 \times H_2$ onto G if and only if

   (a)
$$G = H_1 H_2$$

   (b)
$$H_1 \cap H_2 = \{e\}$$

7. Let H be a proper closed subgroup of an ALS-group G. Prove that if $a \in G \setminus H$, then there exists γ in $\widehat{G}$ such that $\gamma(a) \neq 1$ and $\gamma(H) = 1$.

8. Show that the rational numbers $\mathbb{Q}$ with the relative topology from $\mathbb{R}$ is compactly generated. ($\mathbb{Q}$ is not locally compact by *Exercise* 14, p. 61.)

8.4 A Duality Theorem

The main theorem in this section is that all ALS-groups satisfy duality. Although ALS-groups are a large interesting class of groups, this theorem is not the best duality theorem, which is reflected in the choice of the article "A" in the section title. The more important and prominent result is that all locally compact Abelian Hausdorff topological groups (LCA-groups) satisfy duality, which is known as the Pontryagin - van Kampen Duality Theorem. The proof presented in this section for ALS-groups roughly follows the final stage of the proof of the Pontryagin van Kampen Duality Theorem in Morris [11]. (There is also an overview with references to other proofs of the Pontryagin - van Kampen Duality Theorem in the literature in [11].) The strategy of the proof is to use the duality results already obtained to ratchet our way up to all ALS-groups. The proofs, however, are easier for ALS-groups than for LCA-groups.

We have a wealth of information about character groups available to us. In many instances there is more than one reason why the next step in a proof is valid. Our preference has been for the simpler reasons, leaving the readers to prefer others as they like.

Recall that if G_1 and G_2 are ALS-groups and $\varphi : G_1 \to G_2$ is a homomorphism, then $\widehat{\varphi}(\chi) = \chi \circ \varphi$ defines a homomorphism of $\widehat{G_2}$ to $\widehat{G_1}$ (*Exercise* 6, p. 261). This construction and it properties will play an essential role in this section.

Proposition 8.4.1 *Let G_1 and G_2 be ALS-groups, and let $\varphi : G_1 \to G_2$ be a homomorphism.*

(a) If φ is onto, then $\widehat{\varphi}$ is one-to-one.

(b) If G_1 is compact and φ is one-to-one, then $\widehat{\varphi}$ is onto.

(c) If φ is one-to-one and open, then $\widehat{\varphi}$ is onto.

Proof. For the first part, let χ be an element of $\widehat{G_2}$ that is not equal to the identity of $\widehat{G_2}$. So there exists a in G_2 such that $\chi(a) \neq 1$. Because φ is onto, $a = \varphi(b)$ for some b in G_1 and $\chi(\varphi(b)) \neq 1$. Since $\widehat{\varphi}(\chi)(b) = \chi(\varphi(b)) \neq 1$, it follows that $\widehat{\varphi}(\chi)$ does not equal the identity in $\widehat{G_1}$ and that the kernel of $\widehat{\varphi}$ can only contain the identity element. Thus $\widehat{\varphi}$ is one-to-one.

For the second part, assume G_1 is compact and φ is one-to-one. If x and y are distinct elements of G_1, then $\varphi(x) \neq \varphi(y)$ and by *Proposition 8.3.3* there exists χ in $\widehat{G_2}$ such that $\chi(\varphi(x)) \neq \chi(\varphi(y))$ or equivalently $\widehat{\varphi}(\chi)(x) \neq \widehat{\varphi}(\chi)(y)$. Thus the subgroup $\widehat{\varphi}(\widehat{G_2})$ separates points of G_1. Since G_1 is compact, it satisfies duality and $\widehat{G_1}$ is discrete. Moreover, if H is a proper closed subgroup of $\widehat{G_1}$, then $\widehat{G_1}/H$ is a discrete ALS-group and has a nontrivial character group by *Exercise* 7, p. 336. Therefore, *Proposition 8.2.8* implies that $\widehat{\varphi}(\widehat{G_2})$ is dense in $\widehat{G_1}$. It follows that $\widehat{\varphi}(\widehat{G_2}) = \widehat{G_1}$ because $\widehat{G_1}$ is discrete.

For the third part, let χ be an element of $\widehat{G_1}$. Because φ is one-to-one and open, it is an isomorphism of G_1 onto $\varphi(G_1)$, an open subgroup of G_2. Then, $G_2/\varphi(G_1)$ is a discrete ALS-group. Furthermore, $\chi \circ \varphi^{-1}$ is a character of $\varphi(G_1)$. Because $\mathbb{K}$ is divisible, *Proposition 8.1.4* can now be applied to extend $\chi \circ \varphi^{-1}$ to an algebraic homomorphism γ of G_2 to $\mathbb{K}$, which is clearly continuous at the identity, and hence a character. If x is in G_1, then $\gamma \circ \varphi(x) = \chi \circ \varphi^{-1} \circ \varphi(x) = \chi(x)$. Therefore, γ is a character of G_2 such that $\widehat{\varphi}(\gamma) = \gamma \circ \varphi = \chi$, proving that $\widehat{\varphi}$ is onto. □

ALS-groups are locally compact by definition and are σ-compact by *Proposition 6.2.1*. Thus the Open Homomorphism Criterion applies whenever $\varphi : G_1 \to G_2$ is a homomorphism of an ALS-group onto an ALS-group. This simple observation is particularly useful for proving that ALS-groups satisfy duality. For LCA-groups, additional work is required to prove that key homomorphisms are in fact open.

Exact sequences provide a convenient algebraic setting for the arguments in this section. Suppose G_1, G_2, and G_3 are groups. For convenience, we will use e to denote the identity of G_1, G_2, and G_3 rather than e_1 etc. The finite sequence of algebraic homomorphisms

$$G_1 \xrightarrow{\varphi} G_2 \xrightarrow{\theta} G_3$$

is an *exact sequence* at G_2 provided that $\varphi(G_1) = \theta^{-1}(e)$, that is, the image of φ equals the kernel of θ.

When G_1, G_2, and G_3 are ALS-groups and

$$G_1 \xrightarrow{\varphi} G_2 \xrightarrow{\theta} G_3$$

is a sequence of homomorphisms, we have a dual sequence

$$\widehat{G_3} \xrightarrow{\widehat{\theta}} \widehat{G_2} \xrightarrow{\widehat{\varphi}} \widehat{G_1}$$

Assuming the original sequence is exact at G_2, we want to know when the dual sequence is exact at $\widehat{G_2}$. The following result provides an answer that will met our needs.

Proposition 8.4.2 *Let G_1, G_2, and G_3 are ALS-groups, and let $\varphi : G_1 \to G_2$ and $\theta : G_2 \to G_3$ be homomorphisms. If θ is onto and*

$$G_1 \xrightarrow{\varphi} G_2 \xrightarrow{\theta} G_3$$

is exact at G_2, then

$$\widehat{G_3} \xrightarrow{\widehat{\theta}} \widehat{G_2} \xrightarrow{\widehat{\varphi}} \widehat{G_1}$$

is exact at $\widehat{G_2}$.

Proof. Observe that $\theta \circ \varphi(x) = e$ for all x in G_1 because the sequence is exact at G_2. Consequently, $\widehat{\theta \circ \varphi}(\chi) = \chi \circ \theta \circ \varphi$ is the identity character in $\widehat{G_1}$, and $\widehat{\theta}(\widehat{G_3})$ is contained in the kernel of $\widehat{\varphi}$ because $\widehat{\theta \circ \varphi} = \widehat{\varphi} \circ \widehat{\theta}$.

To prove that the kernel of $\widehat{\varphi}$ is contained $\widehat{\theta}(\widehat{G_3})$, let χ be in the kernel of $\widehat{\varphi}$, that is, χ is an element of $\widehat{G_2}$ such that $\chi(\varphi(x)) = 1$ for all x in G_1. It follows by exactness that

$$\{x \in G_2 : \theta(x) = e\} = \varphi(G_1) \subset \{x \in G_2 : \chi(x) = 1\},$$

and the kernel of θ is also contained in the kernel of χ.

To construct a character γ in $\widehat{G_3}$ such that $\widehat{\theta}(\gamma) = \gamma \circ \theta = \chi$, let y be an element of G_3. Using the fact that θ is onto, set $\gamma(y) = \chi(x)$ for x such that $\theta(x) = y$. If $\theta(x') = y$, then $\theta(x'x^{-1}) = e$ and $\chi(x'x^{-1}) = e$ or $\chi(x') = \chi(x)$. So the definition of χ is unambiguous. It is easily verified that γ is an algebraic homomorphism such that $\chi = \gamma \circ \theta$.

If U is an open subset of $\mathbb{K}$, then $\gamma^{-1}(U) = \theta(\chi^{-1}(U))$, which is an open subset of G_3 because θ is an open function by the Open Homomorphism Criterion. Therefore, γ is continuous and an element of $\widehat{G_3}$ such that $\widehat{\theta}(\gamma) = \chi$. (The construction of γ is just a slight modification of the proof of *Exercise* 15, p. 51, applied to χ.) □

The trivial group consisting of one element will be denoted by 1. For any group G, there is a unique algebraic homomorphism of 1 to G and a unique algebraic homomorphism of G onto 1. They will be denoted simply as $1 \longrightarrow G$ and $G \longrightarrow 1$. When G is a metric group, 1 will have the discrete topology and these algebraic homomorphisms will be continuous.

The sequence of algebraic homomorphisms

$$1 \longrightarrow G_1 \xrightarrow{\varphi} G_2 \xrightarrow{\theta} G_3 \longrightarrow 1$$

is called a *short exact sequence* provide that it is exact at G_1, G_2, and G_3. *This equivalent to saying that φ is one-to-one, the image of φ is the kernel of θ, and θ is onto.*

Next, we prove two propositions showing when the dual of a short exact sequence is a short exact sequence. They are complementary results in the sense that they can be applied one after the other in either order.

Proposition 8.4.3 *Let G_1, G_2, and G_3 be ALS-groups. If*

$$1 \longrightarrow G_1 \xrightarrow{\varphi} G_2 \xrightarrow{\theta} G_3 \longrightarrow 1$$

is a short exact sequence of homomorphisms such that G_1 *is compact, then*

$$1 \longrightarrow \widehat{G_3} \xrightarrow{\widehat{\theta}} \widehat{G_2} \xrightarrow{\widehat{\varphi}} \widehat{G_1} \longrightarrow 1$$

is a short exact sequence such that $\widehat{\theta}$ *is an open homomorphism.*

Proof. *Proposition 8.4.2* implies that the required sequence is exact at $\widehat{G_2}$. Parts *(a)* and *(b)*, respectively, of *Proposition 8.4.1* imply that $\widehat{\theta}$ and $\widehat{\varphi}$ are one-to-one and onto, respectively. Thus the sequence

$$1 \longrightarrow \widehat{G_3} \xrightarrow{\widehat{\theta}} \widehat{G_2} \xrightarrow{\widehat{\varphi}} \widehat{G_1} \longrightarrow 1$$

is a short exact sequence.

Since G_1 is compact, $\widehat{G_1}$ is discrete and the kernel of $\widehat{\varphi}$ is an open subgroup of $\widehat{G_2}$. Then by exactness $\widehat{\theta}(\widehat{G_3})$ is an open-and-closed subgroup of $\widehat{G_2}$. It follows that $\widehat{\theta}(\widehat{G_3})$ is an ALS-group.

Since ALS-groups are σ-compact and locally compact, the Open Homomorphism Criterion implies that $\widehat{\theta} : \widehat{G_3} \to \widehat{\theta}(\widehat{G_3})$ is an open homomorphism. An open set in $\widehat{\theta}(\widehat{G_3})$ is open in $\widehat{G_2}$ because $\widehat{\theta}(\widehat{G_3})$ is an open set in $\widehat{G_2}$. Consequently, $\widehat{\theta}$ is also an open homomorphism of $\widehat{G_3}$ to $\widehat{G_2}$. □

Proposition 8.4.4 *Let* G_1, G_2, *and* G_3 *ALS-groups. If*

$$1 \longrightarrow G_1 \xrightarrow{\varphi} G_2 \xrightarrow{\theta} G_3 \longrightarrow 1$$

is a short exact sequence such that φ *is open homomorphisms, then*

$$1 \longrightarrow \widehat{G_3} \xrightarrow{\widehat{\theta}} \widehat{G_2} \xrightarrow{\widehat{\varphi}} \widehat{G_1} \longrightarrow 1$$

is a short exact sequence such that $\widehat{G_3}$ *is compact.*

Proof. The exactness of the sequence at $\widehat{G_3}$ and $\widehat{G_1}$ follows from parts *(a)* and *(c)* of *Proposition 8.4.1*, respectively. As in the preceding proof, the sequence is exact at $\widehat{G_2}$ by *Proposition 8.4.2*. Thus the required sequence is a short exact sequence.

Since φ is an open homomorphism by hypothesis, $\varphi(G_1)$ is an open-and-closed subgroup of G_2 and $G_2/\varphi(G_1)$ is a discrete group. Moreover, $G_2/\varphi(G_1)$ is isomorphic to G_3 by *Corollary 1.5.19* because $\varphi(G_1)$ equals the kernel of θ by exactness. Therefore, G_3 is a discrete ALS-group and $\widehat{G_3}$ is compact. $\square$

Observe that the conclusion of *Proposition 8.4.3* matches the hypothesis of *Proposition 8.4.4* and visa versa. Consequently, *Proposition 8.4.4* can be applied to the result of *Proposition 8.4.3* and visa versa to obtain short exact sequences of the form

$$1 \longrightarrow G_1^* \xrightarrow{\varphi^*} G_2^* \xrightarrow{\theta^*} G_3^* \longrightarrow 1$$

where $\varphi^* = \widehat{\widehat{\varphi}}$ and $\theta^* = \widehat{\widehat{\theta}}$. The strategy is to choose the right G_1 and G_3 so this observation can be used to prove that G_2 satisfies duality.

Suppose X, X', Y Y' are sets and $f : X \to Y$, $f' : X' \to Y'$, $g : X \to X'$, and $h : Y \to Y'$ are functions. This situation can be conveniently represented by the following square diagram:

$$\begin{array}{ccc} X & \xrightarrow{f} & Y \\ {\scriptstyle g}\downarrow & & \downarrow{\scriptstyle h} \\ X' & \xrightarrow{f'} & Y' \end{array}$$

Such a diagram is called a *commutative square* provided that $h \circ f = f' \circ g$. For example, the heart of the proof of *Proposition 8.2.4* was showing that $\widehat{\beta} \circ \eta^* = \beta' \circ (\eta_1^* \times \eta_2^*)$ or in other words

$$\begin{array}{ccc} G_1 \times G_2 & \xrightarrow{\eta_1^* \times \eta_2^*} & G_1^* \times G_2^* \\ {\scriptstyle \eta^*}\downarrow & & \downarrow{\scriptstyle \beta'} \\ (G_1 \times G_2)^* & \xrightarrow{\widehat{\beta}} & \Gamma \end{array}$$

was a commutative square.

In preparation for proving that ALS-groups satisfy duality, we need two lemmas.

Lemma 8.4.5 *Let G_1 and G_2 be ALS-groups. If $\varphi : G_1 \to G_2$ is a homomorphism, then*

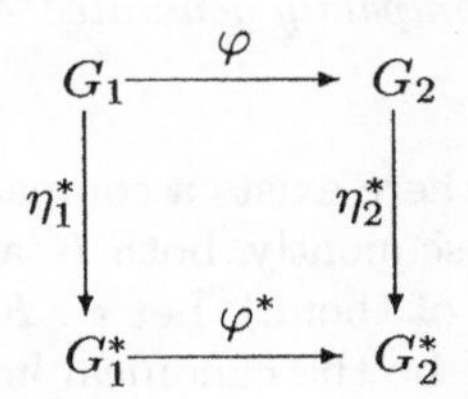

is a commutative square.

Proof. Let x be in G_1. To show that $\varphi^* \circ \eta_1^*(x) = \eta_2^* \circ \varphi(x)$ it must be shown that for every χ in $\widehat{G_2}$

$$\varphi^* \circ \eta_1^*(x)(\chi) = \eta_2^* \circ \varphi(x)(\chi).$$

First,

$$\varphi^* \circ \eta_1^*(x)(\chi) = (\eta_1^*(x) \circ \widehat{\varphi})(\chi) = \eta_1^*(x)(\chi \circ \varphi) = \chi(\varphi(x)).$$

Second,

$$\eta_2^* \circ \varphi(x)(\chi) = \eta_2^*(\varphi(x))(\chi) = \chi(\varphi(x))$$

to complete the proof. □

The second lemma is a special case of the five lemma from algebraic topology.

Lemma 8.4.6 *Suppose*

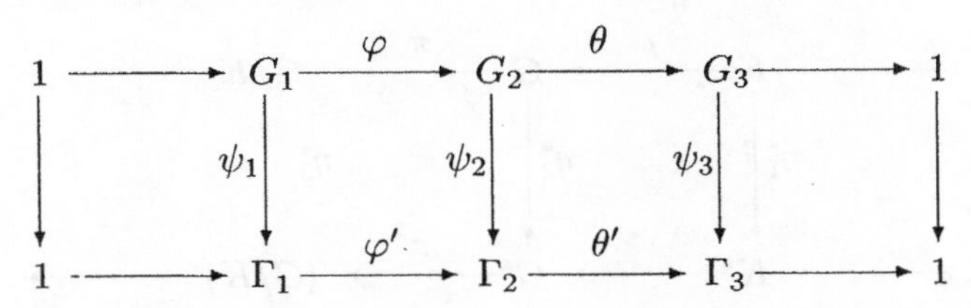

is a diagram of groups and algebraic homomorphisms such that the top and bottom sequences are short exact sequences and every square is commutative. If ψ_1 and ψ_3 are algebraic isomorphisms, then ψ_2 is also an algebraic isomorphism.

Proof. To prove that ψ_2 is onto, let x be an element of Γ_2. Because θ is onto and ψ_3 is an isomorphism, there exists y in G_2 such that $\theta'(x) = \psi_3(\theta(y))$. Set $z = \psi_2(y)$ and check that $\theta'(xz^{-1}) = e$. By exactness, there exists u in Γ_1 such that $\varphi'(u) = xz^{-1}$. Set $v = \varphi(\psi_1^{-1}(u))$ and check that $\psi_2(v) = xz^{-1}$. It follows that $\psi_2(vy) = xz^{-1}z = x$ and φ is onto.

The fact that ψ_2 is also one-to-one will not be used because the characters of ALS-groups separate points, implying that η^* is one-to-one. So the proof will be left to the reader. □

Theorem 8.4.7 *If G is a compactly generated ALS-group, then G satisfies duality.*

Proof. By *Theorem 8.3.9*, there exists a compact subgroup K such that G/K is an elementary group. Consequently, both K and G/K satisfy duality. (*Theorem 8.2.14* applies to both of them.) Let $\iota : K \to G$ be the inclusion homomorphism and $\pi : G \to G/K$ be the canonical homomorphism. Then,

$$1 \longrightarrow K \xrightarrow{\ \iota\ } G \xrightarrow{\ \pi\ } G/K \longrightarrow 1$$

is a short exact sequence. Since K is compact, *Proposition 8.4.3* implies that

$$1 \longrightarrow \widehat{G/K} \xrightarrow{\ \widehat{\pi}\ } \widehat{G} \xrightarrow{\ \widehat{\iota}\ } \widehat{K} \longrightarrow 1$$

is a short exact sequence of homomorphisms such that $\widehat{\pi}$ is open. Now *Proposition 8.4.4* applies and

$$1 \longrightarrow K^* \xrightarrow{\ \iota^*\ } G^* \xrightarrow{\ \pi^*\ } (G/K)^* \longrightarrow 1$$

is a short exact sequence of homomorphisms.

It follows from *Lemma 8.4.5* that every square in the diagram

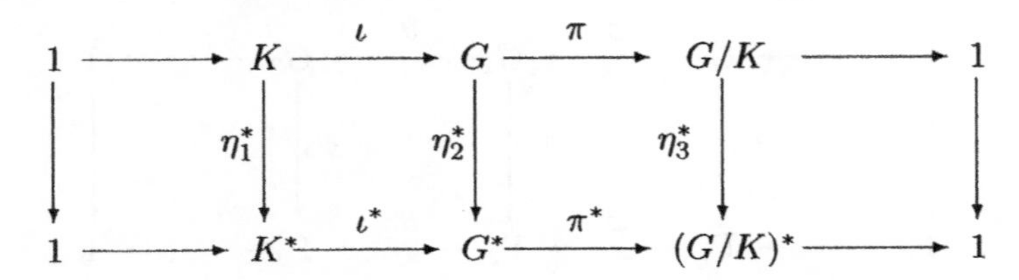

of ALS-groups and homomorphisms is commutative. The homomorphisms η_1^* and η_3^* are isomorphisms because K and G/K satisfy duality. Consequently, *Lemma 8.4.6* applies and η_2^* onto. The homomorphism η_2^* is one-to-one because characters separate points in ALS-groups (*Proposition 8.3.3*). To prove that η_2^* is an isomorphism of G onto G^*, it remains only to show that it is open. Once again the Open Homomorphism Criterion applies. □

Theorem 8.4.8 *If G is an ALS-group, then G satisfies duality.*

Proof. Let U be compact symmetric neighborhood of e in G and set $H = \langle U \rangle$. Then H is a compactly generated open subgroup of G, and G/H is a discrete ALS-group. Thus H and G/H satisfy duality by *Theorems 8.4.7 and 8.2.13*, respectively. Moreover, the proof of *Theorem 8.4.7* with only slight modification can be used again.

Consider the exact sequence

$$1 \longrightarrow H \xrightarrow{\iota} G \xrightarrow{\pi} G/H \longrightarrow 1$$

and note that ι is open because H is an open subgroup of G. So *Proposition 8.4.4* applies first, and then *Proposition 8.4.3* instead of in the reverse order as above. Now the second paragraph of the proof of *Theorem 8.4.7* can be repeated verbatim with H replacing K. □

The following elementary consequences of duality are now obvious:

Corollary 8.4.9 *If G is an ALS-group, then there exists an ALS-group H such that G is isomorphic to $\widehat{H}$.*

Corollary 8.4.10 *If G_1 and G_2 are ALS-groups such that $\widehat{G_1}$ and $\widehat{G_2}$ are isomorphic, then G_1 and G_2 are isomorphic.*

Corollary 8.4.11 *The following hold for an ALS-group G:*

(a) G is compact if and only if $\widehat{G}$ is discrete.

(b) G is discrete if and only if $\widehat{G}$ is compact.

(c) G is an elementary group if and only if $\widehat{G}$ is an elementary group.

The proofs of *Propositions 8.4.3 and 8.4.4* and *Theorems 8.4.7 and 8.4.8* could use the Open Homomorphism Criterion to show easily that critical homomorphisms were open because ALS-groups are locally compact and σ-compact. Since LCA-groups are not always σ-compact, the proofs of the same results in the more general setting of LCA-groups require a more careful analysis to prove these homomorphisms are open. These proofs can be found in [11] on pages 79-85.

The remainder of the section is devoted to using duality to obtain structural information about ALS-groups. Although these results illustrate the power of duality theorems, they represent only a small fraction of the results that can be obtained using Pontryagin - van Kampen Duality.

Theorem 8.4.12 *If G is a compactly generated ALS-group, then there exists a compact Abelian metric group C and non-negative integers p and n such that G is isomorphic to $\mathbb{R}^p \times \mathbb{Z}^n \times C$.*

Proof. It follows from *Theorem 8.3.9* that there exist a compact subgroup K of G such that G/K is isomorphic to a group of the form $\mathbb{R}^p \times \mathbb{K}^q \times \mathbb{Z}^n \times F$, where $p \geq 0$, $q \geq 0$, $n \geq 0$, and F is a finite Abelian group. As in the proof of *Theorem 8.4.7*

$$1 \longrightarrow \widehat{G/K} \xrightarrow{\ \widehat{\pi}\ } \widehat{G} \xrightarrow{\ \widehat{\iota}\ } \widehat{K} \longrightarrow 1$$

is a short exact sequence with $\widehat{\pi}$ open.

The character group of $\mathbb{R}^p \times \mathbb{K}^q \times \mathbb{Z}^n \times F$ is $\mathbb{R}^p \times \mathbb{Z}^q \times \mathbb{K}^n \times F$ and $\mathbb{R}^p \times \mathbb{K}^n$ is an open subgroup of $\mathbb{R}^p \times \mathbb{Z}^q \times \mathbb{K}^n \times F$. Hence, the group $\widehat{G}$ contains an open subgroup H isomorphic to the divisible group $\mathbb{R}^p \times \mathbb{K}^n$. It follows that $D = \widehat{G}/H$ is a discrete ALS-group. So *Proposition 8.3.4* implies that $\widehat{G}$ is isomorphic to $\mathbb{R}^p \times \mathbb{K}^n \times D$. Therefore, G^* and G by duality are isomorphic to $\mathbb{R}^p \times \mathbb{Z}^n \times \widehat{D}$ and $\widehat{D}$ is compact. □

Theorem 8.4.13 *Let G be an ALS-group. Then the following hold:*

(a) There exists an open subgroup H of G such that H is isomorphic to $\mathbb{R}^p \times C$, where $p \geq 0$ and C is a compact Abelian metric group.

(b) If G is connected, then G is isomorphic to $\mathbb{R}^p \times C$, where $p \geq 0$ and C is a compact connected Abelian metric group.

Proof. Let U be a compact symmetric neighborhood of e. Then $\langle U \rangle$ is an open compactly generated subgroup of G. By *Theorem 8.4.12*, there exists a compact Abelian metric group C and non-negative integers p and r such that $\langle U \rangle$ is isomorphic to $\mathbb{R}^p \times \mathbb{Z}^r \times C$. Since $\mathbb{R}^p \times C$ is an open subgroup of $\mathbb{R}^p \times \mathbb{Z}^r \times C$, it follows that $\langle U \rangle$ contains an open subgroup H isomorphic to $\mathbb{R}^p \times C$. Clearly, H is also open in G.

If G is connected, then the only open nontrivial subgroup of G is G itself. Thus $H = G$, forcing $\mathbb{R}^p \times C$ and C to be connected. □

Theorem 8.4.14 *Let G be an ALS-group. If there exists a neighborhood of the identity in G that contains no subgroup other than the identity itself (no small subgroups), then G is isomorphic to $\mathbb{R}^p \times \mathbb{K}^m \times D$, where $p \geq 0$, $m \geq 0$, and D is discrete ALS-group.*

Proof. By *Theorem 8.4.13*, G contains an open subgroup H isomorphic to $\mathbb{R}^p \times C$, where $p \geq 0$ and C is a compact Abelian metric group. Clearly, *Corollary 8.3.7* applies to C, and C is isomorphic to $\mathbb{K}^m \times F$, where $m \geq 0$ and F is a finite Abelian group. It follows that H is isomorphic to $\mathbb{R}^p \times \mathbb{K}^m \times F$. Consequently, G has an open subgroup H' isomorphic to the divisible metric group $\mathbb{R}^p \times \mathbb{K}^m$ and *Proposition 8.3.4* applies. So G is isomorphic to $\mathbb{R}^p \times \mathbb{K}^m \times D$ where $p \geq 0$, $m \geq 0$, and $D = G/H'$. □

Theorem 8.4.15 *An ALS-group G is monothetic if and only if $\widehat{G}$ is isomorphic to $\mathbb{K}$ or to a countable subgroup of $\mathbb{K}_d$.*

Proof. By *Propositions 7.4.12, 8.1.2, and 8.2.15*, $\widehat{G}$ is isomorphic to $\mathbb{K}$ or to a countable subgroup of $\mathbb{K}_d$, when G is a monothetic ALS-group.

By duality, it suffices to prove that $\widehat{H}$ is monothetic, when H is a countable subgroup of $\mathbb{K}_d$. Observe that $\gamma : H \to \mathbb{K}$ defined by $\gamma(z) = z$ is a character of H that separates points of H. Now it follows from *Proposition 8.2.10* that the subgroup $\langle\gamma\rangle$ of $\widehat{H}$ is dense in $\widehat{H}$. Hence, $\varphi(k) = \gamma^k$ is a homomorphism of $\mathbb{Z}$ onto a dense subgroup of $\widehat{H}$. □

Let E be a nonempty subset of an ALS-group G. The *annihilator* of E is defined by

$$\mathfrak{A}(E) = \{\chi \in \widehat{G} : \chi(x) = 1 \text{ for all } x \in E\}.$$

Clearly, $\mathfrak{A}(E)$ is a subgroup of $\widehat{G}$. Because $\eta(\chi, x) = \chi(x)$ is continuous, it is also easy to see that $\mathfrak{A}(E)$ is closed in $\widehat{G}$. If F is a nonempty subset of $\widehat{G}$, then by duality

$$\{\xi \in G^* : \xi(\chi) = 1 \text{ for all } \chi \in F\} = \eta^*(\{x \in G : \chi(x) = 1 \text{ for all } \chi \in F\}).$$

So it makes more sense to define the *annihilator* of a nonempty subset F of $\widehat{G}$ by

$$\mathfrak{A}(F) = \{x \in G : \chi(x) = 1 \text{ for all } \chi \in F\}.$$

This is not our first encounter with annihilators. Every character of $\mathbb{R}^n$ has the form $\gamma_{\mathbf{x}}(\mathbf{u}) = e^{2\pi i \mathbf{u}\cdot\mathbf{x}}$ for a unique $\mathbf{x} \in \mathbb{R}^n$. So $\gamma_{\mathbf{x}}(\mathbf{u}) = 1$ if and only if $\mathbf{x}\cdot\mathbf{u} \in \mathbb{Z}$. Thus the annihilator of E, a nonempty subset of $\mathbb{R}^n$, can be described as

$$\{\mathbf{x} \in \mathbb{R}^n : \mathbf{x}\cdot\mathbf{u} \in \mathbb{Z} \text{ for all } \mathbf{u} \in E\},$$

which is precisely the set denoted by $E^{\#}$ in Section 3.3 and used in the proof of the Kronecker Theorem on page 134.

Proposition 8.4.16 *If H is a closed subgroup of an ALS-group G, then*

$$\mathfrak{A}(\mathfrak{A}(H)) = H.$$

Proof. Clearly, $H \subset \mathfrak{A}(\mathfrak{A}(H))$. To complete the proof it suffices to show that $G \setminus H \subset G \setminus \mathfrak{A}(\mathfrak{A}(H))$. Let a be in $G \setminus H$. By *Exercise* 7, p. 343, there exists $\gamma \in \widehat{G}$ such that $\gamma(a) \neq 1$ and $\gamma(H) = 1$. So γ is in $\mathfrak{A}(H)$ and a is not in $\mathfrak{A}(\mathfrak{A}(H))$. □

The annihilators can be used to describe the character groups of closed subgroups and their quotient groups.

Theorem 8.4.17 *If H is a closed subgroup of an ALS-group G, then the following hold:*

(a) $\widehat{G/H}$ is isomorphic to $\mathfrak{A}(H)$.

(b) $\widehat{H}$ is isomorphic to and $\widehat{G}/\mathfrak{A}(H)$.

(c) Every character of H is the restriction of a character of G to H.

Proof. Consider the short exact sequence

$$1 \longrightarrow H \xrightarrow{\ \iota\ } G \xrightarrow{\ \pi\ } G/H \longrightarrow 1$$

It follows from *Proposition 8.4.1* that $\widehat{\pi}$ is one-to-one and the sequence

$$1 \longrightarrow \widehat{G/H} \xrightarrow{\ \widehat{\pi}\ } \widehat{G} \xrightarrow{\ \widehat{\iota}\ } \widehat{H} \longrightarrow 1$$

is exact at $\widehat{G/H}$. By *Proposition 8.4.2*, it is exact at $\widehat{G}$. The kernel of $\widehat{\iota}$ equals $\mathfrak{A}(H)$ because $\widehat{\iota}(\chi) = \chi|H$. Thus $\widehat{\pi}(\widehat{G/H}) = \mathfrak{A}(H)$. Because $\mathfrak{A}(H)$ is a closed subgroup of $\widehat{G}$, it is an ALS-group. Using the Open Homomorphism Criterion once again, the homomorphism $\widehat{\pi} : \widehat{G/H} \to \mathfrak{A}(H)$ is open and thus an isomorphism of $\widehat{G/H}$ onto $\mathfrak{A}(H)$.

To prove part *(b)*, consider the closed subgroup $\mathfrak{A}(H)$ of $\widehat{G}$. To avoid confusion with the proof of part *(a)*, let $\sigma : \widehat{G} \to \widehat{G}/\mathfrak{A}(H)$ denote the canonical homomorphism. It follow from the proof of part *(a)* that $\widehat{\sigma}$ is an isomorphism of $\Gamma = \widehat{\widehat{G}/\mathfrak{A}(H)}$ onto the closed subgroup $\{\xi \in G^* : \xi(\mathfrak{A}(H)) = 1\}$ of G^*. Using duality and *Proposition 8.4.16*,

$$\begin{aligned} \{\xi \in G^* : \xi(\mathfrak{A}(H)) = 1\} &= \\ \eta^*(\{x \in G : \chi(x)) = 1 \text{ for all } \chi \in \mathfrak{A}(H)\} &= \\ \eta^*(\mathfrak{A}(\mathfrak{A}(H))) &= \eta^*(H) \end{aligned}$$

and $\eta^*(H) = \widehat{\sigma}(\Gamma)$. Therefore, $\varphi(x) = \widehat{\sigma}^{-1} \circ \eta^*$ defines an isomorphism of H onto Γ, and $\widehat{\varphi}$ is an isomorphism of $\widehat{\Gamma} = (\widehat{G}/\mathfrak{A}(H))^*$ onto $\widehat{H}$. Part *(b)* now follows by applying the duality theorem to $\widehat{G}/\mathfrak{A}(H)$.

Part *(c)*, which is equivalent to the dual sequence being exact at $\widehat{H}$, requires a careful calculation of the isomorphism $\varphi = \widehat{\sigma}^{-1} \circ \eta^*$ and builds on the proof of part *(b)*. Consider ξ in $\eta^*(H) = \{\xi \in G^* : \xi(\mathfrak{A}(H)) = 1\}$. Setting $\tilde{\xi}(\mathfrak{A}(H)\chi) = \xi(\chi)$ defines $\tilde{\xi} \in \Gamma$ such that $\tilde{\xi} \circ \sigma = \xi$ by *Exercise* 15, p. 51. Thus $\widehat{\sigma}(\tilde{\xi}) = \xi$ and $\widehat{\sigma}^{-1}(\xi) = \tilde{\xi}$, which implies that

$$\widehat{\sigma}^{-1}(\xi)(\mathfrak{A}(H)\chi) = \xi(\chi).$$

Therefore,

$$\varphi(x)(\mathfrak{A}(H)\chi) = \widehat{\sigma}^{-1}(\eta^*(x))(\mathfrak{A}(H)\chi) = \eta^*(x)(\chi) = \chi(x).$$

Now, let γ be in $\widehat{H}$. Since $\widehat{\varphi}$ is an isomorphism of $\widehat{\Gamma} = (\widehat{G}/\mathfrak{A}(H))^*$ onto $\widehat{H}$, there exists a unique ψ in $(\widehat{G}/\mathfrak{A}(H))^*$ such that $\gamma = \psi \circ \varphi$. By duality, there exists a unique coset $\mathfrak{A}(H)\chi$ in $\widehat{G}/\mathfrak{A}(H)$ such that $\psi(\theta) = \theta(\mathfrak{A}(H)\chi)$ for all θ in Γ. Therefore, using the previous formula,

$$\gamma(x) = \psi(\varphi(x)) = \varphi(x)(\mathfrak{A}(H)\chi) = \chi(x)$$

for all x in H, and $\gamma = \chi|H$. □

Corollary 8.4.18 *A subgroup H of an ALS-group G is open if and only if $\mathfrak{A}(H)$ is compact.*

Proof. Observe that H is an open subgroup of G if and only if G/H is a discrete group if and only if $\widehat{G/H}$, which is isomorphic to $\mathfrak{A}(H)$, is compact. □

We will finish with a study of the relationship between connectivity properties of an ALS-group and properties of its character group. This leads to interesting structural results about ALS-groups. Recall that $C(e)$, the component of the identity of a topological group G, is a closed subgroup of G.

An element x of a metric group G is said to be *compact* provided that $\langle x\rangle^-$ is compact. Let $\mathfrak{K}(G)$ denote the subset of all compact elements of G. Obviously, every compact subgroup of G including $\{e\}$ is contained in $\mathfrak{K}(G)$. In particular, $\mathfrak{K}(G) = \{e\}$ if and only if the only compact subgroup of G is $\{e\}$.

Theorem 8.4.19 *The following hold for an ALS-group G:*

(a) G is connected if and only if $\mathfrak{K}(\widehat{G}) = \{\zeta\}$.

(b) G is totally disconnected if and only if $\mathfrak{K}(\widehat{G}) = \widehat{G}$.

(c) $\mathfrak{A}(C(e)) = \mathfrak{K}(\widehat{G})$.

Proof. Because G is connected if and only if G itself is the only open subgroup of G, part *(a)* is an immediate consequence of *Corollary 8.4.18* and *Exercise* 5, p. 357.

Next, suppose G is totally disconnected and consider γ, an element of $\widehat{G}$. Let r be a real number such that $0 < r < 1/4$. There exists a neighborhood V of e in G such that $\gamma(V) \subset U_r = \{e^{2\pi i x} : |x| < r\}$. By *Theorem 5.4.7* there exists an open subgroup H of G such $H \subset V$. Because U_r contains no subgroups of $\mathbb{K}$ except $\{1\}$, it follows that $\gamma(H) = 1$ and γ is in $\mathfrak{A}(H)$, which is compact because H is open. Thus $\widehat{G} \subset \mathfrak{K}(\widehat{G})$ and $\mathfrak{K}(\widehat{G}) = \widehat{G}$.

Continuing the proof of part *(b)*, suppose $\mathfrak{K}(\widehat{G}) = \widehat{G}$. Given a in G, there exists γ in $\widehat{G}$ such that $\gamma(a) \neq 1$. Since $\widehat{G} = \mathfrak{K}(\widehat{G})$, the group $\Gamma = \langle\gamma\rangle^-$ is compact. Because $\mathfrak{A}(\mathfrak{A}(\Gamma)) = \Gamma$ (*Exercise* 5, p. 357), it follows from *Corollary 8.4.18* that $\mathfrak{A}(\Gamma)$ is open and closed. Therefore, a is not in $C(e)$ because a is not in $\mathfrak{A}(\Gamma)$. It follows that $C(e) = \{e\}$ and completes the proof of part *(b)*

To start the proof of part *(c)*, consider γ in $\mathfrak{K}(\widehat{G})$ and let $\Gamma = \langle\gamma\rangle^-$. Then as above $\mathfrak{A}(\Gamma)$ is an open-and-closed subgroup of G. It follows that $C(e) \subset \mathfrak{A}(\Gamma)$ and $\Gamma = \mathfrak{A}(\mathfrak{A}(\Gamma) \subset \mathfrak{A}(C(e))$. Therefore, $\mathfrak{K}(\widehat{G}) \subset \mathfrak{A}(C(e))$.

Finally, consider γ in $\mathfrak{A}(C(e))$. Then $\widehat{\pi}$ is an isomorphism of $\widehat{G/C(e)}$ onto $\mathfrak{A}(C(e))$ by the proof of part *(a)* of *Theorem 8.4.17*. So there exists $\widetilde{\gamma}$ in $\widehat{G/C(e)}$ such that $\widehat{\pi}(\widetilde{\gamma}) = \gamma$. Observe that $\widetilde{\gamma}$ is compact by part *(b)* because $G/C(e)$ is totally disconnected (*Theorem 5.1.10*). Then, $\widehat{\pi}(\langle\widetilde{\gamma}\rangle^{-}) = \langle\gamma\rangle^{-}$ and γ is compact, proving that $\mathfrak{A}(C(e)) \subset \mathfrak{K}(\widehat{G})$ □

Corollary 8.4.20 *If G is an ALS-group, then $\mathfrak{K}(G)$ is a closed subgroup of G.*

Proof. It follows from part *(c)* of the theorem that $\mathfrak{K}(\widehat{H})$ is a closed subgroup of $\widehat{H}$ for any ALS-group H. By *Corollary 8.4.9*, there exists an ALS-group H such that G is isomorphic to $\widehat{H}$. □

Corollary 8.4.21 *A compact Abelian metric group G is connected if and only if $\widehat{G}$ is torsion-free.*

Proof. An element of a discrete group is compact if and only if it is of finite order. □

Theorem 8.4.22 *For a compact Abelian metric group G, the following are equivalent:*

(a) G is connected.

(b) $\widehat{G}$ is torsion-free.

(c) G is divisible.

Proof. It was just shown (*Corollary 8.4.21*) that *(a)* and *(b)* are equivalent. The equivalence of *(b)* and *(c)* hinges on the following simple equation for $x \in G$, $\chi \in \widehat{G}$, and $n \in \mathbb{Z}^{+}$:

$$\chi(x^n) = \chi(x)^n = \chi^n(x).$$

The function $\theta_n(x) = x^n$ is a homomorphism of G to G for n in $\mathbb{Z}^{+}$. Observe that G is divisible if and only if $\theta_n(G) = G$ for all $n \in \mathbb{Z}^{+}$. It follows from the above equation that χ is in $\mathfrak{A}(\theta_n(G))$ if and only if $\chi^n = \zeta$, the identity of $\widehat{G}$. in particular, χ is in the torsion group $T_{\widehat{G}}$ if and only if χ is in $\mathfrak{A}(\theta_n(G))$ for some $n \in \mathbb{Z}^{+}$.

Suppose that G is divisible. Then, $\mathfrak{A}(\theta_n(G)) = \mathfrak{A}(G) = \{\zeta\}$ for all $n \in \mathbb{Z}^{+}$. It follows that $T_{\widehat{G}} = \{\zeta\}$ and *(c)* implies *(b)*.

Now suppose that $T_{\widehat{G}} = \{\zeta\}$. Then, $\mathfrak{A}(\theta_n(G)) = \{\zeta\} = \mathfrak{A}(G)$ for all $n \in \mathbb{Z}^{+}$. Since G is compact, $\theta_n(G)$ is compact and a closed subgroup of G. So *Proposition 8.4.16* applies, and $\theta_n(G) = \mathfrak{A}(\{\zeta\}) = G$ for all $n \in \mathbb{Z}^{+}$. □

Theorem 8.4.23 *If G is a connected ALS-group, then G is divisible and isomorphic to $\mathbb{R}^p \times C$, where C is a compact, connected, and divisible Abelian metric group.*

Proof. By the second part of *Theorem 8.4.13*, there exists a compact connected Abelian metric group C and a non-negative integer p such that G is isomorphic to $\mathbb{R}^p \times C$. So C is divisible by the *Theorem 8.4.22*, and $\mathbb{R}^p \times C$ is divisible by *Exercise* 5, p. 328. Therefore, G is a divisible group. □

Theorem 8.4.24 *Let G be an ALS-group. If G has a connected neighborhood of the identity, then there exist a compact connected divisible Abelian metric group, a countable discrete Abelian group D, and a non-negative integer p such that G is isomorphic to $\mathbb{R}^p \times C \times D$.*

Proof. Because G has a connected neighborhood of the identity, the component of the identity, $C(e)$, must be an open-and-closed connected subgroup of G. *Theorem 8.4.23* implies that $C(e)$ is divisible. Consequently, G is isomorphic to $C(e) \times G/C(e)$ by a now familiar argument. Clearly, $D = G/C(e)$ is a countable discrete Abelian group. *Theorem 8.4.23* also implies that $C(e)$ is isomorphic to $\mathbb{R}^p \times C$, where C is a compact, connected, and divisible Abelian metric group and completes the proof. □

EXERCISES

1. Prove that the homomorphism ψ_2 in *Lemma 8.4.6* is one-to-one.
2. Show that G is an elementary group if and only if G and $\widehat{G}$ are compactly generated.
3. Let G be an ALS-group. Show that G is finitely generated if and only if $\widehat{G}$ is isomorphic to a closed subgroup of $\mathbb{K}^n$ for some $n \geq 0$.
4. Show that if H is a subgroup of an ALS-group G, then $\mathfrak{A}(\mathfrak{A}(H)) = H^-$.
5. Let G be an ALS-group. Show that if Γ is a closed subgroup of $\widehat{G}$, then $\mathfrak{A}(\mathfrak{A}(\Gamma)) = \Gamma$.
6. Prove that a subgroup H of an ALS-group G is compact if and only if $\mathfrak{A}(H)$ is an open subgroup of $\widehat{G}$.
7. Give an example of a noncompact ALS-group G such that $\mathfrak{K}(G) = G$.
8. Prove that an ALS-group G is isomorphic to $\mathbb{R}^n$ for some $n \geq 0$ if and only if both G and $\widehat{G}$ are connected.
9. Show that the **a**-adic groups Γ_m and Γ_n are isomorphic if and only if m and n have the same prime divisors, for example, 2 and 3 are the prime divisors of both 6 and 72 have.
10. Give an example of a divisible ALS-group that is not compactly generated.
11. Prove that if G is an ALS-group and A is a subgroup of $\widehat{G}$ that separates points in G, then A is dense in $\widehat{G}$.

12. Let G be a compact Abelian metric group. Prove that G is a solenoidal group if and only if $\widehat{G}$ is isomorphic to a countable subgroup of $\mathbb{R}_d$.

13. Show that $\widehat{\mathbb{Q}_d}$ is a compact divisible group.

14. Let G be a compact Abelian metric group and let H be a noncompact ALS-group. Prove that there exists a homomorphism $\varphi : H \to G$ such that $\varphi(H)^- = G$ if and only if $\widehat{G}$ is algebraically isomorphic to a countable subgroup of $\widehat{H}$.

15. Consider the group $L = \{z \in \mathbb{K} : z^k = 1 \text{ for some } k \in \mathbb{Z}^+\}$ with the discrete topology from *Corollary 7.4.14*. Show that $\widehat{L}$ is a totally disconnected compact monothetic metric group. Prove that if G is a totally disconnected compact monothetic metric group, then there exists an onto homomorphism $\varphi : \widehat{L} \to G$.

Bibliography

[1] L. V. Ahlfors. *Complex Analysis.* International Series on Pure and Applied Mathematics. McGraw-Hill Book Company, NY, second edition, 1966.

[2] N. Bourbaki. *General Topology Part 2.* Addison-Wesley Publishing Company, Reading, MA, 1966.

[3] B. C. Hall. *Lie Groups, Lie Algebras and Representations*, volume 222 of *Graduate Texts in Mathematics.* Springer, NY, 2000.

[4] E. Hewitt and K. A. Ross. *Abstract Harmonic Analysis I*, volume 115 of *Der Grundlehren Mathematischen Wissenschaften.* Springer-Verlag, Berlin, 1963.

[5] P. J. Higgins. *An Introduction to Topological Groups.* Number 15 in London Mathematical Society Lecture Note Series. Cambridge University Press, Cambridge, MA, 1974.

[6] J. L. Kelley. The tychonoff product theorem implies the axiom of choice. *Fund. Math.*, 37:75–76, 1950.

[7] J. L. Kelley. *General Topology.* University Series in Higher Mathematics. D. Van Nostrand, Princeton, NJ, 1955.

[8] J. M. Lee. *Introduction to Topological Manifolds.* Number 202 in Graduate Texts in Mathematics. Springer, NY, 2000.

[9] J. M. Lee. *Introduction to Smooth Manifolds.* Number 218 in Graduate Texts in Mathematics. Springer, NY, 2006.

[10] N. G. Markley. *Principles of Differential Equations.* Pure and Applied Mathematics. John Wiley & Sons, Inc., Hoboken, NJ, 2004.

[11] S. A. Morris. *Pontryagin Duality and the Structure of Locally Compact Abelian Groups.* Number 29 in London Mathematical Society Lecture Note Series. Cambridge University Press, Cambridge, MA, 1977.

[12] L. Nachbin. *The Haar Integral.* The University Series in Higher Mathematics. D. Van Nostrand Company, Inc, NY, 1965.

[13] L. S. Pontryagin. *Topological Groups*, volume XXIX of *Russian Monographs and Texts on Advanced Mathematics and Physics*. Gordon and Breach, NY, second edition, 1966.

[14] H. L. Royden. *Real Analysis*. Macmillan Publishing Company, NY, third edition, 1988.

[15] G. F. Simmons. *Introduction to Topology and Modern Analysis*. International Series in Pure and Applied Mathematics. McGraw-Hill Book Company, Inc, NY, 1963.

[16] S. Willard. *General Topology*. Addison-Wesley, Reading, MA, 1970.

Index of Special Symbols

Index

Printed and bound by CPI Group (UK) Ltd, Croydon, CR0 4YY

16/04/2025

14658517-0003